INTRODUCTORY ALGEBRAIC NUMBER THEORY

Algebraic number theory is a subject that came into being through the attempts of mathematicians to try to prove Fermat's last theorem and that now has a wealth of applications to Diophantine equations, cryptography, factoring, primality testing, and public-key cryptosystems.

This book provides an introduction to the subject suitable for senior undergraduate and beginning graduate students in mathematics. The material is presented in a straightforward, clear, and elementary fashion, and the approach is hands on, with an explicit computational flavor. Prerequisites are kept to a minimum, and numerous examples illustrating the material occur throughout the text. References to suggested readings and to the biographies of mathematicians who have contributed to the development of algebraic number theory are given at the end of each chapter. There are more than 320 exercises, an extensive index, and helpful location guides to theorems and lemmas in the text.

Şaban Alaca is Lecturer in Mathematics at Carleton University, where he has been honored by three teaching awards: Faculty of Science Teaching Award, Professional Achievement Award, and Students Choice Award. His main research interest is in algebraic number theory.

Kenneth S. Williams is Professor Emeritus and Distinguished Research Professor of Mathematics at Carleton University. Dr. Williams has published more than 240 research papers in number theory, linear algebra, algebra, and analysis. This is his seventh book.

INTRODUCTORY ALGEBRAIC NUMBER THEORY

ŞABAN ALACA

Carleton University, Ottawa

KENNETH S. WILLIAMS

Carleton University, Ottawa

CAMBRIDGE UNIVERSITY PRESS
Cambridge, New York, Melbourne, Madrid, Cape Town, Singapore,
São Paulo, Delhi, Dubai, Tokyo

Cambridge University Press
The Edinburgh Building, Cambridge CB2 8RU, UK

Published in the United States of America by Cambridge University Press, New York

www.cambridge.org
Information on this title: www.cambridge.org/9780521540117

First published 2004

A catalogue record for this publication is available from the British Library

Library of Congress Cataloguing in Publication data
Alaca, Saban, 1964–
Introductory algebraic number theory / Saban Alaca, Kenneth S. Williams.
p.   cm.
Includes bibliographical references and index.
ISBN 0-521-83250-0 (hb.) – ISBN 0-521-54011-9 (pbk.)
1. Algebraic number theory.   I. Williams, Kenneth S.   II. Title.
QA247 .A43   2003
512'.74 – dc21   2003051243

ISBN 978-0-521-83250-2 Hardback
ISBN 978-0-521-54011-7 Paperback

Transferred to digital printing 2009

To our wives
Ayşe and Carole

Contents

List of Tables

Notation

$\mathbb{N} = \{1, 2, 3, \ldots\}$

$\mathbb{Z} = \{0, \pm 1, \pm 2, \ldots\}$

$\mathbb{Q}$ = field of rational numbers

$\mathbb{R}$ = field of real numbers

$\mathbb{C}$ = field of complex numbers

ϕ = empty set

$\left(\dfrac{m}{p}\right)$ = Legendre symbol = $\begin{cases} 1, & \text{if } p \nmid m \text{ and } x^2 \equiv m \text{ (mod } p) \text{ is solvable,} \\ -1, & \text{if } p \nmid m \text{ and } x^2 \equiv m \text{ (mod } p) \text{ is insolvable,} \\ 0, & \text{if } p \mid m, \end{cases}$

where $m \in \mathbb{Z}$ and p is a prime

$[x]$ = greatest integer less than or equal to the real number x

$\dbinom{m}{n}$ = binomial coefficient = $\dfrac{m!}{(m-n)!\,n!}$, where m and n are integers such that $0 \le n \le m$

If A is a set containing 0 then $A^* = A \setminus \{0\}$

$\mathbb{Z}_n$ = cyclic group of order n

$\text{card}(S)$ = cardinality of the set S

$O_n = n \times n$ zero matrix

$I_n = n \times n$ identity matrix

$O_{r,s} = r \times s$ zero matrix

Introduction

This book is intended as an introductory text for senior undergraduate and beginning graduate students wishing to learn the fundamentals of algebraic number theory. It is based upon a course in algebraic number theory given by the second author at Carleton University for more than thirty years. Keeping in mind that this is an introductory text, the authors have strived to present the material in as straightforward, clear, and elementary fashion as possible. Throughout the text many numerical examples are given to illustrate the theory. Each chapter closes with a set of exercises on the material covered in the chapter, as well as some suggested further reading. References cited in each chapter are listed under suggested reading. Biographical references for some of the mathematicians mentioned in the text are also given at the end of each chapter. For the convenience of the reader, the book concludes with page references for the definitions, theorems, and lemmas in the text. In addition an extensive bibliography of books on algebraic number theory is provided.

The main aim of the book is to present to the reader a detailed self-contained development of the classical theory of algebraic numbers. This theory is one of the crowning achievements of nineteenth-century mathematics. It came into being through the attempts of mathematicians of that century to prove Fermat's last theorem, namely, that the equation $x^n + y^n = z^n$ has no solutions in nonzero integers x, y, z, where n is an integer ≥ 3. A wonderful achievement of the twentieth century was the proof of Fermat's last theorem by Andrew Wiles of Princeton University. Although the proof of Fermat's last theorem is beyond the scope of this book, we will show how algebraic number theory can be used to find the solutions in integers (if any) of other equations.

The contents of the book are divided into fourteen chapters. Chapter 1 serves as an introduction to the basic properties of integral domains. Chapters 2 and 3 are devoted to Euclidean domains and Noetherian domains respectively. In Chapter 4 the reader is introduced to algebraic numbers and algebraic integers. Algebraic number fields are introduced in Chapter 6 after a discussion of algebraic extensions of fields in Chapter 5. Chapter 7 is devoted to the study of integral bases. Minimal integers are introduced as a tool for finding integral bases and many numerical examples are given. Chapter 8 is concerned with Dedekind domains. The ring of integers of an algebraic number field is the prototype of a Dedekind domain. Chapters 9 and 10 discuss the factorization of ideals into prime ideals. The structure of the unit group of a real quadratic field is determined in Chapter 11. In Chapter 12 the classic theorems of Minkowski in the geometry of numbers are proved and are used to show that the ideal class group is finite. Dirichlet's determination of the units in an arbitrary algebraic number field is presented in Chapter 13 using the approach given by van der Waerden. Finally, in Chapter 14, the algebraic number-theoretic tools developed in earlier chapters are used to discuss the solvability of certain equations in integers.

The prerequisites for this book are a basic course in linear algebra (systems of linear equations, vector spaces over a field), a basic course in modern algebra (groups, rings, and fields including Eisenstein's irreducibility criterion), and a basic course in elementary number theory (the Legendre symbol, quadratic residues, and the law of quadratic reciprocity.) No Galois theory is needed.

A possible outline for a one-semester course (three hours of lectures per week for twelve weeks) together with an approximate breakdown of lecture time is as follows:

Chapter 1 (excluding Theorem 1.2.2)	2 hours
Chapter 2 (excluding Sections 2.3, 2.4)	2 hours
Chapter 3	3 hours
Chapter 4	3 hours
Chapter 5	3 hours
Chapter 6	5 hours
Chapter 7 (Section 7.1 only)	3 hours
Chapter 8	3 hours
Chapter 9	3 hours
Chapter 10 (excluding Sections 10.4, 10.5, 10.6)	2 hours
Chapter 11	3 hours
Chapter 12 (excluding Section 12.7)	2 hours
Chapter 14 (Section 14.2 only)	2 hours

It is planned to provide solutions to selected questions, as well as corrections to any errors, on the website

http://mathstat.carleton.ca/~williams/books.html
or
http://www.math.carleton.ca/~williams/books.html.

The authors would like to thank their colleagues John D. Dixon, James G. Huard, Pierre Kaplan, Blair K. Spearman, and P. Gary Walsh for helpful suggestions in connection with the writing of this book. The second author would like to thank the many students who have taken the course Mathematics 70.436*/70.536 Algebraic Number Theory with him at Carleton University over the years. Special thanks go to the class of 2000–1 (Yaroslav Bezverkhnyev, Joanne Charlebois, Colette Haley, Mathieu Lemire, Rima Rahal, Fabien Roche, Tom Wiley, and Benjamin Young) for their suggestions for improvement to the preliminary draft of this book used in class. Finally, the authors would like to thank Austin Behne for his help in translating van der Waerden's paper on Dirichlet's unit theorem from German into English.

1

Integral Domains

1.1 Integral Domains

In this chapter we recall the definition and properties of an integral domain and develop the concept of divisibility in such a domain. We expect the reader to be familiar with the elementary properties of groups, rings, and fields and to have a basic knowledge of both elementary number theory and linear algebra over a field.

Definition 1.1.1 (Integral domain) *An integral domain is a commutative ring that has a multiplicative identity but no divisors of zero.*

An integral domain D is called a field if for each $a \in D$, $a \neq 0$, there exists $b \in D$ with $ab = 1$.

Example 1.1.1 *The ring $\mathbb{Z} = \{0, \pm 1, \pm 2, \ldots\}$ of all integers is an integral domain.*

Example 1.1.2 *$\mathbb{Z} + \mathbb{Z}i = \{a + bi \mid a, b \in \mathbb{Z}\}$ is an integral domain. The elements of $\mathbb{Z} + \mathbb{Z}i$ are called Gaussian integers after the famous mathematician Carl Friedrich Gauss (1777–1855), who developed their properties in his work on biquadratic reciprocity. $\mathbb{Z} + \mathbb{Z}i$ is called the Gaussian domain.*

Example 1.1.3 *$\mathbb{Z} + \mathbb{Z}\omega = \{a + b\omega \mid a, b \in \mathbb{Z}\}$, where ω is the complex cube root of unity given by $\omega = (-1 + \sqrt{-3})/2$, is an integral domain. The elements of $\mathbb{Z} + \mathbb{Z}\omega$ are called Eisenstein integers after Gotthold Eisenstein (1823–1852), who introduced them in his pioneering work on the law of cubic reciprocity. $\mathbb{Z} + \mathbb{Z}\omega$ is called the Eisenstein domain. The other complex cube root of unity is $\omega^2 = \overline{\omega} = (-1 - \sqrt{-3})/2$. Note that $\mathbb{Z} + \mathbb{Z}\omega = \mathbb{Z} + \mathbb{Z}\omega^2$ as $\omega^2 = -\omega - 1$. Also $\mathbb{Z} + \mathbb{Z}\omega = \mathbb{Z} + \mathbb{Z}\left(\frac{1+\sqrt{-3}}{2}\right)$.*

Example 1.1.4 *$\mathbb{Z} + \mathbb{Z}\sqrt{m} = \{a + b\sqrt{m} \mid a, b \in \mathbb{Z}\}$, where m is a positive or negative integer that is not a perfect square, is an integral domain. As $\sqrt{m}$ is a root of an irreducible quadratic polynomial (namely $x^2 - m$), $\mathbb{Z} + \mathbb{Z}\sqrt{m}$ is called* *a quadratic domain. If k is a nonzero integer such that k^2 divides m then*

$$\mathbb{Z} + \mathbb{Z}\sqrt{m} \subseteq \mathbb{Z} + \mathbb{Z}\sqrt{m/k^2}$$

with equality if and only if $k^2 = 1$. $\mathbb{Z} + \mathbb{Z}\sqrt{m}$ is called a subdomain of $\mathbb{Z} + \mathbb{Z}\sqrt{m/k^2}$. Thus $\mathbb{Z} + 2\mathbb{Z}i \subset \mathbb{Z} + \mathbb{Z}i$.

Example 1.1.5 $\mathbb{Z} + \mathbb{Z}\left(\frac{1+\sqrt{m}}{2}\right) = \{a + b\left(\frac{1+\sqrt{m}}{2}\right) \mid a, b \in \mathbb{Z}\}$, *where m is a non-square integer (positive or negative), which is congruent to 1 modulo 4, is an integral domain. We emphasize that $\mathbb{Z} + \mathbb{Z}\left(\frac{1+\sqrt{m}}{2}\right)$ is not an integral domain if $m \not\equiv 1 \pmod 4$ since in this case it is not closed under multiplication as*

$$\left(\frac{1 + \sqrt{m}}{2}\right)\left(1 - \left(\frac{1 + \sqrt{m}}{2}\right)\right) = \left(\frac{1 + \sqrt{m}}{2}\right)\left(\frac{1 - \sqrt{m}}{2}\right) = \frac{1 - m}{4} \notin \mathbb{Z}.$$

Again as $\frac{1+\sqrt{m}}{2}$ is a root of an irreducible quadratic polynomial (namely $x^2 - x + \left(\frac{1-m}{4}\right)$), $\mathbb{Z} + \mathbb{Z}\left(\frac{1+\sqrt{m}}{2}\right)$ is called a quadratic domain. We note that the elements of the integral domain $\mathbb{Z} + \mathbb{Z}\left(\frac{1+\sqrt{m}}{2}\right)$ can also be written in the form $\frac{1}{2}(x + y\sqrt{m})$, where x and y are integers such that $x \equiv y \pmod 2$. Clearly the domain $\mathbb{Z} + \mathbb{Z}\sqrt{m}$ is a subdomain of $\mathbb{Z} + \mathbb{Z}\left(\frac{1+\sqrt{m}}{2}\right)$.

Example 1.1.6 $F[x] =$ *the ring of polynomials in the indeterminate x with coefficients from a field F is an integral domain.*

Example 1.1.7 $\mathbb{Z}[x] =$ *the ring of polynomials in the indeterminate x with integral coefficients is an integral domain.*

Example 1.1.8 $D[x] =$ *the ring of polynomials in the indeterminate x with coefficients from the integral domain D is an integral domain.*

Example 1.1.9 $F[x, y] =$ *the ring of polynomials in the two indeterminates x and y with coefficients from the field F is an integral domain.*

Example 1.1.10 $\mathbb{Z} + \mathbb{Z}\theta + \mathbb{Z}\theta^2 = \{a + b\theta + c\theta^2 \mid a, b, c \in \mathbb{Z}\}$, *where θ is a root of the cubic equation $\theta^3 + \theta + 1 = 0$, is an integral domain. It is called a cubic domain.*

Example 1.1.11 $D = \{a + b\sqrt{2} + ci + di\sqrt{2} \mid a, c \text{ integers; } b, d \text{ both integers or both halves of odd integers}\}$ *is an integral domain. Clearly $\mathbb{Z} + \mathbb{Z}\sqrt{2} \subset D$, $\mathbb{Z} + \mathbb{Z}i \subset D$, $\mathbb{Z} + \mathbb{Z}i\sqrt{2} \subset D$.*

Properties of an Integral Domain

Let D be an integral domain. Then the following properties hold.

(a) The identity element of D is unique, for if 1 and $1'$ are two identities for D then

$$1 = 1 \cdot 1' \text{ (as } 1' \text{ is an identity)} = 1' \text{ (as } 1 \text{ is an identity)}.$$

(b) D possesses a left cancellation law, that is,

$$ab = ac, \ a \neq 0 \Longrightarrow b = c \ (a, b, c \in D)$$

as well as a right cancellation law

$$ac = bc, \ c \neq 0 \Longrightarrow a = b \ (a, b, c \in D).$$

(c) It is well known that if D is an integral domain then there exists a field F, called the field of quotients of D or the quotient field of D, that contains an isomorphic copy D' of D (see, for example, Fraleigh [3]). In practice it is usual to identify D with D' and so consider D as a subdomain of F. The quotient field of $\mathbb{Z}$ is the field of rational numbers $\mathbb{Q}$. The quotient field of the polynomial domain $F[X]$ (where F is a field) is the field $F(X)$ of rational functions in X.

Definition 1.1.2 (Divisor) *Let a and b belong to the integral domain D. The element a is said to be a divisor of b (or a divides b) if there exists an element c of D such that $b = ac$. If a is a divisor of b, we write $a \mid b$. If a is not a divisor of b, we write $a \nmid b$.*

Example 1.1.12 $1 + i \mid 2$ *in* $\mathbb{Z} + \mathbb{Z}i$ *as* $2 = (1 + i)(1 - i)$.

Example 1.1.13 $x^2 + x + 1 \mid x^4 + x^2 + 1$ *in* $\mathbb{Z}[x]$ *as* $x^4 + x^2 + 1 = (x^2 + x + 1)(x^2 - x + 1)$.

Example 1.1.14 $(1 - \omega)^2 \mid 3$ *in* $\mathbb{Z} + \mathbb{Z}\omega$ *as* $3 = (1 - \omega)^2(1 + \omega)$ *(see Example 1.1.3)*.

Example 1.1.15 $1 + \theta - \theta^2 \mid -\theta - 2\theta^2$ *in* $\mathbb{Z} + \mathbb{Z}\theta + \mathbb{Z}\theta^2$ *as* $-\theta - 2\theta^2 = (1 + \theta - \theta^2)(1 - \theta)$ *(see Example 1.1.10)*.

Example 1.1.16 $2 + \sqrt{2} \nmid 3$ *in* $\mathbb{Z} + \mathbb{Z}\sqrt{2}$ *as* $3/(2 + \sqrt{2}) = 3 - \frac{3}{2}\sqrt{2} \notin \mathbb{Z} + \mathbb{Z}\sqrt{2}$.

Properties of Divisors

Let $a, b, c \in D$, where D is an integral domain. Then the following properties hold.

(a) $a \mid a$ (reflexive property).
(b) $a \mid b$ and $b \mid c$ implies $a \mid c$ (transitive property).

(c) $a \mid b$ and $a \mid c$ implies $a \mid xb + yc$ for any $x \in D$ and $y \in D$.
(d) $a \mid b$ implies $ac \mid bc$.
(e) $ac \mid bc$ and $c \neq 0$ implies $a \mid b$.
(f) $1 \mid a$.
(g) $a \mid 0$.
(h) $0 \mid a$ implies $a = 0$.

Definition 1.1.3 (Unit) *An element a of an integral domain D is called a unit if $a \mid 1$. The set of units of D is denoted by $U(D)$.*

Properties of Units

Let D be an integral domain. Then $U(D)$ has the following properties.

(a) $\pm 1 \in U(D)$.
(b) If $a \in U(D)$ then $-a \in U(D)$.
(c) If $a \in U(D)$ then $a^{-1} \in U(D)$.
(d) If $a \in U(D)$ and $b \in U(D)$ then $ab \in U(D)$.
(e) If $a \in U(D)$ then $\pm a^n \in U(D)$ for any $n \in \mathbb{Z}$.

Example 1.1.17

(a) $i \in U(\mathbb{Z} + \mathbb{Z}i)$.
(b) $\omega \in U(\mathbb{Z} + \mathbb{Z}\omega)$ (*see Example 1.1.3*).
(c) $\theta \in U(\mathbb{Z} + \mathbb{Z}\theta + \mathbb{Z}\theta^2)$ as $1 = \theta(-1 - \theta^2)$ (*see Example 1.1.10*).

Theorem 1.1.1 *If D is an integral domain then $U(D)$ is an Abelian group with respect to multiplication.*

Proof: $U(D)$ is closed under multiplication by property (d). Multiplication of elements of $U(D)$ is both associative and commutative as D is an integral domain. $U(D)$ possesses an identity element, namely 1, by property (a). Every element of $U(D)$ has a multiplicative inverse by property (c). Thus $U(D)$ is an Abelian group with respect to multiplication. ∎

Abelian groups are named after the Norwegian mathematician Niels Henrik Abel (1802–1829), who proved in 1824 the impossibility of solving the general quintic equation by means of radicals.

Example 1.1.18 *Let $\mathbb{Z}_n$ denote the cyclic group of order n.*

(a) $U(\mathbb{Z}) = \{\pm 1\} \simeq \mathbb{Z}_2$.
(b) $U(\mathbb{Z} + \mathbb{Z}i) = \{\pm 1, \pm i\} \simeq \mathbb{Z}_4$.
(c) $U(F[x]) = F^*$, *where F is a field and $F^* = F \setminus \{0\}$.*

(d) $U(\mathbb{Z}[x]) = \{\pm 1\} \simeq \mathbb{Z}_2$.
(e) $\pm(1 + \sqrt{2})^n \in U(\mathbb{Z} + \mathbb{Z}\sqrt{2})$, for all $n \in \mathbb{Z}$.
(f) $\frac{1}{2}\sqrt{2} + \frac{1}{2}i\sqrt{2} \in U(D)$, where D is defined in Example 1.1.11.

We remark that in Chapter 11 we will show that

$$U(\mathbb{Z} + \mathbb{Z}\sqrt{2}) = \{\pm(1 + \sqrt{2})^n \mid n \in \mathbb{Z}\} \simeq \mathbb{Z}_2 \times \mathbb{Z}.$$

Definition 1.1.4 (Associate) *Two nonzero elements a and b of an integral domain D are called associates, or said to be associated, if each divides the other. If a and b are associates we write a $\sim$ b. If a and b are not associates we write a $\not\sim$ b.*

Properties of Associates

Let $a, b, c \in D^* = D \setminus \{0\}$, where D is an integral domain. The following properties hold.

(a) $a \sim a$ (reflexive property).
(b) $a \sim b$ implies $b \sim a$ (symmetric property).
(c) $a \sim b$ and $b \sim c$ imply $a \sim c$ (transitive property).
(d) $a \sim b$ if and only if $ab^{-1} \in U(D)$.
(e) $a \sim 1$ if and only if a is a unit.

Properties (a), (b), and (c) show that $\sim$ is an equivalence relation. The equivalence class containing $a \in D$ is just the set $\{ua \mid u \in U(D)\}$.

Example 1.1.19

(a) *In $\mathbb{Z}$, $a \sim b$ if and only if $a = \pm b$, equivalently $|a| = |b|$.*
(b) *In $\mathbb{Z} + \mathbb{Z}i$ we have $1 + i \sim 1 - i$ as $\frac{1+i}{1-i} = i \in U(\mathbb{Z} + \mathbb{Z}i)$.*
(c) *In $\mathbb{Z} + \mathbb{Z}\sqrt{2}$ we have $1 + 3\sqrt{2} \sim 5 - 2\sqrt{2}$ as $\frac{1+3\sqrt{2}}{5-2\sqrt{2}} = 1 + \sqrt{2} \in U(\mathbb{Z} + \mathbb{Z}\sqrt{2})$.*

1.2 Irreducibles and Primes

In $\mathbb{Z}$ an integer p (≥ 2) that is divisible only by the positive integers 1 and p is called a prime. Each prime p in $\mathbb{Z}$ has the following two properties:

$$p = ab \ (a, b \in \mathbb{Z}) \Longrightarrow a \text{ or } b = \pm 1 \tag{1.2.1}$$

and

$$p \mid ab \ (a, b \in \mathbb{Z}) \Longrightarrow p \mid a \text{ or } p \mid b. \tag{1.2.2}$$

Our next definition generalizes property (1.2.1) to an arbitrary integral domain D, and an element of D with this property is called an irreducible element.

Definition 1.2.1 (Irreducible) *A nonzero, nonunit element a of an integral domain D is called an irreducible, or said to be irreducible, if $a = bc$, where $b, c \in D$, implies that either b or c is a unit.*

   A nonzero, nonunit element that is not irreducible is called reducible.

Example 1.2.1   *2 is irreducible in $\mathbb{Z}$, for if $2 = ab$ with $a \in \mathbb{Z}$ and $b \in \mathbb{Z}$ then either $a = \pm 1$ or $b = \pm 1$.*

Example 1.2.2   *2 is irreducible in $\mathbb{Z} + \mathbb{Z}\sqrt{-5}$. To show this, suppose that $2 = (a + b\sqrt{-5})(c + d\sqrt{-5})$, where $a, b, c, d \in \mathbb{Z}$. Taking the modulus of both sides of this equation, we obtain $4 = (a^2 + 5b^2)(c^2 + 5d^2)$. Thus $a^2 + 5b^2$ is a positive integral divisor of 4 and so we must have*

$$a^2 + 5b^2 = 1, 2, \text{ or } 4.$$

Hence we see that

$$(a, b) = (\pm 1, 0) \text{ or } (\pm 2, 0)$$

so that

$$a + b\sqrt{-5} = \pm 1 \text{ or } \pm 2.$$

In the former case $a + b\sqrt{-5}$ is a unit of $\mathbb{Z} + \mathbb{Z}\sqrt{-5}$. In the latter case

$$c + d\sqrt{-5} = \frac{2}{a + b\sqrt{-5}} = \frac{2}{\pm 2} = \pm 1$$

is a unit of $\mathbb{Z} + \mathbb{Z}\sqrt{-5}$. Hence 2 is irreducible in $\mathbb{Z} + \mathbb{Z}\sqrt{-5}$.

Example 1.2.3   *$7 + \sqrt{-5}$ is reducible in $\mathbb{Z} + \mathbb{Z}\sqrt{-5}$ because*

$$7 + \sqrt{-5} = (1 + \sqrt{-5})(2 - \sqrt{-5})$$

and neither $1 + \sqrt{-5}$ nor $2 - \sqrt{-5}$ is a unit of $\mathbb{Z} + \mathbb{Z}\sqrt{-5}$.

Our next definition generalizes property (1.2.2) to an arbitrary integral domain, and an element with this property is called a prime element.

Definition 1.2.2 (Prime) *A nonzero, nonunit element p of an integral domain D is called a prime if $p \mid ab$, where $a, b \in D$, implies that $p \mid a$ or $p \mid b$.*

Example 1.2.4   *2 is a prime in $\mathbb{Z}$. Suppose $2 \mid ab$, where $a, b \in \mathbb{Z}$, so that ab is even. Since the product of two odd integers is odd, at least one of a and b must be even, that is, $2 \mid a$ or $2 \mid b$, showing that 2 is prime.*

Example 1.2.5 *2 is not a prime in* $\mathbb{Z} + \mathbb{Z}\sqrt{-5}$ *as* $2 \mid (1 + \sqrt{-5})(1 - \sqrt{-5})$ *yet* $2 \nmid 1 \pm \sqrt{-5}$.

Example 1.2.6 $1 + i$ *is a prime in* $\mathbb{Z} + \mathbb{Z}i$. *To show this, suppose that* $1 + i \mid (a + bi)(c + di)$, *where* $a, b, c, d \in \mathbb{Z}$. *Then there exist integers* x *and* y *such that*

$$(a + bi)(c + di) = (1 + i)(x + yi).$$

Taking the modulus of both sides of this equation, we obtain

$$(a^2 + b^2)(c^2 + d^2) = 2(x^2 + y^2).$$

As 2 is a prime in $\mathbb{Z}$, *we have either* $2 \mid a^2 + b^2$ *or* $2 \mid c^2 + d^2$. *Interchanging* $a + bi$ *and* $c + di$, *if necessary, we may suppose that* $2 \mid a^2 + b^2$. *Thus, either* a *and* b *are both even or they are both odd. In the former case* $a = 2r$ *and* $b = 2s$, *where* r *and* s *are integers, and*

$$a + bi = 2(r + si) = (1 + i)((r + s) + (-r + s)i),$$

so that $1 + i \mid a + bi$. *In the latter case* $a = 2r + 1$ *and* $b = 2s + 1$, *where* r *and* s *are integers, and*

$$a + bi = 2(r + si) + (1 + i) = (1 + i)((r + s + 1) + (-r + s)i),$$

so that $1 + i \mid a + bi$. *Hence* $1 + i$ *is a prime in* $\mathbb{Z} + \mathbb{Z}i$.

Theorem 1.2.1 *In any integral domain* D *a prime is irreducible.*

Proof: Let $p \in D$ be a prime and suppose that $p = ab$, where $a, b \in D$. As $ab = p \cdot 1$ we have $p \mid ab$, and so, as p is prime, we deduce that $p \mid a$ or $p \mid b$, that is, $a/p \in D$ or $b/p \in D$. Since $1 = a/p \cdot b$ or $1 = a \cdot b/p$, either b is a unit or a is a unit of D. This proves that p is an irreducible element of D. ∎

The converse of Theorem 1.2.1 is not true. From Examples 1.2.2 and 1.2.5 we see that the element 2 of $\mathbb{Z} + \mathbb{Z}\sqrt{-5}$ is irreducible but not prime.

Waterhouse [6] has recently given a class of integral domains in which every irreducible is prime.

Theorem 1.2.2 *Let* D *be an integral domain that has the following property:*

Every quadratic polynomial in $D[X]$ *having roots in the quotient field* F *of* D *is a product of linear polynomials in* $D[X]$. (1.2.3)

Then every irreducible in D *is prime.*

Proof: Let p be an irreducible element in D, which is not prime. Then there exist $a, b \in D$ such that

$$p \mid ab, \quad p \nmid a, \quad p \nmid b.$$

Let $r = ab/p \in D$, and consider the quadratic polynomial

$$f(X) = pX^2 - (a+b)X + r.$$

In $F[X]$ we have

$$f(X) = p(X - a/p)(X - b/p).$$

We show that $f(X)$ does not factor into linear factors in $D[X]$. Indeed, suppose on the contrary that

$$f(X) = (cX + s)(dX + t)$$

in $D[X]$. Then $cd = p$. As p is irreducible, one of c and d is a unit of D, say d, so that $c = d^{-1}p$. Then the roots of $f(X)$ in F are $-ds/p$ and $-d^{-1}t$. But $-d^{-1}t \in D$, while neither a/p nor b/p is in D. Thus no such factorization can exist. Hence every irreducible in D is prime. ∎

1.3 Ideals

Subsets of an integral domain D that are closed under addition and under multiplication by elements of D play a special role and are called ideals.

Definition 1.3.1 (Ideal) *An ideal I of an integral domain D is a nonempty subset of D having the following two properties:*

$$a \in I, \ b \in I \Longrightarrow a + b \in I,$$

$$a \in I, r \in D \Longrightarrow ra \in I.$$

It is clear that if $a_1, \ldots, a_n \in I$ then $r_1 a_1 + \cdots + r_n a_n \in I$ for all $r_1, \ldots, r_n \in D$. In particular if $a \in I$ and $b \in I$ then $-a \in I$ and $a - b \in I$. Also $0 \in I$, and if $1 \in I$ then $I = D$.

Example 1.3.1 *If $\{a_1, \ldots, a_n\}$ is a set of elements of the integral domain D then the set of all finite linear combinations of $a_1, \ldots, a_n$*

$$\left\{ \sum_{i=1}^{n} r_i a_i \mid r_1, \ldots, r_n \in D \right\}$$

is an ideal of D, which we denote by $\langle a_1, \ldots, a_n \rangle$.

Definition 1.3.2 (Principal ideal) *An ideal I of an integral domain D is called a principal ideal if there exists an element $a \in I$ such that $I = \langle a \rangle$. The element a is called a generator of the ideal I.*

If D is an integral domain the principal ideal $\langle a \rangle$ generated by $a \in D$ is just the set $\{ra \mid r \in D\}$. Clearly the principal ideal $\langle 0 \rangle$ is just the singleton set $\{0\}$ and the principal ideal $\langle 1 \rangle$ is D.

Definition 1.3.3 (Proper ideal) *An ideal I of an integral domain D is called a proper ideal of D if $I \neq \langle 0 \rangle, \langle 1 \rangle$.*

Thus a proper ideal of an integral domain D is an ideal I such that $\{0\} \subset I \subset D$.

Example 1.3.2 *For any positive integer k, the set*

$$k\mathbb{Z} = \{0, \pm k, \pm 2k, \ldots\}$$

is an ideal of $\mathbb{Z}$. Indeed $k\mathbb{Z}$ is a principal ideal generated by k (or $-k$) so that

$$k\mathbb{Z} = \langle k \rangle = \langle -k \rangle.$$

Example 1.3.3 *Let*

$$I = \{f(x) \in \mathbb{Z}[x] \mid f(0) = 0\}.$$

Then I is an ideal of $\mathbb{Z}[x]$ and $I = \langle x \rangle$.

Example 1.3.4 *Let*

$$J = \{f(x) \in \mathbb{Z}[x] \mid f(0) \equiv 0 \pmod{2}\}.$$

Then J is an ideal of $\mathbb{Z}[x]$ and $J = \langle 2, x \rangle$. However, J is not a principal ideal.

Theorem 1.3.1 *Let D be an integral domain and let $a, b \in D^* = D \setminus \{0\}$. Then*

$$\langle a \rangle = \langle b \rangle \text{ if and only if } a/b \in U(D).$$

Proof: If $a/b \in U(D)$ then $a = bu$ for some $u \in U(D)$. Let $x \in \langle a \rangle$. Then $x = ac$ for some $c \in D$. Hence $x = buc$ with $uc \in D$. Thus $x \in \langle b \rangle$. We have shown that $\langle a \rangle \subseteq \langle b \rangle$. As $a/b \in U(D)$ and $U(D)$ is a group with respect to multiplication, we have $b/a = (a/b)^{-1} \in U(D)$. Then, proceeding exactly as before with the roles of a and b interchanged, we find that $\langle b \rangle \subseteq \langle a \rangle$. Thus $\langle a \rangle = \langle b \rangle$.

Conversely, suppose that $\langle a \rangle = \langle b \rangle$. Then $a = bc$ for some $c \in D$ and $b = ad$ for some $d \in D$. Hence $b = bcd$. As $b \neq 0$ we deduce that $1 = cd$ so that $c \in U(D)$. Thus $a/b = c \in U(D)$. ∎

1.4 Principal Ideal Domains

An important class of integral domains are those in which every ideal is principal.

Definition 1.4.1 (Principal ideal domain) *An integral domain D is called a principal ideal domain if every ideal in D is principal.*

We begin by giving an example of an integral domain in which every ideal is principal.

Theorem 1.4.1 $\mathbb{Z}$ *is a principal ideal domain.*

Proof: Let I be an ideal of $\mathbb{Z}$. If $I = \{0\}$ then $I = \langle 0 \rangle$ is a principal ideal. Thus we may suppose that $I \neq \{0\}$. Hence I contains a nonzero element a. As both a and $-a$ belong to I, we can suppose that $a > 0$. Hence I contains at least one positive integer, namely a.

We let m denote the least positive integer in I. Dividing a by m, we obtain integers q and r such that $a = mq + r$ and $0 \leq r < m$. As $a \in I$ and $m \in I$, we have $r = a - mq \in I$. This contradicts the minimality of m unless $r = 0$, in which case $a = mq$; that is, $I = \langle m \rangle = m\mathbb{Z}$. ∎

Theorems 1.3.1 and 1.4.1 show that the set of ideals of $\mathbb{Z}$ is $\{k\mathbb{Z} \mid k \in \{0, 1, 2, \ldots\}\}$. Moreover, if I is an ideal of $\mathbb{Z}$ then it is generated by the least positive integer in I.

Other examples of principal ideal domains will be given in Chapter 2 where we discuss Euclidean domains.

Theorem 1.4.2 *In a principal ideal domain, an irreducible element is prime.*

Proof: Let p be an irreducible element in a principal ideal domain D. Suppose that $p \mid ab$, where $a, b \in D$. If $p \nmid a$ we let I be the ideal $\langle p, a \rangle$ of D. As D is a principal ideal domain there is an element $c \in D$ such that $I = \langle c \rangle$. As $a \in I$ and $p \in I$ we must have $c \mid a$ and $c \mid p$. If $c \sim p$ then $p \mid a$, contradicting $p \nmid a$. Hence $c \not\sim p$, and as p is irreducible, c must be a unit. Thus there exists $d \in D$ such that $cd = 1$. Now $c \in \langle a, p \rangle$ so there exist $x, y \in D$ such that $c = xa + yp$. Hence

$$1 = cd = dxa + dyp,$$

and so

$$b = (dx)ab + (bdy)p.$$

Since $p \mid ab$ this shows that $p \mid b$. Thus $p \mid a$ or $p \mid b$ and p is a prime element of D. ∎

Theorem 1.4.3 *In a principal ideal domain, an element is irreducible if and only if it is prime.*

Proof: This follows immediately from Theorems 1.2.1 and 1.4.2. ∎

Example 1.4.1 *It was noted in Section 1.2 that 2 is irreducible but not prime in $\mathbb{Z} + \mathbb{Z}\sqrt{-5}$. Hence, by Theorem 1.4.3, the integral domain $\mathbb{Z} + \mathbb{Z}\sqrt{-5}$ is not a principal ideal domain. Indeed the ideal $\langle 2, 1 + \sqrt{-5}\rangle$ of $\mathbb{Z} + \mathbb{Z}\sqrt{-5}$ is not principal. This can be shown directly as follows. Suppose, on the contrary, that the ideal $\langle 2, 1 + \sqrt{-5}\rangle$ is principal, that is, $\langle 2, 1 + \sqrt{-5}\rangle = \langle \alpha \rangle$ for some $\alpha \in \mathbb{Z} + \mathbb{Z}\sqrt{-5}$. Hence $2 \in \langle \alpha \rangle$ and $1 + \sqrt{-5} \in \langle \alpha \rangle$ so that $\alpha \mid 2$ and $\alpha \mid 1 + \sqrt{-5}$. From the first of these, as 2 is irreducible in $\mathbb{Z} + \mathbb{Z}\sqrt{-5}$, it must be the case that $\alpha \sim 1$ or $\alpha \sim 2$. If $\alpha \sim 2$ then $2 \mid 1 + \sqrt{-5}$, which is impossible as $\frac{1+\sqrt{-5}}{2} = \frac{1}{2} + \frac{1}{2}\sqrt{-5} \notin \mathbb{Z} + \mathbb{Z}\sqrt{-5}$. Hence $\alpha \sim 1$, and so $\langle 2, 1 + \sqrt{-5}\rangle = \langle 1 \rangle$. This shows that 1 is a linear combination of 2 and $1 + \sqrt{-5}$ with coefficients from $\mathbb{Z} + \mathbb{Z}\sqrt{-5}$; that is, there exist $x, y, z, w \in \mathbb{Z}$ such that*

$$1 = (x + y\sqrt{-5})2 + (z + w\sqrt{-5})(1 + \sqrt{-5}).$$

Equating coefficients of 1 and $\sqrt{-5}$, we obtain

$$1 = 2x + z - 5w, \ 0 = 2y + z + w.$$

The difference of these equations yields

$$1 = 2(x - y - 3w),$$

which is clearly impossible as the left-hand side is an odd integer and the right-hand side is an even integer. Hence the ideal $\langle 2, 1 + \sqrt{-5}\rangle$ is not principal in $\mathbb{Z} + \mathbb{Z}\sqrt{-5}$.

Definition 1.4.2 (Greatest common divisor) *Let D be a principal ideal domain and let $\{a_1, \ldots, a_n\}$ be a set of elements of D. Then the ideal $\langle a_1, \ldots, a_n \rangle$ is a principal ideal. A generator of this ideal is called a greatest common divisor of $a_1, \ldots, a_n$.*

Let D be a principal ideal domain. If a and b are greatest common divisors of $a_1, \ldots, a_n \in D$ then

$$\langle a \rangle = \langle a_1, \ldots, a_n \rangle = \langle b \rangle,$$

so that, by Theorem 1.3.1, $a \sim b$. We write $(a_1, \ldots, a_n)$ for a greatest common divisor of $a_1, \ldots, a_n$, understanding that $(a_1, \ldots, a_n)$ is only defined up to a unit. We note that $(a_1, \ldots, a_n) = 0$ if $a_1 = \cdots = a_n = 0$. Also $(a_1, \ldots, a_n) = (a_1, \ldots, a_{n-1})$ if $a_n = 0$. Furthermore,

$$a \in \langle a \rangle = \langle a_1, \ldots, a_n \rangle,$$

so that

$$a = r_1 a_1 + \cdots + r_n a_n$$

for some $r_1, \ldots, r_n \in D$. Thus if $c \in D$ is such that

$$c \mid a_j \ (j = 1, 2, \ldots, n)$$

then

$$c \mid a.$$

Moreover, for $j = 1, 2, \ldots, n$, we have

$$a_j \in \langle a_1, \ldots, a_n \rangle = \langle a \rangle$$

so that

$$a \mid a_j.$$

This justifies calling a "a greatest common divisor" of $a_1, \ldots, a_n$. The elements $a_1, \ldots, a_n$ are called relatively prime if $(a_1, \ldots, a_n)$ is a unit, that is,

$$\langle a_1, \ldots, a_n \rangle = \langle 1 \rangle = D.$$

It is easy to verify that

$$(a_1, \ldots, a_{n-1}, a_n) = ((a_1, \ldots, a_{n-1}), a_n),$$

so that a greatest common divisor can be obtained by finding a succession of greatest common divisors of pairs of elements, that is, if $(a_1, a_2) = b$ then $(a_1, a_2, a_3) = (b, a_3)$, etc.

In the next theorem we use our knowledge of primes and irreducibles in a principal ideal domain to give conditions under which a prime p can be expressed as $u^2 - mv^2$ or $mv^2 - u^2$ for some integers u and v, where m is a given nonsquare integer.

Theorem 1.4.4 *Let m be a nonsquare integer such that $\mathbb{Z} + \mathbb{Z}\sqrt{m}$ is a principal ideal domain. Let p be an odd prime for which the Legendre symbol*

$$\left(\frac{m}{p} \right) = 1.$$

Then there exist integers u and v such that

$$p = u^2 - mv^2 \ \text{if} \ m < 0, \ \text{or if} \ m > 0,$$

and there are integers T, U such that $T^2 - mU^2 = -1$,

$$p = u^2 - mv^2 \ \text{or} \ mv^2 - u^2, \ \text{if} \ m > 0,$$

and there are no integers T, U with $T^2 - mU^2 = -1$.

Proof: As $\left(\frac{m}{p}\right) = 1$, there exists an integer x such that $x^2 \equiv m \pmod{p}$. Thus

$$p \mid (x + \sqrt{m})(x - \sqrt{m})$$

in $\mathbb{Z} + \mathbb{Z}\sqrt{m}$. Clearly $\frac{x \pm \sqrt{m}}{p} = \frac{x}{p} \pm \frac{1}{p}\sqrt{m} \notin \mathbb{Z} + \mathbb{Z}\sqrt{m}$ so that

$$p \nmid x \pm \sqrt{m}.$$

Hence p is not a prime in $\mathbb{Z} + \mathbb{Z}\sqrt{m}$. As $\mathbb{Z} + \mathbb{Z}\sqrt{m}$ is a principal ideal domain, by Theorem 1.4.3 p is not irreducible in $\mathbb{Z} + \mathbb{Z}\sqrt{m}$. Hence

$$p = (u + v\sqrt{m})(w + t\sqrt{m}) \tag{1.4.1}$$

for some $u + v\sqrt{m} \in \mathbb{Z} + \mathbb{Z}\sqrt{m}$ and $w + t\sqrt{m} \in \mathbb{Z} + \mathbb{Z}\sqrt{m}$, where neither $u + v\sqrt{m}$ nor $w + t\sqrt{m}$ is a unit in $\mathbb{Z} + \mathbb{Z}\sqrt{m}$. From (1.4.1) we deduce that

$$p - (uw + tvm) = (ut + vw)\sqrt{m}.$$

As m is not a square, $\sqrt{m} \notin \mathbb{Q}$, so that

$$p - (uw + tvm) = ut + vm = 0.$$

Then

$$p^2 = (uw + tvm)^2 = (uw + tvm)^2 - m(ut + vm)^2$$

so that

$$p^2 = (u^2 - mv^2)(w^2 - mt^2). \tag{1.4.2}$$

As $m, u, v, w, t \in \mathbb{Z}$ and $m \in \mathbb{N}$, we see that $u^2 - mv^2 \in \mathbb{Z}$ and $w^2 - mt^2 \in \mathbb{Z}$. Moreover, $u^2 - mv^2 \neq \pm 1$ and $w^2 - mt^2 \neq \pm 1$, as $u + v\sqrt{m}$ and $w + t\sqrt{m}$ are not units in $\mathbb{Z} + \mathbb{Z}\sqrt{m}$. Thus, from (1.4.2), as p is a prime, we must have $\pm p = u^2 - mv^2 = w^2 - mt^2$. Hence there are integers u and v such that $p = u^2 - mv^2$ or $-(u^2 - mv^2)$.

If $m < 0$ then $u^2 - mv^2 > 0$, so we must have $p = u^2 - mv^2$.

If $m > 0$, $p = -(u^2 - mv^2)$, and there exist integers T and U such that $T^2 - mU^2 = -1$ then $p = u'^2 - mv'^2$ with $u' = Tu + mUv$, $v' = Uu + Tv$. ∎

In Chapter 2 we give some nonsquare values of m for which $\mathbb{Z} + \mathbb{Z}\sqrt{m}$ is a principal ideal domain. Then, by Theorem 1.4.4, we know that for those odd primes p for which $\left(\frac{m}{p}\right) = 1$ there are integers u and v such that $p = u^2 - mv^2$ or $mv^2 - u^2$. For a general positive integer m it is a difficult problem to decide which primes are expressible as $u^2 - mv^2$ with $u, v \in \mathbb{Z}$. The reader interested in knowing more about this problem should consult Cox [2].

In the next theorem we give conditions that ensure that a prime p can be expressed in the form $u^2 + uv + \frac{1}{4}(1 - m)v^2$ or $-(u^2 + uv + \frac{1}{4}(1 - m)v^2)$

for some integers u and v, where m is a given nonsquare integer with $m \equiv 1 \pmod 4$.

Theorem 1.4.5 *Let* $m \equiv 1 \pmod 4$ *be a nonsquare integer such that* $\mathbb{Z} + \mathbb{Z}\left(\frac{1+\sqrt{m}}{2}\right)$ *is a principal ideal domain. Let p be an odd prime for which* $\left(\frac{m}{p}\right) = 1$. *Then there exist integers u and v such that*

$$p = u^2 + uv + \frac{1}{4}(1-m)v^2,$$

if $m < 0$, or if $m > 0$, and there are integers T, U such that

$$T^2 + TU + \frac{1}{4}(1-m)U^2 = -1,$$

and

$$p = u^2 + uv + \frac{1}{4}(1-m)v^2 \text{ or } -(u^2 + uv + \frac{1}{4}(1-m)v^2) \text{ if } m > 0,$$

and there are no integers T, U with

$$T^2 + TU + \frac{1}{4}(1-m)U^2 = -1.$$

Proof: As $\left(\frac{m}{p}\right) = 1$ there exists an integer z such that $z^2 \equiv -m \pmod p$. Set

$$y = \begin{cases} z, & \text{if } z \text{ is odd,} \\ p - z, & \text{if } z \text{ is even,} \end{cases}$$

so that y is an odd integer satisfying $y^2 \equiv m \pmod p$. Now let $x = \frac{1}{2}(y-1) \in \mathbb{Z}$. Clearly $4\left(x^2 + x + \frac{1}{4}(1-m)\right) = (2x+1)^2 - m = y^2 - m \equiv 0 \pmod p$ so that $p \mid x^2 + x + \frac{1}{4}(1-m)$. Hence $p \mid \left(x + \frac{1+\sqrt{m}}{2}\right)\left(x + \frac{1-\sqrt{m}}{2}\right)$ in $\mathbb{Z} + \mathbb{Z}\left(\frac{1+\sqrt{m}}{2}\right)$. Clearly

$$\frac{x + \frac{1 \pm \sqrt{m}}{2}}{p} \notin \mathbb{Z} + \mathbb{Z}\left(\frac{1+\sqrt{m}}{2}\right)$$

so that

$$p \nmid x + \frac{1 \pm \sqrt{m}}{2}.$$

Hence p is not a prime in $\mathbb{Z} + \mathbb{Z}\left(\frac{1+\sqrt{m}}{2}\right)$.

As $\mathbb{Z} + \mathbb{Z}\left(\frac{1+\sqrt{m}}{2}\right)$ is a principal ideal domain, by Theorem 1.4.3 p is not irreducible in $\mathbb{Z} + \mathbb{Z}\left(\frac{1+\sqrt{m}}{2}\right)$. Hence

$$p = \left(u + v\left(\frac{1+\sqrt{m}}{2}\right)\right)\left(w + t\left(\frac{1+\sqrt{m}}{2}\right)\right) \tag{1.4.3}$$

for some $u + v \left(\frac{1+\sqrt{m}}{2} \right) \in \mathbb{Z} + \mathbb{Z}\sqrt{m}$ and $w + t \left(\frac{1+\sqrt{m}}{2} \right) \in \mathbb{Z} + \mathbb{Z} \left(\frac{1+\sqrt{m}}{2} \right),$ where neither $u + v \left(\frac{1+\sqrt{m}}{2} \right)$ nor $w + t \left(\frac{1+\sqrt{m}}{2} \right)$ is a unit in $\mathbb{Z} + \mathbb{Z} \left(\frac{1+\sqrt{m}}{2} \right)$. From (1.4.3) we have

$$p = \left(u + \frac{v}{2} \right) \left(w + \frac{t}{2} \right) + \frac{vt}{4}m + \frac{(vw + ut)}{2}\sqrt{m}.$$

As m is not a square, $\sqrt{m} \notin \mathbb{Q}$, so that 1 and $\sqrt{m}$ are linearly independent over $\mathbb{Q}$. Hence

$$p = \left(u + \frac{v}{2} \right) \left(w + \frac{t}{2} \right) + \frac{vt}{4}m, \quad vw + ut = 0.$$

Thus

$$p = \left(u + v \left(\frac{1 - \sqrt{m}}{2} \right) \right) \left(w + t \left(\frac{1 - \sqrt{m}}{2} \right) \right). \qquad (1.4.4)$$

Multiplying (1.4.3) and (1.4.4) together we obtain

$$p^2 = \left(u^2 + uv + \frac{1}{4}(1 - m)v^2 \right) \left(w^2 + wt + \frac{1}{4}(1 - m)t^2 \right) \qquad (1.4.5)$$

since

$$\left(x + y \left(\frac{1 + \sqrt{m}}{2} \right) \right) \left(x + y \left(\frac{1 - \sqrt{m}}{2} \right) \right) = x^2 + xy + \frac{1}{4}(1 - m)y^2.$$

As $m \equiv 1 \pmod 4$, $u^2 + uv + \frac{1}{4}(1 - m)v^2 \in \mathbb{Z}$ and $w^2 + wt + \frac{1}{4}(1 - m)t^2 \in \mathbb{Z}$. Moreover $u^2 + uv + \frac{1}{4}(1 - m)v^2 \neq \pm 1$ and $w^2 + wt + \frac{1}{4}(1 - m)t^2 \neq \pm 1$ as $u + v \left(\frac{1+\sqrt{m}}{2} \right)$ and $w + t \left(\frac{1+\sqrt{m}}{2} \right)$ are not units in $\mathbb{Z} + \mathbb{Z} \left(\frac{1+\sqrt{m}}{2} \right)$. Thus from (1.4.5) we deduce that

$$\pm p = u^2 + uv + \frac{1}{4}(1 - m)v^2 = w^2 + wt + \frac{1}{4}(1 - m)t^2$$

as p is a prime. Hence there are integers u and v such that

$$p = u^2 + uv + \frac{1}{4}(1 - m)v^2 \text{ or } -\left(u^2 + uv + \frac{1}{4}(1 - m)v^2 \right).$$

If $m < 0$ then $u^2 + uv + \frac{1}{4}(1 - m)v^2 > 0$ so that $p = u^2 + uv + \frac{1}{4}(1 - m)v^2$. If $m > 0$, $p = -\left(u^2 + uv + \frac{1}{4}(1 - m)v^2 \right)$, and there exist integers T and U such that $T^2 + TU + \frac{1}{4}(1 - m)U^2 = -1$ then $p = u'^2 + u'v' + \frac{1}{4}(1 - m)v'^2$ with $u' = uT + \frac{1}{4}(1 - m)vU$ and $v' = uU + vT + vU$. ∎

Examples illustrating Theorems 1.4.4 and 1.4.5 are given in Section 2.5.

1.5 Maximal Ideals and Prime Ideals

In this section we give the basic properties of maximal and prime ideals. These will be important when we discuss Dedekind domains in Chapter 8.

Definition 1.5.1 (Maximal ideal) *A proper ideal M of an integral domain D is called a maximal ideal if whenever I is an ideal of D such that $M \subseteq I \subseteq D$ then $I = M$ or $I = D$.*

Example 1.5.1 *The ideal $\langle x^2 + 1 \rangle$ is maximal in $\mathbb{R}[x]$. To show this, assume that I is an ideal of $\mathbb{R}[x]$ such that $\langle x^2 + 1 \rangle \subset I \subset \mathbb{R}[x]$. As $\langle x^2 + 1 \rangle$ is properly contained in I, there exists $f(x) \in I$ and $f(x) \notin \langle x^2 + 1 \rangle$. Dividing $f(x)$ by $x^2 + 1$, we obtain*

$$f(x) = (x^2 + 1)q(x) + r(x),$$

where $r(x) \neq 0$ and $\deg(r(x)) < 2$. Thus $r(x) = ax + b$, where $a \in \mathbb{R}$ and $b \in \mathbb{R}$ are not both 0, and

$$ax + b = r(x) = f(x) - q(x)(x^2 + 1) \in I.$$

Thus

$$a^2 x^2 - b^2 = (ax + b)(ax - b) \in I$$

and

$$a^2(x^2 + 1) \in I.$$

Hence

$$a^2 + b^2 = (a^2(x^2 + 1)) - (a^2 x^2 - b^2) \in I.$$

Thus I contains a nonzero real number, that is, I contains a unit of $\mathbb{R}[x]$. This proves that $I = \mathbb{R}[x]$, a contradiction. Hence no such ideal I exists, and consequently $\langle x^2 + 1 \rangle$ is a maximal ideal of $\mathbb{R}[x]$.

Example 1.5.2 *$\langle 5 \rangle$ is not a maximal ideal of $\mathbb{Z} + \mathbb{Z}i$ as*

$$\langle 5 \rangle \subset \langle 1 + 2i \rangle \subset \mathbb{Z} + \mathbb{Z}i.$$

Theorem 1.5.1 *Let D be an integral domain. Let $a \in D$ be such that $a \neq 0$ and $a \notin U(D)$. Then*

$$\langle a \rangle \text{ is a maximal ideal of } D \Longrightarrow a \text{ is irreducible in } D.$$

Proof: Suppose that a is not an irreducible element of D. Then, as a is neither 0 nor a unit, it must be reducible. Hence there exist $b \in D$ and $c \in D$ such that $a = bc$

and neither b nor c is a unit or 0. Thus

$$\langle a \rangle \subset \langle b \rangle \subset D$$

so that $\langle a \rangle$ is not a maximal ideal. Hence we have shown that

$$\langle a \rangle \text{ is a maximal ideal} \Longrightarrow a \text{ is irreducible,}$$

as asserted. ∎

The next example shows that the converse of Theorem 1.5.1 is not true in general.

Example 1.5.3

(a) x is an irreducible element of $\mathbb{Z}[x]$ but $\langle x \rangle$ is not a maximal ideal of $\mathbb{Z}[x]$ as

$$\langle x \rangle \subset \langle 2, x \rangle \subset \mathbb{Z}[x].$$

(b) $1 + \sqrt{-5}$ is an irreducible element of $\mathbb{Z} + \mathbb{Z}\sqrt{-5}$ but $\langle 1 + \sqrt{-5} \rangle$ is not a maximal ideal of $\mathbb{Z} + \mathbb{Z}\sqrt{-5}$ as

$$\langle 1 + \sqrt{-5} \rangle \subset \langle 2, 1 + \sqrt{-5} \rangle \subset \mathbb{Z} + \mathbb{Z}\sqrt{-5}.$$

However, the converse of Theorem 1.5.1 is true in a principal ideal domain.

Theorem 1.5.2 *Let D be a principal ideal domain. Let $a \in D$ be such that $a \neq 0$ and $a \notin U(D)$. Then*

$$\langle a \rangle \text{ is a maximal ideal of } D \Longleftrightarrow a \text{ is irreducible in } D.$$

Proof: In view of Theorem 1.5.1 we have only to show that

$$a \text{ is irreducible} \Longrightarrow \langle a \rangle \text{ is maximal.} \tag{1.5.1}$$

Suppose that a is irreducible but that $\langle a \rangle$ is not a maximal ideal. Then there exists an ideal I such that

$$\langle a \rangle \subset I \subset D.$$

As D is a principal ideal domain, $I = \langle b \rangle$ for some $b \in D$. Hence

$$\langle a \rangle \subset \langle b \rangle \subset D$$

and so

$$a = bc,$$

for some $c \in D$. Since $\langle b \rangle \subset D$, b is not a unit, and since $\langle a \rangle \subset \langle b \rangle$, c is not a unit. Thus a is reducible, which is a contradiction. This completes the proof of (1.5.1). ∎

Theorem 1.5.3 *Let D be an integral domain and let I be an ideal of D. Then*

$$D/I \text{ is a field} \iff I \text{ is maximal.}$$

Proof: Suppose that D/I is a field and that J is an ideal of D with

$$I \subset J \subseteq D.$$

Thus there exists $b \in J$ with $b \notin I$. Then $b + I$ is a nonzero element of D/I and therefore, as D/I is a field, there exists an element $c + I \in D/I$ such that

$$(b + I)(c + I) = 1 + I.$$

Thus

$$bc + I = 1 + I$$

and so

$$bc - 1 \in I \subset J.$$

Since $b \in J$ and $c \in D$ we have

$$bc \in J.$$

Hence

$$1 = bc - (bc - 1) \in J,$$

so that $J = \langle 1 \rangle = D$. This proves that I is maximal.

Now suppose that I is maximal. To show that D/I is a field we have only to show that $b + I \neq 0 + I$ has a multiplicative inverse, as all the other field properties follow trivially. As $b + I \neq 0 + I$ we have $b \notin I$. Consider

$$B = \{x \in D \mid x = by + w \text{ for some } y \in D \text{ and some } w \in I\}.$$

It is easy to check that B is an ideal of D such that $I \subset B$ (Exercise 12). Since I is maximal we must have $B = D$. Thus $1 \in B$ so that $1 = by' + w'$ for some $y' \in D$ and some $w' \in I$. Then

$$(b + I)(y' + I) = by' + I = 1 - w' + I = 1 + I$$

so that $(b + I)^{-1}$ exists and is equal to $y' + I$.   ∎

Definition 1.5.2 (Prime ideal) *A proper ideal P of an integral domain D is called a prime ideal if*

$$a, b \in D \text{ and } ab \in P \text{ implies } a \in P \text{ or } b \in P.$$

Example 1.5.4 *The principal ideal $I = \langle x^2 + 1 \rangle$ is not a prime ideal of $\mathbb{C}[x]$ as*

$$x \pm i \in \mathbb{C}[x], \ (x + i)(x - i) = x^2 + 1 \in I \ but \ x \pm i \notin I.$$

Example 1.5.5 *The ideal $I = \langle 1 + i \rangle$ is a prime ideal of $\mathbb{Z} + \mathbb{Z}i$. To see this, suppose that $a + bi \in \mathbb{Z} + \mathbb{Z}i$ and $c + di \in \mathbb{Z} + \mathbb{Z}i$ are such that*

$$(a + bi)(c + di) \in \langle 1 + i \rangle.$$

Then there exists $x + yi \in \mathbb{Z} + \mathbb{Z}i$ such that

$$(a + bi)(c + di) = (1 + i)(x + yi).$$

Equating real and imaginary parts we obtain

$$ac - bd = x - y, \ \ ad + bc = x + y.$$

Adding these two equations, we have

$$ac + ad + bc - bd = 2x,$$

so that

$$(a + b)(c + d) = ac + ad + bc + bd \equiv ac + ad + bc - bd = 2x \equiv 0 \, (\mathrm{mod} \ 2).$$

Hence either $a + b$ or $c + d$ is even. Without loss of generality we may suppose that $a + b$ is even. Hence there exist $u \in \mathbb{Z}$ and $v \in \mathbb{Z}$ such that $a + b = 2u$ and $a - b = 2v$. Then

$$a + bi = (u + v) + (u - v)i = (1 + i)(u - vi)$$

and thus $a + bi \in \langle 1 + i \rangle$, proving that $\langle 1 + i \rangle$ is a prime ideal.

We next determine which principal ideals of an integral domain are prime.

Theorem 1.5.4 *Let D be an integral domain. Let $a \in D$ be such that $a \neq 0$ and $a \notin U(D)$. Then*

$$\langle a \rangle \ is \ a \ prime \ ideal \ of \ D \Longleftrightarrow a \ is \ prime \ in \ D.$$

Proof: Suppose that $\langle a \rangle$ is a prime ideal of D. Let $b, c \in D$ be such that $a \mid bc$ so that $bc \in \langle a \rangle$. As $\langle a \rangle$ is a prime ideal, we must have $b \in \langle a \rangle$ or $c \in \langle a \rangle$; that is, $a \mid b$ or $a \mid c$, showing that a is prime.

Now suppose that a is a prime in D. Let $b \in D$ and $c \in D$ be such that $bc \in \langle a \rangle$. Hence there exists $d \in D$ such that $bc = ad$, so that $a \mid bc$. As a is prime we have $a \mid b$ or $a \mid c$. Without loss of generality we may suppose that $a \mid b$. Hence there exists $e \in D$ such that $b = ae$ and so $b \in \langle a \rangle$. This proves that $\langle a \rangle$ is a prime ideal. ∎

Theorem 1.5.5 *Let D be an integral domain and let I be an ideal of D. Then*

$$D/I \text{ is an integral domain} \Longleftrightarrow I \text{ is prime.}$$

Proof: Suppose first that D/I is an integral domain and that $a, b \in D$ are such that $ab \in I$. Then $(a + I)(b + I) = ab + I = 0 + I$, the zero element of the integral domain D/I. Because an integral domain has no divisors of zero, we have $a + I = 0 + I$ or $b + I = 0 + I$; that is, we have either $a \in I$ or $b \in I$, so that I is prime.

Now suppose that I is a prime ideal of D. As I is a proper ideal of D, D/I is a commutative ring with identity $1 + I$. Thus we have only to check that when I is prime, D/I has no divisors of zero. Suppose that $a + I \in D/I$ and $b + I \in D/I$ are such that $(a + I)(b + I) = 0 + I$. Then $ab + I = I$, so that $ab \in I$. As I is prime, either $a \in I$ or $b \in I$; that is, $a + I = 0 + I$ or $b + I = 0 + I$, so D/I has no zero divisors. ∎

Theorem 1.5.6 *Let D be an integral domain. Let I be a maximal ideal of D. Then I is a prime ideal of D.*

Proof: Let I be a maximal ideal of D. Then, by Theorem 1.5.3, D/I is a field. But a field is always an integral domain, so D/I is an integral domain. Then, by Theorem 1.5.5, I is a prime ideal of D. ∎

The next example shows that the converse of Theorem 1.5.6 is not true in general.

Example 1.5.6 $\langle x \rangle$ *is a prime ideal of $\mathbb{Z}[x]$, but it is not a maximal ideal of $\mathbb{Z}[x]$.*

The converse of Theorem 1.5.6 is true in a principal ideal domain.

Theorem 1.5.7 *Let D be a principal ideal domain. Let I be a proper ideal of D. Then*

$$I \text{ is maximal} \Longleftrightarrow I \text{ is prime.}$$

Proof: In view of Theorem 1.5.6 we have only to show that if I is a prime ideal of D then I is a maximal ideal.

Suppose that I is a prime ideal of D that is not maximal. Then there exists an ideal J of D such that

$$I \subset J \subset D.$$

As D is a principal ideal domain, we have $I = \langle a \rangle$ and $J = \langle b \rangle$ for some $a, b \in D$. As $\langle a \rangle \subset \langle b \rangle$ we have $a = bc$ for some $c \in D$. Now $bc = a \in \langle a \rangle = I$, and I is prime, so that either $b \in I$ or $c \in I$. If $b \in I$ then $J = \langle b \rangle \subseteq I \subset J$, which is a contradiction. Hence $c \in I$. Thus $c = ad$ for some $d \in D$, and so $a = bda$. But

$a \neq 0$ so $bd = 1$. Thus b is a unit and $J = \langle b \rangle = D \supset J$, a contradiction. Hence I is maximal. ∎

1.6 Sums and Products of Ideals

In this section we show how to add and multiply ideals to obtain further ideals. First we define the sum of two ideals.

Definition 1.6.1 (Sum of ideals) *Let I and J be ideals in an integral domain D. The sum of I and J, written $I + J$, is defined by*

$$I + J = \{i + j \mid i \in I, j \in J\}.$$

It is readily checked that $I + J$ is also an ideal and that it is the minimal ideal containing both I and J. The following properties are also easily checked: For ideals I, J, K of the integral domain D

$$I + J = J + I,$$
$$(I + J) + K = I + (J + K),$$
$$I + \langle 0 \rangle = I,$$
$$I + \langle 1 \rangle = \langle 1 \rangle.$$

Further, if $I = \langle i \rangle$ and $J = \langle j \rangle$ are principal ideals, then $I + J = \langle i, j \rangle$. It is easy to extend Definition 1.6.1 to the sum of a finite number of ideals.

Next we define the product of two ideals.

Definition 1.6.2 (Product of ideals) *Let I and J be ideals in an integral domain D. The product of I and J, written IJ, is defined by*

$$IJ = \{x \in D \mid x = i_1 j_1 + \cdots + i_r j_r \text{ for some } r \in \mathbb{N},$$
$$\text{some } i_1, \ldots, i_r \in I, \text{ and some } j_1, \ldots, j_r \in J\}.$$

Clearly IJ is the set of all finite sums of products of elements of I and J, and it is easily checked that IJ is an ideal. The following properties are also easily verified: For ideals I, J, K of the integral domain D

$$IJ = JI,$$
$$(IJ)K = I(JK),$$
$$I\langle 0 \rangle = \langle 0 \rangle,$$
$$I\langle 1 \rangle = I.$$

Further, if $I = \langle i \rangle$ and $J = \langle j \rangle$ are principal ideals, then $IJ = \langle ij \rangle$. We leave it to the reader to extend Definition 1.6.2 to a product of a finite number of ideals.

Addition and multiplication of ideals are related by the distributive law

$$(I + J)K = IK + JK.$$

Example 1.6.1 *Let m and n be integers that are not both zero. Set $d = (m, n)$, the greatest common divisor of m and n. We show that*

$$\langle m \rangle + \langle n \rangle = \langle d \rangle.$$

Let $a \in \langle m \rangle + \langle n \rangle$. Then there exist integers r and s such that $a = rm + sn$. As $d = (m, n)$ there exist coprime integers m_1 and n_1 such that $m = dm_1$, $n = dn_1$. Thus $a = rdm_1 + sdn_1 = (rm_1 + sn_1)d \in \langle d \rangle$. This shows that $\langle m \rangle + \langle n \rangle \subseteq \langle d \rangle$.

Now let $a \in \langle d \rangle$, so that there exists an integer e such that $a = de$. As $d = (m, n)$ there exist integers x and y such that $d = xm + yn$. Hence $a = (xm + yn)e = (xe)m + (ye)n \in \langle m \rangle + \langle n \rangle$. This proves that $\langle d \rangle \subseteq \langle m \rangle + \langle n \rangle$.

The two inclusions show that $\langle m \rangle + \langle n \rangle = \langle d \rangle$.

Next we give another necessary and sufficient condition for a proper ideal to be a prime ideal.

Theorem 1.6.1 *Let P be a proper ideal of an integral domain D. Then P is a prime ideal if and only if for any two ideals A and B of D satisfying $AB \subseteq P$ either $A \subseteq P$ or $B \subseteq P$.*

Proof: Suppose that P is a proper ideal of D with the property

$$AB \subseteq P \implies A \subseteq P \text{ or } B \subseteq P \ (A, B \text{ ideals of } D). \tag{1.6.1}$$

Let $a, b \in D$ be such that $ab \in P$. Set $A = \langle a \rangle$, $B = \langle b \rangle$ so that $AB = \langle a \rangle \langle b \rangle = \langle ab \rangle \subseteq P$. Hence $\langle a \rangle \subseteq P$ or $\langle b \rangle \subseteq P$. Thus $a \in P$ or $b \in P$, showing that P is a prime ideal.

Now suppose that P does not satisfy (1.6.1). Then there exist ideals A and B of D with

$$A \not\subseteq P, \ B \not\subseteq P, \ AB \subseteq P.$$

Let $a \in A$, $a \notin P$ and $b \in B$, $b \notin P$. Then $ab \in AB \subseteq P$ but $a \notin P$, $b \notin P$, so P is not a prime ideal.   ■

Our final theorem of this chapter shows that a prime ideal P of an integral domain D_1 remains prime when restricted to a subdomain D of D_1.

Theorem 1.6.2 *Let D and D_1 be integral domains satisfying $D \subseteq D_1$. Let P be a prime ideal of D_1 such that $P \cap D \neq \{0\}$, D. Then $P \cap D$ is a prime ideal of D.*

Proof: We show first that $P \cap D$ is an ideal of D. Let $a, b \in P \cap D$. Then $a, b \in P$ and $a, b \in D$. From the first of these, as P is an ideal, we see that $a + b \in P$. From the second, as D is an integral domain, it is closed under addition so that $a + b \in D$. Hence $a + b \in P \cap D$. Now suppose that $a \in P \cap D$ and $d \in D$. As $d \in D$, $a \in P$ and P is an ideal of D, we deduce that $da \in P$. As $d \in D$, $a \in D$ and D being an integral domain is closed under multiplication, we see that $da \in D$. Thus $da \in P \cap D$. This proves that $P \cap D$ is an ideal of D. Since $P \cap D \neq \{0\}$, D, by assumption, $P \cap D$ is a proper ideal of D.

Finally, we show that $P \cap D$ is a prime ideal. Let $a, b \in D$ be such that $ab \in P \cap D$. Then $a, b \in D_1$ and $ab \in P$. As P is a prime ideal of D_1, we deduce that $a \in P$ or $b \in P$. This completes the proof that $P \cap D$ is a prime ideal of D. ∎

Exercises

1. Prove that $U(\mathbb{Z} + \mathbb{Z}i) = \{\pm 1, \pm i\}$.
2. Prove that $U(\mathbb{Z} + \mathbb{Z}\omega) = \{\pm 1, \pm\omega, \pm\omega^2\}$.
3. Let m be an integer with $m < -1$. Prove that

$$U(\mathbb{Z} + \mathbb{Z}\sqrt{m}) = \{\pm 1\}.$$

4. Let m be an integer with $m \equiv 1 \pmod 4$ and $m < -3$. Prove that

$$U\left(\mathbb{Z} + \mathbb{Z}\left(\frac{1 + \sqrt{m}}{2}\right)\right) = \{\pm 1\}.$$

5. Let

$$D = \{f(x) \in \mathbb{Q}[x] \mid f(0) \in \mathbb{Z}\}.$$

   Prove that D is a subdomain of $\mathbb{Q}[x]$.
6. Determine $U(D)$ for D as given in Exercise 5.
7. Let D be an integral domain. Let $u \in U(D)$. Let I be an ideal of D that contains u. Prove that $I = D$.
8. In Example 1.3.4 prove that J is not a principal ideal.
9. Let

$$S = \{a + bi \in \mathbb{Z} + \mathbb{Z}i \mid b \equiv 0 \pmod 2\}.$$

   Is S an ideal of $\mathbb{Z} + \mathbb{Z}i$?
10. If A and B are ideals of an integral domain D, prove that $A \cap B$ is also an ideal of D.
11. Give an example to show that if A and B are ideals of an integral domain D then $A \cup B$ may not be an ideal of D.
12. Prove that the set B defined in the proof of Theorem 1.5.3 is an ideal.
13. Let A and B be ideals of an integral domain D. Prove that $AB \subseteq A \cap B$.
14. Let A and B be ideals of an integral domain D. Show that $(A \cap B)(A + B) \subseteq AB$. Give an example to show that equality does not always hold.
15. Give an example to show that an integral domain may not contain any irreducible elements.

16. Prove that $\langle x \rangle$ is a prime ideal of $\mathbb{Z}[x]$.

17. Let m be a positive integer that is not a perfect square. Let $\alpha = a + b\sqrt{m} \in \mathbb{Z} + \mathbb{Z}\sqrt{m}$. Prove that if

$$a^2 - mb^2 = \pm 1$$

then $\alpha \in U(\mathbb{Z} + \mathbb{Z}\sqrt{m})$.

18. Let m be a positive integer with $m \equiv 1 \pmod 4$ that is not a perfect square. Let $\alpha = a + b\left(\frac{1+\sqrt{m}}{2}\right) \in \mathbb{Z} + \mathbb{Z}\left(\frac{1+\sqrt{m}}{2}\right)$. Prove that if

$$a^2 + ab + \left(\frac{1-m}{4}\right)b^2 = \pm 1$$

then $\alpha \in U\left(\mathbb{Z} + \mathbb{Z}\left(\frac{1+\sqrt{m}}{2}\right)\right)$.

19. Prove that $\langle 1 - 3i,\ 3 - i \rangle$ is a principal ideal in $\mathbb{Z} + \mathbb{Z}i$ by finding a generator for this ideal.

20. Prove that $\langle 2, 1 + \sqrt{-5} \rangle = \langle 2, 1 - \sqrt{-5} \rangle$, $\langle 3, 1 + \sqrt{-5} \rangle \neq \langle 3, 1 - \sqrt{-5} \rangle$, $\langle 2, 1 + \sqrt{-5} \rangle \neq \langle 3, 1 + \sqrt{-5} \rangle$, and $\langle 2, 1 + \sqrt{-5} \rangle \neq \langle 3, 1 - \sqrt{-5} \rangle$ in $\mathbb{Z} + \mathbb{Z}\sqrt{-5}$.

21. Prove that $\langle 2, 1 + \sqrt{-5} \rangle$, $\langle 3, 1 + \sqrt{-5} \rangle$, and $\langle 3, 1 - \sqrt{-5} \rangle$ are prime ideals of $\mathbb{Z} + \mathbb{Z}\sqrt{-5}$. Determine $\langle 2, 1 + \sqrt{-5} \rangle \cap \mathbb{Z}$, $\langle 3, 1 + \sqrt{-5} \rangle \cap \mathbb{Z}$, and $\langle 3, 1 - \sqrt{-5} \rangle \cap \mathbb{Z}$.

22. Let D be an integral domain. Let $a, b, c \in D$ be such that $\langle a, c \rangle = D$. Prove that $\langle a, bc \rangle = \langle a, b \rangle$.

23. Prove that $17 - 3\sqrt{3} \sim 83 + 47\sqrt{3}$ in $\mathbb{Z} + \mathbb{Z}\sqrt{3}$.

24. Give an example of an integral domain satisfying (1.2.3).

25. Express $2 + 8\sqrt{-5}$ as a product of irreducibles in $\mathbb{Z} + \mathbb{Z}\sqrt{-5}$. In how many ways can this be done?

26. Prove that $\sqrt{-6}$ is not a prime in $\mathbb{Z} + \mathbb{Z}\sqrt{-6}$.

27. Prove that $\sqrt{-6}$ is an irreducible in $\mathbb{Z} + \mathbb{Z}\sqrt{-6}$.

28. Prove that $\mathbb{Z} + \mathbb{Z}\sqrt{-6}$ is not a principal ideal domain.

29. Give an example of an ideal in $\mathbb{Z} + \mathbb{Z}\sqrt{-6}$ that is not principal.

30. Prove that $\sqrt{10}$ is not a prime in $\mathbb{Z} + \mathbb{Z}\sqrt{10}$.

31. Prove that $\sqrt{10}$ is an irreducible in $\mathbb{Z} + \mathbb{Z}\sqrt{10}$.

32. Prove that $\mathbb{Z} + \mathbb{Z}\sqrt{10}$ is not a principal ideal domain.

33. Give an example of an ideal in $\mathbb{Z} + \mathbb{Z}\sqrt{10}$ that is not principal.

34. Let P be a prime ideal of an integral domain D. Let $A_1, \ldots, A_k$ be ideals of D such that $P \supseteq A_1 \cdots A_k$. Prove that $P \supseteq A_i$ for some $i \in \{1, 2, \ldots, k\}$.

35. Let $r \in \mathbb{Z} \setminus \{-2, 0\}$. Prove that

$$D = \{a + b\theta + c\theta^2 \mid a, b, c \in \mathbb{Z}\},$$

where

$$\theta^3 + r\theta + 1 = 0$$

is an integral domain. Prove that $\theta \in U(D)$.

36. Let p be a prime. Let m be an integer with $m \leq -(p + 1)$. Prove that p is irreducible in $\mathbb{Z} + \mathbb{Z}\sqrt{m}$.

37. Let p be a prime. Let m be an integer with $m \equiv 1 \pmod 4$ and $m \leq -(4p+1)$. Prove that p is irreducible in $\mathbb{Z} + \mathbb{Z}\left(\frac{1+\sqrt{m}}{2}\right)$.

Suggested Reading

1. P. M. Cohn, *Rings of fractions,* American Mathematical Monthly 78 (1971), 596–615.

   The author won the Lester R. Ford award for expository writing for this paper. The paper reviews Ore's work on embedding certain non-commutative rings in skew fields, a generalization of the corresponding standard result for integral domains mentioned in Section 1.1.

2. D. A. Cox, *Primes of the Form $x^2 + my^2$,* Wiley, New York, 1989.

   The main theorem of the book (Theorem 9.2, p. 180) asserts (with some details omitted) that if m is a positive integer then there is a polynomial $f_m(x) \in \mathbb{Z}[x]$ (of a certain degree depending only on m) such that if p is an odd prime satisfying $\left(\frac{-m}{p}\right) = 1$ then $p = u^2 + mv^2$ for integers u and v if and only if the congruence $f_m(x) \equiv 0 \pmod p$ is solvable.

   For example if p is an odd prime such that $\left(\frac{-36}{p}\right) = 1$, that is, $p \equiv 1 \pmod 4$, then

   $$p = u^2 + 36v^2 \iff x^4 + 3 \equiv 0 \pmod p \text{ is solvable.}$$

3. J. B. Fraleigh, *A First Course in Abstract Algebra,* Addison-Wesley, Reading, Massachusetts, 1968.

   Chapter 26 is devoted to constructing the field of quotients of an integral domain.

4. L. Kinkade and J. Wagner, *When polynomial rings are principal ideal rings,* Journal of Undergraduate Mathematics 23 (1991), 59–62.

   In this paper it is shown that $R[x]$ is a principal ideal ring if and only if $R \simeq R_1 \oplus R_2 \oplus \cdots \oplus R_n$, where $R_1, \ldots, R_n$ are fields.

5. D. E. Rowe, *Gauss, Dirichlet and the law of biquadratic reciprocity,* The Mathematical Intelligencer 10 (1988), 13–26.

   This paper gives a discussion of the relationship between Gauss and Dirichlet, mainly concerning their contributions to number theory including their work on biquadratic reciprocity.

6. W. C. Waterhouse, *Quadratic polynomials and unique factorization,* American Mathematical Monthly 109 (2002), 70–72.

   Theorem 1.2.2 is taken from this paper.

Biographies

1. E. T. Bell, *Men of Mathematics,* Simon and Schuster, New York, 1937.

   Chapters 14 and 17 are devoted to Gauss and Abel respectively.

2. W. K. Bühler, *Gauss: A Biographical Study,* Springer-Verlag, Berlin, Heildelberg, New York, 1981.

   This book provides a comprehensive discussion of Gauss's life and work.

3. G. Eisenstein, *Mathematische Werke,* Bände I, II, Chelsea Publishing Co., New York, 1989.

   The foreword to Eisenstein's Collected Papers comprises an interesting discussion of Eisenstein's work by André Weil.

4. O. Ore, *Niels Henrik Abel, Mathematician Extraordinary,* University of Minnesota Press, 1957; Chelsea, New York, 1974.

   A nontechnical biography is presented here in an easy-to-read fashion.

5. A. Stubhaug, *Niels Henrik Abel and His Times,* Springer-Verlag, New York, 2000.

   Stubhaug presents a very readable account of a remarkable mathematician.

6. The website

                    http://www-groups.dcs.st-and.ac.uk/~history/

   has biographies of Abel, Gauss, and Eisenstein.

2

Euclidean Domains

In the proof of Theorem 1.4.1 we made use of the following property of $\mathbb{Z}$: Given $a, b \in \mathbb{Z}$ with $b > 0$ then there exist $q, r \in \mathbb{Z}$ such that

$$a = qb + r, \; 0 \le r < b. \qquad (2.0.1)$$

In fact the integers q and r are uniquely determined by a and b. We have

$$q = [a/b], \; r = a - b[a/b], \qquad (2.0.2)$$

where $[x]$ denotes the greatest integer less than or equal to the real number x. The integer q is called the quotient and the integer r the remainder. An important class of integral domains are those possessing a property analogous to (2.0.1). Such domains are called Euclidean domains. In Theorem 2.1.2 we show that Euclidean domains are principal ideal domains.

2.1 Euclidean Domains

To define a Euclidean domain we must first define a Euclidean function.

Definition 2.1.1 (Euclidean function) *Let D be an integral domain. A mapping $\phi : D \to \mathbb{Z}$ is called a Euclidean function on D if it has the following two properties:*

$$\phi(ab) \ge \phi(a), \; \text{for all } a, b \in D \text{ with } b \ne 0, \qquad (2.1.1)$$

$$\text{if } a, b \in D \text{ with } b \ne 0 \text{ then there exist } q, r \in D \qquad (2.1.2)$$
$$\text{such that } a = qb + r \text{ and } \phi(r) < \phi(b).$$

Example 2.1.1 $\phi(a) = |a| \; (a \in \mathbb{Z})$ *is a Euclidean function on $\mathbb{Z}$.*

Example 2.1.2 *Let $D = F[x]$, where F is a field. D is the domain of polynomials in x with coefficients in F. Let $p(x) \in D$. Then*

$$\phi(p(x)) = \begin{cases} \deg (p(x)), & \text{if } p(x) \ne 0, \\ -1, & \text{if } p(x) = 0, \end{cases}$$

is a Euclidean function on D.

In general the elements q and r in (2.1.2) are not uniquely determined. If D is an integral domain that is not a field and that possesses a Euclidean function ϕ for which the quotient and remainder r in (2.1.2) are always uniquely determined by a and b then $D = F[x]$ for some field F. This result is due to Rhai [14]; see also Jodeit [12].

Theorem 2.1.1 (Properties of a Euclidean function) *Let D be an integral domain that possesses a Euclidean function ϕ. Let $a, b \in D$. Then*

(a) $a \sim b \Longrightarrow \phi(a) = \phi(b)$,
(b) $a \mid b$ and $\phi(a) = \phi(b) \Longrightarrow a \sim b$,
(c) $a \in U(D) \Longleftrightarrow \phi(a) = \phi(1)$,
(d) $\phi(a) > \phi(0)$, *if $a \neq 0$.*

Proof: **(a)** As $a \sim b$ there exists $u \in U(D)$ such that $a = ub$. Then by (2.1.1) we have $\phi(a) = \phi(ub) \geq \phi(b)$. As $u \in U(D)$, we have $u^{-1} \in U(D)$ and $b = u^{-1}a$, so again by (2.1.1) we have $\phi(b) = \phi(u^{-1}a) \geq \phi(a)$. From these two inequalities, we deduce that $\phi(a) = \phi(b)$.

   (b) By (2.1.2) there exist $q, r \in D$ such that $a = qb + r$ and $\phi(r) < \phi(b) = \phi(a)$. Now $a \mid b$ so that we have $a \mid r$. Suppose $r \neq 0$. Then by (2.1.1) we have $\phi(r) \geq \phi(a)$, which is a contradiction. Hence $r = 0$. Thus $a = qb$. But $a \mid b$ so $q \in U(D)$ and thus $a \sim b$.

   (c) First we have

$$a \in U(D) \Longrightarrow a \sim 1 \Longrightarrow \phi(a) = \phi(1)$$

by part (a). Second, we have

$$1 \mid a, \ \phi(1) = \phi(a) \Longrightarrow 1 \sim a \Longrightarrow a \in U(D)$$

by part (b).

   (d) By (2.1.2) there exist $q, r \in D$ such that

$$0 = qa + r, \ \phi(r) < \phi(a).$$

Suppose $r \neq 0$. Then $q \neq 0$ and by (2.1.1) we have

$$\phi(r) = \phi((-q)a) \geq \phi(a),$$

which is a contradiction. Hence $r = 0$ and $\phi(0) < \phi(a)$. ∎

Definition 2.1.2 (Euclidean domain) *Let D be an integral domain. If D possesses a Euclidean function ϕ then D is called a Euclidean domain with respect to ϕ.*

If D is a Euclidean domain with respect to some Euclidean function ϕ and it is not important to specify ϕ, we just call D a Euclidean domain. Before giving examples of Euclidean domains in the next section, we prove the fundamental theorem that every Euclidean domain is a principal ideal domain.

Theorem 2.1.2 *A Euclidean domain is a principal ideal domain.*

Proof: Let D be a Euclidean domain. Hence D possesses a Euclidean function, say ϕ. Let I be an ideal in D. If $I = \{0\}$ then $I = \langle 0 \rangle$ is a principal ideal. If $I \neq \{0\}$ we consider the set S of integers defined by

$$S = \{\phi(x) \mid x \in I, \ x \neq 0\}.$$

As $I \neq \{0\}$, S is a nonempty set. By Theorem 2.1.1(d), S is bounded below. Hence S has a least element, say $\phi(a)$, $a \in I$, $a \neq 0$. If $b \in I$ then, as ϕ is a Euclidean function, there exist $q, r \in D$ such that

$$b = qa + r, \ \phi(r) < \phi(a).$$

Now, as I is an ideal, $r = b - qa \in I$, and so, as $\phi(a)$ is the least element of S, we have $r = 0$. Hence $b = qa$ and so $I = \langle a \rangle$. Thus every ideal in D is principal and so D is a principal ideal domain. $\blacksquare$

The integral domain $\mathbb{Z} + \mathbb{Z}\left(\frac{1+\sqrt{-19}}{2}\right)$ is a principal ideal domain. This will be proved in Chapter 12 (see Example 12.6.1). However, it is not a Euclidean domain (Theorem 2.3.8), so the converse of Theorem 2.1.2 is not true.

In a Euclidean domain D a greatest common divisor of two elements a and b of D (see Definition 1.4.2) can be obtained by means of the Euclidean algorithm. Since $(c, 0) = (0, c) = c$ for all $c \ (\neq 0) \in D$, it suffices to consider only elements a and b that are not zero.

Theorem 2.1.3 (Euclidean algorithm) *Let a and b be nonzero elements of a Euclidean domain D with Euclidean function ϕ. Define elements $q_1, q_2, \ldots$ and $r_{-1}, r_0, r_1, r_2, \ldots$ of D recursively by*

$$r_{-1} = a, \ r_0 = b, \tag{2.1.3}$$

and

$$r_j = q_{j+2} r_{j+1} + r_{j+2}, \ \phi(r_{j+2}) < \phi(r_{j+1}), \tag{2.1.4}$$

for $j = -1, 0, 1, 2, \ldots, k$, where k is the least integer ≥ -1 such that

$$r_{k+2} = 0.$$

Then

$$(a, b) = r_{k+1}.$$

Proof: By property (2.1.2) of the Euclidean function ϕ, the relations (2.1.3) and (2.1.4) define $q_1, q_2, \ldots, q_{k+2}$ and $r_{-1}, r_0, r_1, \ldots, r_{k+2}$, and since the sequence $\phi(r_1), \phi(r_2), \ldots$ is a decreasing sequence of integers bounded below by $\phi(0)$ (Theorem 2.1.1(d)) it must terminate after a finite number of steps (say $k + 2$ steps) so that $r_{k+2} = 0$. From (2.1.4) we deduce that

$$\langle r_j, r_{j+1} \rangle = \langle q_{j+2} r_{j+1} + r_{j+2}, r_{j+1} \rangle = \langle r_{j+2}, r_{j+1} \rangle = \langle r_{j+1}, r_{j+2} \rangle$$

for $j = -1, 0, 1, 2, \ldots, k$. Hence

$$\langle a, b \rangle = \langle r_{-1}, r_0 \rangle = \langle r_0, r_1 \rangle = \cdots = \langle r_k, r_{k+1} \rangle$$
$$= \langle r_{k+1}, r_{k+2} \rangle = \langle r_{k+1}, 0 \rangle = \langle r_{k+1} \rangle$$

so that

$$\langle a, b \rangle = r_{k+1}. \qquad \blacksquare$$

2.2 Examples of Euclidean Domains

In view of Examples 2.1.1 and 2.1.2 we have

Theorem 2.2.1

(a) $\mathbb{Z}$ *is a Euclidean domain.*
(b) *Let F be a field. Then $F[x]$ is a Euclidean domain.*

From Theorems 2.1.2 and 2.2.1 we see that $\mathbb{Z}$ and $F[x]$ are principal ideal domains. In the remainder of this section we investigate when the integral domains $\mathbb{Z} + \mathbb{Z}\sqrt{m}$ ($m \equiv 2, 3 \pmod 4$) and $\mathbb{Z} + \mathbb{Z}\left(\frac{1+\sqrt{m}}{2}\right)$ ($m \equiv 1 \pmod 4$) are Euclidean with respect to the function that maps $r + s\sqrt{m}$ to $|r^2 - ms^2|$. In this section we denote this function by ϕ_m. Later in Section 9.2 we recognize ϕ_m as the absolute value of the norm of the element $r + s\sqrt{m}$. Integral domains that are Euclidean with respect to the absolute value of the norm are called norm-Euclidean.

Definition 2.2.1 (Function ϕ_m) *Let m be a squarefree integer. The function ϕ_m : $\mathbb{Q}(\sqrt{m}) \to \mathbb{Q}$ is defined by*

$$\phi_m(r + s\sqrt{m}) = |r^2 - ms^2|$$

for all $r, s \in \mathbb{Q}$.

The basic properties of ϕ_m are given in the next lemma.

Lemma 2.2.1 *Let m be a squarefree integer.*

(a) $\phi_m : \mathbb{Z} + \mathbb{Z}\sqrt{m} \to \mathbb{N} \cup \{0\}$.
(b) *If $m \equiv 1 \pmod 4$ then $\phi_m : \mathbb{Z} + \mathbb{Z}\left(\frac{1+\sqrt{m}}{2}\right) \to \mathbb{N} \cup \{0\}$.*

(c) *Let $\alpha \in \mathbb{Q}(\sqrt{m})$. Then $\phi_m(\alpha) = 0 \iff \alpha = 0$.*

(d) *$\phi_m(\alpha\beta) = \phi_m(\alpha)\phi_m(\beta)$ for all $\alpha, \beta \in \mathbb{Q}(\sqrt{m})$.*

(e) *$\phi_m(\alpha\beta) \geq \phi_m(\alpha)$ for all $\alpha, \beta \in \mathbb{Z} + \mathbb{Z}\sqrt{m}$ with $\beta \neq 0$.*

(f) *If $m \equiv 1 \pmod 4$, then $\phi_m(\alpha\beta) \geq \phi_m(\alpha)$ for all $\alpha, \beta \in \mathbb{Z} + \mathbb{Z}\left(\frac{1+\sqrt{m}}{2}\right)$ with $\beta \neq 0$.*

Proof: **(a)** Let $\alpha \in \mathbb{Z} + \mathbb{Z}\sqrt{m}$ so that $\alpha = x + y\sqrt{m}$ for some $x, y \in \mathbb{Z}$. Then $x^2 - my^2 \in \mathbb{Z}$ and $|x^2 - my^2| \geq 0$ so that

$$\phi_m(\alpha) = \phi_m(x + y\sqrt{m}) = |x^2 - my^2| \in \mathbb{N} \cup \{0\}.$$

(b) If $m \equiv 1 \pmod 4$ then $\mathbb{Z} + \mathbb{Z}\left(\frac{1+\sqrt{m}}{2}\right)$ is an integral domain (Example 1.1.5). Let $\alpha \in \mathbb{Z} + \mathbb{Z}\left(\frac{1+\sqrt{m}}{2}\right)$ so that $\alpha = x + y\left(\frac{1+\sqrt{m}}{2}\right) = \left(x + \frac{y}{2}\right) + \frac{y}{2}\sqrt{m}$ for some $x, y \in \mathbb{Z}$. Then

$$\phi_m(\alpha) = \phi_m\left(\left(x + \frac{y}{2}\right) + \frac{y}{2}\sqrt{m}\right)$$
$$= \left|\left(x + \frac{y}{2}\right)^2 - m\left(\frac{y}{2}\right)^2\right|$$
$$= \left|x^2 + xy + \frac{1}{4}(1 - m)y^2\right| \in \mathbb{N} \cup \{0\}, \text{ as } \frac{1}{4}(1 - m) \in \mathbb{Z}.$$

(c) Let $\alpha \in \mathbb{Q}(\sqrt{m})$ so that $\alpha = r + s\sqrt{m}$ for some $r, s \in \mathbb{Q}$. Then, as m is squarefree, we have

$$\phi_m(\alpha) = 0 \iff \phi_m(r + s\sqrt{m}) = 0$$
$$\iff |r^2 - ms^2| = 0$$
$$\iff r^2 = ms^2$$
$$\iff r = s = 0$$
$$\iff r + s\sqrt{m} = 0$$
$$\iff \alpha = 0.$$

(d) Let $\alpha, \beta \in \mathbb{Q}(\sqrt{m})$. Then $\alpha = x + y\sqrt{m}$ and $\beta = u + v\sqrt{m}$ for some $x, y, u, v \in \mathbb{Z}$. Thus

$$\phi_m(\alpha\beta) = \phi_m((x + y\sqrt{m})(u + v\sqrt{m}))$$
$$= \phi_m((xu + myv) + (xv + yu)\sqrt{m})$$
$$= |(xu + myv)^2 - m(xv + yu)^2|$$
$$= |x^2u^2 + m^2y^2v^2 - mx^2v^2 - my^2u^2|$$
$$= |(x^2 - my^2)(u^2 - mv^2)|$$
$$= |x^2 - my^2|\,|u^2 - mv^2|$$
$$= \phi_m(\alpha)\phi_m(\beta).$$

(e) Let $\alpha, \beta \in \mathbb{Z} + \mathbb{Z}\sqrt{m}$ with $\beta \neq 0$. By part (c), we have $\phi_m(\beta) \neq 0$. Then, by part (a), we deduce that $\phi_m(\alpha) \geq 0$ and $\phi_m(\beta) \geq 1$. Thus, by part (d), we have

$$\phi_m(\alpha\beta) = \phi_m(\alpha)\phi_m(\beta) \geq \phi_m(\alpha).$$

(f) This follows in exactly the same way as part (e) except that we use part (b) in place of part (a). ∎

Our next theorem uses the properties of ϕ_m given in Lemma 2.2.1 to give a convenient necessary and sufficient condition for $\mathbb{Z} + \mathbb{Z}\sqrt{m}$ to be Euclidean with respect to ϕ_m, that is, norm-Euclidean.

Theorem 2.2.2 *Let m be a squarefree integer. Then the integral domain $\mathbb{Z} + \mathbb{Z}\sqrt{m}$ is Euclidean with respect to ϕ_m if and only if for all $x, y \in \mathbb{Q}$ there exist $a, b \in \mathbb{Z}$ such that*

$$\phi_m((x + y\sqrt{m}) - (a + b\sqrt{m})) < 1. \tag{2.2.1}$$

Proof: Suppose first that $\mathbb{Z} + \mathbb{Z}\sqrt{m}$ is Euclidean with respect to ϕ_m. Let $x, y \in \mathbb{Q}$. Then $x + y\sqrt{m} = (r + s\sqrt{m})/t$ for integers r, s, t with $t \neq 0$. As ϕ_m is a Euclidean function on $\mathbb{Z} + \mathbb{Z}\sqrt{m}$ there exist $a + b\sqrt{m}, c + d\sqrt{m} \in \mathbb{Z} + \mathbb{Z}\sqrt{m}$ such that

$$r + s\sqrt{m} = t(a + b\sqrt{m}) + (c + d\sqrt{m}), \quad \phi_m(c + d\sqrt{m}) < \phi_m(t).$$

Hence

$$\phi_m((x + y\sqrt{m}) - (a + b\sqrt{m})) = \phi_m\left(\frac{r + s\sqrt{m}}{t} - (a + b\sqrt{m})\right)$$

$$= \phi_m\left(\frac{r + s\sqrt{m} - t(a + b\sqrt{m})}{t}\right)$$

$$= \phi_m\left(\frac{c + d\sqrt{m}}{t}\right)$$

$$= \frac{\phi_m(c + d\sqrt{m})}{\phi_m(t)} < 1,$$

by Lemma 2.2.1(d).

Now suppose that (2.2.1) holds. To show that $\mathbb{Z} + \mathbb{Z}\sqrt{m}$ is Euclidean with respect to ϕ_m, we must show that (2.1.1) and (2.1.2) hold. The inequality (2.1.1) holds in view of Lemma 2.2.1(e). We now show that (2.1.2) holds. Let $r + s\sqrt{m}$, $t + u\sqrt{m} \in \mathbb{Z} + \mathbb{Z}\sqrt{m}$ with $t + u\sqrt{m} \neq 0$. Then

$$\frac{r + s\sqrt{m}}{t + u\sqrt{m}} = x + y\sqrt{m},$$

where

$$x = \frac{rt - msu}{t^2 - mu^2} \in \mathbb{Q}, \quad y = \frac{st - ru}{t^2 - mu^2} \in \mathbb{Q}.$$

Note that $t + u\sqrt{m} \neq 0$ ensures that $t^2 - mu^2 \neq 0$. By (2.2.1) there exists $a + b\sqrt{m} \in \mathbb{Z} + \mathbb{Z}\sqrt{m}$ such that

$$\phi_m((x + y\sqrt{m}) - (a + b\sqrt{m})) < 1.$$

Set $c = r - at - bum \in \mathbb{Z}$, $d = s - au - bt \in \mathbb{Z}$, so that

$$c + d\sqrt{m} = (r + s\sqrt{m}) - (a + b\sqrt{m})(t + u\sqrt{m}) \in \mathbb{Z} + \mathbb{Z}\sqrt{m}.$$

Hence

$$r + s\sqrt{m} = (a + b\sqrt{m})(t + u\sqrt{m}) + (c + d\sqrt{m})$$

and

$$
\begin{aligned}
\phi_m(c + d\sqrt{m}) &= \phi_m((r + s\sqrt{m}) - (a + b\sqrt{m})(t + u\sqrt{m})) \\
&= \phi_m((x + y\sqrt{m})(t + u\sqrt{m}) - (a + b\sqrt{m})(t + u\sqrt{m})) \\
&= \phi_m((t + u\sqrt{m})((x + y\sqrt{m}) - (a + b\sqrt{m}))) \\
&= \phi_m(t + u\sqrt{m})\phi_m((x + y\sqrt{m}) - (a + b\sqrt{m})) \\
&< \phi_m(t + u\sqrt{m}),
\end{aligned}
$$

by Lemma 2.2.1(d), which completes the proof of (2.1.2). ■

Theorem 2.2.2 enables us to determine those negative squarefree integers m for which $\mathbb{Z} + \mathbb{Z}\sqrt{m}$ is Euclidean with respect to ϕ_m.

Theorem 2.2.3 *Let m be a negative squarefree integer. Then the integral domain $\mathbb{Z} + \mathbb{Z}\sqrt{m}$ is Euclidean with respect to ϕ_m if and only if $m = -1, -2$.*

Proof: First we show that $\mathbb{Z} + \mathbb{Z}\sqrt{m}$ is Euclidean with respect to ϕ_m for $m = -1$ and $m = -2$. Let $x, y \in \mathbb{Q}$. We can choose $a, b \in \mathbb{Z}$ such that

$$|x - a| \leq \frac{1}{2}, \quad |y - b| \leq \frac{1}{2}.$$

Then

$$
\begin{aligned}
\phi_m((x + y\sqrt{m}) - (a + b\sqrt{m})) &= \phi_m((x - a) + (y - b)\sqrt{m})) \\
&= |(x - a)^2 - m(y - b)^2| \\
&\leq |x - a|^2 + |m||y - b|^2 \\
&\leq \frac{1}{4} + 2 \cdot \frac{1}{4} \\
&= \frac{3}{4} < 1
\end{aligned}
$$

and, appealing to Theorem 2.2.2, we deduce that $\mathbb{Z} + \mathbb{Z}\sqrt{m}$ is Euclidean with respect to ϕ_m for $m = -1$ and $m = -2$.

Now suppose that $\mathbb{Z} + \mathbb{Z}\sqrt{m}$ is Euclidean with respect to ϕ_m. Then, by Theorem 2.2.2, there exist $a, b \in \mathbb{Z}$ such that

$$\phi_m \left((\tfrac{1}{2} + \tfrac{1}{2}\sqrt{m}) - (a + b\sqrt{m}) \right) < 1;$$

that is (as $-m = |m|$),

$$\left(\tfrac{1}{2} - a \right)^2 + |m| \left(\tfrac{1}{2} - b \right)^2 < 1.$$

But for any integer x, we have

$$\left| \tfrac{1}{2} - x \right| \geq \tfrac{1}{2}, \quad \left(\tfrac{1}{2} - x \right)^2 \geq \tfrac{1}{4},$$

so

$$\tfrac{1}{4} + \tfrac{|m|}{4} < 1;$$

that is, $|m| < 3$. Hence $m = -1$ and $m = -2$ are the only possibilities.   ∎

In an exactly similar way to the proof of Theorem 2.2.2, we can prove the following result.

Theorem 2.2.4 *Let m be a squarefree integer with $m \equiv 1 \pmod 4$. Then the integral domain $\mathbb{Z} + \mathbb{Z} \left(\frac{1+\sqrt{m}}{2} \right)$ is Euclidean with respect to ϕ_m if and only if for all $x, y \in \mathbb{Q}$ there exist $a, b \in \mathbb{Z}$ such that*

$$\phi_m \left((x + y\sqrt{m}) - \left(a + b \left(\frac{1 + \sqrt{m}}{2} \right) \right) \right) < 1.$$

From Theorem 2.2.4, exactly as we proved Theorem 2.2.3, we can determine those negative squarefree integers $m \equiv 1 \pmod 4$ for which $\mathbb{Z} + \mathbb{Z} \left(\frac{1+\sqrt{m}}{2} \right)$ is Euclidean with respect to ϕ_m.

Theorem 2.2.5 *Let m be a negative squarefree integer with $m \equiv 1 \pmod 4$. Then the integral domain $\mathbb{Z} + \mathbb{Z} \left(\frac{1+\sqrt{m}}{2} \right)$ is Euclidean with respect to ϕ_m if and only if $m = -3, -7, -11$.*

The determination of the positive squarefree integers m for which $\mathbb{Z} + \mathbb{Z}\sqrt{m}$ ($m \equiv 2, 3 \pmod 4$) and $\mathbb{Z} + \mathbb{Z} \left(\frac{1+\sqrt{m}}{2} \right)$ ($m \equiv 1 \pmod 4$) are Euclidean with respect to ϕ_m is much more difficult and was the culmination of the efforts of numerous mathematicians including E. S. Barnes (1874–1953), H. Behrbohm, E. Berg, A. T. Brauer (1894–1985), H. Chatland, H. Davenport (1907–1969), L. E. Dickson (1874–1954), P. Erdös (1913–1996), H. A. Heilbronn (1908–1975), N. Hofreiter,

L. K. Hua, K. Inkeri, J. F. Keston, C. Ko, S. H. Min, A. Oppenheim, O. Perron (1880–1975), L. Rédei, R. Remak (1888–1942), L. Schuster, W. T. Sheh, and H. P. F. Swinnerton-Dyer.

The final step was taken in 1950 by Chatland and Davenport [4], who established the following two theorems.

Theorem 2.2.6 *Let m be a positive squarefree integer with* $m \equiv 2, 3 \pmod 4$. *Then the integral domain* $\mathbb{Z} + \mathbb{Z}\sqrt{m}$ *is Euclidean with respect to* ϕ_m *if and only if* $m = 2, 3, 6, 7, 11, 19, 57$.

Theorem 2.2.7 *Let m be a positive squarefree integer with* $m \equiv 1 \pmod 4$. *Then the integral domain* $\mathbb{Z} + \mathbb{Z}\left(\frac{1+\sqrt{m}}{2}\right)$ *is Euclidean with respect to* ϕ_m *if and only if* $m = 5, 13, 17, 21, 29, 33, 37, 41, 73$.

We will not prove these two theorems here. We will just prove the following result.

Theorem 2.2.8 *The integral domain* $\mathbb{Z} + \mathbb{Z}\sqrt{m}$ *is Euclidean with respect to* ϕ_m *for* $m = 2, 3, 6$.

Proof: $m = 2, 3$. Let $x, y \in \mathbb{Q}$. We choose $a, b \in \mathbb{Z}$ such that

$$|x - a| \le \frac{1}{2}, \quad |y - b| \le \frac{1}{2}.$$

As $(x - a)^2 \ge 0$ and $m(y - b)^2 \ge 0$, we have

$$|(x - a)^2 - m(y - b)^2| \le \max(|x - a|^2, m|y - b|^2) \le \frac{3}{4}.$$

Thus

$$\phi_m((x + y\sqrt{m}) - (a + b\sqrt{m})) = |(x - a)^2 - m(y - b)^2| < 1,$$

and the result follows by Theorem 2.2.2.

$m = 6$. Suppose that $\mathbb{Z} + \mathbb{Z}\sqrt{6}$ is not Euclidean with respect to ϕ_6. Then, by Theorem 2.2.2, there exist $r, s \in \mathbb{Q}$ such that

$$\phi_6((r + s\sqrt{6}) - (x + y\sqrt{6})) \ge 1 \text{ for all } x, y \in \mathbb{Z};$$

that is,

$$|(r - x)^2 - 6(s - y)^2| \ge 1 \text{ for all } x, y \in \mathbb{Z}.$$

We can choose $\epsilon_1 = \pm 1$ and $u_1 \in \mathbb{Z}$ such that

$$0 \le \epsilon_1 r + u_1 \le \frac{1}{2}$$

and $\epsilon_2 = \pm 1$ and $u_2 \in \mathbb{Z}$ such that

$$0 \le \epsilon_2 s + u_2 \le \frac{1}{2}.$$

Set

$$r_1 = \epsilon_1 r + u_1 \in \mathbb{Q}, \quad x_1 = \epsilon_1 x + u_1 \in \mathbb{Z},$$
$$s_1 = \epsilon_2 s + u_2 \in \mathbb{Q}, \quad y_1 = \epsilon_2 y + u_2 \in \mathbb{Z},$$

so that

$$0 \le r_1 \le \frac{1}{2}, \ 0 \le s_1 \le \frac{1}{2}, \tag{2.2.2}$$

and

$$|(r_1 - x_1)^2 - 6(s_1 - y_1)^2| \ge 1 \text{ for all } x_1, y_1 \in \mathbb{Z}. \tag{2.2.3}$$

Taking $(x_1, y_1) = (0,0),\ (1,0),$ and $(-1,0)$ in (2.2.3), we obtain the inequalities

$$\begin{cases} |r_1^2 - 6s_1^2| \ge 1, \\ |(1-r_1)^2 - 6s_1^2| \ge 1, \\ |(1+r_1)^2 - 6s_1^2| \ge 1. \end{cases} \tag{2.2.4}$$

From (2.2.2) we deduce that

$$\begin{cases} -\dfrac{3}{2} \le r_1^2 - 6s_1^2 \le \dfrac{1}{4}, \\ -\dfrac{5}{4} \le (1-r_1)^2 - 6s_1^2 \le 1, \\ -\dfrac{1}{2} \le (1+r_1)^2 - 6s_1^2 \le \dfrac{9}{4}. \end{cases} \tag{2.2.5}$$

From (2.2.4) and (2.2.5), we deduce that

$$-\frac{3}{2} \le r_1^2 - 6s_1^2 \le -1, \tag{2.2.6}$$

$$\text{(i) } (1-r_1)^2 - 6s_1^2 = 1 \text{ or (ii) } -\frac{5}{4} \le (1-r_1)^2 - 6s_1^2 \le -1, \tag{2.2.7}$$

$$1 \le (1+r_1)^2 - 6s_1^2 \le \frac{9}{4}. \tag{2.2.8}$$

From (2.2.6) and (2.2.8), we obtain

$$1 \le 1 + 2r_1 + (r_1^2 - 6s_1^2) \le 2r_1,$$

so that $r_1 \geq \frac{1}{2}$. But $r_1 \leq \frac{1}{2}$ so we must have $r_1 = \frac{1}{2}$. Then (2.2.7)(i) gives $\frac{1}{4} - 6s_1^2 = 1$, which is impossible, and (2.2.7)(ii) gives $\frac{1}{4} - 6s_1^2 \leq -1$, so that $s_1^2 \geq \frac{5}{24}$. But from (2.2.8) we have $6s_1^2 \leq (1 + r_1)^2 - 1 = \frac{5}{4}$; that is, $s_1^2 \leq \frac{5}{24}$, so that $s_1^2 = \frac{5}{24}$, which is impossible. This completes the proof that $\mathbb{Z} + \mathbb{Z}\sqrt{6}$ is Euclidean with respect to ϕ_6. ∎

2.3 Examples of Domains That are Not Euclidean

We begin by giving a class of values of m for which $\mathbb{Z} + \mathbb{Z}\sqrt{m}$ is not Euclidean with respect to ϕ_m.

Theorem 2.3.1 *Let m be a positive squarefree integer. If there exist distinct odd primes p and q such that*

$$\left(\frac{m}{p}\right) = \left(\frac{m}{q}\right) = -1,$$

and positive integers t and u such that

$$pt + qu = m, \quad p \nmid t, \quad q \nmid u,$$

and an integer r such that

$$r^2 \equiv pt \pmod{m},$$

then $\mathbb{Z} + \mathbb{Z}\sqrt{m}$ is not Euclidean with respect to ϕ_m.

Proof: Suppose that $\mathbb{Z} + \mathbb{Z}\sqrt{m}$ is Euclidean with respect to ϕ_m. Then there exist $\gamma, \delta \in \mathbb{Z} + \mathbb{Z}\sqrt{m}$ such that

$$r\sqrt{m} = m\gamma + \delta, \quad \phi_m(\delta) < \phi_m(m).$$

Setting $\gamma = x + y\sqrt{m}$ $(x, y \in \mathbb{Z})$ we obtain

$$\phi_m(r\sqrt{m} - m(x + y\sqrt{m})) < \phi_m(m);$$

that is

$$|m^2 x^2 - m(r - my)^2| < m^2,$$

so that

$$|mx^2 - (my - r)^2| < m.$$

Since

$$mx^2 - (my - r)^2 \equiv -r^2 \equiv -pt \pmod{m}$$

and

$$0 < pt < pt + qu = m,$$

we must have

$$mx^2 - (my - r)^2 = -pt \text{ or } m - pt;$$

that is

$$mX^2 - Y^2 = -pt \text{ or } qu$$

for integers $X (= x)$ and $Y (= my - r)$. Suppose that $mX^2 - Y^2 = -pt$. As $\left(\frac{m}{p}\right) = -1$ we have $p \nmid m$. Also, as $p \nmid t$ we have $p \mid\mid -pt$. Hence $p \nmid X$ and $p \nmid Y$. Thus

$$\left(\frac{m}{p}\right) = \left(\frac{mX^2}{p}\right) = \left(\frac{Y^2}{p}\right) = 1,$$

contradicting $\left(\frac{m}{p}\right) = -1$. Now suppose that $mX^2 - Y^2 = qu$. As $\left(\frac{m}{q}\right) = -1$ we have $q \nmid m$. Also, as $q \nmid u$ we have $q \mid\mid qu$. Hence $q \nmid X$ and $q \nmid Y$. Thus

$$\left(\frac{m}{q}\right) = \left(\frac{mX^2}{q}\right) = \left(\frac{Y^2}{q}\right) = 1,$$

contradicting $\left(\frac{m}{q}\right) = -1$. This proves that $\mathbb{Z} + \mathbb{Z}\sqrt{m}$ is not Euclidean with respect to ϕ_m. ∎

We next use Theorem 2.3.1 to give some explicit, small, positive, squarefree values of m for which $\mathbb{Z} + \mathbb{Z}\sqrt{m}$ is not Euclidean with respect to ϕ_m.

Theorem 2.3.2 $\mathbb{Z} + \mathbb{Z}\sqrt{m}$ *is not Euclidean with respect to* ϕ_m *for* $m = 23, 47, 59, 83$.

Proof: This follows immediately from Theorem 2.3.1 and the following table.

m	p	q	t	u	r
23	3	5	1	4	7
47	3	5	4	7	23
59	3	7	15	2	24
83	3	5	1	16	13

∎

The corresponding result to Theorem 2.3.1 for $\mathbb{Z} + \mathbb{Z}\left(\frac{1+\sqrt{m}}{2}\right)$ $(m \equiv 1 \pmod 4)$ is not quite so elegant.

Theorem 2.3.3 *Let m be a positive squarefree integer with $m \equiv 1 \pmod 4$. If there exist distinct odd primes p and q such that*

$$\left(\frac{m}{p}\right) = \left(\frac{m}{q}\right) = -1$$

and an odd integer r such that

$$p \parallel (m-1)r^2 - 4m \left[\frac{(m-1)r^2}{4m}\right],$$

$$q \parallel (m-1)r^2 - 4m \left[\frac{(m-1)r^2}{4m}\right] - 4m,$$

then $\mathbb{Z} + \mathbb{Z}\left(\frac{1+\sqrt{m}}{2}\right)$ is not Euclidean with respect to ϕ_m.

Proof: As m and r are both odd, we have $\frac{m-r}{2} \in \mathbb{Z}$. Hence

$$\frac{m + r\sqrt{m}}{2} = \frac{m-r}{2} + r\left(\frac{1+\sqrt{m}}{2}\right) \in \mathbb{Z} + \mathbb{Z}\left(\frac{1+\sqrt{m}}{2}\right).$$

Suppose that $\mathbb{Z} + \mathbb{Z}\left(\frac{1+\sqrt{m}}{2}\right)$ is Euclidean with respect to ϕ_m. Then there exist γ, $\delta \in \mathbb{Z} + \mathbb{Z}\left(\frac{1+\sqrt{m}}{2}\right)$ such that

$$\frac{m + r\sqrt{m}}{2} = m\gamma + \delta, \quad \phi_m(\delta) < \phi_m(m).$$

As $\gamma \in \mathbb{Z} + \mathbb{Z}\left(\frac{1+\sqrt{m}}{2}\right)$ there exist x, $y \in \mathbb{Z}$ such that $\gamma = x + y\left(\frac{1+\sqrt{m}}{2}\right)$, and thus

$$\phi_m\left(\frac{m + r\sqrt{m}}{2} - m\left(x + y\left(\frac{1+\sqrt{m}}{2}\right)\right)\right) < \phi_m(m).$$

Hence

$$\left|\left(\frac{m}{2} - mx - \frac{my}{2}\right)^2 - m\left(\frac{r}{2} - \frac{my}{2}\right)^2\right| < m^2.$$

Multiplying both sides of this inequality by $4/m$, we obtain

$$|m(1 - 2x - y)^2 - (r - my)^2| < 4m.$$

Set $X = 1 - 2x - y \in \mathbb{Z}$ and $Y = r - my \in \mathbb{Z}$ so that

$$|mX^2 - Y^2| < 4m.$$

As $m \equiv 1 \pmod 4$ and $(u + 2v)^2 \equiv u^2 \pmod 4)$ for any u, $v \in \mathbb{Z}$, we deduce that

$$mX^2 - Y^2 \equiv (1 - y)^2 - (1 - y)^2 \equiv 0 \pmod 4.$$

Also

$$mX^2 - Y^2 \equiv -Y^2 \equiv -r^2 \pmod{m}.$$

Hence

$$mX^2 - Y^2 \equiv (m-1)r^2 \pmod{4m}.$$

Thus

$$mX^2 - Y^2 = (m-1)r^2 - 4m \left[\frac{(m-1)r^2}{4m} \right]$$

or

$$mX^2 - Y^2 = (m-1)r^2 - 4m \left[\frac{(m-1)r^2}{4m} \right] - 4m.$$

In the first case we have $p \,\|\, mX^2 - Y^2$. As $\left(\frac{m}{p} \right) = -1$ we have $p \nmid m$. Thus $p \nmid X$ and $p \nmid Y$. Then

$$\left(\frac{m}{p} \right) = \left(\frac{mX^2}{p} \right) = \left(\frac{Y^2}{p} \right) = 1,$$

contradicting $\left(\frac{m}{p} \right) = -1$.

The second case can be treated similarly.   ∎

We now use Theorem 2.3.3 to show that $\mathbb{Z} + \mathbb{Z} \left(\frac{1+\sqrt{53}}{2} \right)$ is not Euclidean with respect to ϕ_{53}.

Theorem 2.3.4 $\mathbb{Z} + \mathbb{Z} \left(\frac{1+\sqrt{53}}{2} \right)$ *is not Euclidean with respect to* ϕ_{53}.

Proof: We choose

$$m = 53, \ p = 5, \ q = 19, \ r = 29.$$

Clearly

$$\left(\frac{m}{p} \right) = \left(\frac{53}{5} \right) = \left(\frac{3}{5} \right) = -1,$$

$$\left(\frac{m}{q} \right) = \left(\frac{53}{19} \right) = \left(\frac{-4}{19} \right) = \left(\frac{-1}{19} \right) = -1,$$

$$
\begin{aligned}
(m-1)r^2 - 4m \left[\frac{(m-1)r^2}{4m} \right] &= 52 \cdot 29^2 - 4 \cdot 53 \cdot \left[\frac{52 \cdot 29^2}{4 \cdot 53} \right] \\
&= 43732 - 212 \cdot 206 \\
&= 43732 - 43672 \\
&= 60 = 5 \cdot 2^2 \cdot 3,
\end{aligned}
$$

and

$$(m-1)r^2 - 4m\left[\frac{(m-1)r^2}{4m}\right] - 4m = 60 - 212 = -152 = -19 \cdot 2^3,$$

so the result follows from Theorem 2.3.3   ∎

In the next theorem we show that $\mathbb{Z} + \mathbb{Z}\sqrt{m}$ ($m \equiv 2, 3 \pmod 4$) is not Euclidean with respect to ϕ_m if m is sufficiently large. The same result is also true for $\mathbb{Z} + \mathbb{Z}\left(\frac{1+\sqrt{m}}{2}\right)$ ($m \equiv 1 \pmod 4$) but the proof is more complicated and we will not give it here.

Theorem 2.3.5 *Let m be a positive squarefree integer.*

(a) *If $m \equiv 2 \pmod 4$ and $m \geq 42$ then $\mathbb{Z} + \mathbb{Z}\sqrt{m}$ is not Euclidean with respect to ϕ_m.*

(b) *If $m \equiv 3 \pmod 4$ and $m \geq 91$ then $\mathbb{Z} + \mathbb{Z}\sqrt{m}$ is not Euclidean with respect to ϕ_m.*

Proof: **(a)** As $m \geq 42$ we have $m > 20 + 8\sqrt{6} = 4(\sqrt{3} + \sqrt{2})^2$ so that $\sqrt{m} > 2(\sqrt{3} + \sqrt{2})$ and thus

$$\left(\frac{\sqrt{3m}-1}{2}\right) - \left(\frac{\sqrt{2m}-1}{2}\right) = \left(\frac{\sqrt{3}-\sqrt{2}}{2}\right)\sqrt{m}$$

$$> \frac{(\sqrt{3}-\sqrt{2})}{2}2(\sqrt{3}+\sqrt{2}) = 1.$$

Hence there exists an integer u satisfying

$$\frac{\sqrt{2m}-1}{2} < u < \frac{\sqrt{3m}-1}{2}.$$

Set $t = 2u + 1$ so that t is an odd integer satisfying

$$2m < t^2 < 3m.$$

Now suppose that $\mathbb{Z} + \mathbb{Z}\sqrt{m}$ is Euclidean with respect to ϕ_m. Then there exist $\gamma, \delta \in \mathbb{Z} + \mathbb{Z}\sqrt{m}$ such that

$$t\sqrt{m} = m\gamma + \delta, \quad \phi_m(\delta) < \phi_m(m).$$

As $\gamma \in \mathbb{Z} + \mathbb{Z}\sqrt{m}$ there exist $x, y \in \mathbb{Z}$ such that $\gamma = x + y\sqrt{m}$, and

$$\phi_m(t\sqrt{m} - m(x + y\sqrt{m})) < \phi_m(m);$$

that is,

$$|m^2x^2 - m(t - my)^2| < m^2,$$

and thus

$$|mx^2 - (t - my)^2| < m.$$

Set $X = my - t \in \mathbb{Z}$ and $Y = x \in \mathbb{Z}$ so that

$$|X^2 - mY^2| < m$$

and

$$X^2 - mY^2 \equiv X^2 \equiv t^2 \ (\text{mod } m).$$

Since $2m < t^2 < 3m$ we have

$$X^2 - mY^2 = t^2 - 2m$$

or

$$X^2 - mY^2 = t^2 - 3m.$$

In the first case, as $t^2 \equiv 1 \ (\text{mod } 8)$ (since t is odd) and $m \equiv 2 \ (\text{mod } 4)$, we have

$$X^2 - mY^2 \equiv 5 \ (\text{mod } 8).$$

Thus X is odd, so $X^2 \equiv 1 \ (\text{mod } 8)$ and

$$mY^2 \equiv 4 \ (\text{mod } 8).$$

This is clearly impossible as

$$mY^2 \equiv \begin{cases} 0 \ (\text{mod } 8), & \text{if } Y \equiv 0 \ (\text{mod } 2), \\ 2 \ (\text{mod } 4), & \text{if } Y \equiv 1 \ (\text{mod } 2). \end{cases}$$

In the second case, as $t^2 \equiv 1 \ (\text{mod } 8)$, we have

$$X^2 - mY^2 \equiv 1 - 3m \ (\text{mod } 8).$$

As m is even we see that X is odd. Hence $X^2 \equiv 1 \ (\text{mod } 8)$ and

$$m(Y^2 - 3) \equiv 0 \ (\text{mod } 8).$$

Hence, as $2 \| m$, we have

$$Y^2 \equiv 3 \ (\text{mod } 4),$$

which is impossible.

(b) As $m \geq 91$ we have $m > 44 + 8\sqrt{30} = 4(\sqrt{6} + \sqrt{5})^2$ so that $\sqrt{m} > 2(\sqrt{6} + \sqrt{5})$ and thus

$$\left(\frac{\sqrt{6m} - 1}{2}\right) - \left(\frac{\sqrt{5m} - 1}{2}\right) = \frac{(\sqrt{6} - \sqrt{5})}{2}\sqrt{m}$$

$$> \frac{(\sqrt{6} - \sqrt{5})}{2} 2(\sqrt{6} + \sqrt{5}) = 1.$$

Hence there is an integer u satisfying

$$\frac{\sqrt{5m} - 1}{2} < u < \frac{\sqrt{6m} - 1}{2}.$$

Set $t = 2u + 1$ so that t is an odd integer satisfying

$$5m < t^2 < 6m.$$

Now suppose that $\mathbb{Z} + \mathbb{Z}\sqrt{m}$ is Euclidean with respect to ϕ_m. Then there exist $\gamma, \delta \in \mathbb{Z} + \mathbb{Z}\sqrt{m}$ such that

$$t\sqrt{m} = m\gamma + \delta, \quad \phi_m(\delta) < \phi_m(m).$$

As $\gamma \in \mathbb{Z} + \mathbb{Z}\sqrt{m}$ there exist $x, y \in \mathbb{Z}$ such that

$$\gamma = x + y\sqrt{m},$$

and

$$\phi_m(t\sqrt{m} - m(x + y\sqrt{m})) < \phi_m(m);$$

that is,

$$|m^2 x^2 - m(t - my)^2| < m^2,$$

and thus

$$|mx^2 - (t - my)^2| < m.$$

Set $X = my - t \in \mathbb{Z}$ and $Y = x \in \mathbb{Z}$ so that

$$|X^2 - mY^2| < m$$

and

$$X^2 - mY^2 \equiv X^2 \equiv t^2 \pmod{m}.$$

Since $5m < t^2 < 6m$ we have

$$X^2 - mY^2 = t^2 - 5m$$

or

$$X^2 - mY^2 = t^2 - 6m.$$

In the first case, as $t^2 \equiv 1 \pmod 8$ (since t is odd) and $m \equiv 3 \pmod 4$, we have

$$X^2 - mY^2 = t^2 - 5m \equiv 1 - 15 = -14 \equiv 2 \pmod 4$$

so that

$$X \equiv Y \equiv 1 \pmod 2.$$

Thus $X^2 \equiv Y^2 \equiv 1 \pmod 8$ so that

$$1 - 5m \equiv t^2 - 5m = X^2 - mY^2 \equiv 1 - m \pmod 8,$$

giving $4m \equiv 0 \pmod 8$, which is clearly impossible. In the second case, as $t^2 \equiv 1 \pmod 8$ and $m \equiv 3 \pmod 4$, we have

$$X^2 - mY^2 = t^2 - 6m \equiv 1 - 18 = -17 \equiv 7 \pmod 8.$$

If X is odd, so $X^2 \equiv 1 \pmod 8$, then

$$mY^2 \equiv 2 \pmod 8,$$

which is impossible. If X is even, so $X^2 \equiv 0 \pmod 4$, then $-3Y^2 \equiv 3 \pmod 4$, so that $Y^2 \equiv 3 \pmod 4$, which is impossible. ∎

It is a consequence of Theorem 2.2.5 that the domain $\mathbb{Z} + \mathbb{Z}\left(\frac{1+\sqrt{-19}}{2}\right)$ is not Euclidean with respect to ϕ_{-19}. But could it be Euclidean with respect to some other function? In fact it is not. How do we see this? One way of showing that an integral domain is not Euclidean with respect to any function is to show that it does not possess certain distinguished elements called universal side divisors, since a domain that has no universal side divisors is not Euclidean with respect to any function. Indeed as we shall see $\mathbb{Z} + \mathbb{Z}\left(\frac{1+\sqrt{-19}}{2}\right)$ has no universal divisors and therefore is not Euclidean with respect to any function.

We now define a universal side divisor. For any integral domain D it is convenient to set

$$\tilde{D} = U(D) \cup \{0\}$$

so that

$$D - \tilde{D} = \phi \text{ if and only if } D \text{ is a field.}$$

Definition 2.3.1 (Universal side divisor) *Let D be an integral domain that is not a field so that $D - \tilde{D} \neq \phi$. An element $u \in D - \tilde{D}$ is called a universal side divisor if for any $x \in D$ there exists some $z \in \tilde{D}$ such that $u \mid x - z$.*

Theorem 2.3.6 *Let D be an integral domain that is not a field. If D has no universal side divisors then D is not Euclidean.*

Proof: Suppose that D is Euclidean with respect to the Euclidean function ϕ and has no universal side divisors. Consider the set of integers defined by

$$S = \{\phi(v) \mid v \in D - \tilde{D}\}.$$

As D is not a field, $D - \tilde{D} \neq \phi$ and so S is nonempty. By Theorem 2.1.1(d), S is bounded below. Thus S possesses a least element, say $\phi(u)$, $u \in D - \tilde{D}$. As D is Euclidean with respect to ϕ, for any $x \in D$ there exist $y, z \in D$ such that $x = uy + z$ and $\phi(z) < \phi(u)$. If $z = 0$ then $x = uy$ and $u \mid x$. If $z \neq 0$ then by the minimality of $\phi(u)$, $z \in U(D)$. Thus in both cases $u \mid x - z$ for some $z \in \tilde{D}$, and so u is a universal side divisor, which is a contradiction. ∎

If m is a negative squarefree integer with $m \equiv 2, 3 \,(\text{mod } 4)$ then $\mathbb{Z} + \mathbb{Z}\sqrt{m}$ is Euclidean with respect to ϕ_m for $m = -1$ and $m = -2$ and is not Euclidean with respect to ϕ_m for $m < -2$ (Theorem 2.2.3). We now use Theorem 2.3.6 to show that $\mathbb{Z} + \mathbb{Z}\sqrt{m}$ is not Euclidean with respect to any function for $m < -2$.

Theorem 2.3.7 *Let m be a negative squarefree integer with $m \equiv 2, 3$ (mod 4) and $m < -2$. Then $\mathbb{Z} + \mathbb{Z}\sqrt{m}$ is not Euclidean.*

Proof: Let $D = \mathbb{Z} + \mathbb{Z}\sqrt{m}$. As $m \neq -1$, $U(D) = \{1, -1\}$ (see Exercise 3 of Chapter 1) so that $\tilde{D} = \{0, 1, -1\}$. Suppose that u is a universal side divisor in D. Then u must divide one of $2 - 1$, $2 + 0$, or $2 + 1$, that is, one of 1, 2, or 3. But u being a universal side divisor is not a unit, so $u \nmid 1$. Hence $u \mid 2$ or $u \mid 3$. Since $m \equiv 2, 3 \,(\text{mod } 4)$ and $m < -2$ we have $m \leq -5$ so that both 2 and 3 are irreducible in D (Exercise 36 of Chapter 1). Hence the only possible universal side divisors are $2, -2, 3$, and -3. However, none of these divides any of the three elements of $\mathbb{Z} + \mathbb{Z}\sqrt{m}$:

$$\sqrt{m} - 1, \ \sqrt{m}, \ \sqrt{m} + 1,$$

so that no such universal side divisor can exist. Hence, by Theorem 2.3.6, D is not Euclidean. ∎

If m is a negative squarefree integer with $m \equiv 1 \,(\text{mod } 4)$ then $\mathbb{Z} + \mathbb{Z}\left(\frac{1+\sqrt{m}}{2}\right)$ is Euclidean with respect to ϕ_m for $m = -3, -7, -11$ and is not Euclidean with respect to ϕ_m for $m < -11$ (Theorem 2.2.5). We use Theorem 2.3.6 to show that $\mathbb{Z} + \mathbb{Z}\left(\frac{1+\sqrt{m}}{2}\right)$ is not Euclidean with respect to any function for $m < -11$.

Theorem 2.3.8 *Let m be a squarefree negative integer with $m \equiv 1$ (mod 4) and $m < -11$. Then $\mathbb{Z} + \mathbb{Z}\left(\frac{1+\sqrt{m}}{2}\right)$ is not Euclidean.*

Proof: Let $D = \mathbb{Z} + \mathbb{Z}\left(\frac{1+\sqrt{m}}{2}\right)$. As $m \neq -3$ we have $U(D) = \{1, -1\}$ (Exercise 4 of Chapter 1) so that $\tilde{D} = \{0, 1, -1\}$. Suppose that u is a universal side divisor in D. Then u must divide one of $2 - 1$, $2 + 0$, or $2 + 1$, that is, one of 1, 2, or 3. As u is not a unit, u must divide 2 or 3. Since $m \leq -15$, both 2 and 3 are irreducible in $\mathbb{Z} + \mathbb{Z}\left(\frac{1+\sqrt{m}}{2}\right)$ (Exercise 37 of Chapter 1). Therefore the only possible side divisors are 2, -2, 3, and -3. However, none of these divides any of the following three elements of $\mathbb{Z} + \mathbb{Z}\left(\frac{1+\sqrt{m}}{2}\right)$,

$$\frac{1}{2}(-1+\sqrt{m}) = \frac{1}{2}(1+\sqrt{m}) - 1, \; \frac{1}{2}(1+\sqrt{m}), \; \frac{1}{2}(3+\sqrt{m}) = \frac{1}{2}(1+\sqrt{m}) + 1,$$

so that no such universal side divisor can exist. Hence, by Theorem 2.3.6, D is not Euclidean. ∎

When m is a positive squarefree integer very little is known. Clark [5] has shown that $\mathbb{Z} + \mathbb{Z}\left(\frac{1+\sqrt{69}}{2}\right)$ is Euclidean with respect to the function

$$\phi\left(a + b\left(\frac{1+\sqrt{69}}{2}\right)\right) = \begin{cases} |a^2 + ab - 17b^2|, & \text{if } (a, b) \neq (10, 3), \\ 26, & \text{if } (a, b) = (10, 3). \end{cases}$$

By Theorem 2.2.7 we know that $\mathbb{Z} + \mathbb{Z}\left(\frac{1+\sqrt{69}}{2}\right)$ is not Euclidean with respect to ϕ_{69}. This is the first example of a real quadratic domain that is Euclidean but not norm-Euclidean. Since the 26 in the definition of ϕ can be replaced by any integer greater than 25, $\mathbb{Z} + \mathbb{Z}\left(\frac{1+\sqrt{69}}{2}\right)$ is Euclidean with respect to infinitely many different functions. Samuel [16] suggests that $\mathbb{Z} + \mathbb{Z}\sqrt{14}$ may be Euclidean with respect to some function different from ϕ_{14}, and this has recently been proved by Harper [9].

2.4 Almost Euclidean Domains

In this section we introduce the concept of an "almost Euclidean domain" and show that such a domain must be a principal ideal domain. In Chapter 3 we show that a principal ideal domain is an almost Euclidean domain (see Theorem 3.3.3). Thus principal ideal domains are domains that are almost Euclidean in a certain sense. We first define an "almost Euclidean function" analogously to that of a Euclidean function (Definition 2.1.1).

Definition 2.4.1 (Almost Euclidean function) *Let D be an integral domain. A mapping $\phi : D \rightarrow \mathbb{N} \cup \{0\}$ is called an almost Euclidean function on D if it has* *the following properties:*

$$\phi(0) = 0, \tag{2.4.1}$$

$$\phi(a) > 0, \text{ for all } a \in D \text{ with } a \neq 0, \tag{2.4.2}$$

$$\phi(ab) \geq \phi(a), \text{ for all } a, b \in D \text{ with } b \neq 0, \tag{2.4.3}$$

$$\text{if } a, b \in D \text{ with } b \neq 0 \text{ then either} \tag{2.4.4}$$

$$(i)\ a = bq \text{ for some } q \in D \text{ or}$$

$$(ii)\ 0 < \phi(ax + by) < \phi(b) \text{ for some } x,\ y \in D.$$

It is clear from Definition 2.1.1 and Theorem 2.1.1(d) that if ϕ is a Euclidean function satisfying $\phi(0) = 0$ then ϕ is an almost Euclidean function.

The concept of an almost Euclidean domain occurs in the work of Cámpoli [2] and Greene [8].

Definition 2.4.2 (Almost Euclidean domain) *Let D be an integral domain. If D possesses an almost Euclidean function ϕ then D is called an almost Euclidean domain with respect to ϕ.*

As for Euclidean domains, if it is not important to specify the almost Euclidean function ϕ, we just call D an almost Euclidean domain.

Theorem 2.4.1 *An almost Euclidean domain is a principal ideal domain.*

Proof: Let D be an almost Euclidean domain. Let ϕ be an almost Euclidean function defined on D. Let I be a nonzero ideal of D. Among the elements x of I, let b be an element with a minimal positive value of $\phi(x)$. Given $a \in I$, for any $x, y \in D$, the element $ax + by$ is in I. By the definition of b, we cannot have $0 < \phi(ax + by) < \phi(b)$ so that as D is almost Euclidean with respect to ϕ, we must have $a = bq$ for some $q \in D$. Thus $I = \langle b \rangle$ and D is a principal ideal domain. ∎

For the converse of this theorem, see Theorem 3.3.3.

2.5 Representing Primes by Binary Quadratic Forms

Expressions of the type $ax^2 + bxy + cy^2$ $(a, b, c \in \mathbb{Z})$ are called binary quadratic forms. The integer n is said to be represented by the binary quadratic form $ax^2 + bxy + cy^2$ if there are integers x and y such that $n = ax^2 + bxy + cy^2$. Thus for example 31 is represented by the form $x^2 + xy + 3y^2$ as $31 = 1^2 + 1 \cdot 3 + 3 \cdot 3^2$, but 2 is not represented by the form $x^2 + 5y^2$.

As $\mathbb{Z} + \mathbb{Z}\sqrt{m}$ $(m = -1, -2)$ and $\mathbb{Z} + \mathbb{Z}\left(\frac{1+\sqrt{m}}{2}\right)$ $(m = -3, -7, -11)$ are Euclidean domains (Theorems 2.2.3 and 2.2.5), we can apply Theorems 1.4.4 and 1.4.5 to determine when an odd prime p is represented by each of the forms $x^2 + y^2$,

$x^2 + 2y^2$, $x^2 + xy + y^2$, $x^2 + xy + 2y^2$, and $x^2 + xy + 3y^2$. To do this we begin by recalling the following Legendre symbol evaluations from elementary number theory. For an odd prime p

$$\left(\frac{-1}{p}\right) = 1 \iff p \equiv 1 \ (\text{mod } 4), \tag{2.5.1}$$

$$\left(\frac{-2}{p}\right) = 1 \iff p \equiv 1, 3 \ (\text{mod } 8), \tag{2.5.2}$$

$$\left(\frac{-3}{p}\right) = 1 \iff p \equiv 1 \ (\text{mod } 3), \tag{2.5.3}$$

$$\left(\frac{-7}{p}\right) = 1 \iff p \equiv 1, 2, 4 \ (\text{mod } 7), \tag{2.5.4}$$

$$\left(\frac{-11}{p}\right) = 1 \iff p \equiv 1, 3, 4, 5, 9 \ (\text{mod } 11). \tag{2.5.5}$$

Theorem 2.5.1 *Let p be a prime such that $p \equiv 1$ (mod 4). Then there exist integers x and y such that $p = x^2 + y^2$.*

Proof: As $p \equiv 1$ (mod 4), by (2.5.1) we have $\left(\frac{-1}{p}\right) = 1$. Since $\mathbb{Z} + \mathbb{Z}\sqrt{-1}$ is a Euclidean domain, by Theorem 2.1.2 it is a principal ideal domain. Thus by Theorem 1.4.4, there are integers x and y such that $p = x^2 + y^2$. ∎

Theorem 2.5.1 is called the Girard–Fermat theorem. Heath-Brown [10] gave an interesting new proof of this theorem in 1984. Varouchas [18] and Williams [20] have given presentations of Heath-Brown's proof. Zagier [22] has given a one-sentence proof.

Theorem 2.5.2 *Let p be a prime such that $p \equiv 1, 3$ (mod 8). Then there exist integers x and y such that $p = x^2 + 2y^2$.*

Proof: The proof is the same as that of Theorem 2.5.1 except that (2.5.2) is used in place of (2.5.1) and $\mathbb{Z} + \mathbb{Z}\sqrt{-2}$ in place of $\mathbb{Z} + \mathbb{Z}\sqrt{-1}$. ∎

Jackson [11] has given a short proof of Theorem 2.5.2 when $p \equiv 3$ (mod 8). Similarly using (2.5.3)–(2.5.5) and Theorem 1.4.5, we obtain the following three theorems.

Theorem 2.5.3 *Let p be a prime such that $p \equiv 1$ (mod 3). Then there exist integers x and y such that $p = x^2 + xy + y^2$.*

Theorem 2.5.4 *Let p be a prime such that $p \equiv 1, 2, 4$ (mod 7). Then there exist integers x and y such that $p = x^2 + xy + 2y^2$.*

Theorem 2.5.5 *Let p be a prime such that $p \equiv 1, 3, 4, 5, 9 \pmod{11}$. Then there exist integers x and y such that $p = x^2 + xy + 3y^2$.*

In Theorem 2.2.8 we showed that $\mathbb{Z} + \mathbb{Z}\sqrt{m}$ is Euclidean for $m = 2, 3, 6$. Recall from elementary number theory that for an odd prime p

$$\left(\frac{2}{p}\right) = 1 \iff p \equiv 1, 7 \pmod{8},$$

$$\left(\frac{3}{p}\right) = 1 \iff p \equiv 1, 11 \pmod{12},$$

$$\left(\frac{6}{p}\right) = 1 \iff p \equiv 1, 5, 19, 23 \pmod{24}.$$

Hence, by Theorem 1.4.4, we obtain the following three theorems.

Theorem 2.5.6 *Let p be a prime such that $p \equiv 1, 7 \pmod{8}$. Then there exist integers x and y such that $p = x^2 - 2y^2$.*

We used the fact that $T^2 - 2U^2 = -1$ for $T = U = 1$.

Theorem 2.5.7 *Let p be a prime such that $p \equiv 1, 11 \pmod{12}$. Then there exist integers x and y such that either $p = x^2 - 3y^2$ or $p = 3y^2 - x^2$.*

In this case there are no integers T and U such that $T^2 - 3U^2 = -1$.

Theorem 2.5.8 *Let p be a prime such that $p \equiv 1, 5, 19, 23 \pmod{24}$. Then there exist integers x and y such that either $p = x^2 - 6y^2$ or $p = 6y^2 - x^2$.*

There are no integers T and U such that $T^2 - 6U^2 = -1$.

Exercises

1. Let D be an integral domain possessing a Euclidean function ϕ. Give an example to show that

$$\phi(a) = \phi(b) \ (a, b \in D) \not\Longrightarrow a \sim b.$$

2. Prove Theorem 2.2.4.
3. Prove Theorem 2.2.5.
4. Give an example to show that q and r in (2.1.2) are not necessarily unique.
5. Prove that $\mathbb{Z} + \mathbb{Z}\sqrt{7}$ is Euclidean with respect to ϕ_7 using the method of the proof of Theorem 2.2.8.

6. Prove that $\mathbb{Z} + \mathbb{Z}\left(\frac{1+\sqrt{5}}{2}\right)$ is Euclidean with respect to ϕ_5 using the method of the proof of Theorem 2.2.8.

7. Use Theorem 2.3.1 to show that $\mathbb{Z} + \mathbb{Z}\sqrt{26}$ is not Euclidean with respect to ϕ_{26}.

8. Prove a modification of Theorem 2.3.1 that allows one of the primes p and q to be the prime 2.

9. Prove an extension of Theorem 2.3.1 that replaces p and q in the equation $pt + qu = m$ by powers of p and q with odd exponents.

10. Use Theorem 2.3.3 to prove that $\mathbb{Z} + \mathbb{Z}\left(\frac{1+\sqrt{77}}{2}\right)$ is not Euclidean with respect to ϕ_{77}.

11. Prove that if p is a prime with $p \equiv 3 \pmod 4$ then there do not exist integers x and y such that $p = x^2 + y^2$.

12. Let p be a prime. Use Theorem 2.5.1 and Exercise 11 to deduce that

$$p = x^2 + y^2 \iff p = 2 \text{ or } p \equiv 1 \pmod 4.$$

13. Prove that if p is a prime with $p \equiv 5, \ 7 \pmod 8$ then there do not exist integers x and y such that $p = x^2 + 2y^2$.

14. Let p be a prime. Use Theorem 2.5.2 and Exercise 13 to deduce that

$$p = x^2 + 2y^2 \iff p = 2 \text{ or } p \equiv 1, \ 3 \pmod 8.$$

15. Prove that if p is a prime with $p \equiv 2 \pmod 3$ then there do not exist integers x and y such that $p = x^2 + xy + y^2$.

16. Let p be a prime. Use Theorem 2.5.3 and Exercise 15 to deduce that

$$p = x^2 + xy + y^2 \iff p = 3 \text{ or } p \equiv 1 \pmod 3.$$

17. Prove that if p is a prime with $p \equiv 3, \ 5, \ 6 \pmod 7$ then there do not exist integers x and y such that $p = x^2 + xy + 2y^2$.

18. Let p be a prime. Use Theorem 2.5.4 and Exercise 17 to deduce that

$$p = x^2 + xy + 2y^2 \iff p = 7 \text{ or } p \equiv 1, \ 2, \ 4 \pmod 7.$$

19. Prove that if m is a positive integer possessing a prime divisor $q \equiv 3 \pmod 4$ then there are no integers T and U such that $T^2 - mU^2 = -1$.

20. Let p be a prime with $p \equiv 1, 11 \pmod{12}$. Deduce from Theorem 2.5.7 that

$$p = x^2 - 3y^2, \text{ if } p \equiv 1 \pmod{12},$$
$$p = 3y^2 - x^2, \text{ if } p \equiv 11 \pmod{12},$$

for some integers x and y.

21. Let p be a prime with $p \equiv 1, 5, 19, 23 \pmod{24}$. Deduce from Theorem 2.5.8 that

$$p = x^2 - 6y^2, \text{ if } p \equiv 1, 19 \pmod{24},$$
$$p = 6y^2 - x^2, \text{ if } p \equiv 5, 23 \pmod{24},$$

for some integers x and y.

22. Prove that the subdomain $\mathbb{Z} + 3\mathbb{Z}\sqrt{-2}$ of the Euclidean domain $\mathbb{Z} + \mathbb{Z}\sqrt{-2}$ is not Euclidean.

23. Prove that the subdomain $\mathbb{Z} + 7\mathbb{Z}\sqrt{2}$ of the Euclidean domain $\mathbb{Z} + \mathbb{Z}\sqrt{2}$ is not Euclidean.

24. Prove that the subdomain $\mathbb{Z} + 2\mathbb{Z}\sqrt{3}$ of the Euclidean domain $\mathbb{Z} + \mathbb{Z}\sqrt{3}$ is not Euclidean.

25. Prove that the subdomain $\mathbb{Z} + 5\mathbb{Z}\sqrt{6}$ of the Euclidean domain $\mathbb{Z} + \mathbb{Z}\sqrt{6}$ is not Euclidean.

26. Let m be a positive integer with $m \equiv 1 \pmod 4$. Show that the solvability of the equation $T^2 + TU + \frac{1}{4}(1 - m)U^2 = -1$ in integers T and U (see Theorem 1.4.5) is equivalent to the solvability of the equation $X^2 - mY^2 = -4$ in integers X and Y.

27. Let m be an integer with $m \equiv 1 \pmod 4$ that possesses a prime divisor $q \equiv 3 \pmod 4$. Prove that there are no integers T and U such that $T^2 + TU + \frac{1}{4}(1 - m)U^2 = -1$.

28. Prove that if p is a prime with $p \equiv 1, 4 \pmod 5$ then there are integers x and y such that $p = x^2 + xy - y^2$. [Hint: Use Theorems 1.4.5 and 2.2.7.]

29. Use Exercise 12 to show that the irreducibles in $\mathbb{Z} + \mathbb{Z}i$ are $1 + i$ and its associates, $x \pm iy$, where $x^2 + y^2 = p$ (prime) $\equiv 1 \pmod 4$, and their associates, and q (prime) $\equiv 3 \pmod 4$ and its associates.

30. Use Exercise 14 to determine the irreducibles in $\mathbb{Z} + \mathbb{Z}\sqrt{-2}$.

Suggested Reading

1. P. J. Arpaia, *A note on quadratic Euclidean domains*, American Mathematical Monthly 75 (1968), 864–865.

   Examples of quadratic Euclidean domains that possess subdomains that are not Euclidean are given. For example the Gaussian domain $\mathbb{Z} + \mathbb{Z}\sqrt{-1}$ is Euclidean but its subdomain $\mathbb{Z} + 2\mathbb{Z}\sqrt{-1}$ is not Euclidean.

2. O. A. Cámpoli, *A principal ideal domain that is not a Euclidean domain*, American Mathematical Monthly 95 (1988), 868–871.

   It is shown in an elementary fashion that $\mathbb{Z} + \mathbb{Z}(\frac{1+\sqrt{-19}}{2})$ is a principal ideal domain but not a Euclidean domain. The idea of a domain being almost Euclidean is introduced (p. 870).

3. H. Chatland, *On the Euclidean algorithm in quadratic number fields*, Bulletin of the American Mathematical Society 55 (1949), 948–953.

   This paper is a valuable source of references to work on the Euclidean algorithm in quadratic domains. It should be noted that $\mathbb{Z} + \mathbb{Z}(\frac{1+\sqrt{97}}{2})$ is not Euclidean, contrary to the claim by Rédei. This was established by Barnes and Swinnerton-Dyer in 1952.

4. H. Chatland and H. Davenport, *Euclid's algorithm in real quadratic fields*, Canadian Journal of Mathematics 2 (1950), 289–296.

   This is where the final steps in the proofs of Theorems 2.2.6 and 2.2.7 are given.

5. D. A. Clark, *A quadratic field which is Euclidean but not norm-Euclidean*, Manuscripta Mathematica 83 (1994), 327–330.

   It is shown that $\mathbb{Z} + \mathbb{Z}(\frac{1+\sqrt{69}}{2})$ is Euclidean but not norm-Euclidean.

6. D. A. Cox, *Primes of the form $x^2 + ny^2$*, Wiley, New York, 1989.

   This book presents a comprehensive treatment of the problem of deciding which primes are represented by $x^2 + ny^2$.

7. D. W. Dubois and A. Steger, *A note on division algorithms in imaginary quadratic number fields*, Canadian Journal of Mathematics 10 (1958), 285–286.

   Let m be a negative squarefree integer. The authors prove that if $\mathbb{Z} + \mathbb{Z}\sqrt{m}$ ($m \equiv 2, 3 \pmod 4$) and $\mathbb{Z} + \mathbb{Z}(\frac{1+\sqrt{m}}{2})$ ($m \equiv 1 \pmod 4$) are Euclidean they must be Euclidean with respect to ϕ_m.

8. J. Greene, *Principal ideal domains are almost Euclidean*, American Mathematical Monthly 104 (1997), 154–156.

   The author proves that an integral domain is a principal ideal domain if and only if it is almost Euclidean.

9. M. Harper, *A proof that $\mathbb{Z}[\sqrt{14}]$ is a Euclidean domain*, Ph.D. thesis, McGill University, Montréal, Canada, 2000.

   It is shown that $\mathbb{Z} + \mathbb{Z}\sqrt{14}$ is Euclidean.

10. D. R. Heath-Brown, *Fermat's two-squares theorem*, Invariant (1984), 3–5.

    A beautifully simple proof of the Girard-Fermat theorem is given.

11. T. Jackson, *A short proof that every prime $p \equiv 3 \pmod 8$ is of the form $x^2 + 2y^2$*, American Mathematical Monthly 107 (2000), 447.

    Heath-Brown's ideas are used to prove Euler's result that every prime $p \equiv 3 \pmod 8$ is represented by $x^2 + 2y^2$.

12. M. A. Jodeit, *Uniqueness in the division algorithm*, American Mathematical Monthly 74 (1967), 835–836.

    Let D be an integral domain. Suppose that ϕ is a Euclidean function on D such that the quotient q and remainder r in (2.1.2) are always unique. Then D is either a field or a polynomial domain $F[x]$, where F is a field.

13. Th. Motzkin, *The Euclidean algorithm*, Bulletin of the American Mathematical Society 55 (1949), 1142–1146.

    In this classic paper on Euclidean domains, it is shown that $\mathbb{Z} + \mathbb{Z}(\frac{1+\sqrt{-19}}{2})$ is a principal ideal domain but not a Euclidean domain. Universal side divisors are introduced.

14. T.-S. Rhai, *A characterization of polynomial domains over a field*, American Mathematical Monthly 69 (1962), 984–986.

    Let D be an integral domain. Suppose that ϕ is a Euclidean function on D such that the quotient q and remainder r in (2.1.2) are always unique. Then D is either a field or a polynomial domain $F[x]$, where F is a field.

15. K. Rogers, *The axioms for Euclidean domains*, American Mathematical Monthly 78 (1971), 1127–1128.

    The role of the condition (2.1.1) in the definition of a Euclidean function is discussed.

16. P. Samuel, *About Euclidean rings*, Journal of Algebra 19 (1971), 282–301.

    In this classic paper on Euclidean rings, the author suggests (p. 294) that $\mathbb{Z} + \mathbb{Z}\sqrt{14}$ may be Euclidean but not norm-Euclidean. This has recently been established by Harper [9].

17. S. Singh, *Non-Euclidean domains: An example*, Mathematics Magazine 49 (1976), 243.

    It is shown that $\mathbb{Z} + \mathbb{Z}\sqrt{m}$ is not Euclidean for negative squarefree integers m with $m < -2$ and $m \equiv 2, 3 \pmod 4$.

18. Y. Varouchas, *Une démonstration élémentaire du théorème des deux carrés*, La Caverne, I. R. E. M. de Lorraine, France, Bulletin No. 6, février 1984, pp. 31–39.

    A presentation in French of Heath-Brown's proof of the Girard–Fermat theorem is given.

19. K. S. Williams, *Note on non-Euclidean principal ideal domains*, Mathematics Magazine 48 (1975), 176–177.

    It is shown that the domains $\mathbb{Z} + \mathbb{Z}(\frac{1+\sqrt{m}}{2})$ ($m = -19, -43, -67, -163$) are not Euclidean.

20. K. S. Williams, *Heath–Brown's elementary proof of the Girard–Fermat theorem*, Carleton Coordinates, Department of Mathematics and Statistics, Carleton University, Ottawa, Ontario, Canada, January 1985, pp. 4–5.

    A presentation of Heath-Brown's proof of the Girard–Fermat theorem is given.

21. J. C. Wilson, *A principal ideal ring that is not a Euclidean ring*, Mathematics Magazine 46 (1973), 34–38.

    The author shows that $\mathbb{Z} + \mathbb{Z}(\frac{1+\sqrt{-19}}{2})$ is a principal ideal domain that is not Euclidean.

22. D. Zagier, *A one-sentence proof that every prime $p \equiv 1 \pmod 4$ is a sum of two squares*, American Mathematical Monthly 97 (1990), 144.

    A one-sentence rendition of Heath-Brown's proof of the Girard-Fermat theorem is given.

Biographies

1. K. Barner, *Pierre de Fermat* (1601?–1665)—*His life beside mathematics*, Canadian Mathematical Society Notes 34 (2002), 3–4, 26–30.

   The author relates an interesting account of the nonmathematical life of Fermat.

2. P. Hoffman, *The Man Who Loved Only Numbers*, Hyperion, New York, 1998.

   The story of Paul Erdös, one of the most prolific and eccentric mathematicians of the twentieth century, is told.

3. M. S. Mahoney, *The Mathematical Career of Pierre de Fermat* (1601–1605), Princeton University Press, Princeton, New Jersey, 1973.

   For two completely different reviews of this book, see Isis 65 (1974), 398–400 and Bulletin of the American Mathematical Society 79 (1973), 1138–1149.

4. C. A. Rogers, *Harold Davenport*, Bulletin of the London Mathematical Society 4 (1972), 66–99.

   A memoir on the life and mathematics of Davenport is presented.

5. The website

   http://www-groups.dcs.st-and.ac.uk/~history/

   has biographies of A. T. Brauer, L. E. Dickson, P. Erdös, H. A. Heilbronn, O. Perron, and R. Remak.

3

Noetherian Domains

3.1 Noetherian Domains

Let I_1 be a nonzero ideal in the domain $\mathbb{Z}$. We consider ideals I such that $I_1 \subseteq I$. As $\mathbb{Z}$ is a principal ideal domain (Theorem 1.4.1), there are nonzero integers m and n such that $I_1 = \langle m \rangle$, $I = \langle n \rangle$, and $n \mid m$. Now m has only finitely many divisors n so there exist only finitely many ideals I with $I_1 \subseteq I$. Thus there cannot exist infinitely many ideals I_k $(k = 2, 3, \ldots)$ such that

$$I_1 \subset I_2 \subset I_3 \subset I_4 \subset \ldots. \tag{3.1.1}$$

The importance of domains such as $\mathbb{Z}$ that do not contain infinite ascending chains of ideals of the type (3.1.1) was first recognized by the German mathematician Emmy Noether (1882–1935). Such domains are now called Noetherian domains in her honor. We note that some domains do contain infinite chains of ideals of the type (3.1.1). For example, if F is a field, the domain $F[X_1, X_2, \ldots]$ contains the infinite chain of ideals

$$\langle X_1 \rangle \subset \langle X_1, \ X_2 \rangle \subset \langle X_1, \ X_2, \ X_3 \rangle \subset \ldots.$$

Definition 3.1.1 (Ascending chain of ideals) *An infinite sequence of ideals $\{I_n : n = 1, 2, \ldots\}$ in an integral domain is said to be an ascending chain if*

$$I_1 \subseteq I_2 \subseteq \ldots \subseteq I_n \subseteq \ldots.$$

The chain is said to be a strictly ascending chain if

$$I_1 \subset I_2 \subset \ldots \subset I_n \subset \ldots.$$

Definition 3.1.2 (Terminating ascending chain) *An ascending chain of ideals*

$$I_1 \subseteq I_2 \subseteq \ldots \subseteq I_n \subseteq \ldots$$

in an integral domain is said to terminate if there exists a positive integer n_0 such that

$$I_n = I_{n_0} \quad \text{for all } n \geq n_0.$$

Definition 3.1.3 (Ascending chain condition) *An integral domain D is said to satisfy the ascending chain condition if every ascending chain of ideals in D terminates or, equivalently, if D does not contain a strictly ascending chain of ideals.*

Definition 3.1.4 (Noetherian domain) *An integral domain that satisfies the ascending chain condition is called a Noetherian domain.*

More generally we define a Noetherian ring to be a ring R in which every ascending chain of (two-sided) ideals in R terminates.

From the remarks preceding the definitions, we have the following two examples.

Example 3.1.1 $\mathbb{Z}$ *is a Noetherian domain.*

Example 3.1.2 *If F is a field, the domain $F[X_1, X_2, \ldots]$ is not Noetherian.*

The next theorem gives a necessary and sufficient condition for an integral domain to be a Noetherian domain.

Theorem 3.1.1 *Let D be an integral domain. Then D is Noetherian if and only if every ideal of D is finitely generated.*

Proof: Let D be a Noetherian domain. Suppose that not every ideal of D is finitely generated. Let I be an ideal of D that is not finitely generated. Thus $I \neq \langle 0 \rangle$, and so there exists $a_1 \in I$ with $a_1 \neq 0$. Let A_1 be the ideal given by $A_1 = \langle a_1 \rangle$. Clearly $A_1 \subseteq I$. Moreover, $I \neq A_1$ as A_1 is finitely generated and I is not. Hence $A_1 \subset I$. Take $a_2 \in I$, $a_2 \notin A_1$, and let A_2 be the ideal given by $A_2 = \langle a_1, a_2 \rangle$. Clearly $A_1 \subset A_2 \subset I$. Continuing in this way, we obtain an infinite strictly increasing sequence of ideals $A_1 \subset A_2 \subset \ldots$, contradicting that D is a Noetherian domain. Hence every ideal of a Noetherian domain must be finitely generated.

Now let D be an integral domain in which every ideal is finitely generated. Let

$$I_1 \subseteq I_2 \subseteq I_3 \subseteq \ldots$$

be an ascending chain of ideals in D. It is easy to check that $\bigcup_{n=1}^{\infty} I_n$ is an ideal of D. Hence $\bigcup_{n=1}^{\infty} I_n$ is finitely generated, so there exist finitely many elements

$a_1, a_2, \ldots, a_m$ of D such that

$$\bigcup_{n=1}^{\infty} I_n = \langle a_1, a_2, \ldots, a_m \rangle.$$

For each $i = 1, 2, \ldots, m$, $a_i \in \bigcup_{n=1}^{\infty} I_n$, say, $a_i \in I_{n_i}$. Set $l = \max(n_1, n_2, \ldots, n_m)$. Clearly $I_l \subseteq \bigcup_{n=1}^{\infty} I_n$. As $n_i \leq l$ we have $I_{n_i} \subseteq I_l$, and thus $a_i \in I_l$ for $i = 1, 2, \ldots, m$. Hence $\langle a_1, \ldots, a_m \rangle \subseteq I_l$ so that $\bigcup_{n=1}^{\infty} I_n \subseteq I_l$. This proves that $\bigcup_{n=1}^{\infty} I_n = I_l$, and thus $I_n = I_l$ for $n \geq l$. Hence D is Noetherian. ■

From Theorem 3.1.1 we see that principal ideal domains are Noetherian.

Theorem 3.1.2 *Let D be a principal ideal domain. Then D is a Noetherian domain.*

Proof: As D is a principal ideal domain, every ideal in D is principal and therefore finitely generated. Hence, by Theorem 3.1.1, D is Noetherian. ■

Example 3.1.3 *By Theorems 2.1.2 and 3.1.2 a Euclidean domain is always Noetherian. Thus*

$$\mathbb{Z} \ (Theorem\ 2.2.1(a)),$$
$$\mathbb{Z} + \mathbb{Z}\sqrt{-1}, \ \mathbb{Z} + \mathbb{Z}\sqrt{-2} \ (Theorem\ 2.2.3),$$
$$\mathbb{Z} + \mathbb{Z}\left(\frac{1 + \sqrt{-3}}{2}\right), \ \mathbb{Z} + \mathbb{Z}\left(\frac{1 + \sqrt{-7}}{2}\right),$$
$$\mathbb{Z} + \mathbb{Z}\left(\frac{1 + \sqrt{-11}}{2}\right) \ (Theorem\ 2.2.5),$$
$$\mathbb{Z} + \mathbb{Z}\sqrt{2}, \ \mathbb{Z} + \mathbb{Z}\sqrt{3}, \ \mathbb{Z} + \mathbb{Z}\sqrt{6} \ (Theorem\ 2.2.8)$$

are all examples of Noetherian domains.

Our next objective is to give another condition (called the maximal condition) that allows us to recognize when an integral domain is Noetherian.

Definition 3.1.5 (Maximal condition) *An integral domain D is said to satisfy the maximal condition if every nonempty set S of ideals of D contains an ideal that is not properly contained in any other ideal of the set S; that is, S possesses an ideal I such that if J is an ideal in S with $I \subseteq J$ then $J = I$.*

We show that satisfying the maximal condition is equivalent to the domain being Noetherian.

Theorem 3.1.3 *Let D be an integral domain. Then D is Noetherian if and only if D satisfies the maximal condition.*

Proof: Suppose that D is a Noetherian domain that does not satisfy the maximal condition. Then D possesses a nonempty set S of ideals with the property that for every ideal I of S there exists an ideal J of S with $I \subset J$. This property enables us to construct inductively an infinite strictly ascending chain of ideals in S, which contradicts D being a Noetherian domain. Hence every Noetherian domain must satisfy the maximal condition.

Now let D be an integral domain that satisfies the maximal condition. Let $I_1 \subseteq I_2 \subseteq I_3 \subseteq \ldots$ be an ascending chain of ideals of D. Set $S = \{I_n \mid n = 1, 2, 3, \ldots\}$. As D satisfies the maximal condition, S contains an ideal I_m, which is not properly contained in any other ideal of S. As $I_m \subseteq I_j$ for $j \geq m$ we must have $I_j = I_m$ for $j \geq m$. Hence the ascending chain $I_1 \subseteq I_2 \subseteq I_3 \subseteq \ldots$ terminates and D is Noetherian. ∎

A famous theorem of David Hilbert (1862–1943) asserts that if D is a Noetherian domain then the polynomial domain $D[X_1, \ldots, X_n]$ is also Noetherian. This is the celebrated Hilbert basis theorem. We will not prove this theorem here; a proof can be found for example in [8, pp. 201–202].

Example 3.1.4 $\mathbb{Z}[X_1, \ldots, X_n]$ *is a Noetherian domain. We can see this as follows. By Theorem 2.2.1(a) $\mathbb{Z}$ is a Euclidean domain. Thus, by Theorem 2.1.2, $\mathbb{Z}$ is a principal ideal domain. Hence, by Theorem 3.1.2, $\mathbb{Z}$ is a Noetherian domain. Finally, by the Hilbert basis theorem, $\mathbb{Z}[X_1, \ldots, X_n]$ is a Noetherian domain.*

Example 3.1.5 $F[X_1, X_2, \ldots, X_n]$ $(n \geq 1)$, *where F is a field, is a Noetherian domain. We can prove this as follows. By Theorem 2.2.1(b), $F[X_1]$ is a Euclidean domain. Hence, by Theorem 2.1.2, $F[X_1]$ is a principal ideal domain, and so, by Theorem 3.1.2, is a Noetherian domain. Then, by the Hilbert basis theorem, $(F[X_1])[X_2, \ldots, X_n]$ is a Noetherian domain; that is, $F[X_1, X_2, \ldots, X_n]$ is a Noetherian domain.*

3.2 Factorization Domains

Let D be an integral domain that is not a field so that D contains nonzero, nonunit elements. It may be the case that all of these elements are reducible so that D contains no irreducibles. The next example illustrates this.

Example 3.2.1 *Let D be the domain of polynomials in positive rational powers of x over $\mathbb{C}$, that is,*

$$D = \{a_1 x^{r_1} + \cdots + a_n x^{r_n} \mid n \in \mathbb{N}, \ a_1, \ldots, a_n \in \mathbb{C}, \ r_1, \ldots, r_n \in \mathbb{Q},$$
$$0 \leq r_1 < \cdots < r_n\}.$$

Clearly $U(D) = \mathbb{C}^*$. We show that D does not possesss any irreducible elements. Suppose that

$$f(x) = a_1 x^{r_1} + \cdots + a_n x^{r_n}$$

is an irreducible element of D. As $f(x)$ is a nonzero element of D, we may suppose that $a_n \neq 0$. If $n = 1$ and $r_1 = 0$ then $f(x) = a_1 \neq 0$ is a unit of D, a contradiction. If $n = 1$ and $r_1 > 0$ then $f(x) = a_1 x^{r_1} = a_1 (x^{r_1/2})^2$ is reducible in D, a contradiction. Hence $n \geq 2$ and $r_n > 0$. Let

$$t = \text{least common multiple of the (positive) denominators of the}$$
$$\text{rationals } r_1, \ldots, r_n$$

so that t is a positive integer such that $r_1 t, \ldots, r_n t$ are integers with

$$0 \leq r_1 t < r_2 t < \cdots < r_n t.$$

Then

$$f(x^t) = a_1 x^{r_1 t} + \cdots + a_n x^{r_n t} \in \mathbb{C}[x].$$

Hence there exist $b_1, \ldots, b_{r_n t} \in \mathbb{C}$ such that

$$a_1 x^{r_1 t} + \cdots + a_n x^{r_n t} = a_n (x - b_1) \cdots (x - b_{r_n t}).$$

Thus

$$f(x) = a_n (x^{1/t} - b_1) \cdots (x^{1/t} - b_{r_n t}).$$

Since

$$x^{1/t} - b_1 = (x^{1/2t} - b_1^{1/2})(x^{1/2t} + b_1^{1/2}),$$

where $x^{1/2t} \pm b_1^{1/2}$ are nonzero, nonunit elements of D, $f(x)$ is reducible, a contradiction.

Thus D does not possess any irreducible elements.

We show next that a Noetherian domain always contains irreducibles.

Theorem 3.2.1 *Let D be an integral domain that is not a field. If D is Noetherian then D contains elements that are irreducible.*

Proof: Suppose that the integral domain D does not contain any irreducibles. As we are assuming that D is not a field, D has nonzero, nonunit elements. Let a be one of these. Then a is not an irreducible. Hence a is reducible. Thus there exists a nonzero, nonunit element a_1 of D such that $a_1 \mid a$ and $a_1 \nsim a$. Clearly $\langle a \rangle \subset \langle a_1 \rangle$. As a_1 is not an irreducible, a_1 is reducible, and we can repeat the preceding argument to obtain a nonzero, nonunit element a_2 of D such that $a_2 \mid a_1$ and $a_2 \nsim a_1$. Thus $\langle a_1 \rangle \subset \langle a_2 \rangle$. Continuing in this way we obtain an ascending chain of principal ideals

$$\langle a \rangle \subset \langle a_1 \rangle \subset \langle a_2 \rangle \subset \cdots,$$

contradicting that D is a Noetherian domain. Hence D contains irreducibles. ■

By Theorem 3.2.1 the domain D in Example 3.2.1 cannot be Noetherian. This is easily seen directly as it contains the infinite ascending chain of ideals

$$\langle x \rangle \subset \langle x^{1/2} \rangle \subset \langle x^{1/4} \rangle \subset \langle x^{1/8} \rangle \subset \cdots.$$

Clearly for this domain it is not possible to express each nonzero, nonunit element as a finite product of irreducibles. Domains in which this is possible are called factorization domains. The main result of this section is that a Noetherian domain is always a factorization domain; that is, in a Noetherian domain every nonzero, nonunit element can be expressed as a finite product of irreducibles. The converse of this result however is not true as we demonstrate in Example 3.2.2.

Definition 3.2.1 (Factorization domain) *Let D be an integral domain. Then D is said to be a factorization domain if every nonzero, nonunit element of D can be expressed as a finite product of irreducible elements of D.*

Our next result shows that a Noetherian domain is always a factorization domain.

Theorem 3.2.2 *Let D be a Noetherian domain. Then D is a factorization domain.*

Proof: Let D be a Noetherian domain and suppose that D is not a factorization domain. Then D contains at least one nonzero, nonunit element that is not a finite product of irreducible elements of D. Let A be the set of all such elements, so A is not empty. Let

$$S = \{\langle a \rangle \mid a \in A\}.$$

Clearly S is a nonempty set of principal ideals of D. As D is a Noetherian domain, by the maximal condition (Theorem 3.1.3), S has a maximal element, say $\langle b \rangle$. As $\langle b \rangle \in S$, $b \in A$ so that b is a nonzero, nonunit element of D that is not a product of irreducibles. Hence b is not irreducible. Thus we can write b in the form $b = cd$, where c and d are nonzero, nonunit elements of D. Hence $\langle b \rangle = \langle cd \rangle \subseteq \langle c \rangle$ and $\langle b \rangle = \langle cd \rangle \subseteq \langle d \rangle$. Moreover, as d is not a unit, b and c are not associates; thus $\langle b \rangle \neq \langle c \rangle$, and so $\langle b \rangle \subset \langle c \rangle$. Similarly, $\langle b \rangle \subset \langle d \rangle$. By the maximality of $\langle b \rangle$, we have $\langle c \rangle \notin S$ and $\langle d \rangle \notin S$. Hence c and d are products of irreducible elements of D. Thus $b = cd$ is also a product of irreducible elements of D, contradicting our assumption. Thus D is a factorization domain. ■

Example 3.2.2 *From Examples 3.1.1 and 3.1.3, and Theorem 3.2.2, we see that* $\mathbb{Z}$, $\mathbb{Z} + \mathbb{Z}\sqrt{-1}$, $\mathbb{Z} + \mathbb{Z}\sqrt{-2}$, $\mathbb{Z} + \mathbb{Z}\left(\frac{1+\sqrt{-3}}{2}\right)$, $\mathbb{Z} + \mathbb{Z}\left(\frac{1+\sqrt{-7}}{2}\right)$, $\mathbb{Z} + \mathbb{Z}\left(\frac{1+\sqrt{-11}}{2}\right)$, $\mathbb{Z} + \mathbb{Z}\sqrt{2}$, $\mathbb{Z} + \mathbb{Z}\sqrt{3}$, and $\mathbb{Z} + \mathbb{Z}\sqrt{6}$ are all examples of factorization domains.*

The next example shows that a factorization domain is not always a Notherian domain.

Example 3.2.3 *Let F be a field. We show that $F[X_1, X_2, \ldots]$ is a factorization domain. Let a be a nonzero, nonunit element of $F[X_1, X_2, \ldots]$. Then a is a polynomial in finitely many of the indeterminates $X_1, X_2, \ldots$ with coefficients in F. Thus there is a positive integer m such that $a \in F[X_1, X_2, \ldots, X_m]$. From Example 3.1.5 we know that $F[X_1, \ldots, X_m]$ is a Noetherian domain. Therefore, by Theorem 3.2.2, it is a factorization domain. Hence every nonzero, nonunit element of $F[X_1, X_2, \ldots, X_m]$ can be expressed as a product of irreducible elements. Thus a is a product of irreducible elements of $F[X_1, X_2, \ldots, X_m]$. But an irreducible element of $F[X_1, X_2, \ldots, X_m]$ is an irreducible element of $F[X_1, X_2, \ldots]$, so a is a product of irreducible elements of $F[X_1, X_2, \ldots]$. Hence $F[X_1, X_2, \ldots]$ is a factorization domain. By Example 3.1.2, $F[X_1, X_2, \ldots]$ is not Noetherian.*

Theorem 3.2.3 *Let D be a principal ideal domain. Then D is a factorization domain.*

Proof: This result follows from Theorems 3.1.2 and 3.2.2.   ■

3.3 Unique Factorization Domains

Let D be a factorization domain. Let a be a nonzero, nonunit element of D. Then there exist irreducible elements of D such that

$$a = h_1 h_2 \cdots h_k.$$

If h_1 and h_2 are associates, say $h_2 = v h_1$, where v is a unit of D, then

$$a = v h_1^2 h_3 \cdots h_k.$$

Repeating this process we eventually obtain a factorization

$$a = w l_1^{k_1} \cdots l_m^{k_m},$$

where w is a unit of D, the k_i are positive integers, and the l_i are irreducible elements of D with no two distinct ones being associates. Suppose

$$a = w' l_1'^{k_1'} \cdots l_{m'}'^{k_{m'}'}$$

is another such factorization of a into powers of nonassociated irreducible elements of D. If $m = m'$ and after a possible rearragement of $l_1, \ldots, l_m$, we have

$$l_1 \sim l'_1, \ldots, l_m \sim l'_m \text{ and } k_1 = k'_1, \ldots, k_m = k'_m,$$

we say that a has a unique factorization as a product of irreducible elements of D.

Definition 3.3.1 (Unique factorization domain) *Let D be a factorization domain. Suppose that every nonzero, nonunit element a of D has a unique factorization as a product of irreducible elements of D. Then D is called a unique factorization domain.*

Theorem 3.3.1 *Let D be a principal ideal domain. Then D is a unique factorization domain.*

Proof: Suppose that D is a principal ideal domain. By Theorem 3.2.3 D is a factorization domain. Suppose however that D is not a unique factorization domain. Then there exists at least one nonzero, nonunit element a, which has at least two different factorizations as a product of irreducible elements of D. Let A be the set of all such elements a, and let

$$S = \{\langle a \rangle \mid a \in A\}.$$

As A is a nonempty set, so is S. Now D is a principal ideal domain, so by Theorem 3.1.2, D is a Noetherian domain. Hence, by the maximal condition (Theorem 3.1.3), S contains a maximal element, say $\langle b \rangle$. Thus $b \in A$ and b has two essentially different factorizations as a product of irreducibles, say,

$$b = u l_1^{k_1} \cdots l_m^{k_m} = v h_1^{j_1} \cdots h_n^{j_n}, \tag{3.3.1}$$

where u and v are units of D, $l_1, \ldots, l_m, h_1, \ldots, h_n$ are irreducible elements of D, $k_1, \ldots, k_m, j_1, \ldots, j_n$ are positive integers, $l_i \not\sim l_j$ $(i \neq j)$, and $h_i \not\sim h_j$ $(i \neq j)$. As $k_1 > 0$, we see that $l_1 \mid b$, and thus $l_1 \mid v h_1^{d_1} \ldots h_n^{j_n}$. As D is a principal ideal domain and l_1 is irreducible, by Theorem 1.4.2, l_1 is prime. Thus $l_1 \mid h_s$ for some integer s with $1 \leq s \leq n$. After relabeling the h's, if necessary, we may suppose that $l_1 \mid h_1$. Since l_1 and h_1 are both irreducibles this means that $l_1 \sim h_1$, say, $h_1 = l_1 w$, where w is a unit of D. Replacing h_1 by $l_1 w$ in (3.3.1), we obtain

$$b/l_1 = u l_1^{k_1-1} \cdots l_m^{k_m} = v w h_1^{d_1-1} \cdots h_n^{j_n}.$$

As l_1 is not a unit we have $\langle b \rangle \subset \langle b/l_1 \rangle$. Hence, by the maximality of $\langle b \rangle$, we have after suitable rearrangement of the h's

$$k_1 - 1 = j_1 - 1, \ k_2 = j_2, \ldots, k_m = j_m, \ m = n,$$
$$l_1 \sim h_1, \ l_2 \sim h_2, \ldots, l_m \sim h_m.$$

This contradicts the assumption that b has two essentially different factorizations. This completes the proof that a principal ideal domain is always a unique factorization domain. ∎

Clearly from Theorems 2.1.2 and 3.3.1 we see that a Euclidean domain is always a unique factorization domain. Thus the domains listed in Example 3.1.3 are all unique factorization domains.

Example 3.3.1 $\mathbb{Z}$, $\mathbb{Z} + \mathbb{Z}\sqrt{-1}$, $\mathbb{Z} + \mathbb{Z}\sqrt{-2}$, $\mathbb{Z} + \mathbb{Z}\left(\frac{1+\sqrt{-3}}{2}\right)$, $\mathbb{Z} + \mathbb{Z}\left(\frac{1+\sqrt{-7}}{2}\right)$, $\mathbb{Z} + \mathbb{Z}\left(\frac{1+\sqrt{-11}}{2}\right)$, $\mathbb{Z} + \mathbb{Z}\sqrt{2}$, $\mathbb{Z} + \mathbb{Z}\sqrt{3}$, and $\mathbb{Z} + \mathbb{Z}\sqrt{6}$ *are unique factorization domains.*

Example 3.3.2 *If F is a field then, by Theorem 2.2.1(b), $F[X]$ is a Euclidean domain and thus a unique factorization domain.*

It is a well-known theorem that if D is a unique factorization domain so is $D[X]$. A proof is given in [2, Theorem 7, p. 305]. Appealing to this result we obtain

Example 3.3.3 $\mathbb{Z}[X]$ *is a unique factorization domain.*

Thus $(\mathbb{Z}[X])[Y] = \mathbb{Z}[X, Y]$ is a unique factorization domain and generally we have

Example 3.3.4 $\mathbb{Z}[X_1, \ldots, X_n]$ *is a unique factorization domain.*

Similarly, as $F[X]$ (F a field) is a unique factorization domain, we have

Example 3.3.5 $F[X_1, \ldots, X_n]$ *(F a field) is a unique factorization domain.*

The next example shows that the converse of Theorem 3.3.1 is not true.

Example 3.3.6 *The unique factorization domain $\mathbb{Z}[X]$ is not a principal ideal domain as it contains the nonprincipal ideal $\langle 2, X \rangle$.*

Theorem 3.3.2 *Let D be a unique factorization domain. Then an element of D is irreducible if and only if it is prime.*

Proof: Let p be an irreducible element of D. Suppose that $p \mid ab$, where a and b are elements of D. Hence there exists an element c of D such that $ab = pc$. Since

D is a factorization domain, we have

$$a = p_1 \cdots p_l, \quad b = q_1 \cdots q_m, \quad c = r_1 \cdots r_n,$$

where $p_1, \ldots, p_l, q_1, \ldots, q_m, r_1, \ldots, r_n$ are irreducible elements of D, which are not necessarily distinct. Then

$$p_1 \cdots p_l q_1 \cdots q_m = p r_1 \cdots r_n.$$

As D is a unique factorization domain, p must be an associate of one of the p_i or q_j. Hence $p \mid a$ or $p \mid b$, showing that p is a prime.

   This completes the proof of the theorem as a prime is always an irreducible by Theorem 1.2.1. ∎

   Let $a_1, \ldots, a_n$ be nonzero elements of a unique factorization domain D. Let $\{\pi_1, \ldots, \pi_k\}$ be a set of irreducibles such that

(i) each π_i $(i = 1, 2, \ldots, k)$ divides at least one of $a_1, \ldots, a_n$,
(ii) $\pi_i \not\sim \pi_j$ if $i \neq j$, and
(iii) if π is an irreducible that divides at least one of $a_1, \ldots, a_n$ then $\pi \sim \pi_i$ for some
    $i \in \{1, 2, \ldots, k\}$.

We remark that if $a_1, \ldots, a_n$ are all units then $\{\pi_1, \ldots, \pi_k\} = \phi$. From (i), (ii), and (iii), we see that

$$a_i = \epsilon_i \prod_{j=1}^{k} \pi_j{}^{e_{ij}}, \quad i = 1, 2, \ldots, n,$$

where $\epsilon_i \in U(D)$ and the e_{ij} are nonnegative integers. (e_{ij} is positive if and only if $\pi_j \mid a_i$.) Set

$$e_j = \min_{1 \le i \le n} e_{ij}, \quad j = 1, 2, \ldots, k,$$

and

$$a = \prod_{j=1}^{k} \pi_j^{e_j} \in D.$$

Clearly

$$a \mid a_i, \quad i = 1, 2, \ldots, n,$$

and if $b \in D$ is such that

$$b \mid a_i, \quad i = 1, 2, \ldots, n,$$

then

$$b \mid a.$$

We call a "a greatest common divisor" of $a_1, \ldots, a_n$, a quantity we had previously defined in a principal ideal domain (Definition 1.4.2). If D is a unique factorization domain that is also a principal ideal domain then it is easily checked that this notion of a greatest common divisor coincides with that of Definition 1.4.2. If the set $\{\pi_1, \ldots, \pi_k\}$ of irreducibles is changed to any other set of irreducibles with properties (i), (ii), and (iii) then a is changed by at most a unit. Thus a greatest common divisor in a unique factorization domain is only defined up to a unit.

In a principal ideal domain D a greatest common divisor a of $a_1, \ldots, a_n$ is a linear combination of $a_1, \ldots, a_n$ with coefficients from D. This is not necessarily the case in a domain D that is a unique factorization domain but not a principal ideal domain. For example in $\mathbb{Z}[x]$ a greatest common divisor of $2x$ and x^2 is x but

$$x \neq f(x)2x + g(x)x^2$$

for any $f(x), \ g(x) \in \mathbb{Z}[x]$.

Theorem 3.3.3 *A principal ideal domain is an almost Euclidean domain.*

Proof: Let D be a principal ideal domain. By Theorem 3.3.1, D is a unique factorization domain. Hence we may define a function $\phi : D \to \mathbb{N} \cup \{0\}$ as follows:

$\phi(0) = 0,$

$\phi(a) = 1, \ \text{if } a \in U(D),$

$\phi(a) = 2^n, \ \text{if } a \in D - \tilde{D} \ \text{and } a = i_1 i_2 \cdots i_n,$

$\qquad\qquad\qquad\qquad \text{where } i_1, \ldots, i_n \text{ are irreducibles.}$

Clearly ϕ satisfies (2.4.1), (2.4.2), and (2.4.3). We show that ϕ satisfies (2.4.4). Let $a, b \in D$ with $b \neq 0$. Let $I = \langle a, b \rangle$. Since I is an ideal in D, $I = \langle r \rangle$ for some $r \in D$ with $r \neq 0$. If $a = bq$ for some $q \in D$ then $I = \langle bq, b \rangle = \langle b \rangle$. Otherwise $I \neq \langle b \rangle$. Since $b \in I, b = xr$ for some $x \in D$, so $\phi(b) \geq \phi(r)$. As $I \neq \langle b \rangle$, x is not a unit. Thus $\phi(x) > 1$ so $\phi(r) < \phi(b)$. Now $r = ax_0 + by_0$ for some $x_0, y_0 \in D$ so $0 < \phi(ax_0 + by_0) < \phi(b)$ and (2.4.4) is satisfied by ϕ. Thus ϕ is an almost Euclidean function on D and D is an almost Euclidean domain. ∎

From Theorems 2.4.1 and 3.3.3 we deduce Greene's theorem [3].

Theorem 3.3.4 *An integral domain is a principal ideal domain if and only if it is almost Euclidean.*

3.4 Modules

Analogous to the concept of a vector space over a field is that of a module over a ring. All rings are assumed to possess an identity.

Definition 3.4.1 (*R*-action) *Let R be a ring with identity and M an additive Abelian group. A function $\alpha : R \times M \to M$ is called an R-action on M if α has the following properties:*

$$\alpha(r + s, m) = \alpha(r, m) + \alpha(s, m), \tag{3.4.1}$$
$$\alpha(r, m + n) = \alpha(r, m) + \alpha(r, n), \tag{3.4.2}$$
$$\alpha(r, \alpha(s, m)) = \alpha(rs, m), \tag{3.4.3}$$
$$\alpha(1, m) = m, \tag{3.4.4}$$

for all $r, s \in R$ and all $m, n \in M$.

Definition 3.4.2 (*R*-module) *Let R be a ring with identity. An additive Abelian group M together with an R-action on M is called an R-module.*

It would be more accurate to call what we have just defined a left *R*-module. There is a similar definition of a right *R*-module in which the elements of *R* are written on the right. In this book we will keep to left modules throughout.

If *M* is an *R*-module with *R*-action α on *M* we write $\alpha(r, m)$ as rm to keep the notation as simple as possible. With this convention (3.4.1)–(3.4.4) become

$$(r + s)m = rm + sm, \tag{3.4.5}$$
$$r(m + n) = rm + rn, \tag{3.4.6}$$
$$r(sm) = (rs)m, \tag{3.4.7}$$
$$1m = m, \tag{3.4.8}$$

valid for all $r, s \in R$, $m, n \in M$. Taking $n = 0$ in (3.4.6) and $s = 0$ in (3.4.5), we deduce that $r0 = 0$ ($r \in R$) and $0m = 0$ ($m \in M$). The reader can easily check from the axioms that

$$(-r)m = -(rm) = r(-m)$$

for all $r \in R$ and all $m \in M$.

Example 3.4.1 *If F is a field then an F-module is the same thing as a vector space over F.*

Example 3.4.2 *Any additive Abelian group A can be thought of as a $\mathbb{Z}$-module in a natural way. The $\mathbb{Z}$-action on A is just the map $(n, a) \to na$ from $\mathbb{Z} \times A$ to A.*

Example 3.4.3 *Any ring R with identity can be thought of as a module over itself in a natural way. We just take M to be the additive group $\langle R, + \rangle$ of R and define a map $R \times R \to R$ by $(r, s) \to rs$ (the product of r and s in R).*

A submodule of an *R*-module *M* is just a subset *N* of *M* such that the operations of *M*, when restricted to *N*, make *N* into an *R*-module. These operations are the

Abelian group operations $+$ and $-$, and the operation of "multiplying on the left" by elements of R.

Definition 3.4.3 (Submodule) *Let R be a ring with identity. Let M be an R-module. A subgroup N of M is called a submodule of M if $rn \in N$ for all $r \in R$ and $n \in N$.*

Example 3.4.4 *Any R-module M has the submodules M and $\{0\}$.*

Example 3.4.5 *If A is an additive Abelian group considered as a $\mathbb{Z}$-module, then the submodules of A are precisely the subgroups of A.*

Example 3.4.6 *Let V be a vector space over a field F considered as an F-module. Then the submodules of V are its subspaces.*

Example 3.4.7 *If R is a commutative ring with identity, then the submodules of R considered as an R-module are the ideals of R.*

Definition 3.4.4 (Submodule generated by a set) *If X is a subset of an R-module M then the submodule generated by X is the smallest submodule of M containing X.*

This definition is a valid one because the intersection of all the submodules of M containing X is such a submodule and is thus the smallest such module. Indeed if X is a nonempty subset of M then it is not difficult to show that the set

$$\left\{ \sum_{i=1}^{n} r_i x_i \mid r_i \in R, \ x_i \in X, \ n \geq 1 \right\}$$

of all finite sums of elements of the form rx with $r \in R$ and $x \in X$ is the smallest submodule of M containing X, and so it is the submodule of M generated by X.

Definition 3.4.5 (Finitely generated module) *An R-module M is called finitely generated if M is generated by some finite set of elements of M.*

Thus an R-module M is finitely generated if and only if there exist finitely many elements $x_1, \ldots, x_n \in M$ such that each $x \in M$ can be expressed as a "linear combination" $\sum_{i=1}^{n} r_i x_i$ of the x_i with coefficients $r_i \in R$.

Definition 3.4.6 (Factor module) *Let N be a submodule of the R-module M. Then the factor module M/N is the quotient group M/N of cosets $\{m + N \mid m \in M\}$ together with the R-action given by $r(m + N) = rm + N$ for each $r \in R$ and each coset $m + N$ in M/N.*

If $m + N = m' + N$ then we have $m - m' \in N$. Hence $r(m - m') \in N$ since N is a submodule and thus $rm + N = rm' + N$. This shows that the action of R on M/N in Definition 3.4.6 is well defined. The axioms are easily verified, so M/N under this R-action is an R-module. We often write $\overline{m}$ for $m + N$.

Definition 3.4.7 (Module homomorphism) *Let M and N be R-modules. A module homomorphism from M to N is a map $\theta : M \to N$ such that*

$$\theta(m_1 + m_2) = \theta(m_1) + \theta(m_2),$$
$$\theta(rm) = r\theta(m),$$

for all $m, m_1, m_2 \in M$ and $r \in R$. A module homomorphism that is bijective is called a module isomorphism. Two modules M and N having a module isomorphism between them are called module isomorphic, and we write $M \simeq N$.

If K and L are submodules of an R-module M then $K + L$ is a submodule of M, K is a submodule of $K + L$, $K \cap L$ is a submodule of L, and $K + L/K \simeq L/K \cap L$ since $\theta : K + L/K \to L/K \cap L$ defined by $\theta(k + l + K) = l + K \cap L$ ($k \in K$, $l \in L$) is a module isomorphism.

3.5 Noetherian Modules

A Noetherian domain is an integral domain in which every ascending chain of ideals terminates. Analogously we define a Noetherian R-module to be an R-module in which every ascending chain of submodules terminates.

Definition 3.5.1 (Noetherian module) *Let R be a ring with identity. An R-module M is called Noetherian if every ascending chain of submodules of M terminates.*

Theorem 3.5.1 *Let R be a ring with identity. Let M be an R-module and let N be a submodule of M. Then M is Noetherian if and only if both N and M/N are Noetherian.*

Proof: Suppose that M is Noetherian. Let

$$N_1 \subseteq N_2 \subseteq \ldots$$

be an ascending chain of submodules of N. As N is a submodule of M this chain is also a chain of submodules of M. But M is Noetherian so this chain terminates. Hence N is also Noetherian. Now let

$$\overline{M_1} \subseteq \overline{M_2} \subseteq \ldots$$

be an ascending chain of submodules of the factor module M/N. For $i = 1, 2, \ldots$ let

$$M_i = \{m \in M \mid \overline{m} \in \overline{M_i}\}.$$

It is easy to check that M_i is a submodule of M and that $M_i \subseteq M_{i+1}$. Hence

$$M_1 \subseteq M_2 \subseteq \ldots$$

is an ascending chain of submodules of M. As M is Noetherian this chain terminates and thus the original chain terminates too. This proves that the R-module M/N is Noetherian.

Now suppose that both N and M/N are Noetherian. Let

$$M_1 \subseteq M_2 \subseteq \ldots$$

be an ascending chain of submodules of M. For $i = 1, 2, \ldots$ set

$$\overline{M_i} = \{\overline{m} \mid m \in M_i\}.$$

Again it is easy to check that $\overline{M_i}$ is a submodule of the R-module M/N and that $\overline{M_i} \subseteq \overline{M_{i+1}}$. Hence

$$\overline{M_1} \subseteq \overline{M_2} \subseteq \ldots$$

is an ascending chain of submodules of M/N. As M/N is Noetherian this chain terminates and there is a positive integer l_1 such that

$$\overline{M_i} \subseteq \overline{M_{l_1}} \text{ for } i \geq l_1.$$

Now $M_i \cap N$ is a submodule of N and $M_i \cap N \subseteq M_{i+1} \cap N$ so that

$$M_1 \cap N \subseteq M_2 \cap N \subseteq \ldots$$

is an ascending chain of submodules of N. As N is Noetherian this chain terminates and there exists a positive integer l_2 such that

$$M_i \cap N = M_{l_2} \cap N \text{ for } i \geq l_2.$$

Set $l = \max(l_1, l_2)$. Thus for $i \geq l$ we have

$$\overline{M_i} = \overline{M_{i+1}}, \ M_i \cap N = M_{i+1} \cap N.$$

Suppose that the original chain $M_1 \subseteq M_2 \subseteq \ldots$ of submodules of M does not terminate. Then $M_i \subset M_{i+1}$ for some $i \geq l$. We can choose $m_{i+1} \in M_{i+1}$, $m_{i+1} \notin M_i$. Hence $\overline{m}_{i+1} \in \overline{M_{i+1}} = \overline{M_i}$ and so there exist $m_i \in M_i$ and $n \in N$ such that $m_{i+1} = m_i + n$. Thus $m_{i+1} - m_i = n \in N$. Also as $M_i \subseteq M_{i+1}$ we have $m_{i+1} - m_i \in M_{i+1}$. Thus $m_{i+1} - m_i \in M_{i+1} \cap N = M_i \cap N \subseteq M_i$ so $m_{i+1} \in M_i$, which is a contradiction. Hence the chain $M_1 \subseteq M_2 \subseteq \ldots$ must terminate and M is Noetherian. ∎

Theorem 3.5.2 *If R is a Noetherian ring, any finitely generated R-module M is Noetherian.*

Proof: Let M be a finitely generated R-module. Then there exist $m_1, m_2, \ldots, m_n \in M$ such that

$$M = Rm_1 + Rm_2 + \cdots + Rm_n.$$

Each Rm_i is an R-module.

For $k = 1, 2, \ldots, n$ we define the R-module M_k by

$$M_k = Rm_1 + \cdots + Rm_k$$

so that $M_n = M$.

We first show that each R-module Rm_i $(i = 1, \ldots, n)$ is Noetherian. Let

$$N_i = \{r \in R \mid rm_i = 0\}.$$

Clearly N_i is a submodule of R. Since the submodules of R are ideals of R and R is a Noetherian ring, R is a Noetherian module. Hence, by Theorem 3.5.1, the factor module R/N_i is Noetherian. But $R/N_i \simeq Rm_i$ so Rm_i is Noetherian. In particular $M_1 = Rm_1$ is Noetherian.

Now suppose that $M_1, \ldots, M_{k-1}$ $(2 \le k \le n)$ are Noetherian. We show that M_k is Noetherian. As Rm_k is Noetherian, we see by Theorem 3.5.1 that the factor module

$$Rm_k / Rm_k \cap M_{k-1}$$

is Noetherian. Hence

$$M_k / M_{k-1} = M_{k-1} + Rm_k / M_{k-1} \simeq Rm_k / Rm_k \cap M_{k-1}$$

is Noetherian. Then, by Theorem 3.5.1, M_k is Noetherian. Hence $M_1, \ldots, M_n$ are Noetherian so $M = M_n$ is a Noetherian module. ∎

The consequence of Theorem 3.5.2 that we use in Chapter 6 is the following result (see Theorem 6.5.3).

Theorem 3.5.3 *Let D and E be integral domains with $D \subseteq E$. If D is a Noetherian domain and E is a finitely generated D-module then E is a Noetherian domain.*

Proof: Let $I_1 \subseteq I_2 \subseteq \ldots$ be an ascending chain of ideals in the domain E. By Theorem 3.5.2, as D is a Noetherian domain and E is a finitely generated D-module, E is a Noetherian D-module. But each I_i is a D-submodule of E so the chain $I_1 \subseteq I_2 \subseteq \ldots$ must terminate. Hence we have shown that E is a Noetherian domain. ∎

In Example 3.1.3 we saw that the integral domain $\mathbb{Z} + \mathbb{Z}\sqrt{m}$ is Noetherian for $m = -1, -2, 2, 3$, and 6. In fact this is true for an arbitrary integer m that is not a perfect square.

Theorem 3.5.4 *Let m be a nonsquare integer. Then $\mathbb{Z} + \mathbb{Z}\sqrt{m}$ is a Noetherian domain (and thus a factorization domain by Theorem 3.2.2).*

Proof: We take $D = \mathbb{Z}$ and $E = \mathbb{Z} + \mathbb{Z}\sqrt{m}$ in Theorem 3.5.3. As $\mathbb{Z}$ is Noetherian (Example 3.1.3) and $\mathbb{Z} + \mathbb{Z}\sqrt{m}$ is a finitely generated $\mathbb{Z}$-module (generated by 1 and $\sqrt{m}$) the theorem follows from Theorem 3.5.3.   ∎

Similarly, taking $D = \mathbb{Z}$ and $E = \mathbb{Z} + \mathbb{Z}\left(\frac{1+\sqrt{m}}{2}\right)$, where m is a nonsquare integer with $m \equiv 1 \pmod{4}$, in Theorem 3.5.3, we obtain

Theorem 3.5.5 *Let m be a nonsquare integer with $m \equiv 1 \pmod{4}$. Then $\mathbb{Z} + \mathbb{Z}\left(\frac{1+\sqrt{m}}{2}\right)$ is a Noetherian domain (and thus a factorization domain by Theorem 3.2.2).*

Example 3.5.1 *$\mathbb{Z} + \mathbb{Z}\sqrt{-5}$ is a factorization domain by Theorem 3.5.4. However, it is not a unique factorization domain as 6 has two different factorizations into irreducibles in $\mathbb{Z} + \mathbb{Z}\sqrt{-5}$, namely*

$$6 = 2 \cdot 3 = (1 + \sqrt{-5})(1 - \sqrt{-5}).$$

The fact that 2 and 3 are irreducibles in $\mathbb{Z} + \mathbb{Z}\sqrt{-5}$ follows from Exercise 36 of Chapter 1. To see that $1 + \sqrt{-5}$ is an irreducible in $\mathbb{Z} + \mathbb{Z}\sqrt{-5}$ suppose that

$$1 + \sqrt{-5} = (a + b\sqrt{-5})(c + d\sqrt{-5})$$

for some $a, b, c, d \in \mathbb{Z}$. Then

$$6 = (a^2 + 5b^2)(c^2 + 5d^2)$$

so that (as $a^2 + 5b^2$ is a nonnegative integer)

$$a^2 + 5b^2 = 1, 2, 3, \text{ or } 6.$$

Clearly $a^2 + 5b^2 \neq 2, 3$. If $a^2 + 5b^2 = 1$ then $a = \pm 1$, $b = 0$, and $a + b\sqrt{-5} = \pm 1$ is a unit of $\mathbb{Z} + \mathbb{Z}\sqrt{-5}$. If $a^2 + 5b^2 = 6$ then $c^2 + 5d^2 = 1$ and $c + d\sqrt{-5} = \pm 1$ is a unit of $\mathbb{Z} + \mathbb{Z}\sqrt{-5}$. This proves that $1 + \sqrt{-5}$ is irreducible in $\mathbb{Z} + \mathbb{Z}\sqrt{-5}$. Similarly we can show that $1 - \sqrt{-5}$ is also irreducible. The irreducibles $2, 3, 1 + \sqrt{-5}$, and $1 - \sqrt{-5}$ are not associates of one another because the quotient of any two of them $\notin U(\mathbb{Z} + \mathbb{Z}\sqrt{-5}) = \{-1, +1\}$.

Example 3.5.2 *$\mathbb{Z} + \mathbb{Z}\sqrt{10}$ is a factorization domain by Theorem 3.5.4. It is not a unique factorization domain as*

$$10 = (\sqrt{10})^2 = 2 \cdot 5,$$

where 2, 5, and $\sqrt{10}$ are nonassociated irreducibles in $\mathbb{Z} + \mathbb{Z}\sqrt{10}$. We just show that 2 is an irreducible in $\mathbb{Z} + \mathbb{Z}\sqrt{10}$. Suppose that

$$2 = (a + b\sqrt{10})(c + d\sqrt{10})$$

for some $a, b, c, d \in \mathbb{Z}$. Then

$$4 = (a^2 - 10b^2)(c^2 - 10d^2).$$

Hence, as $a^2 - 10b^2 \in \mathbb{Z}$, we deduce that

$$a^2 - 10b^2 = -4, -2, -1, 1, 2, \ or \ 4.$$

If $a^2 - 10b^2 = \pm 1$ then $a + b\sqrt{10}$ is a unit of $\mathbb{Z} + \mathbb{Z}\sqrt{10}$. If $a^2 - 10b^2 = \pm 4$ then $c^2 - 10d^2 = \pm 1$ and $c + d\sqrt{10}$ is a unit of $\mathbb{Z} + \mathbb{Z}\sqrt{10}$. If $a^2 - 10b^2 = \pm 2$ then $a^2 \equiv \pm 2 \,(\mathrm{mod}\ 5)$, which is a contradiction as a square is congruent to 0, 1, or 4 (mod 5). Thus this case cannot occur. Hence 2 is irreducible in $\mathbb{Z} + \mathbb{Z}\sqrt{10}$. We leave it to the reader to show that 5 and $\sqrt{10}$ are also irreducible in $\mathbb{Z} + \mathbb{Z}\sqrt{10}$ and that 2, 5, and $\sqrt{10}$ are not associates of one another (Exercise 13).

In Example 3.5.2 we showed that the equation $x^2 - 10y^2 = 2$ (or -2) has no solutions in integers x and y. Here this was very easy to do: We just considered the equation modulo 5 and got a contradiction. In general one cannot show that an equation of the type $x^2 - my^2 = N$ has no solutions in integers x and y by congruence considerations alone. We show how to determine the solvability or insolvability of the equation $x^2 - my^2 = N$ ($m, n \in \mathbb{Z}$ with m positive and nonsquare and $0 < |N| < \sqrt{m}$) in Section 11.7.

Exercises

1. Let F be a field. If M is an F-module prove that M is a vector space over F. Conversely show that if M is a vector space over F then M is an F-module.
2. Considering $\mathbb{Z}$ as a $\mathbb{Z}$-module, where the $\mathbb{Z}$-action on $\mathbb{Z}$ is just multiplication, determine all the $\mathbb{Z}$-submodules of $\mathbb{Z}$.
3. Let $I_1 \subseteq I_2 \subseteq \ldots$ be an ascending chain of ideals in an integral domain D. Prove that $\bigcup_{n=1}^{\infty} I_n$ is an ideal in D.
4. Let F be a field. Is the domain $F[X]$ Noetherian?
5. Prove that the ideal $\langle 2, X \rangle$ in $\mathbb{Z}[X]$ is not principal (Example 3.3.6).
6. Prove that a subset N of an R-module M is a submodule of M if and only if
   (i) $0 \in N$,
   (ii) $n_1, n_2 \in N \Longrightarrow n_1 - n_2 \in N$, and
   (iii) $n \in N, r \in R \Longrightarrow rn \in N$.
7. Prove that the intersection of any nonempty collection of submodules of an R-module is itself a submodule of M.

8. If $M_1, \ldots, M_n$ are $n(\geq 1)$ nonempty subsets of an R-module M, we define

$$M_1 + \cdots + M_n = \{m_1 + \cdots + m_n \mid m_i \in M_i\}.$$

If $M_1, \ldots, M_n$ are submodules of M prove that $M_1 + \cdots + M_n$ is a submodule of M.

9. Let M and N be R-modules. Let $\theta : M \to N$ be an R-homomorphism. Define

$$\ker \theta = \{m \in M \mid \theta(m) = 0\},$$
$$\operatorname{im} \theta = \{n \in N \mid n = \theta(m) \text{ for some } m \in M\}.$$

Prove the following:
   (i) $\ker \theta$ is a submodule of M.
   (ii) $\operatorname{im} \theta$ is a submodule of N.
   (iii) $M/\ker \theta \simeq \operatorname{im} \theta$.

10. If K and L are submodules of an R-module M with $K \subseteq L$, prove that

$$M/K \,/\, L/K \simeq M/L.$$

11. Suppose that D is a unique factorization domain and $a(\neq 0)$ and $b(\neq 0)$ are coprime nonunits in D. Prove that if $ab = c^n$ for some $c \in D$ and some $n \in \mathbb{N}$ then there is a unit $e \in D$ such that ea and $e^{-1}b$ are nth powers in D.

12. Let D be a unique factorization domain. Give an example to show that the following assertion is not true in general: If a is an irreducible element of D then $\langle a \rangle$ is a maximal ideal of D.

13. Prove that 5 and $\sqrt{10}$ are irreducible elements of $\mathbb{Z} + \mathbb{Z}\sqrt{10}$ and that 2, 5, and $\sqrt{10}$ are not associates of one another, as asserted in Example 3.5.2.

14. Let p be a prime and m be a positive nonsquare integer such that the Legendre symbol $\left(\frac{\pm p}{q}\right) = -1$ for some odd prime factor q of m. Prove that the equation $x^2 - my^2 = \pm p$ has no solution in integers x and y. Deduce that p is an irreducible element of $\mathbb{Z} + \mathbb{Z}\sqrt{m}$.

15. Prove that $\mathbb{Z} + \mathbb{Z}\sqrt{-6}$ is not a unique factorization domain by exhibiting an element of $\mathbb{Z} + \mathbb{Z}\sqrt{-6}$ that has two different factorizations into irreducibles.

16. Prove that $\mathbb{Z} + \mathbb{Z}\sqrt{-10}$ is not a unique factorization domain.

17. Prove that $\mathbb{Z} + \mathbb{Z}\sqrt{15}$ is not a unique factorization domain.

Suggested Reading

1. P. M. Cohn, *Unique factorization domains*, American Mathematical Monthly 80 (1973), 1–18 (correction, American Mathematical Monthly 80 (1973), 1115).

   A brief survey of unique factorization domains is given in the first six sections of the article.

2. D. S. Dummit and R. M. Foote, *Abstract Algebra*, second edition, Prentice Hall, Upper Saddle River, New Jersey, 1999.

   In Section 9.3 it is shown that D is a unique factorization domain if and only if $D[X]$ is a unique factorization domain, and that if D is a unique factorization domain so is $D[X_1, X_2, \ldots]$.

3. J. Greene, *Principal ideal domains are almost Euclidean*, American Mathematical Monthly 104 (1997), 154–156.

   This paper is where Theorem 3.3.4 was first proved.

4. B. Hartley and T. O. Hawkes, *Rings, Modules and Linear Algebra*, Chapman and Hall, London, New York, 1974.

   Chapters 5 and 6 give a very nice introduction to modules.

5. N. Jacobson, *Lectures in Abstract Algebra*, Volume I, van Nostrand, Princeton, New Jersey, 1955.

   Chapter VI contains a proof of the Hilbert basis theorem.

6. W. Rudin, *Unique factorization of Gaussian integers*, American Mathematical Monthly 68 (1961), 907–908.

   A very simple and short proof is given that the Gaussian domain is a unique factorization domain.

7. P. Samuel, *Unique factorization*, American Mathematical Monthly 75 (1968), 945–952.

   A classic overview of unique factorization is given.

8. O. Zariski and P. Samuel, *Commutative Algebra*, Volume 1, van Nostrand Company, Princeton, New Jersey, 1967.

   Chapter 4 gives a proof of the Hilbert basis theorem.

Biographies

1. A. Dick, *Emmy Noether, 1882–1935*, Birkhäuser, Boston, Massachusetts, 1981.

   This biography of Emny Noether includes obituaries by B. L. van der Waerden, H. Weyl, and P. S. Alexandrov as well as a list of her publications.

2. C. H. Kimberling, *Emmy Noether*, American Mathematical Monthly 79 (1972), 136–149 (addendum, American Mathematical Monthly 79 (1972), 755).

   This personal biography has excerpts from articles on Noether by P. S. Aleksandrov and H. Weyl.

3. C. Reid, *Hilbert*, Springer-Verlag, Berlin, Heidelberg, New York, 1970.

   This wonderful book covers the life of David Hilbert, a man deeply devoted to the world of logic and mathematics.

4. The website

   http://www-groups.dcs.st-and.ac.uk/~history/

   has biographies of both Hilbert and Noether.

4

Elements Integral over a Domain

4.1 Elements Integral over a Domain

Let A be an integral domain and let B be an integral domain containing A. We are interested in those elements of B that are roots of monic polynomials with coefficients in A.

Definition 4.1.1 (Element integral over a domain) *Let A and B be integral domains with $A \subseteq B$. The element $b \in B$ is said to be integral over A if it satisfies a polynomial equation*

$$x^n + a_{n-1}x^{n-1} + \cdots + a_1 x + a_0 = 0,$$

where $a_0, a_1, \ldots, a_{n-1} \in A$.

Note that every element $a \in A$ is integral over A as it is a root of $x - a \in A[x]$.

Definition 4.1.2 (Algebraic integer) *A complex number which is integral over $\mathbb{Z}$ is called an algebraic integer.*

Example 4.1.1 *$\sqrt{2}$ is an algebraic integer as it satisfies the equation $x^2 - 2 = 0$.*

Example 4.1.2 *$\frac{1}{2}(-1 + i\sqrt{3})$ is an algebraic integer as it satisfies the equation $x^2 + x + 1 = 0$.*

Example 4.1.3 *$\sqrt[3]{2} - \sqrt[3]{4}$ is an algebraic integer as it satisfies the equation $x^3 + 6x + 2 = 0$.*

Example 4.1.4 *$\frac{1}{4}(1 + \sqrt{21} + \sqrt{33} - \sqrt{77})$ is an algebraic integer as it satisfies the equation $x^4 - x^3 - 16x^2 + 37x - 17 = 0$.*

Example 4.1.5 *A root of unity is an algebraic integer as it is a root of $x^n - 1 \in \mathbb{Z}[x]$ for some $n \in \mathbb{N}$.*

Example 4.1.6 $\frac{3\sqrt{2}-i\sqrt{2}}{2} \in D$ *(see Example 1.1.11) is integral over* $\mathbb{Z} + \mathbb{Z}i$ *as it is a root of the polynomial* $x^2 - (4 - 3i) \in (\mathbb{Z} + \mathbb{Z}i)[x]$.

Example 4.1.7 $1/\sqrt{2}$ *is not an algebraic integer. Suppose on the contrary that* $1/\sqrt{2}$ *is an algebraic integer. Then there exists a positive integer n and integers* $a_0, a_1, \ldots, a_{n-1}$ *such that*

$$\left(\frac{1}{\sqrt{2}}\right)^n + a_{n-1}\left(\frac{1}{\sqrt{2}}\right)^{n-1} + \cdots + a_0 = 0.$$

Multiplying both sides of this equation by $(\sqrt{2})^n$ *we obtain*

$$1 + a_{n-1}\sqrt{2} + a_{n-2}(\sqrt{2})^2 + \cdots + a_0(\sqrt{2})^n = 0.$$

Thus

$$(1 + 2a_{n-2} + 4a_{n-4} + \cdots) + \sqrt{2}(a_{n-1} + 2a_{n-3} + \cdots) = 0.$$

If $a_{n-1} + 2a_{n-3} + \cdots \neq 0$ *then*

$$\sqrt{2} = \frac{-(1 + 2a_{n-2} + 4a_{n-4} + \cdots)}{(a_{n-1} + 2a_{n-3} + \cdots)}$$

is the quotient of two integers and thus a rational number, a contradiction. Hence, $a_{n-1} + 2a_{n-3} + \cdots = 0$ *and so* $1 + 2a_{n-2} + 4a_{n-4} + \cdots = 0$. *This is a contradiction as the integer* $1 + 2a_{n-2} + 4a_{n-4} + \cdots$ *is clearly odd.*

Definition 4.1.3 (Element algebraic over a field) *Let A and B be integral domains with* $A \subseteq B$. *Suppose that A is a field and* $b \in B$ *is integral over A; then b is said to be algebraic over A.*

Definition 4.1.4 (Algebraic number) *A complex number that is algebraic over* $\mathbb{Q}$ *is called an algebraic number.*

Example 4.1.8 $1/\sqrt{2}$ *is an algebraic number as it satisfies the equation* $x^2 - 1/2 = 0$.

Example 4.1.9 *Let* $A = \{a + b\sqrt{2} \mid a, b \in \mathbb{Q}\}$ *and* $B = \{x + yi + z\sqrt{2} + wi\sqrt{2} \mid x, y, z, w \in \mathbb{Q}\}$ *so that A and B are fields with* $A \subset B$. *Then* $b = \frac{1}{2}(1 + i + \sqrt{2}) \in B$ *is algebraic over A as b satisfies the equation*

$$x^2 - (1 + \sqrt{2})x + \left(1 + \frac{1}{\sqrt{2}}\right) = 0.$$

Definition 4.1.5 (Domain integral over a subdomain) *Let A and B be integral domains with* $A \subseteq B$. *If every* $b \in B$ *is integral over A we say that B is integral over A.*

Example 4.1.10 *The quadratic domain* $\mathbb{Z} + \mathbb{Z}\sqrt{m}$, *where* m *is a nonsquare integer, is integral over* $\mathbb{Z}$ *as every element* $u + v\sqrt{m}$ *of* $\mathbb{Z} + \mathbb{Z}\sqrt{m}$ *is integral over* $\mathbb{Z}$ *as it satisfies the polynomial equation*

$$x^2 - 2ux + u^2 - mv^2 = 0,$$

where $-2u \in \mathbb{Z}$ *and* $u^2 - mv^2 \in \mathbb{Z}$.

Example 4.1.11 *The quadratic domain* $\mathbb{Z} + \mathbb{Z}\left(\frac{1+\sqrt{m}}{2}\right)$, *where* m *is a nonsquare integer with* $m \equiv 1 \pmod 4$, *is integral over* $\mathbb{Z}$ *as every element* $\alpha = u + v\left(\frac{1+\sqrt{m}}{2}\right) \in \mathbb{Z} + \mathbb{Z}\left(\frac{1+\sqrt{m}}{2}\right)$ *is integral over* $\mathbb{Z}$ *because it satisfies the polynomial equation*

$$x^2 - (2u + v)x + \left(u^2 + uv + \frac{1}{4}(1 - m)v^2\right) = 0,$$

where $-(2u + v) \in \mathbb{Z}$ *and* $u^2 + uv + \frac{1}{4}(1 - m)v^2 \in \mathbb{Z}$.

Example 4.1.12 *Let* A *and* B *be integral domains with* $A \subseteq B$. *Let* $a \in A$. *Let* $b \in B$ *be integral over* A. *We show that* ab *is integral over* A.
 As $b \in B$ *is integral over* A *there exist* $a_0, a_1, \ldots, a_{n-1} \in A$ *such that*

$$b^n + a_{n-1}b^{n-1} + \cdots + a_1 b + a_0 = 0.$$

Let $a \in A$. *Then* $ab \in B$ *and*

$$(ab)^n + a_{n-1}a(ab)^{n-1} + \cdots + a_1 a^{n-1}(ab) + a_0 a^n = 0.$$

As $a_{n-1}a, \ldots, a_1 a^{n-1}, a_0 a^n \in A$ *we deduce that* ab *is integral over* A.

Theorem 4.1.1 *Let* $A \subseteq B \subseteq C$ *be a tower of integral domains. If* $c \in C$ *is integral over* A *then* c *is integral over* B.

Proof: As $c \in C$ is integral over A there exist $a_0, a_1, \ldots, a_{n-1} \in A$ such that

$$c^n + a_{n-1}c^{n-1} + \cdots + a_1 c + a_0 = 0.$$

As $A \subseteq B$, $a_0, a_1, \ldots, a_{n-1} \in B$ and so c is integral over B. ∎

Theorem 4.1.2 *Let* $A \subseteq B \subseteq C$ *be a tower of integral domains. If* C *is integral over* A *then* C *is integral over* B.

Proof: Let $c \in C$. As C is integral over A, c is integral over B. Thus c is integral over B, so that C is integral over B. ∎

Theorem 4.1.3 *Let A and B be integral domains with $A \subseteq B$. Let $b \in B$. Then b is integral over A if and only if $A[b]$ is a finitely generated A-module.*

Proof: Suppose that b is integral over A. Then there exist $a_0, a_1, \ldots, a_{n-1} \in A$ such that

$$b^n - a_{n-1}b^{n-1} - a_{n-2}b^{n-2} - \cdots - a_1 b - a_0 = 0.$$

Hence

$$b^n = a_{n-1}b^{n-1} + a_{n-2}b^{n-2} + \cdots + a_1 b + a_0 \in Ab^{n-1} + Ab^{n-2} + \cdots + Ab + A.$$

Also

$$\begin{aligned}
b^{n+1} &= a_{n-1}b^n + a_{n-2}b^{n-1} + \cdots + a_1 b^2 + a_0 b \\
&\in Ab^n + Ab^{n-1} + \cdots + Ab^2 + Ab \\
&\subseteq Ab^{n-1} + \cdots + Ab + A.
\end{aligned}$$

By induction we see that

$$b^k \in Ab^{n-1} + \cdots + Ab + A$$

for all nonnegative integers k. This shows that the integral domain $A[b]$ of polynomials in b with coefficients in A is a finitely generated A-module.

Conversely suppose that $A[b]$ is a finitely generated A-module. Then there exist $u_1, u_2, \ldots, u_n \in A[b]$ such that

$$A[b] = Au_1 + \cdots + Au_n.$$

Clearly $u_1, \ldots, u_n$ are not all zero. Now each $u_i \in A[b]$ and so $bu_i \in A[b]$ for $i = 1, 2, \ldots, n$. Thus there exist $a_{ij} \in A$ $(i, j = 1, 2, \ldots, n)$ such that

$$\begin{cases}
bu_1 = a_{11}u_1 + \cdots + a_{1n}u_n, \\
\quad \cdots \\
bu_n = a_{n1}u_1 + \cdots + a_{nn}u_n.
\end{cases}$$

Thus the homogeneous system of n equations in the n unknowns $x_1, \ldots, x_n$,

$$\begin{cases}
(b - a_{11})x_1 - a_{12}x_2 - \cdots - a_{1n}x_n = 0, \\
-a_{21}x_1 + (b - a_{22})x_2 - \cdots - a_{2n}x_n = 0, \\
\quad \cdots \\
-a_{n1}x_1 - a_{n2}x_2 - \cdots + (b - a_{nn})x_n = 0,
\end{cases}$$

has a nontrivial solution $(x_1, x_2, \ldots, x_n) = (u_1, u_2, \ldots, u_n)$ in the integral domain $A[b]$ and so in its quotient field. But this can only happen if the determinant of the coefficient matrix is zero. Hence

$$\begin{vmatrix} b - a_{11} & -a_{12} & \cdots & -a_{1n} \\ -a_{21} & b - a_{22} & \cdots & -a_{2n} \\ \vdots & \vdots & & \vdots \\ -a_{n1} & -a_{n2} & \cdots & b - a_{nn} \end{vmatrix} = 0.$$

When this determinant is expanded, we obtain an equation

$$b^n + a_{n-1}b^{n-1} + \cdots + a_1 b + a_0 = 0,$$

where $a_0, a_1, \ldots, a_{n-1} \in A$. Hence b is integral over A.   ■

The proof of our next theorem follows closely that of the previous theorem.

Theorem 4.1.4 *Let A and B be integral domains with $A \subseteq B$. Let $b \in B$. If there exists an integral domain C such that*

$$A[b] \subseteq C \subseteq B$$

and C is a finitely generated A-module then b is integral over A and $A[b]$ is a finitely generated A-module.

Proof: As C is a finitely generated A-module, there exist $c_1, \ldots, c_n \in C$ such that

$$C = Ac_1 + \cdots + Ac_n.$$

Clearly $c_1, \ldots, c_n$ are not all zero. Now $b \in A[b]$ and $A[b] \subseteq C$ so that $b \in C$. But C is an integral domain so that $bc_1, \ldots, bc_n \in C$. Hence there exist $a_{ij} \in A$ ($i, j = 1, 2, \ldots, n$) such that

$$\begin{cases} bc_1 = a_{11}c_1 + \cdots + a_{1n}c_n, \\ \quad \cdots \\ bc_n = a_{n1}c_1 + \cdots + a_{nn}c_n. \end{cases}$$

Thus the homogeneous system of n equations in the n unknowns $x_1, \ldots, x_n$,

$$\begin{cases} (b - a_{11})x_1 - a_{12}x_2 - \cdots - a_{1n}x_n = 0, \\ -a_{21}x_1 + (b - a_{22})x_2 - \cdots - a_{2n}x_n = 0, \\ \quad \cdots \\ -a_{n1}x_1 - a_{n2}x_2 - \cdots + (b - a_{nn})x_n = 0, \end{cases}$$

has a nontrivial solution $(x_1, \ldots, x_n) = (c_1, \ldots, c_n)$ in the integral domain C and thus in its quotient field. Hence the determinant of its coefficient matrix is zero;

that is,

$$\begin{vmatrix} b - a_{11} & -a_{12} & \cdots & -a_{1n} \\ -a_{21} & b - a_{22} & \cdots & -a_{2n} \\ \vdots & \vdots & & \vdots \\ -a_{n1} & -a_{n2} & \cdots & b - a_{nn} \end{vmatrix} = 0.$$

Expanding this determinant we obtain an equation

$$b^n + a_{n-1}b^{n-1} + \cdots + a_1 b + a_0 = 0,$$

where $a_0, a_1, \ldots, a_{n-1} \in A$. Hence b is integral over A and, by Theorem 4.1.3, $A[b]$ is a finitely generated A-module. ∎

The special case $C = B$ in Theorem 4.1.4 shows that if A and B are integral domains with $A \subseteq B$ and B is a finitely generated A-module then B is integral over A.

Theorem 4.1.5 *Let $A \subseteq B \subseteq C$ be a tower of integral domains. If B is a finitely generated A-module and C is a finitely generated B-module then C is a finitely generated A-module.*

Proof: As B is a finitely generated A-module there exist $b_1, \ldots, b_m \in B$ such that

$$B = Ab_1 + \cdots + Ab_m.$$

As C is a finitely generated B-module there exist $c_1, \ldots, c_n \in C$ such that

$$C = Bc_1 + \cdots + Bc_n.$$

Let $c \in C$. Then

$$c = \sum_{j=1}^{n} x_j c_j,$$

where $x_1, \ldots, x_n \in B$. Moreover, for $j = 1, \ldots, n$ we have

$$x_j = \sum_{i=1}^{m} a_{ij} b_i,$$

where $a_{11}, \ldots, a_{mn} \in A$. Hence

$$c = \sum_{j=1}^{n} \sum_{i=1}^{m} a_{ij} b_i c_j$$

so that

$$C = Ab_1 c_1 + \cdots + Ab_m c_n$$

is a finitely generated A-module. ∎

Theorem 4.1.6 *Let A and B be integral domains with $A \subseteq B$. Let $b_1, b_2 \in B$ be integral over A. Then $b_1 + b_2$, $b_1 - b_2$, and $b_1 b_2$ are integral over A.*

Proof: As b_1 is integral over A, by Theorem 4.1.3, $A[b_1]$ is a finitely generated A-module. Moreover, b_2 is integral over A and so by Theorem 4.1.1 b_2 is integral over $A[b_1]$. Hence, by Theorem 4.1.3, $(A[b_1])[b_2] = A[b_1, b_2]$ is a finitely generated $A[b_1]$-module. Thus $A[b_1, b_2]$ is a finitely generated A-module by Theorem 4.1.5. Let λ denote any one of $b_1 + b_2$, $b_1 - b_2$, $b_1 b_2$. Then we have

$$A \subseteq A[\lambda] \subseteq A[b_1, b_2] \subseteq B,$$

where the integral domain $A[b_1, b_2]$ is a finitely generated A-module. Hence, by Theorem 4.1.4, λ is integral over A. ∎

The next theorem is an immediate consequence of Theorem 4.1.6.

Theorem 4.1.7 *Let A and B be integral domains with $A \subseteq B$. Then the set of all elements of B that are integral over A is a subdomain of B containing A.*

Taking $A = \mathbb{Z}$ and $B = \mathbb{C}$ in Theorem 4.1.7, we obtain

Theorem 4.1.8 *The set of all algebraic integers is an integral domain.*

The domain of all algebraic integers is denoted by Ω.

Theorem 4.1.9 *Let A and B be integral domains with $A \subseteq B$. Let $b_1, \ldots, b_n \in B$ be integral over A. Then $A[b_1, \ldots, b_n]$ is a finitely generated A-module.*

Proof: We prove the theorem by induction on n. If $b_1 \in B$ is integral over A then $A[b_1]$ is a finitely generated A-module by Theorem 4.1.3, so the theorem is true for $n = 1$.

Now assume that $A[b_1, \ldots, b_{n-1}]$ ($n \geq 2$) is a finitely generated A-module, where $b_1, \ldots, b_{n-1} \in B$ are integral over A. Let $b_n \in B$ be integral over A. Then, by Theorem 4.1.1, b_n is integral over $A[b_1, \ldots, b_{n-1}]$. Hence, by Theorem 4.1.3, $(A[b_1, \ldots, b_{n-1}])[b_n] = A[b_1, \ldots, b_n]$ is a finitely generated A-module. This completes the inductive step and the theorem follows by the principle of mathematical induction. ∎

Theorem 4.1.10 *Let A and B be integral domains with $A \subseteq B$. If each of $b_1, \ldots, b_n \in B$ is integral over A then $A[b_1, \ldots, b_n]$ is integral over A.*

Proof: We prove the theorem by induction on n.

Suppose first that $b_1 \in B$ is integral over A. Then, by Theorem 4.1.6 and Example 4.1.12, we deduce that $a_0 + a_1 b_1 + \cdots + a_n b_1^n$ is integral over A for all $a_0, a_1, \ldots, a_n \in A$. This proves that $A[b_1]$ is integral over A.

Next let $b_1, \ldots, b_{n-1} \in B$ be integral over A and suppose that $A[b_1, \ldots, b_{n-1}]$ is integral over A. Let $b_n \in B$ be integral over A. Let f be any element of $A[b_1, \ldots, b_n]$. Then

$$f = f_0 + f_1 b_n + \cdots + f_m b_n^m,$$

where $f_0, f_1, \ldots, f_m \in A[b_1, \ldots, b_{n-1}]$. By the inductive hypothesis $f_0, f_1, \ldots, f_m$ are all integral over A. Then, as b_n is integral over A, we deduce by Theorem 4.1.6 that $f_0 + f_1 b_n + \cdots + f_m b_n^m$ is integral over A. Hence every element f of $A[b_1, \ldots, b_n]$ is integral over A, proving that $A[b_1, \ldots, b_n]$ is integral over A. ∎

Theorem 4.1.11 *Let $A \subseteq B \subseteq C$ be a tower of integral domains. If B is integral over A and $c \in C$ is integral over B then c is integral over A.*

Proof: As $c \in C$ is integral over B there exist $b_0, b_1, \ldots, b_{n-1} \in B$ such that

$$c^n + b_{n-1} c^{n-1} + \cdots + b_1 c + b_0 = 0.$$

This shows that c is integral over $A[b_0, b_1, \ldots, b_{n-1}]$. As each $b_i \in B$ and B is integral over A, each b_i is integral over A. Thus, by Theorem 4.1.9, $A[b_0, b_1, \ldots, b_{n-1}]$ is a finitely generated A-module. As c is integral over $A[b_0, b_1, \ldots, b_{n-1}]$, by Theorem 4.1.3 we see that $(A[b_0, b_1, \ldots, b_{n-1}])[c] = A[b_0, b_1, \ldots, b_{n-1}, c]$ is a finitely generated A-module. Hence, by Theorem 4.1.4, c is integral over A. ∎

We can now prove that "integral over" is a transitive relation.

Theorem 4.1.12 *Let $A \subseteq B \subseteq C$ be a tower of integral domains. If C is integral over B and B is integral over A then C is integral over A.*

Proof: Let c be any element of C. Then c is integral over B. As B is integral over A, by Theorem 4.1.11 c is integral over A. Hence C is integral over A. ∎

4.2 Integral Closure

Let A and B be integral domains with $A \subseteq B$. In Theorem 4.1.7 we showed that the set of all elements of B that are integral over A is a subdomain of B containing A. We now give this domain a name.

Definition 4.2.1 (Integral closure) *Let A and B be integral domains with $A \subseteq B$. The integral closure of A in B is the subdomain of B consisting of all elements* of B that are integral over A. The integral closure of A in B is denoted by A^B.

From Theorem 4.1.7 we have

Theorem 4.2.1 *Let A and B be integral domains with $A \subseteq B$. Then the integral closure A^B of A in B is an integral domain satisfying*

$$A \subseteq A^B \subseteq B.$$

Clearly $A^A = A$ for any integral domain A.

Our next theorem determines the integral closure of $\mathbb{Z}$ in the field $\mathbb{Q}(i) = \{x + yi \mid x, y \in \mathbb{Q}\}$.

Theorem 4.2.2 *The integral closure of $A = \mathbb{Z}$ in $B = \mathbb{Q}(i)$ is*

$$A^B = \mathbb{Z} + \mathbb{Z}i.$$

Proof: We first show that $\mathbb{Z} + \mathbb{Z}i \subseteq A^B$. Let $\alpha \in \mathbb{Z} + \mathbb{Z}i$. Then $\alpha = m + ni$, where $m, n \in \mathbb{Z}$. Hence α is a root of the quadratic polynomial

$$x^2 - 2mx + (m^2 + n^2) \in \mathbb{Z}[x].$$

This shows that α is an algebraic integer. Clearly $\alpha \in \mathbb{Q}(i)$. Thus $\alpha \in A^B$ so that $\mathbb{Z} + \mathbb{Z}i \subseteq A^B$.

We now show that $A^B \subseteq \mathbb{Z} + \mathbb{Z}i$. Let $\alpha \in A^B$. Hence $\alpha \in \mathbb{Q}(i)$ is algebraic over $\mathbb{Z}$. As $\alpha \in \mathbb{Q}(i)$ we have $\alpha = r + si$, where $r, s \in \mathbb{Q}$. We just treat the case $s \neq 0$. The case $s = 0$ can be treated in a similar and easier manner. Clearly α is a root of $g(x) = x^2 - 2rx + (r^2 + s^2) \in \mathbb{Q}[x]$. As α is algebraic over $\mathbb{Z}$, there exists a monic polynomial $f(x) \in \mathbb{Z}[x]$ with $f(\alpha) = 0$. Since $f(x), g(x) \in \mathbb{Q}[x]$, by the division algorithm there exist polynomials $q(x), r(x) \in \mathbb{Q}[x]$ such that

$$f(x) = q(x)g(x) + r(x), \quad \deg r(x) < \deg g(x).$$

As $\deg g(x) = 2$ we see that $r(x) = r_0 + r_1 x$, where $r_0, r_1 \in \mathbb{Q}$. Hence

$$f(x) = q(x)g(x) + r_0 + r_1 x.$$

Taking $x = \alpha$ we obtain (as $f(\alpha) = g(\alpha) = 0$)

$$r_0 + r_1 \alpha = 0$$

so that

$$(r_0 + r_1 r) + i r_1 s = 0.$$

Equating real and imaginary parts, we obtain

$$r_0 + r_1 r = r_1 s = 0.$$

As $s \neq 0$ we deduce that $r_1 = r_0 = 0$. Hence

$$f(x) = q(x)g(x),$$

where $q(x)$, $g(x) \in \mathbb{Q}[x]$. Let a be the least common multiple of the denominators of the coefficients of $q(x)$ and b the least common multiple of the denominators of $g(x)$. Then $abf(x) = aq(x)bg(x)$, where $aq(x)$ and $bg(x) \in \mathbb{Z}[x]$. Let c be the content of $aq(x)$ and d the content of $bg(x)$. (Recall that the content of a nonzero polynomial $a_n x^n + \cdots + a_1 x + a_0 \in \mathbb{Z}[x]$ is the greatest common divisor of the integers $a_n, \ldots, a_1, a_0$ and that a primitive polynomial is a polynomial of $\mathbb{Z}[x]$ with content 1.) Then we have $aq(x) = cq_1(x)$ and $bg(x) = dg_1(x)$, where $q_1(x) \in \mathbb{Z}[x]$ and $g_1(x) \in \mathbb{Z}[x]$ are both primitive polynomials. Also $abf(x) = cq_1(x)dg_1(x)$. Since $f(x) \in \mathbb{Z}[x]$ is monic the content of $abf(x)$ is ab. By a theorem of Gauss, the product of two primitive polynomials is primitive. Hence $q_1(x)g_1(x)$ is primitive and the content of $cq_1(x)dg_1(x)$ is cd. Thus $ab = cd$ and $f(x) = q_1(x)g_1(x)$, where

$$q_1(x) = \frac{a}{c}q(x) \in \mathbb{Z}[x], \quad g_1(x) = \frac{b}{d}g(x) \in \mathbb{Z}[x].$$

Suppose that

$$f(x) = x^n + a_{n-1}x^{n-1} + \cdots + a_0,$$
$$q_1(x) = b_{n-2}x^{n-2} + \cdots + b_0,$$
$$g_1(x) = c_2 x^2 + c_1 x + c_0,$$

where $a_0, \ldots, a_{n-1}, b_0, \ldots, b_{n-2}, c_0, c_1, c_2 \in \mathbb{Z}$. Equating coefficients of x^n in $f(x) = q_1(x)g_1(x)$, we obtain $b_{n-2}c_2 = 1$. As $b_{n-2}, c_2 \in \mathbb{Z}$ we have $b_{n-2} = c_2 = \pm 1$. Changing $q_1(x)$ to $-q_1(x)$ and $g_1(x)$ to $-g_1(x)$, if necessary, we may suppose that $c_2 = 1$. Then $g_1(x)$ and $g(x)$ are both monic so from $g_1(x) = (b/d)g(x)$, we deduce that $b = d$ and

$$x^2 - 2rx + (r^2 + s^2) = g(x) = g_1(x) \in \mathbb{Z}[x].$$

Thus $2r \in \mathbb{Z}$ and $r^2 + s^2 \in \mathbb{Z}$. If $2r \in 2\mathbb{Z} + 1$ then $2s \in 2\mathbb{Z} + 1$ and $4r^2 + 4s^2 \in 4\mathbb{Z} + 2$, contradicting $4r^2 + 4s^2 \in 4\mathbb{Z}$. Hence $2r \in 2\mathbb{Z}$ so that $r \in \mathbb{Z}$ and $s \in \mathbb{Z}$, that is, $r + si \in \mathbb{Z} + \mathbb{Z}i$, proving that $A^B \subseteq \mathbb{Z} + \mathbb{Z}i$. ∎

Theorem 4.2.3 *Let D be a unique factorization domain. Let F be the field of quotients of D. Then $c \in F$ is integral over D if and only if $c \in D$.*

Proof: If $c \in D$ then c satisfies the equation $x - c = 0$ and so is integral over D.

Conversely, suppose that $c \in F$ is integral over D. Then c satisfies a polynomial equation

$$x^n + a_{n-1}x^{n-1} + \cdots + a_1 x + a_0 = 0,$$

where $a_0, a_1, \ldots, a_{n-1} \in D$. As $c \in F$ we can express c in the form $c = r/s$, where $r \in D$, $s(\neq 0) \in D$, and $\gcd(r, s) = 1$. Hence

$$r^n + a_{n-1}r^{n-1}s + \cdots + a_1 r s^{n-1} + a_0 s^n = 0. \qquad (4.2.1)$$

If s is not a unit in D then it is divisible by some prime p. From (4.2.1) we see that $p \mid r^n$, and thus, as p is prime, $p \mid r$. This contradicts that $\gcd(r, s) = 1$. Hence s must be a unit and $c = rs^{-1} \in D$. ∎

Theorem 4.2.4 $\mathbb{Q} \cap \Omega = \mathbb{Z}$.

Proof: By Example 3.3.1 $\mathbb{Z}$ is a unique factorization domain. Choose $D = \mathbb{Z}$ in Theorem 4.2.3 so that $F = \mathbb{Q}$. Then $c \in \mathbb{Q} \cap \Omega$ if and only if $c \in \mathbb{Z}$. Hence $\mathbb{Q} \cap \Omega = \mathbb{Z}$. ∎

Theorem 4.2.4 tells us that a rational algebraic integer must be an ordinary integer. We will use this result on a number of occasions.

If D is an integral domain and F its field of quotients then it may happen that the integral closure D^F of D in F is equal to D. If this happens we say that D is integrally closed. Apparently the term "integrally closed" was first defined by Ernst Steinitz (1871–1928) in 1912, but the importance of the concept was already known to Richard Dedekind (1831–1916).

Definition 4.2.2 (Integrally closed domain) *An integral domain D is said to be integrally closed if the only elements of its quotient field that are integral over D are those of D itself.*

Theorem 4.2.5 *Let D be a unique factorization domain. Then D is integrally closed.*

Proof: Let F be the field of quotients of D. By Theorem 4.2.3 we have $D^F = D$ so that D is integrally closed. ∎

Example 4.2.1 $\mathbb{Z} + \mathbb{Z}\sqrt{-3}$ *is not integrally closed. The quotient field of $\mathbb{Z} + \mathbb{Z}\sqrt{-3}$ is $\mathbb{Q}(\sqrt{-3}) = \{x + y\sqrt{-3} \mid x, y \in \mathbb{Q}\}$. Set $\alpha = \frac{1}{2}(1 + \sqrt{-3})$. Clearly $\alpha \in \mathbb{Q}(\sqrt{-3})$ but $\alpha \notin \mathbb{Z} + \mathbb{Z}\sqrt{-3}$. Moreover, α is integral over $\mathbb{Z}$ as it satisfies the equation $\alpha^2 - \alpha + 1 = 0$. Hence α is integral over $\mathbb{Z} + \mathbb{Z}\sqrt{-3}$. This shows that $\mathbb{Z} + \mathbb{Z}\sqrt{-3}$ is not integrally closed. Further, by Theorem 4.2.5, we see that $\mathbb{Z} + \mathbb{Z}\sqrt{-3}$ is not a unique factorization domain. For example, 4 has two quite different* *factorizations into irreducibles in* $\mathbb{Z} + \mathbb{Z}\sqrt{-3}$, *namely* $4 = 2 \cdot 2 = (1 + \sqrt{-3})(1 - \sqrt{-3})$.

Example 4.2.2 $\mathbb{Z} + \mathbb{Z}\left(\frac{1+\sqrt{-3}}{2}\right)$ *is integrally closed. By Example 3.3.1,* $\mathbb{Z} + \mathbb{Z}\left(\frac{1+\sqrt{-3}}{2}\right)$ *is a unique factorization domain. The assertion then follows from Theorem 4.2.5.*

Theorem 4.2.6 *Every algebraic number is of the form* a/b, *where* a *is an algebraic integer and* b *is a nonzero ordinary integer.*

Proof: Let c be an algebraic number. Then there exist $a_0, a_1, \ldots, a_{n-1} \in \mathbb{Q}$ such that

$$c^n + a_{n-1}c^{n-1} + \cdots + a_1c + a_0 = 0. \tag{4.2.2}$$

Let b be the least common multiple of the denominators of $a_0, a_1, \ldots, a_{n-1}$. Thus $b \in \mathbb{N}$ and $ba_i \in \mathbb{Z}$ for $i = 0, 1, 2, \ldots, n - 1$. From (4.2.2) we obtain

$$(bc)^n + (ba_{n-1})(bc)^{n-1} + \cdots + (b^{n-1}a_1)(bc) + (b^na_0) = 0.$$

This shows that bc is a root of a monic polynomial in $\mathbb{Z}$. Thus bc is an algebraic integer, say a. Then $c = a/b$, where $a \in \Omega$ and $b \in \mathbb{Z}$. ∎

Example 4.2.3 *Let*

$$c = \frac{1}{3^{4/3}} - \frac{1}{3^{2/3}}.$$

Then

$$c^3 = \left(\frac{1}{3^{4/3}} - \frac{1}{3^{2/3}}\right)^3 = \frac{1}{3^4} - 3\frac{1}{3^{8/3}}\frac{1}{3^{2/3}} + 3\frac{1}{3^{4/3}}\frac{1}{3^{4/3}} - \frac{1}{3^2}$$

$$= \left(\frac{1}{3^4} - \frac{1}{3^2}\right) - \frac{3}{3^2}\left(\frac{1}{3^{4/3}} - \frac{1}{3^{2/3}}\right) = -\frac{8}{81} - \frac{1}{3}c,$$

so that

$$c^3 + \frac{1}{3}c + \frac{8}{81} = 0.$$

Thus c *is an algebraic number. The least common multiple of the denominators of the coefficients* $0/1, 1/3, 8/81$ *of* c^2, c, 1 *respectively is* $b = 81$. *Then* $a = bc = 81c$ *is a root of*

$$a^3 + 2187a + 52488 = 0.$$

Thus

$$a = 3^{8/3} - 3^{10/3}$$

is an algebraic integer. Hence

$$c = \frac{3^{8/3} - 3^{10/3}}{81}$$

is the quotient of an algebraic integer and an ordinary integer.

Exercises

1. Prove that

$$\frac{1}{3}\left(1 + 10^{1/3} + 10^{2/3}\right)$$

is an algebraic integer.

2. Prove that

$$\frac{10^{2/3} - 1}{\sqrt{-3}}$$

is an algebraic integer.

3. Let m and n be distinct squarefree integers such that $m \equiv n \equiv 3 \pmod 4$. Let $l = (m, n)$ and set $m = lm_1$, $n = ln_1$ so that $(m_1, n_1) = 1$. If x_0, x_1, x_2, x_3 are integers such that

$$x_0 \equiv x_3 \pmod 2, \quad x_1 \equiv x_2 \pmod 2,$$

prove that

$$\frac{1}{2}\left(x_0 + x_1\sqrt{m} + x_2\sqrt{n} + x_3\sqrt{m_1}\sqrt{n_1}\right)$$

is an algebraic integer.

4. Express the algebraic number

$$\left(\frac{1 + \sqrt{2}}{9}\right)^{1/3} + \left(\frac{1 - \sqrt{2}}{9}\right)^{1/3}$$

as the quotient of an algebraic integer and an ordinary integer.

5. Express the algebraic number

$$\sqrt[3]{\frac{1 + \sqrt{17}}{5}} + \sqrt[3]{\frac{1 - \sqrt{17}}{5}}$$

as the quotient of an algebraic integer and an ordinary integer.

6. Let D be a principal ideal domain. Prove that D is integrally closed.

7. Let m be a nonsquare integer, which is congruent to 1 modulo 4. Prove that the domain $\mathbb{Z} + \mathbb{Z}\sqrt{m}$ is not integrally closed.

8. Let θ be a root of $x^3 + 6x + 34$. Prove that the domain $\mathbb{Z} + \mathbb{Z}\theta + \mathbb{Z}\theta^2$ is not integrally closed. [Hint: Consider $\phi = (1 + \theta)/3$.]

9. Let $A = \mathbb{Z}$ and $B = \mathbb{Q}(\sqrt{2}) = \{a + b\sqrt{2} \mid a, b \in \mathbb{Q}\}$. Prove that $A^B = \mathbb{Z} + \mathbb{Z}\sqrt{2}$.

10. Prove that the domain $\mathbb{Z} + \mathbb{Z}i + \mathbb{Z}\sqrt{2} + \mathbb{Z}i\sqrt{2}$ is not integrally closed.

11. If $A \subseteq B \subseteq C$ is a tower of integral domains, prove that $A^B \subseteq A^C \subseteq B^C$.

12. Prove that $\mathbb{Z} + \mathbb{Z}\sqrt{2} + \mathbb{Z}\sqrt{5} + \mathbb{Z}\sqrt{10}$ is not integrally closed in its quotient field. [Hint: Consider $\left(\frac{1+\sqrt{5}}{\sqrt{2}}\right)^2$.]

13. Prove that the integral closure of $\mathbb{Z} + \mathbb{Z}\sqrt{5}$ in the field $\mathbb{Q}(\sqrt{5}, i) = \{a + bi + c\sqrt{5} + di\sqrt{5} \mid a, b, c, d \in \mathbb{Q}\}$ is

$$\left\{ \alpha + i\beta \mid \alpha, \beta \in \mathbb{Z} + \mathbb{Z}\left(\frac{1+\sqrt{5}}{2}\right) \right\}.$$

14. Prove that the integral closure of $\mathbb{Z} + \mathbb{Z}\sqrt{5}$ in the field $\mathbb{Q}(\sqrt{5}, \omega) = \{a + b\omega + c\sqrt{5} + d\omega\sqrt{5} \mid a, b, c, d \in \mathbb{Q}\}$, where ω is a primitive cube root of unity, is

$$\left\{ \alpha + \beta\omega \mid \alpha, \beta \in \mathbb{Z} + \mathbb{Z}\left(\frac{1+\sqrt{5}}{2}\right) \right\}.$$

15. Let A and B be integral domains with $A \subseteq B$ and B integral over A. If I is a nonzero ideal of B, prove that $I \cap A$ is a nonzero ideal of A.

Suggested Reading

1. T. W. Atterton, *A note on certain subsets of algebraic integers*, Bulletin of the Australian Mathematical Society 1 (1969), 345–352.

The author shows for example that the integral closure of $\mathbb{Z} + \mathbb{Z}\sqrt{5}$ in $\mathbb{Q}(\sqrt{5}, i) = \{a + bi + c\sqrt{5} + di\sqrt{5} \mid a, b, c, d \in \mathbb{Q}\}$ is

$$\left\{ \alpha + i\beta \mid \alpha, \beta \in \mathbb{Z} + \mathbb{Z}\left(\frac{1+\sqrt{5}}{2}\right) \right\}.$$

2. N. Bourbaki, *Éléments d'histoire des mathématiques*, second edition, Hermann, Paris, 1974.

On page 141 it is mentioned that Steinitz showed how a small number of abstract ideas, such as an irreducible ideal, chain conditions, and an integrally closed ring, could be used to prove general results characterizing Dedekind rings and that the last two of these ideas had already been introduced by Dedekind.

Biographies

1. The website

http://www-groups.dcs.st-and.ac.uk/~history/

has biographies of both Richard Dedekind (1831–1916) and Ernst Steinitz (1871–1928).

5

Algebraic Extensions of a Field

5.1 Minimal Polynomial of an Element Algebraic over a Field

Let K be a subfield of the field $\mathbb{C}$ of complex numbers. Let $\alpha \in \mathbb{C}$ be algebraic over K (see Definition 4.1.3). As α is algebraic over K, there exists a nonzero polynomial $g(x) \in K[x]$ such that $g(\alpha) = 0$. We let $I_K(\alpha)$ denote the set of all polynomials in $K[x]$ having α as a root, that is,

$$I_K(\alpha) = \{f(x) \in K[x] \mid f(\alpha) = 0\}. \tag{5.1.1}$$

Clearly the set $I_K(\alpha)$ contains the zero polynomial. It is easy to check that $I_K(\alpha)$ is an ideal of $K[x]$. Moreover, $I_K(\alpha) \neq \langle 0 \rangle$ as $g(x) \in I_K(\alpha)$.

As K is a field, by Theorem 2.2.1(b) we know that $K[x]$ is a Euclidean domain and thus, by Theorem 2.1.2, a principal ideal domain. Hence there exists $p(x) \in K[x]$ such that

$$I_K(\alpha) = \langle p(x) \rangle. \tag{5.1.2}$$

Suppose $p_1(x) \in K[x]$ is another polynomial that generates $I_K(\alpha)$, that is,

$$I_K(\alpha) = \langle p_1(x) \rangle.$$

Then

$$\langle p(x) \rangle = \langle p_1(x) \rangle$$

and so, by Theorem 1.3.1, we have

$$p_1(x) = u(x)p(x),$$

where $u(x)$ is a unit in $K[x]$. However, from Example 1.1.18(c), we have

$$U(K[x]) = K^*,$$

so that

$$u(x) \in K^*.$$

88

This shows that we may take the polynomial $p(x)$ to be monic, in which case $p(x)$ is uniquely determined by (5.1.2).

Definition 5.1.1 (Minimal polynomial of α over K) *Let K be a subfield of $\mathbb{C}$. Let $\alpha \in \mathbb{C}$ be algebraic over K. Then the unique monic polynomial $p(x) \in K[x]$ such that*

$$I_K(\alpha) = \langle p(x) \rangle$$

is called the minimal polynomial of α over K and is denoted by $\mathrm{irr}_K(\alpha)$.

Definition 5.1.2 (Degree of α over K) *Let K be a subfield of $\mathbb{C}$. Let $\alpha \in \mathbb{C}$ be algebraic over K. Then the degree of α over K, written $\deg_K(\alpha)$, is defined by*

$$\deg_K(\alpha) = \deg(\mathrm{irr}_K(\alpha)).$$

When $K = \mathbb{Q}$ we write $\deg(\alpha)$ for $\deg_{\mathbb{Q}}(\alpha)$.

Theorem 5.1.1 *Let K be a subfield of $\mathbb{C}$. Let $\alpha \in \mathbb{C}$ be algebraic over K. Then $\mathrm{irr}_K(\alpha)$ is irreducible in $K[x]$.*

Proof: Suppose that $\mathrm{irr}_K(\alpha)$ is reducible in $K[x]$. Then there exist nonzero polynomials $r(x) \in K[x]$ and $s(x) \in K[x]$ such that

$$\mathrm{irr}_K(\alpha) = r(x)s(x) \tag{5.1.3}$$

with $r(x) \notin U(K[x])$ and $s(x) \notin U(K[x])$. Hence $r(x) \notin K$ and $s(x) \notin K$ so that $\deg r(x) \geq 1$ and $\deg s(x) \geq 1$. Thus

$$\deg(\mathrm{irr}_K(\alpha)) = \deg r(x) + \deg s(x) > \max(\deg r(x), \deg s(x)). \tag{5.1.4}$$

As α is a root of $\mathrm{irr}_K(\alpha)$, from (5.1.3) we have $r(\alpha)s(\alpha) = 0$, so that either $r(\alpha) = 0$ or $s(\alpha) = 0$. Without loss of generality we may suppose that $r(\alpha) = 0$. Hence

$$r(x) \in I_K(\alpha) = \langle \mathrm{irr}_K(\alpha) \rangle$$

so that

$$\mathrm{irr}_K(\alpha) \mid r(x)$$

and thus

$$\deg(\mathrm{irr}_K(\alpha)) \leq \deg r(x),$$

which contradicts (5.1.4). Hence $\mathrm{irr}_K(\alpha)$ is irreducible in $K[x]$. ■

Example 5.1.1 $\alpha = (1+i)/\sqrt{2} \in \mathbb{C}$ *is a root of* $x^4+1 \in \mathbb{Q}[x]$. *As* x^4+1 *is monic and irreducible in* $\mathbb{Q}[x]$, *we have*

$$\text{irr}_{\mathbb{Q}}\left(\frac{1+i}{\sqrt{2}}\right) = x^4+1, \ \deg\left(\frac{1+i}{\sqrt{2}}\right) = 4.$$

Example 5.1.2 *Let* K *be the field* $\mathbb{Q}(\sqrt{2}) = \{a+b\sqrt{2} \mid a,b \in \mathbb{Q}\}$. *Let* $\alpha = (1+i)/\sqrt{2} \in \mathbb{C}$. *Then* α *is a root of* $x^2 - \sqrt{2}x + 1 \in K[x]$. *As* $x^2 - \sqrt{2}x+1$ *is monic and irreducible in* $K[x]$, *we have*

$$\text{irr}_{\mathbb{Q}(\sqrt{2})}\left(\frac{1+i}{\sqrt{2}}\right) = x^2 - \sqrt{2}x+1, \ \deg_{\mathbb{Q}(\sqrt{2})}\left(\frac{1+i}{\sqrt{2}}\right) = 2.$$

5.2 Conjugates of α over K

We define the conjugates of an element over a subfield of $\mathbb{C}$.

Definition 5.2.1 (Conjugates of α over K) *Let* $\alpha \in \mathbb{C}$ *be algebraic over a subfield* K *of* $\mathbb{C}$. *The conjugates of* α *over* K *are the roots in* $\mathbb{C}$ *of* $\text{irr}_K(\alpha)$.

Example 5.2.1 *We have from Example 5.1.1 that*

$$\text{irr}_{\mathbb{Q}}\left(\frac{1+i}{\sqrt{2}}\right) = x^4+1.$$

As

$$x^4+1 = \left(x - \left(\frac{1+i}{\sqrt{2}}\right)\right)\left(x - \left(\frac{1-i}{\sqrt{2}}\right)\right)\left(x + \left(\frac{1+i}{\sqrt{2}}\right)\right)\left(x + \left(\frac{1-i}{\sqrt{2}}\right)\right)$$

the conjugates of $(1+i)/\sqrt{2}$ *over* $\mathbb{Q}$ *are*

$$\frac{1+i}{\sqrt{2}}, \ \frac{1-i}{\sqrt{2}}, \ \frac{-1-i}{\sqrt{2}}, \ \frac{-1+i}{\sqrt{2}}.$$

Example 5.2.2 *We have from Example 5.1.2 that*

$$\text{irr}_{\mathbb{Q}(\sqrt{2})}\left(\frac{1+i}{\sqrt{2}}\right) = x^2 - \sqrt{2}x+1.$$

As

$$x^2 - \sqrt{2}x+1 = \left(x - \frac{1+i}{\sqrt{2}}\right)\left(x - \frac{1-i}{\sqrt{2}}\right)$$

the conjugates of $(1+i)/\sqrt{2}$ *over* $\mathbb{Q}(\sqrt{2})$ *are*

$$\frac{1+i}{\sqrt{2}}, \ \frac{1-i}{\sqrt{2}}.$$

Example 5.2.3 *Similarly to Example 5.2.2 we find that the conjugates of* $(1 + i)/\sqrt{2}$ *over the field* $\mathbb{Q}(i) = \{x + yi \mid x, y \in \mathbb{Q}\}$ *are*

$$\frac{1+i}{\sqrt{2}}, \quad \frac{-1-i}{\sqrt{2}}$$

and the conjugates of $(1 + i)/\sqrt{2}$ *over the field* $\mathbb{Q}(\sqrt{-2}) = \{x + y\sqrt{-2} \mid x, y \in \mathbb{Q}\}$ *are*

$$\frac{1+i}{\sqrt{2}}, \quad \frac{-1+i}{\sqrt{2}}.$$

Theorem 5.2.1 *Let K be a subfield of $\mathbb{C}$. Let $\alpha \in \mathbb{C}$ be algebraic over K. Then the conjugates of α over K are distinct.*

Proof: Suppose that α has two conjugates over K that are the same. Then $\mathrm{irr}_K(\alpha)$ has a root of order at least 2. Let $\beta \in \mathbb{C}$ be such a multiple root. Then

$$\mathrm{irr}_K(\alpha) = (x - \beta)^2 r(x), \qquad (5.2.1)$$

where $r(x) \in \mathbb{C}[x]$. Differentiating (5.2.1) with respect to x, we obtain

$$\mathrm{irr}_K(\alpha)' = (x - \beta)^2 r'(x) + 2(x - \beta)r(x).$$

Thus β is a root of the derivative $\mathrm{irr}_K(\alpha)'$ of $\mathrm{irr}_K(\alpha)$. As $\mathrm{irr}_K(\alpha)' \in K[x]$ we have

$$\mathrm{irr}_K(\alpha)' \in I_K(\alpha) = \langle \mathrm{irr}_K(\alpha) \rangle$$

so that

$$\mathrm{irr}_K(\alpha) \mid \mathrm{irr}_K(\alpha)'$$

and thus

$$\deg(\mathrm{irr}_K(\alpha)) \leq \deg(\mathrm{irr}_K(\alpha)'),$$

which is impossible. Hence the conjugates of α over K are distinct. ■

5.3 Conjugates of an Algebraic Integer

Theorem 5.3.1 *If α is an algebraic integer then its conjugates over $\mathbb{Q}$ are also algebraic integers.*

Proof: As α is an algebraic integer it is a root of a polynomial

$$h(x) = x^m + a_{m-1}x^{m-1} + \cdots + a_1x + a_0 \in \mathbb{Z}[x].$$

Since $h(x) \in \mathbb{Q}[x]$ and $h(\alpha) = 0$ we have $h(x) \in I_{\mathbb{Q}}(\alpha) = \langle \mathrm{irr}_{\mathbb{Q}}(\alpha) \rangle$ so that

$$h(x) = \mathrm{irr}_{\mathbb{Q}}(\alpha)q(x)$$

for some $q(x) \in \mathbb{Q}[x]$. Let β be a conjugate of α over $\mathbb{Q}$. Then β is also a root of $\mathrm{irr}_{\mathbb{Q}}(\alpha)$. Hence $h(\beta) = 0$ and so β is also an algebraic integer. ∎

We recall that a monic polynomial $f(x) = x^n + a_1 x^{n-1} + \cdots + a_{n-1}x + a_n \in \mathbb{Z}[x]$ is said to be *p-Eisenstein* with respect to the prime p if

$$p \mid a_1, \ldots, p \mid a_{n-1}, \, p \mid a_n, \, p^2 \nmid a_n.$$

Eisenstein's irreducibility criterion asserts that if $f(x)$ is *p*-Eisenstein for some prime p then $f(x)$ is irreducible in $\mathbb{Z}[x]$.

Example 5.3.1 *Let $\alpha = \sqrt[3]{2} - \sqrt[3]{4}$. Then*

$$\alpha^3 = 2 - 6\sqrt[3]{2} + 6\sqrt[3]{4} - 4 = -2 - 6(\sqrt[3]{2} - \sqrt[3]{4}) = -2 - 6\alpha$$

so that α is a root of the monic cubic polynomial $x^3 + 6x + 2 \in \mathbb{Z}[x]$ and is thus an algebraic integer. As $x^3 + 6x + 2$ is 2-Eisenstein it is irreducible in $\mathbb{Z}[x]$. Hence

$$\mathrm{irr}_{\mathbb{Q}}(\alpha) = x^3 + 6x + 2.$$

The other two roots of $\mathrm{irr}_{\mathbb{Q}}(\alpha)$ are

$$\alpha' = \omega\sqrt[3]{2} - \omega^2\sqrt[3]{4}, \ \alpha'' = \omega^2\sqrt[3]{2} - \omega\sqrt[3]{4},$$

where ω is a complex cube root of unity. Thus α' and α'' are also algebraic integers.

Theorem 5.3.2 *If α is an algebraic integer then*

$$\mathrm{irr}_{\mathbb{Q}}(\alpha) \in \mathbb{Z}[x].$$

Proof: Let the conjugates of the algebraic integer α over $\mathbb{Q}$ be $\alpha_1 = \alpha, \, \alpha_2, \ldots, \alpha_n$. Then

$$\begin{aligned}
\mathrm{irr}_{\mathbb{Q}}(\alpha) &= (x - \alpha_1)(x - \alpha_2) \cdots (x - \alpha_n) \\
&= x^n - (\alpha_1 + \alpha_2 + \cdots + \alpha_n)x^{n-1} + (\alpha_1\alpha_2 + \cdots + \alpha_{n-1}\alpha_n)x^{n-2} \\
&\quad + \cdots + (-1)^n \alpha_1\alpha_2 \cdots \alpha_n.
\end{aligned}$$

As $\mathrm{irr}_{\mathbb{Q}}(\alpha) \in \mathbb{Q}[x]$, we have

$$\alpha_1 + \cdots + \alpha_n \in \mathbb{Q},$$
$$\alpha_1\alpha_2 + \cdots + \alpha_{n-1}\alpha_n \in \mathbb{Q},$$
$$\cdots$$
$$\alpha_1\alpha_2 \cdots \alpha_n \in \mathbb{Q}.$$

But, by Theorem 5.3.1, $\alpha_1, \ldots, \alpha_n$ are all algebraic integers. Hence, by Theorem 4.1.8,

$$\alpha_1 + \cdots + \alpha_n, \quad \alpha_1\alpha_2 + \cdots + \alpha_{n-1}\alpha_n, \quad \ldots, \alpha_1\alpha_2 \cdots \alpha_n$$

are all algebraic integers. Since they are all rational, by Theorem 4.2.4 they must in fact be ordinary integers. Hence $\mathrm{irr}_{\mathbb{Q}}(\alpha) \in \mathbb{Z}[x]$. ∎

It is an immediate consequence of Theorem 5.3.2 that if $\alpha \in \mathbb{C}$ satisfies a polynomial of the form

$$x^m + a_{m-1}x^{m-1} + \cdots + a_1 x + a_0 \in \mathbb{Z}[x]$$

then the monic polynomial of least degree in $\mathbb{Q}[x]$ of which α is a root belongs to $\mathbb{Z}[x]$.

We use Theorem 5.3.2 to prove the following result (compare Theorem 4.2.2).

Theorem 5.3.3 *The integral closure of $A = \mathbb{Z} + \mathbb{Z}\sqrt{-3}$ in the field $B = \mathbb{Q}(\sqrt{-3}) = \{a + b\sqrt{-3} \mid a, b \in \mathbb{Q}\}$ is*

$$A^B = \mathbb{Z} + \mathbb{Z}\left(\frac{1 + \sqrt{-3}}{2}\right).$$

Proof: Let $\alpha \in \mathbb{Z} + \mathbb{Z}\left(\frac{1+\sqrt{-3}}{2}\right)$. Then $\alpha = m + n\left(\frac{1+\sqrt{-3}}{2}\right)$ for some $m, n \in \mathbb{Z}$. Clearly $\alpha \in B$. As α is a root of the monic polynomial

$$x^2 - (2m + n)x + (m^2 + mn + n^2) \in A[x],$$

α is integral over A and thus belongs to A^B. Hence $\mathbb{Z} + \mathbb{Z}\left(\frac{1+\sqrt{-3}}{2}\right) \subseteq A^B$.

We now show that $A^B \subseteq \mathbb{Z} + \mathbb{Z}\left(\frac{1+\sqrt{-3}}{2}\right)$. Let $\alpha \in A^B$. Clearly $\alpha \in B$ so that $\alpha = a + b\sqrt{-3}$ for some $a, b \in \mathbb{Q}$. Thus α is a root of the monic polynomial

$$x^2 - 2ax + (a^2 + 3b^2) \in \mathbb{Q}[x].$$

The discriminant of this polynomial is

$$(2a)^2 - 4(a^2 + 3b^2) = -12b^2$$

so that it is reducible in $\mathbb{Q}[x]$ if $b = 0$ and irreducible in $\mathbb{Q}[x]$ if $b \neq 0$. Hence

$$\mathrm{irr}_{\mathbb{Q}}(\alpha) = \begin{cases} x - a, & \text{if } b = 0, \\ x^2 - 2ax + (a^2 + 3b^2), & \text{if } b \neq 0. \end{cases}$$

As $\alpha \in A^B$, α is integral over A and thus is a root of a monic polynomial

$$x^n + \alpha_1 x^{n-1} + \cdots + \alpha_n \in A[x].$$

For $i = 1, 2, \ldots, n$ we have $\alpha_i \in A$ so that $\alpha_i = a_i + b_i\sqrt{-3}$ for some a_i, $b_i \in \mathbb{Z}$. Thus α is a root of the monic polynomial

$$(x^n + a_1 x^{n-1} + \cdots + a_n)^2 + 3(b_1 x^{n-1} + \cdots + b_n)^2 \in \mathbb{Z}[x].$$

Hence α is an algebraic integer and so, by Theorem 5.3.2, $\mathrm{irr}_{\mathbb{Q}}(\alpha) \in \mathbb{Z}[x]$, that is

$$\begin{cases} a \in \mathbb{Z}, & \text{if } b = 0, \\ 2a,\ a^2 + 3b^2 \in \mathbb{Z}, & \text{if } b \neq 0. \end{cases}$$

In the former case $\alpha = a + b\sqrt{-3} = a = a + 0\left(\frac{1+\sqrt{-3}}{2}\right) \in \mathbb{Z} + \mathbb{Z}\left(\frac{1+\sqrt{-3}}{2}\right)$. In the latter case we have $a = m/2$ for some $m \in \mathbb{Z}$. If $m \in 2\mathbb{Z}$ then $a \in \mathbb{Z}$ and $b \in \mathbb{Z}$. If $m \in 2\mathbb{Z} + 1$ then $2b \in 2\mathbb{Z} + 1$. Hence, in both cases, we see that $a = m/2$ and $b = m/2 + n$, where $m, n \in \mathbb{Z}$. Thus

$$\alpha = \frac{m}{2} + \left(\frac{m}{2} + n\right)\sqrt{-3} = -n + (m + 2n)\left(\frac{1+\sqrt{-3}}{2}\right) \in \mathbb{Z} + \mathbb{Z}\left(\frac{1+\sqrt{-3}}{2}\right).$$

Hence $A^B \subseteq \mathbb{Z} + \mathbb{Z}\left(\frac{1+\sqrt{-3}}{2}\right)$.

This completes the proof that

$$A^B = \mathbb{Z} + \mathbb{Z}\left(\frac{1+\sqrt{-3}}{2}\right). \qquad \blacksquare$$

5.4 Algebraic Integers in a Quadratic Field

In this section we determine the algebraic integers in a field $\mathbb{Q}(\alpha)$ obtained by adjoining a root $\alpha(\in \mathbb{C})$ of an irreducible quadratic polynomial $x^2 + ax + b \in \mathbb{Q}[x]$ to $\mathbb{Q}$; that is, $\mathbb{Q}(\alpha)$ is the smallest subfield of $\mathbb{C}$ containing both $\mathbb{Q}$ and α. We note that $\alpha \notin \mathbb{Q}$ as $x^2 + ax + b$ is irreducible in $\mathbb{Q}[x]$. Clearly

$$\mathbb{Q}(\alpha) = \left\{ \frac{a_0 + a_1\alpha + \cdots + a_m\alpha^m}{b_0 + b_1\alpha + \cdots + b_n\alpha^n} \;\middle|\; m, n \text{ (nonnegative integers)}, \right.$$
$$\left. a_0, \ldots, a_m, b_0, \ldots, b_n \in \mathbb{Q},\ b_0 + b_1\alpha + \cdots + b_n\alpha^n \neq 0 \right\}.$$

As $\alpha^2 = -b - a\alpha$, we obtain recursively that $\alpha^k = c_k + d_k\alpha$ ($k = 2, 3, \ldots$), where $c_k, d_k \in \mathbb{Q}$. Thus

$$\mathbb{Q}(\alpha) = \left\{ \frac{e_0 + e_1\alpha}{f_0 + f_1\alpha} \;\middle|\; e_0, e_1, f_0, f_1 \in \mathbb{Q},\ (f_0, f_1) \neq (0, 0) \right\}.$$

As $f_0^2 - af_0f_1 + bf_1^2 \neq 0$ for $(f_0, f_1) \neq (0, 0)$ and

$$\frac{e_0 + e_1\alpha}{f_0 + f_1\alpha} = \left(\frac{e_0 f_0 - ae_0 f_1 + be_1 f_1}{f_0^2 - af_0 f_1 + bf_1^2}\right) + \left(\frac{e_1 f_0 - e_0 f_1}{f_0^2 - af_0 f_1 + bf_1^2}\right)\alpha,$$

we deduce that

$$\mathbb{Q}(\alpha) = \{x + y\alpha \mid x, y \in \mathbb{Q}\},$$

where $\alpha^2 + a\alpha + b = 0$. The field $\mathbb{Q}(\alpha)$ is called a quadratic field or a quadratic extension of $\mathbb{Q}$. Different quadratic polynomials, for example $x^2 + x + 1$ and $x^2 + 6x + 12$, can give rise to the same quadratic field K. Our next theorem gives a unique way of representing a quadratic field.

Theorem 5.4.1 *Let K be a quadratic field. Then there exists a unique squarefree integer m such that $K = \mathbb{Q}(\sqrt{m})$.*

Proof: Suppose that $K = \mathbb{Q}(\alpha)$, where α is a root of the irreducible polynomial $x^2 + ax + b \in \mathbb{Q}[x]$. Then $\alpha = \alpha_1$ or α_2, where

$$\alpha_1 = \frac{-a + \sqrt{a^2 - 4b}}{2}, \quad \alpha_2 = \frac{-a - \sqrt{a^2 - 4b}}{2}.$$

As

$$\alpha_1 + \alpha_2 = -a \in \mathbb{Q}$$

we have

$$\mathbb{Q}(\alpha_1) = \mathbb{Q}(\alpha_2)$$

so that

$$K = \mathbb{Q}(\alpha) = \mathbb{Q}(\alpha_1) = \mathbb{Q}\left(\frac{-a + \sqrt{a^2 - 4b}}{2}\right) = \mathbb{Q}(\sqrt{c}),$$

where $c = a^2 - 4b \in \mathbb{Q}$ is not the square of a rational number as $x^2 + ax + b$ is irreducible in $\mathbb{Q}[x]$. Now

$$c = p/q,$$

where $p, q \in \mathbb{Z}$ are such that

$$q > 0, \ (p, q) = 1.$$

Let r^2 denote the largest square dividing pq. Then $pq = r^2 m$, where m is a square-free integer ($\neq 1$) and

$$K = \mathbb{Q}\left(\sqrt{c}\right) = \mathbb{Q}\left(\sqrt{\frac{p}{q}}\right) = \mathbb{Q}(\sqrt{pq}) = \mathbb{Q}(\sqrt{r^2 m}) = \mathbb{Q}(r\sqrt{m}) = \mathbb{Q}(\sqrt{m}).$$

Now let n be another squarefree integer such that $K = \mathbb{Q}(\sqrt{n})$. Hence

$$\mathbb{Q}(\sqrt{m}) = \mathbb{Q}(\sqrt{n})$$

and so

$$\sqrt{m} = x + y\sqrt{n}$$

for some $x, y \in \mathbb{Q}$. Squaring we obtain

$$m = x^2 + ny^2 + 2xy\sqrt{n}.$$

If $xy \neq 0$ then

$$\sqrt{n} = \frac{m - x^2 - ny^2}{2xy},$$

contradicting that $\sqrt{n} \notin \mathbb{Q}$ as n is squarefree. Hence $xy = 0$. If $y = 0$ then

$$\sqrt{m} = x,$$

contradicting that $\sqrt{m} \notin \mathbb{Q}$ as m is squarefree. Thus $x = 0$ and $\sqrt{m} = y\sqrt{n}$ so that

$$m = y^2 n.$$

As m is squarefree, we deduce that $y^2 = 1$ so that $m = n$. Hence m is uniquely determined by K. ∎

We next determine the algebraic integers in the quadratic field $K = \mathbb{Q}(\sqrt{m}) = \{a + b\sqrt{m} \mid a, b \in \mathbb{Q}\}$, where m is a squarefree integer. The set of algebraic integers in K is denoted by O_K.

Theorem 5.4.2 *Let K be a quadratic field. Let m be the unique squarefree integer such that $K = \mathbb{Q}(\sqrt{m})$. Then the set O_K of algebraic integers in K is given by*

$$O_K = \begin{cases} \mathbb{Z} + \mathbb{Z}\sqrt{m}, & \text{if } m \not\equiv 1 \ (\mathrm{mod}\ 4), \\ \mathbb{Z} + \mathbb{Z}\left(\dfrac{1 + \sqrt{m}}{2}\right), & \text{if } m \equiv 1 \ (\mathrm{mod}\ 4). \end{cases}$$

Proof: It is easily checked that the elements of $\mathbb{Z} + \mathbb{Z}\sqrt{m}$ if $m \not\equiv 1 \ (\mathrm{mod}\ 4)$ and of $\mathbb{Z} + \mathbb{Z}\left(\frac{1+\sqrt{m}}{2}\right)$ if $m \equiv 1 \ (\mathrm{mod}\ 4)$ are algebraic integers in $K = \mathbb{Q}(\sqrt{m})$. Thus

$$O_K \supseteq \begin{cases} \mathbb{Z} + \mathbb{Z}\sqrt{m}, & \text{if } m \not\equiv 1 \ (\mathrm{mod}\ 4), \\ \mathbb{Z} + \mathbb{Z}\left(\dfrac{1 + \sqrt{m}}{2}\right), & \text{if } m \equiv 1 \ (\mathrm{mod}\ 4). \end{cases}$$

We complete the proof by showing the inclusion in the reverse direction. Let $\alpha \in O_K$. Then $\alpha \in K$ and so $\alpha = a + b\sqrt{m}$ for some $a, b \in \mathbb{Q}$. Thus α is a root of the monic polynomial

$$x^2 - 2ax + (a^2 - mb^2) \in \mathbb{Q}[x].$$

The discriminant of this polynomial is

$$(2a)^2 - 4(a^2 - mb^2) = 4mb^2$$

so that it is reducible in $\mathbb{Q}[x]$ if $b = 0$ and irreducible in $\mathbb{Q}[x]$ if $b \neq 0$. Hence

$$\mathrm{irr}_{\mathbb{Q}}(\alpha) = \begin{cases} x - a, & \text{if } b = 0, \\ x^2 - 2ax + (a^2 - mb^2), & \text{if } b \neq 0. \end{cases}$$

As α is an algebraic integer, by Theorem 5.3.2 we have $\mathrm{irr}_{\mathbb{Q}}(\alpha) \in \mathbb{Z}[x]$ so that

$$\begin{cases} a \in \mathbb{Z}, & \text{if } b = 0, \\ 2a, \ a^2 - mb^2 \in \mathbb{Z}, & \text{if } b \neq 0. \end{cases}$$

If $b = 0$ we have $\alpha = a \in \mathbb{Z} \subset \mathbb{Z} + \mathbb{Z}\sqrt{m}$. Now suppose that $b \neq 0$. If $2a \in 2\mathbb{Z}$ then $a \in \mathbb{Z}$ and so $mb^2 \in \mathbb{Z}$. Since m is squarefree we see that $b \in \mathbb{Z}$. In this case $\alpha = a + b\sqrt{m} \in \mathbb{Z} + \mathbb{Z}\sqrt{m}$. If $2a \in 2\mathbb{Z} + 1$ then as $4(a^2 - mb^2) \in \mathbb{Z}$ we deduce that $4mb^2 \in \mathbb{Z}$. As m is squarefree we have $2b \in \mathbb{Z}$. If $2b \in 2\mathbb{Z}$ then $b \in \mathbb{Z}$ and so

$$a^2 = (a^2 - mb^2) + mb^2 \in \mathbb{Z},$$

contradicting that $2a \in 2\mathbb{Z} + 1$. Hence $2b \in 2\mathbb{Z} + 1$. Thus $a = (2u + 1)/2$ and $b = (2v + 1)/2$, where $u, v \in \mathbb{Z}$. Then

$$a^2 - mb^2 = \frac{1}{4}((2u + 1)^2 - m(2v + 1)^2)$$

so that

$$\frac{m - 1}{4} = u^2 + u - m(v^2 + v) - (a^2 - mb^2) \in \mathbb{Z}.$$

Hence $m \equiv 1 \pmod 4$ and

$$\begin{aligned} \alpha = a + b\sqrt{m} &= \frac{2u + 1}{2} + \frac{2v + 1}{2}\sqrt{m} \\ &= (u - v) + (2v + 1)\left(\frac{1 + \sqrt{m}}{2}\right) \\ &\in \mathbb{Z} + \mathbb{Z}\left(\frac{1 + \sqrt{m}}{2}\right). \end{aligned}$$

This completes the proof of the reverse inclusion and thus the proof of the theorem. ∎

The quadratic field $K = \mathbb{Q}(\sqrt{m})$, where m is a squarefree integer, is said to be real if $K \subseteq \mathbb{R}$ and imaginary if $K \not\subseteq \mathbb{R}$. Clearly K is real if $m > 0$ and imaginary if $m < 0$. We close this section by determining the unit group $U(O_K)$ when K is an imaginary quadratic field.

Theorem 5.4.3 *Let K be an imaginary quadratic field. Then*

$$U(O_K) = \begin{cases} \{\pm 1, \pm i\} \simeq \mathbb{Z}_4, & \text{if } K = \mathbb{Q}(\sqrt{-1}), \\ \{\pm 1, \pm \omega, \pm \omega^2\} \simeq \mathbb{Z}_6, & \text{if } K = \mathbb{Q}(\sqrt{-3}), \\ \{\pm 1\} \simeq \mathbb{Z}_2, & \text{otherwise}, \end{cases}$$

where $\omega = (-1 + \sqrt{-3})/2$.

Proof: If $K = \mathbb{Q}(\sqrt{-1})$ then, by Theorem 5.4.2, we have $O_K = \mathbb{Z} + \mathbb{Z}\sqrt{-1} = \mathbb{Z} + \mathbb{Z}i$, and $U(O_K) = \{\pm 1, \pm i\}$ follows from Exercise 1 of Chapter 1.

If $K = \mathbb{Q}(\sqrt{-3})$ then, by Theorem 5.4.2, we have $O_K = \mathbb{Z} + \mathbb{Z}\left(\frac{1+\sqrt{-3}}{2}\right) = \mathbb{Z} + \mathbb{Z}\omega$, and $U(O_K) = \{\pm 1, \pm \omega, \pm \omega^2\}$ follows from Exercise 2 of Chapter 1.

If K is an imaginary quadratic field $\neq \mathbb{Q}(\sqrt{-1})$, $\mathbb{Q}(\sqrt{-3})$ then by Theorem 5.4.1 $K = \mathbb{Q}(\sqrt{m})$ for a unique negative, squarefree, integer $m \neq -1, -3$. If $m \not\equiv 1$ (mod 4) then $O_K = \mathbb{Z} + \mathbb{Z}\sqrt{m}$ and $U(O_K) = \{\pm 1\}$ by Exercise 3 of Chapter 1 as $m < -1$. If $m \equiv 1$ (mod 4) then $O_K = \mathbb{Z} + \mathbb{Z}\left(\frac{1+\sqrt{-3}}{2}\right)$ by Theorem 5.4.2 and $U(O_K) = \{\pm 1\}$ by Exercise 4 of Chapter 1 as $m < -3$. ∎

5.5 Simple Extensions

Definition 5.5.1 (Simple extension) *Let K be a subfield of $\mathbb{C}$ and let $\alpha \in \mathbb{C}$. Let*

$$K(\alpha) = \bigcap_{\substack{F \\ \alpha \in F \\ K \subseteq F \subseteq \mathbb{C}}} F,$$

where the intersection is taken over all subfields F of $\mathbb{C}$, which contain both K and α. The intersection is nonempty as $\mathbb{C}$ itself is such a field. Since the intersection of subfields of $\mathbb{C}$ is again a subfield of $\mathbb{C}$, $K(\alpha)$ is the smallest field containing both K and α. We say that $K(\alpha)$ is formed from K by adjoining a single element α. A subfield L of $\mathbb{C}$ for which there exists $\alpha \in \mathbb{C}$ such that $L = K(\alpha)$ is called a simple extension of K.

Clearly if $\alpha \in K$ then $K(\alpha) = K$.

For $K \subseteq \mathbb{C}$ and $\alpha \in \mathbb{C}$ let

$$L = \left\{ \frac{b_0 + b_1\alpha + \cdots + b_k\alpha^k}{c_0 + c_1\alpha + \cdots + c_h\alpha^h} \,\middle|\, k, h \in \mathbb{N} \cup \{0\}, \begin{array}{l} b_0, \ldots, b_k, c_0, \ldots, c_h \in K, \\ c_0 + c_1\alpha + \cdots + c_n\alpha^h \neq 0 \end{array} \right\}.$$

Then L is a subfield of $\mathbb{C}$ that contains both K and α. Moreover any subfield of $\mathbb{C}$ containing both K and α must contain all the elements of L. Hence L is the smallest subfield of $\mathbb{C}$ containing both K and α, so that $L = K(\alpha)$.

We are interested in simple extensions $K(\alpha)$ of K when α is algebraic over K and the minimal polynomial of α over K has degree n. By the preceding remarks, each element β of $K(\alpha)$ is of the form

$$\beta = \frac{f(\alpha)}{g(\alpha)},$$

where

$$f(x) = b_0 + b_1 x + \cdots + b_k x^k \in K[x],$$
$$g(x) = c_0 + c_1 x + \cdots + c_h x^h \in K[x],$$

and

$$g(\alpha) \neq 0.$$

This implies that $\operatorname{irr}_K(\alpha) \nmid g(x)$ and, since $\operatorname{irr}_K(\alpha)$ is irreducible in $K[x]$, that

$$\langle \operatorname{irr}_K(\alpha), g(x) \rangle = K[x].$$

Thus we can find polynomials $m(x), n(x) \in K[x]$ such that

$$m(x)\operatorname{irr}_K(\alpha) + n(x)g(x) = 1.$$

As $\operatorname{irr}_K(\alpha)$ has the root α, we see that

$$n(\alpha)g(\alpha) = 1$$

so that

$$\frac{1}{g(\alpha)} = n(\alpha)$$

and thus

$$\beta = \frac{f(\alpha)}{g(\alpha)} = f(\alpha)n(\alpha).$$

Hence each element β of $K(\alpha)$ can be expressed as a polynomial in α with coefficients in K, say

$$\beta = d_0 + d_1 \alpha + \cdots + d_l \alpha^l,$$

where l is a nonnegative integer and $d_0, d_1, \ldots, d_l \in K$. Let

$$h(x) = d_0 + d_1 x + \cdots + d_l x^l \in K[x],$$

so that $\beta = h(\alpha)$. As K is a field we can divide $h(x)$ by $\operatorname{irr}_K(\alpha)$ to obtain polynomials $u(x) \in K[x]$ and $v(x) \in K[x]$ such that

$$h(x) = u(x)\operatorname{irr}_K(\alpha) + v(x), \quad \deg v(x) < \deg(\operatorname{irr}_K(\alpha)) = n.$$

Then, as $\mathrm{irr}_K(\alpha)$ has the root α, we have

$$h(\alpha) = v(\alpha),$$

and so

$$\beta = v(\alpha).$$

Hence every element of $K(\alpha)$ is of the form

$$a_0 + a_1\alpha + \cdots + a_{n-1}\alpha^{n-1},$$

where $a_0, a_1, \ldots, a_{n-1} \in K$ and $n = \deg(\mathrm{irr}_K(\alpha))$. Thus we have proved the following result.

Theorem 5.5.1 *Let K be a subfield of $\mathbb{C}$. Let $\alpha \in \mathbb{C}$ be algebraic over K. Let $n = \deg(\mathrm{irr}_K(\alpha))$. Then*

$$K(\alpha) = \{a_0 + a_1\alpha + \cdots + a_{n-1}\alpha^{n-1} \mid a_0, \ldots, a_{n-1} \in K\}.$$

Theorem 5.5.1 shows that $K(\alpha)$ can be viewed as an n-dimensional vector space over K with basis $\{1, \alpha, \ldots, \alpha^{n-1}\}$. The dimension n is called the degree of the extension $K(\alpha)$ over K.

Definition 5.5.2 (Degree of the extension $K(\alpha)$ over K) *Let K be a subfield of $\mathbb{C}$. Let $\alpha \in \mathbb{C}$ be algebraic over K of degree n (so that $n = \deg_K(\alpha) = \deg(\mathrm{irr}_K(\alpha))$). The degree of the extension $K(\alpha)$ over K, written $[K(\alpha) : K]$, is defined by*

$$[K(\alpha) : K] = n.$$

Example 5.5.1 *Let m be a squarefree integer. Then $\sqrt{m} \in \mathbb{C}$ is a root of the polynomial $x^2 - m \in \mathbb{Q}[x]$. Now $x^2 - m = (x - \sqrt{m})(x + \sqrt{m})$, where $\pm\sqrt{m} \notin \mathbb{Q}$ as m is squarefree, so that $x^2 - m$ is irreducible in $\mathbb{Q}[x]$, and thus*

$$\mathrm{irr}_{\mathbb{Q}}(\sqrt{m}) = x^2 - m.$$

By Theorem 5.5.1 we have

$$\mathbb{Q}(\sqrt{m}) = \{a_0 + a_1\sqrt{m} \mid a_0, a_1 \in \mathbb{Q}\}$$

and

$$[\mathbb{Q}(\sqrt{m}) : \mathbb{Q}] = 2,$$

so that $\mathbb{Q}(\sqrt{m})$ is a quadratic extension of $\mathbb{Q}$.

Example 5.5.2 *Let*

$$\alpha = (5 + \sqrt{17})^{1/3} + (5 - \sqrt{17})^{1/3} \in \mathbb{R}.$$

Then

$$\begin{aligned}
\alpha^3 &= (5 + \sqrt{17}) + 3(5 + \sqrt{17})^{2/3}(5 - \sqrt{17})^{1/3} \\
&\quad + 3(5 + \sqrt{17})^{1/3}(5 - \sqrt{17})^{2/3} + (5 - \sqrt{17}) \\
&= 10 + 3(5 + \sqrt{17})^{1/3}(5 - \sqrt{17})^{1/3}((5 + \sqrt{17})^{1/3} + (5 - \sqrt{17})^{1/3}) \\
&= 10 + 3((5 + \sqrt{17})(5 - \sqrt{17}))^{1/3}\alpha \\
&= 10 + 3 \cdot 8^{1/3}\alpha \\
&= 10 + 6\alpha,
\end{aligned}$$

so that α is a root of the monic polynomial $x^3 - 6x - 10 \in \mathbb{Z}[x]$. Hence α is an algebraic integer. Moreover, as $x^3 - 6x - 10$ is 2-Eisenstein, it is irreducible. Hence

$$\operatorname{irr}_{\mathbb{Q}}(\alpha) = x^3 - 6x - 10.$$

Thus, by Theorem 5.5.1,

$$\mathbb{Q}(\alpha) = \{a_0 + a_1\alpha + a_2\alpha^2 \mid a_0, a_1, a_2 \in \mathbb{Q}\}$$

and

$$[\mathbb{Q}(\alpha) : \mathbb{Q}] = \deg(\operatorname{irr}_{\mathbb{Q}}(\alpha)) = 3,$$

so that $\mathbb{Q}(\alpha)$ is a cubic extension of $\mathbb{Q}$.

Example 5.5.3 *Let p be a prime number and let $\omega = e^{2\pi i/p} \in \mathbb{C}$. Clearly $\omega^p = e^{2\pi i} = 1$, so that ω is a root of the monic polynomial $x^p - 1 \in \mathbb{Z}[x]$. Thus ω is an algebraic integer. In $\mathbb{Z}[x]$ we have*

$$x^p - 1 = (x - 1)(x^{p-1} + x^{p-2} + \cdots + x + 1).$$

As ω is not a root of $x - 1$, it must be a root of $f_p(x) = x^{p-1} + x^{p-2} + \cdots + x + 1$. We show that $f_p(x)$ is irreducible in $\mathbb{Z}[x]$. We have

$$f_p(x + 1) = \frac{(x + 1)^p - 1}{(x + 1) - 1} = x^{p-1} + \binom{p}{1}x^{p-2} + \binom{p}{2}x^{p-3} + \cdots + \binom{p}{p-1}.$$

As p is a prime the coefficients $\binom{p}{i}$ ($i = 1, 2, \ldots, p - 1$) of $f_p(x + 1)$ are all divisible by p. Moreover, the constant term $\binom{p}{p-1} = p$ is not divisible by p^2. Hence $f_p(x + 1)$ is p-Eisenstein and therefore irreducible in $\mathbb{Z}[x]$. Thus $f_p(x)$ is irreducible in $\mathbb{Z}[x]$, and thus in $\mathbb{Q}[x]$, proving that

$$\operatorname{irr}_{\mathbb{Q}}(\omega) = x^{p-1} + x^{p-2} + \cdots + x + 1$$

and

$$\deg(\operatorname{irr}_{\mathbb{Q}}(\omega)) = p - 1.$$

Thus $\mathbb{Q}(\omega)$ is an extension of $\mathbb{Q}$ of degree $p - 1$. A field such as $\mathbb{Q}(\omega)$, which is formed by adjoining a root of unity to $\mathbb{Q}$, is called a cyclotomic field.

Definition 5.5.3 (Cyclotomic field) *If K is a subfield of $\mathbb{C}$ such that $K = \mathbb{Q}(\omega)$ for some root of unity ω then K is called a cyclotomic field.*

Example 5.5.4 *The quadratic field $\mathbb{Q}(\sqrt{-3}) = \{a + b\sqrt{-3} \mid a, b \in \mathbb{Q}\}$ is a cyclotomic field as $\mathbb{Q}(\sqrt{-3}) = \mathbb{Q}(\omega)$, where ω is a complex cube root of unity.*

5.6 Multiple Extensions

We now consider the field obtained by adjoining several elements $\alpha_1, \ldots, \alpha_k \in \mathbb{C}$ ($k \geq 2$) to a subfield K of $\mathbb{C}$. We denote this field by $K(\alpha_1, \ldots, \alpha_k)$. It is the smallest subfield of $\mathbb{C}$ that contains both K and the α_i; that is, it is the intersection of all the subfields of $\mathbb{C}$ containing both K and the α_i. The field $K(\alpha_1, \ldots, \alpha_k)$ is called a multiple extension of K. Clearly the order of $\alpha_1, \ldots, \alpha_k$ does not matter. The field $K(\alpha_1, \ldots, \alpha_k)$ can be regarded as the field obtained by a succession of k single adjunctions, namely,

$$K(\alpha_1, \alpha_2) = K(\alpha_1)(\alpha_2),$$
$$K(\alpha_1, \alpha_2, \alpha_3) = K(\alpha_1, \alpha_2)(\alpha_3),$$
$$\cdots$$
$$K(\alpha_1, \alpha_2, \ldots, \alpha_k) = K(\alpha_1, \alpha_2, \ldots, \alpha_{k-1})(\alpha_k).$$

When each α_i ($i = 1, 2, \ldots, k$) is algebraic over K, it is an important result that the multiple extension $K(\alpha_1, \ldots, \alpha_k)$ is in fact a simple extension $K(\alpha)$ for a suitable $\alpha \in \mathbb{C}$ that is algebraic over K. We prove this in Theorem 5.6.2 after treating the case $k = 2$ in Theorem 5.6.1.

Theorem 5.6.1 *Let K be a subfield of $\mathbb{C}$. Let $\alpha \in \mathbb{C}$ and $\beta \in \mathbb{C}$ be algebraic over K. Then there exists $\gamma \in \mathbb{C}$ that is algebraic over K such that*

$$K(\alpha, \beta) = K(\gamma).$$

Proof: Let

$$p(x) = \mathrm{irr}_K(\alpha), \quad q(x) = \mathrm{irr}_K(\beta).$$

Then

$$p(x) = (x - \alpha_1) \cdots (x - \alpha_m) \in K[x],$$

where

$$\alpha_1 = \alpha, \alpha_2, \ldots, \alpha_m$$

are the conjugates of α over K, and

$$q(x) = (x - \beta_1) \cdots (x - \beta_n) \in K[x],$$

where

$$\beta_1 = \beta, \beta_2, \ldots, \beta_n$$

are the conjugates of β over K. By Theorem 5.2.1 we know that the α_i are distinct, as are the β_j. The set

$$S = \left\{ \frac{\alpha_r - \alpha_s}{\beta_t - \beta_u} \mid r, s = 1, \ldots, m; \ t, u = 1, \ldots, n; \ t \neq u \right\}$$

consists of a finite number of complex numbers. We choose a rational number c different from all the members of S. With this choice the mn elements

$$\alpha_i + c\beta_j \ (i = 1, \ldots, m; \ j = 1, \ldots, n)$$

are all distinct. Let

$$\gamma = \alpha_1 + c\beta_1 = \alpha + c\beta$$

and set

$$K_1 = K(\gamma).$$

We also let

$$p_1(x) = p(\gamma - cx) \in K_1[x].$$

As

$$p_1(\beta) = p(\gamma - c\beta) = p(\alpha) = 0$$

and

$$q(\beta) = 0$$

we see that β is a common root of $p_1(x)$ and $q(x)$. We show next that these polynomials have no other common roots. Let $\lambda \in \mathbb{C}$ be a common root of $p_1(x)$ and $q(x)$ with $\lambda \neq \beta$. As λ is a root of $q(x)$ different from β, we have $\lambda = \beta_j$ for some j with $2 \leq j \leq n$. Then, as

$$p(\gamma - c\beta_j) = p_1(\beta_j) = 0,$$

$\gamma - c\beta_j$ must be equal to one of $\alpha_1, \ldots, \alpha_m$, say α_k. Hence

$$\alpha_k + c\beta_j = \gamma = \alpha_1 + c\beta_1$$

so that

$$c = \frac{\alpha_1 - \alpha_k}{\beta_j - \beta_1},$$

contradicting the choice of c. Now let $h(x) = \text{irr}_{K_1}(\beta)$. Then $h(x) \mid p_1(x)$ and $h(x) \mid q(x)$. Since $p_1(x)$ and $q(x)$ have exactly one common root in $\mathbb{C}$, we must have $\deg h(x) = 1$. Thus $h(x) = x + \delta$ for some $\delta \in K_1$. Now $0 = h(\beta) = \beta + \delta$ so that $\beta = -\delta \in K_1$. Then $\alpha = \gamma - c\beta \in K_1$. This shows that

$$K(\alpha, \beta) \subseteq K_1 = K(\gamma).$$

Since $\gamma = \alpha + c\beta \in K(\alpha, \beta)$ we have

$$K(\gamma) \subseteq K(\alpha, \beta)$$

and thus

$$K(\alpha, \beta) = K(\gamma). \qquad \blacksquare$$

Theorem 5.6.2 *Let K be a subfield of $\mathbb{C}$. Let $\alpha_1, \alpha_2, \ldots, \alpha_n$ be algebraic over K. Then there exists $\alpha \in \mathbb{C}$ algebraic over K such that*

$$K(\alpha_1, \alpha_2, \ldots, \alpha_n) = K(\alpha).$$

Proof: The result is trivial if $n = 1$, so we may suppose that $n \geq 2$. By Theorem 5.6.1 there exists $\beta_2 \in \mathbb{C}$ algebraic over K such that $K(\alpha_1, \alpha_2) = K(\beta_2)$. Again by Theorem 5.6.1 there exists $\beta_3 \in \mathbb{C}$ algebraic over K such that $K(\alpha_1, \alpha_2, \alpha_3) = K(\beta_2, \alpha_3) = K(\beta_3)$. Continuing in this way, we obtain a finite sequence $\beta_2, \beta_3, \ldots, \beta_n$ of complex numbers, each algebraic over K, such that

$$\begin{aligned}
K(\alpha_1, \alpha_2, \ldots, \alpha_n) &= K(\beta_2, \alpha_3, \ldots, \alpha_n) \\
&= K(\beta_3, \alpha_4, \ldots, \alpha_n) \\
&= \ldots \\
&= K(\beta_{n-1}, \alpha_n) \\
&= K(\beta_n). \qquad \blacksquare
\end{aligned}$$

If K is a subfield of $\mathbb{C}$, and $\alpha \in \mathbb{C}$ and $\beta \in \mathbb{C}$ are algebraic over K, the proof of Theorem 5.6.1 shows how to find $\gamma \in \mathbb{C}$ algebraic over K such that $K(\alpha, \beta) = K(\gamma)$. We have only to find a rational number c such that the elements $\alpha' + c\beta'$ are all distinct as α' ranges over the conjugates of α over K and β' ranges over the conjugates of β over K. Then $K(\alpha, \beta) = K(\alpha + c\beta)$. We illustrate this in the next two examples.

Example 5.6.1 *We express $\mathbb{Q}(\sqrt{2}, \sqrt{3})$ as a simple extension. The conjugates of $\sqrt{2}$ over $\mathbb{Q}$ are $\sqrt{2}$ and $-\sqrt{2}$. The conjugates of $\sqrt{3}$ over $\mathbb{Q}$ are $\sqrt{3}$ and $-\sqrt{3}$. The four numbers*

$$\sqrt{2} + \sqrt{3}, \ \sqrt{2} - \sqrt{3}, \ -\sqrt{2} + \sqrt{3}, \ -\sqrt{2} - \sqrt{3}$$

are all distinct, so by Theorem 5.6.1 we have

$$\mathbb{Q}(\sqrt{2}, \sqrt{3}) = \mathbb{Q}(\sqrt{2} + \sqrt{3}).$$

Set

$$\alpha = \sqrt{2} + \sqrt{3} \in \mathbb{R}.$$

Squaring α we obtain

$$\alpha^2 = 5 + 2\sqrt{6},$$

so that

$$\alpha^2 - 5 = 2\sqrt{6}.$$

Squaring $\alpha^2 - 5$ we get

$$\alpha^4 - 10\alpha^2 + 25 = 24.$$

Thus α is a root of the monic quartic polynomial

$$f(x) = x^4 - 10x^2 + 1 \in \mathbb{Z}[x].$$

This shows that α is an algebraic integer.

   We now show that $f(x)$ is irreducible in $\mathbb{Z}[x]$ and thus in $\mathbb{Q}[x]$. Since $f(\pm 1) = -8 \neq 0$, $f(x)$ has no linear factors in $\mathbb{Z}[x]$. Thus if $f(x)$ factors in $\mathbb{Z}[x]$, it must factor as a product of two quadratic polynomials in $\mathbb{Z}[x]$, say,

$$x^4 - 10x^2 + 1 = (x^2 + ax + b)(x^2 + cx + d),$$

where $a, b, c, d \in \mathbb{Z}$. Equating coefficients of x^3, x^2, x, and 1, we obtain

$$a + c = 0,$$
$$b + ac + d = -10,$$
$$bc + ad = 0,$$
$$bd = 1.$$

From the first equation we have $c = -a$, so the second equation becomes

$$b + d + 10 = a^2.$$

From the last equation we have $b = d = \pm 1$, so that $b + d = \pm 2$. Hence $a^2 = 8$ or 12, which is impossible. This proves that $x^4 - 10x^2 + 1$ is irreducible in $\mathbb{Q}[x]$ and so

$$\mathrm{irr}_{\mathbb{Q}}(\sqrt{2} + \sqrt{3}) = x^4 - 10x^2 + 1$$

and

$$[\mathbb{Q}(\sqrt{2}, \sqrt{3}) : \mathbb{Q}] = [\mathbb{Q}(\sqrt{2} + \sqrt{3}) : \mathbb{Q}] = 4.$$

Example 5.6.2   *We express $\mathbb{Q}(\sqrt{3}, \sqrt[3]{2})$ as a simple extension. The conjugates of $\sqrt{3}$ are $\sqrt{3}$ and $-\sqrt{3}$. The conjugates of $\sqrt[3]{2}$ are $\sqrt[3]{2}$, $\omega\sqrt[3]{2}$, and $\omega^2\sqrt[3]{2}$, where ω is* *a complex cube root of unity. The six numbers*

$$\sqrt{3} + \sqrt[3]{2}, \ -\sqrt{3} + \sqrt[3]{2}, \ \sqrt{3} + \omega\sqrt[3]{2},$$
$$-\sqrt{3} + \omega\sqrt[3]{2}, \ \sqrt{3} + \omega^2\sqrt[3]{2}, \ -\sqrt{3} + \omega^2\sqrt[3]{2}$$

are all distinct, so by Theorem 5.6.1 we have

$$\mathbb{Q}(\sqrt{3}, \sqrt[3]{2}) = \mathbb{Q}(\sqrt{3} + \sqrt[3]{2}).$$

We conclude this chapter by proving the very important fact that every element of a simple extension $K(\alpha)$ of a subfield K of $\mathbb{C}$, where α is algebraic over K, is algebraic over K.

Theorem 5.6.3 *Let K be a subfield of $\mathbb{C}$. Let $\alpha \in \mathbb{C}$ be algebraic over K. Then every element β of $K(\alpha)$ is algebraic over K, and the degree of β over K is less than or equal to the degree of α over K.*

Proof: Let $\beta \in K(\alpha)$, where α is algebraic over K. Let $n = \deg(\mathrm{irr}_K(\alpha))$. By Theorem 5.5.1 each of the powers β^j, $j = 0, 1, \ldots, n$, of β can be written in the form

$$\beta^j = \sum_{k=0}^{n-1} a_{jk}\alpha^k,$$

where each $a_{jk} \in K$. The homogeneous system of linear equations

$$\sum_{j=0}^{n} a_{jk}x_j = 0, \ k = 0, 1, \ldots, n-1,$$

has a solution $(x_0, x_1, \ldots, x_n) \in K^{n+1}$ with not all of the x_j equal to zero, as the number of unknowns is greater than the number of equations. Then

$$\sum_{j=0}^{n} x_j\beta^j = \sum_{j=0}^{n} x_j \sum_{k=0}^{n-1} a_{jk}\alpha^k = \sum_{k=0}^{n-1} \alpha^k \sum_{j=0}^{n} a_{jk}x_j = 0,$$

proving that β is algebraic over K and that the degree of β over K is less than or equal to the degree of α over K. ∎

Exercises

1. Prove that the set $I_K(\alpha)$ defined in Section 5.1 is an ideal.
2. Prove that $x^4 + 1$ is irreducible in $\mathbb{Q}[x]$ (see Example 5.1.1).
3. Prove that $x^2 - \sqrt{2}x + 1$ is irreducible in $\mathbb{Q}(\sqrt{2})[x]$ (see Example 5.1.2).

4. Determine

$$\text{irr}_{\mathbb{Q}(i)}\left(\frac{1+i}{\sqrt{2}}\right)$$

and

$$\text{irr}_{\mathbb{Q}(\sqrt{-2})}\left(\frac{1+i}{\sqrt{2}}\right).$$

5. Prove that $[\mathbb{Q}(\sqrt{3}+\sqrt[3]{2}):\mathbb{Q}]=6$ (see Example 5.6.2).
6. Prove that $\mathbb{Q}(\sqrt{2},i)=\mathbb{Q}(\sqrt{2}+i)$.
7. Prove that $\mathbb{Q}(\sqrt{2},i\sqrt{2})=\mathbb{Q}(\sqrt{2}+i\sqrt{2})$.
8. Find the minimal polynomial of $2^{1/3}+\omega$ over $\mathbb{Q}(2^{1/3})$, where ω is a complex cube root of unity.
9. Determine $\alpha \in \mathbb{C}$ such that

$$\mathbb{Q}(\sqrt{2},\sqrt{3},\sqrt{5})=\mathbb{Q}(\alpha).$$

10. Prove that

$$[\mathbb{Q}(\sqrt{2},\sqrt{3},\sqrt{5}):\mathbb{Q}]=8.$$

11. Determine the conjugates of $3^{1/3}-3^{2/3}$.
12. Let $\theta \in \mathbb{C}$ be a root of $x^3+11x+4=0$. Prove that $[\mathbb{Q}(\theta):\mathbb{Q}]=3$.
13. Prove that $(-\theta+\theta^2)/2$ is an algebraic integer in $K=\mathbb{Q}(\theta)$, where $\theta^3+11\theta-4=0$.
14. Let $\theta \in \mathbb{C}$ be a root of $x^5+x+1=0$. If $\theta \notin \mathbb{Q}(\sqrt{-3})$, what is $\text{irr}_{\mathbb{Q}}\theta$?
15. Let $\omega=e^{2\pi i/5}$. Prove that

$$\omega=\frac{1}{4}\left(\sqrt{5}-1+i\sqrt{10+2\sqrt{5}}\right).$$

16. Let $\omega=e^{2\pi i/5}$. Show that $\sqrt{5}\in\mathbb{Q}(\omega)$ by expressing $\sqrt{5}$ in the form

$$\sqrt{5}=a\omega+b\omega^2+c\omega^3+d\omega^4$$

for suitable integers a,b,c,d.

17. Prove that

$$\frac{1}{2}\left(i\sqrt{10+2\sqrt{5}}+2i\sqrt{10-2\sqrt{5}}\right)$$

is an algebraic integer in $\mathbb{Q}(e^{2\pi i/5})$.

18. Determine the conjugates of

$$12^{1/5}+54^{1/5}-144^{1/5}+648^{1/5}$$

over $\mathbb{Q}$.

19. Let m be a squarefree integer $\equiv 1 \pmod 4$. Let $A=\mathbb{Z}+\mathbb{Z}\sqrt{m}$ and $B=\mathbb{Q}(\sqrt{m})$. Prove that

$$A^B=\mathbb{Z}+\mathbb{Z}\left(\frac{1+\sqrt{m}}{2}\right).$$

20. Let θ be a nonreal algebraic number. Prove that the complex conjugate $\bar{\theta}$ of θ is one of the conjugates of θ over $\mathbb{Q}$.

21. Let p be an odd prime. Let a and c be integers with $a \equiv 1 \pmod 2$ and $\left(\frac{a^2 - 4c}{p} \right) = -1$. Prove that $x^4 + ax^2 + px + c$ is irreducible in $\mathbb{Z}[x]$.

22. Use Exercise 21 to prove that $[\mathbb{Q}(\theta) : \mathbb{Q}] = 4$, where θ is a root of $x^4 + 7x^2 + 5x + 4 = 0$.

Suggested Reading

1. E. R. Scheinerman, *When close enough is close enough*, American Mathematical Monthly 107 (2000), 489–499.

   A technique is presented for proving identities involving algebraic integers numerically. For example Shanks's identity [2]

   $$\sqrt{5} + \sqrt{22 + 2\sqrt{5}} = \sqrt{11 + 2\sqrt{29}} + \sqrt{16 - 2\sqrt{29} + 2\sqrt{55 - 10\sqrt{29}}}$$

   can be proved using this technique.

2. D. Shanks, *Incredible identities*, Fibonacci Quarterly 12 (1974), 271, 280.

Biographies

1. H. C. Williams, *Daniel Shanks (1917–1996)*, Notices of the American Mathematical Society 44 (1997), 813–816.

2. H. C. Williams, *Daniel Shanks (1917–1996)*, Mathematics of Computation 66 (1997), 929–934.

6

Algebraic Number Fields

6.1 Algebraic Number Fields

An algebraic number field is a field K that is obtained from the field of rational numbers $\mathbb{Q}$ by adjoining a finite number of algebraic numbers.

Definition 6.1.1 (Algebraic number field) *An algebraic number field is a subfield of $\mathbb{C}$ of the form $\mathbb{Q}(\alpha_1, \ldots, \alpha_n)$, where $\alpha_1, \ldots, \alpha_n$ are algebraic numbers.*

Example 6.1.1 *$\mathbb{Q}(\sqrt{2}, \sqrt{3}, \sqrt{7})$; $\mathbb{Q}(\sqrt[7]{1+i}, \theta)$, where θ is a root of the polynomial $x^5 - x + 1$; and $\mathbb{Q}(\sqrt[3]{1 + \sqrt{2}} + \sqrt[3]{1 - \sqrt{2}}, \sqrt{53} + \sqrt[3]{5})$ are all examples of algebraic number fields.*

By Theorem 5.6.2 an algebraic number field can be obtained by adjoining a single algebraic number θ to $\mathbb{Q}$.

Theorem 6.1.1 *If K is an algebraic number field then there exists an algebraic number θ such that $K = \mathbb{Q}(\theta)$.*

Proof: This is the special case $K = \mathbb{Q}$ of Theorem 5.6.2. ∎

In fact the algebraic number θ in Theorem 6.1.1 can always be taken to be an algebraic integer.

Theorem 6.1.2 *If K is an algebraic number field then there is an algebraic integer θ such that $K = \mathbb{Q}(\theta)$.*

Proof: Let K be an algebraic number field. By Theorem 6.1.1 there is an algebraic number ϕ such that $K = \mathbb{Q}(\phi)$. By Theorem 4.2.6 we have $\phi = \theta/b$, where θ is an algebraic integer and b is a nonzero rational integer. Thus

$$K = \mathbb{Q}(\phi) = \mathbb{Q}(\theta/b) = \mathbb{Q}(\theta).$$
∎

Example 6.1.2 *We show that the algebraic number field* $\mathbb{Q}(\sqrt{2}, \sqrt{3})$ *is* $\mathbb{Q}(\theta)$, *where* $\theta = \sqrt{2} + \sqrt{3}$. *This was done in a different way in Example 5.6.1. Clearly* $\mathbb{Q}(\theta) \subseteq \mathbb{Q}(\sqrt{2}, \sqrt{3})$. *As*

$$\sqrt{2} = -\frac{9}{2}\theta + \frac{1}{2}\theta^3 \in \mathbb{Q}(\theta)$$

and

$$\sqrt{3} = \frac{11}{2}\theta - \frac{1}{2}\theta^3 \in \mathbb{Q}(\theta),$$

we see that $\mathbb{Q}(\sqrt{2}, \sqrt{3}) \subseteq \mathbb{Q}(\sqrt{2} + \sqrt{3})$. *Hence* $\mathbb{Q}(\sqrt{2}, \sqrt{3}) = \mathbb{Q}(\sqrt{2} + \sqrt{3})$.

The form of the elements in an algebraic number field follows immediately from Theorem 5.5.1. We have

Theorem 6.1.3 *Let* $K = \mathbb{Q}(\theta)$ *be an algebraic number field, where* θ *is an algebraic number. Let the degree of the polynomial* $\mathrm{irr}_{\mathbb{Q}}(\theta)$ *be* n. *Then every element of* K *is expressible uniquely in the form*

$$c_0 + c_1\theta + \cdots + c_{n-1}\theta^{n-1},$$

where $c_0, c_1, \ldots, c_{n-1} \in \mathbb{Q}$, *and every such quantity* $c_0 + c_1\theta + \cdots + c_{n-1}\theta^{n-1}$ $(c_0, c_1, \ldots, c_{n-1} \in \mathbb{Q})$ *belongs to* K.

Clearly K is an n-dimensional vector space over $\mathbb{Q}$ and the degree of K over $\mathbb{Q}$ is n. K is called a quadratic field if $n = 2$, a cubic field if $n = 3$, a quartic field if $n = 4$, and a quintic field if $n = 5$.

Definition 6.1.2 (The set O_K) *The set of all algebraic integers that lie in the algebraic number field* K *is denoted by* O_K; *that is,*

$$O_K = \Omega \cap K.$$

Theorem 6.1.4 *Let* K *be an algebraic number field. Then* O_K *is an integral domain.*

Proof: By Theorem 4.1.8 we know that Ω is an integral domain ($\subseteq \mathbb{C}$). Hence, as $K (\subseteq \mathbb{C})$ is a field, $O_K = \Omega \cap K$ is an integral domain. ∎

Definition 6.1.3 (Ring of integers of an algebraic number field) O_K *is called the ring of integers of the algebraic number field* K.

Example 6.1.3 *In Theorem 5.4.2 we determined the ring of integers* O_K *of a quadratic field* K. *Taking* $m = -1$ *and* $m = -3$ *in Theorem 5.4.2, we see that*

$$O_{\mathbb{Q}(\sqrt{-1})} = \mathbb{Z} + \mathbb{Z}\sqrt{-1}$$

and

$$O_{\mathbb{Q}(\sqrt{-3})} = \mathbb{Z} + \mathbb{Z} \left(\frac{1 + \sqrt{-3}}{2} \right).$$

Thus the Gaussian domain $\mathbb{Z} + \mathbb{Z}\sqrt{-1}$ (Example 1.1.2) is the ring of integers of the quadratic field $\mathbb{Q}(\sqrt{-1})$ and the Eisenstein domain $\mathbb{Z} + \mathbb{Z}\omega = \mathbb{Z} + \mathbb{Z}\left(\frac{-1+\sqrt{-3}}{2}\right) = \mathbb{Z} + \mathbb{Z}\left(\frac{1+\sqrt{-3}}{2}\right)$ (Example 1.1.3) is the ring of integers of the quadratic field $\mathbb{Q}(\sqrt{-3})$.

When K is not a quadratic field, it is a more difficult problem to determine O_K. We determine O_K for some algebraic number fields of degree > 2 in Chapter 7. Indeed it is an area of current research to determine O_K explicitly for certain classes of algebraic number fields K. See the references at the end of Chapter 7 in this connection.

Theorem 6.1.5 *If K is an algebraic number field then the quotient field of O_K is K.*

Proof: Let F denote the quotient field of O_K, and let $\alpha \in F$. Then $\alpha = b/c$, where $b \in O_K$ and $c \in O_K$ with $c \neq 0$. As $O_K \subseteq K$ we have $b \in K$ and $c \in K$ so that, as K is a field, $\alpha = b/c \in K$. Hence $F \subseteq K$.

Now let $\alpha \in K$. By Theorem 4.2.6 we have $\alpha = b/c$, where b is an algebraic integer and c is a nonzero rational integer. Clearly $b = \alpha c \in K$ so $b \in O_K$. Thus $\alpha = b/c \in F$. Hence $K \subseteq F$.

This proves that $F = K$, so the quotient field of O_K is K. ∎

Theorem 6.1.6 *If K is an algebraic number field then O_K is integrally closed.*

Proof: By Theorem 6.1.5, the quotient field of O_K is K. Let $\beta \in K$ be integral over O_K. As O_K is integral over $\mathbb{Z}$, by Theorem 4.1.11 β is integral over $\mathbb{Z}$, that is, β is an algebraic integer in K. Hence $\beta \in O_K$. This proves that O_K is integrally closed. ∎

For an algebraic number field K, we showed in Theorem 6.1.2 that there is an algebraic integer θ such that $K = \mathbb{Q}(\theta)$. As $\theta \in K$ and $\theta \in \Omega$ we see that $\theta \in \Omega \cap K = O_K$. We now wish to show that θ can be taken from any given nonzero ideal of O_K (Theorem 6.1.8). To prove this we make use of the next result (Theorem 6.1.7), which asserts that if I is a nonzero ideal of O_K then $I \cap \mathbb{Z}$ always contains a nonzero integer.

Theorem 6.1.7 *Let K be an algebraic number field. Then every nonzero ideal in O_K contains a nonzero rational integer.*

Proof: Let $I \neq \{0\}$ be an ideal in O_K. Choose $\alpha \in I$ with $\alpha \neq 0$. As $\alpha \in I \subseteq O_K$, α is an algebraic integer. Let $\mathrm{irr}_{\mathbb{Q}}(\alpha) = x^n + b_1 x^{n-1} + \cdots + b_n$. We show that $b_n \neq 0$. If $n = 1$ then $\mathrm{irr}_{\mathbb{Q}}(\alpha) = x + b_1$ so that $\alpha + b_1 = 0$. Hence $b_1 = -\alpha \neq 0$. If $n \geq 2$ then $b_n \neq 0$ as $\mathrm{irr}_{\mathbb{Q}}(\alpha)$ is irreducible in $\mathbb{Q}[x]$ by Theorem 5.1.1. By Theorem 5.3.2 we know that $\mathrm{irr}_{\mathbb{Q}}(\alpha) \in \mathbb{Z}[x]$ as α is an algebraic integer. Hence $b_1, \ldots, b_n \in \mathbb{Z}$. Thus $b_n = -\alpha^n - b_1 \alpha^{n-1} - \cdots - b_1 \alpha \in I$. Hence b_n is a nonzero rational integer in I. ∎

Theorem 6.1.8 *Let K be an algebraic number field. Let I be a nonzero ideal of O_K. Then there exists $\gamma \in I$ such that $K = \mathbb{Q}(\gamma)$.*

Proof: By Theorem 6.1.2 there exists $\theta \in O_K$ such that $K = \mathbb{Q}(\theta)$. By Theorem 6.1.7 there exists $c \in \mathbb{Z} \cap I$ with $c \neq 0$. Set $\gamma = c\theta$. As $\theta \in O_K$ and $c \in I$ we have $\gamma \in I$. Moreover, as $c \in \mathbb{Z} \setminus \{0\}$, we have $K = \mathbb{Q}(\theta) = \mathbb{Q}(c\theta) = \mathbb{Q}(\gamma)$, where $\gamma \in I$. ∎

6.2 Conjugate Fields of an Algebraic Number Field

Let K be an algebraic number field. In this section we begin by determining the number of monomorphisms $\sigma : K \to \mathbb{C}$. For example, if $K = \mathbb{Q}(\sqrt{2})$ then

$$\sigma_1(x + y\sqrt{2}) = x + y\sqrt{2} \quad (x, y \in \mathbb{Q})$$

and

$$\sigma_2(x + y\sqrt{2}) = x - y\sqrt{2} \quad (x, y \in \mathbb{Q})$$

are two monomorphisms from K to $\mathbb{C}$.

Theorem 6.2.1 *Let K be an algebraic number field of degree n over $\mathbb{Q}$. Then there are exactly n distinct monomorphisms $\sigma_k : K \to \mathbb{C}$ $(k = 1, \ldots, n)$.*

Proof: By Theorem 6.1.1 there exists an algebraic number $\theta \in K$ such that $K = \mathbb{Q}(\theta)$. Let $p(x) = \mathrm{irr}_{\mathbb{Q}}(\theta)$. Then

$$\deg p(x) = \deg (\mathrm{irr}_{\mathbb{Q}}(\theta)) = [\mathbb{Q}(\theta) : \mathbb{Q}] = n,$$

so that θ has n distinct conjugates over $\mathbb{Q}$ (Theorem 5.2.1), say $\theta_1 = \theta, \theta_2, \ldots, \theta_n$, and

$$p(x) = (x - \theta_1)(x - \theta_2) \cdots (x - \theta_n).$$

By Theorem 6.1.3 each element α of K can be expressed uniquely in the form $\alpha = a_0 + a_1 \theta + \cdots + a_{n-1} \theta^{n-1}$, where $a_0, a_1, \ldots, a_{n-1} \in \mathbb{Q}$, so, for $k = 1, 2, \ldots, n$, we can define

$$\sigma_k : K \to \mathbb{C}$$

by

$$\sigma_k(a_0 + a_1\theta + \cdots + a_{n-1}\theta^{n-1}) = a_0 + a_1\theta_k + \cdots + a_{n-1}\theta_k^{n-1}.$$

We show that σ_k $(k = 1, 2, \ldots, n)$ is a field homomorphism.

First we show that σ_k $(k = 1, 2, \ldots, n)$ is additive. Let $\alpha, \beta \in K$. Then

$$\alpha = a_0 + a_1\theta + \cdots + a_{n-1}\theta^{n-1}$$

and

$$\beta = b_0 + b_1\theta + \cdots + b_{n-1}\theta^{n-1},$$

where $a_0, a_1, \ldots, a_{n-1}, b_0, b_1, \ldots, b_{n-1} \in \mathbb{Q}$. Hence

$$\alpha + \beta = (a_0 + b_0) + (a_1 + b_1)\theta + \cdots + (a_{n-1} + b_{n-1})\theta^{n-1}$$

and so

$$\begin{aligned}
\sigma_k(\alpha + \beta) &= (a_0 + b_0) + (a_1 + b_1)\theta_k + \cdots + (a_{n-1} + b_{n-1})\theta_k^{n-1} \\
&= (a_0 + a_1\theta_k + \cdots + a_{n-1}\theta_k^{n-1}) + (b_0 + b_1\theta_k + \cdots + b_{n-1}\theta_k^{n-1}) \\
&= \sigma_k(\alpha) + \sigma_k(\beta).
\end{aligned}$$

Thus σ_k is additive.

Next we show that σ_k $(k = 1, 2, \ldots, n)$ is multiplicative. With the same notation, we let

$$f(x) = a_0 + a_1 x + \cdots + a_{n-1} x^{n-1} \in \mathbb{Q}[x]$$

and

$$g(x) = b_0 + b_1 x + \cdots + b_{n-1} x^{n-1} \in \mathbb{Q}[x]$$

so that

$$f(\theta) = \alpha, \quad g(\theta) = \beta.$$

Dividing $f(x)g(x)$ by $p(x)$ in $\mathbb{Q}[x]$, we obtain a quotient $q(x) \in \mathbb{Q}[x]$ and a remainder $r(x) \in \mathbb{Q}[x]$ such that

$$f(x)g(x) = p(x)q(x) + r(x), \quad \deg r(x) < \deg p(x) = n.$$

Hence, as $p(\theta) = 0$, we have

$$\alpha\beta = f(\theta)g(\theta) = p(\theta)q(\theta) + r(\theta) = r(\theta).$$

Thus, as $p(\theta_k) = 0$, we have

$$\sigma_k(\alpha\beta) = \sigma_k(r(\theta)) = r(\theta_k) = p(\theta_k)q(\theta_k) + r(\theta_k) = f(\theta_k)g(\theta_k) = \sigma_k(\alpha)\sigma_k(\beta),$$

so that σ_k is multiplicative.

Hence we have shown that σ_k $(k = 1, 2, \ldots, n)$ is a homomorphism.

We now show that σ_k $(k = 1, 2, \ldots, n)$ is injective so that it is a monomorphism. Suppose $\alpha = a_0 + a_1\theta + \cdots + a_{n-1}\theta^{n-1} \in K$ and $\beta = b_0 + b_1\theta + \cdots + b_{n-1}\theta^{n-1} \in K$ are such that $\sigma_k(\alpha) = \sigma_k(\beta)$. Then we have

$$a_0 + a_1\theta_k + \cdots + a_{n-1}\theta_k^{n-1} = b_0 + b_1\theta_k + \cdots + b_{n-1}\theta_k^{n-1}$$

so that θ_k is a root of the polynomial

$$(a_0 - b_0) + (a_1 - b_1)x + \cdots + (a_{n-1} - b_{n-1})x^{n-1} \in \mathbb{Q}[x]$$

of degree $< n$. As the $\deg(\mathrm{irr}_{\mathbb{Q}}\theta_k) = \deg p(x) = n$, this polynomial must be the zero polynomial so that

$$a_0 - b_0 = a_1 - b_1 = \cdots = a_{n-1} - b_{n-1} = 0;$$

that is,

$$a_0 = b_0, \ a_1 = b_1, \ldots, a_{n-1} = b_{n-1},$$

and so

$$\alpha = a_0 + a_1\theta + \cdots + a_{n-1}\theta^{n-1} = b_0 + b_1\theta + \cdots + b_{n-1}\theta^{n-1} = \beta,$$

proving that σ_k is injective.

Finally, let $\lambda : K \to \mathbb{C}$ be a monomorphism. Then

$$p(\lambda(\theta)) = \lambda(p(\theta)) = \lambda(0) = 0$$

so that

$$\lambda(\theta) = \theta_k$$

for some $k \in \{1, 2, \ldots, n\}$. Thus

$$\lambda(\theta) = \sigma_k(\theta)$$

and so

$$\lambda(a_0 + a_1\theta + \cdots + a_{n-1}\theta^{n-1}) = a_0 + a_1\theta_k + \cdots + a_{n-1}\theta_k^{n-1}$$
$$= \sigma_k(a_0 + a_1\theta + \cdots + a_{n-1}\theta^{n-1})$$

for all $a_0, a_1, \ldots, a_{n-1} \in \mathbb{Q}$, proving that

$$\lambda = \sigma_k.$$

Hence $\{\sigma_k \mid k = 1, 2, \ldots, n\}$ comprise all the monomorphisms from K to $\mathbb{C}$. ∎

For $k = 1, 2, \ldots, n$, we have

$$\text{range } \sigma_k = \sigma_k(K)$$
$$= \{\sigma_k(a_0 + a_1\theta + \cdots + a_{n-1}\theta^{n-1}) \mid a_0, a_1, \ldots, a_{n-1} \in \mathbb{Q}\}$$
$$= \{a_0 + a_1\theta_k + \cdots + a_{n-1}\theta_k^{n-1} \mid a_0, a_1, \ldots, a_{n-1} \in \mathbb{Q}\}$$
$$= \mathbb{Q}(\theta_k)$$

so that

$$\sigma_k : \mathbb{Q}(\theta) \to \mathbb{Q}(\theta_k)$$

is an isomorphism. Hence all the fields $\mathbb{Q}(\theta_k)$ $(k = 1, 2, \ldots, n)$ are isomorphic.

Definition 6.2.1 (Conjugate fields of an algebraic number field) *Let K be an algebraic number field. Let θ be an algebraic number such that $K = \mathbb{Q}(\theta)$. Let*

$$\theta_1 = \theta, \theta_2, \ldots, \theta_n$$

be the conjugates of θ over $\mathbb{Q}$. Then the fields

$$\mathbb{Q}(\theta_1) = \mathbb{Q}(\theta) = K, \mathbb{Q}(\theta_2), \ldots, \mathbb{Q}(\theta_n)$$

are called the conjugate fields of K.

By the remarks preceding Definition 6.2.1 each of the conjugate fields of K is isomorphic to K.

It appears from the definition that the conjugate fields of K may depend upon the choice of algebraic number θ such that $K = \mathbb{Q}(\theta)$. We show that this is in fact not the case.

Theorem 6.2.2 *Let K be an algebraic number field. Let θ be an algebraic number such that $K = \mathbb{Q}(\theta)$. Let $\theta_1 = \theta, \theta_2, \ldots, \theta_n$ be the conjugates of θ. Let ϕ be another algebraic number such that $K = \mathbb{Q}(\phi)$. Let $c_0, c_1, \ldots, c_{n-1} \in \mathbb{Q}$ be such that*

$$\phi = c_0 + c_1\theta + \cdots + c_{n-1}\theta^{n-1}.$$

For $k = 1, 2, \ldots, n$ set

$$\phi_k = c_0 + c_1\theta_k + \cdots + c_{n-1}\theta_k^{n-1}$$

so that $\phi_1 = \phi$. Then $\phi_1, \phi_2, \ldots, \phi_n$ are the conjugates of ϕ over $\mathbb{Q}$, and

$$\mathbb{Q}(\theta_k) = \mathbb{Q}(\phi_k), \ k = 1, 2, \ldots, n.$$

Proof: Let

$$f(x) = \prod_{k=1}^{n} (x - \phi_k) = \prod_{k=1}^{n} (x - (c_0 + c_1\theta_k + \cdots + c_{n-1}\theta_k^{n-1})) \in K_1[x],$$

where K_1 is the algebraic number field given by

$$K_1 = \mathbb{Q}(\theta_1, \theta_2, \ldots, \theta_n).$$

Clearly $\mathbb{Q} \subseteq K \subseteq K_1 \subseteq \mathbb{C}$. The coefficients of $f(x)$ are (up to sign) the elementary symmetric polynomials in $c_0 + c_1\theta_k + \cdots + c_{n-1}\theta_k^{n-1}$ $(k = 1, 2, \ldots, n)$ and so are polynomials with rational coefficients in the elementary symmetric polynomials in $\theta_1, \theta_2, \ldots, \theta_n$. Since $\theta_1 + \theta_2 + \cdots + \theta_n$, $\theta_1\theta_2 + \cdots + \theta_{n-1}\theta_n, \ldots, \theta_1\theta_2\cdots\theta_n$ are (up to sign) the coefficients of $\text{irr}_{\mathbb{Q}}(\theta) \in \mathbb{Q}[x]$, they are all rational numbers, and so the coefficients of $f(x)$ are all rational. Hence $f(x) \in \mathbb{Q}[x]$. As $f(\phi) = 0$ we have $\text{irr}_{\mathbb{Q}}(\phi) \mid f(x)$, say $f(x) = \text{irr}_{\mathbb{Q}}(\phi)g(x)$, where $g(x) \in \mathbb{Q}[x]$. Then

$$n = \deg f(x) = \deg(\text{irr}_{\mathbb{Q}}(\phi)g(x)) = \deg(\text{irr}_{\mathbb{Q}}(\phi)) + \deg g(x).$$

Now $\deg(\text{irr}_{\mathbb{Q}}(\phi)) = [\mathbb{Q}(\phi) : \mathbb{Q}] = [K : \mathbb{Q}] = n$, so that $\deg g(x) = 0$; that is, $g(x) \in \mathbb{Q}$, say $g(x) = c$. Since $f(x) = c\ \text{irr}_{\mathbb{Q}}(\phi)$ and both $f(x)$ and $\text{irr}_{\mathbb{Q}}(\phi)$ are monic polynomials of degree n, we have $c = 1$. Thus $f(x) = \text{irr}_{\mathbb{Q}}(\phi)$. Hence $\phi_1, \phi_2, \ldots, \phi_n$ are the conjugates of ϕ over $\mathbb{Q}$.

Finally, for $k = 1, 2, \ldots, n$, we have

$$\mathbb{Q}(\phi_k) = \mathbb{Q}(c_0 + c_1\theta_k + \cdots + c_{n-1}\theta_k^{n-1}) \subseteq \mathbb{Q}(\theta_k)$$

and

$$[\mathbb{Q}(\phi_k) : \mathbb{Q}] = [\mathbb{Q}(\theta_k) : \mathbb{Q}]\ (= n)$$

so that

$$\mathbb{Q}(\phi_k) = \mathbb{Q}(\theta_k),\quad k = 1, 2, \ldots, n. \qquad \blacksquare$$

6.3 The Field Polynomial of an Element of an Algebraic Number Field

Let K be an algebraic number field of degree n over $\mathbb{Q}$. Let $\theta \in K$ be such that $K = \mathbb{Q}(\theta)$. Let $\theta_1 = \theta, \theta_2, \ldots, \theta_n$ be the conjugates of θ over $\mathbb{Q}$.

For $\alpha \in K$ there exist unique rational numbers $c_0, c_1, \ldots, c_{n-1}$ such that

$$\alpha = c_0 + c_1\theta + \cdots + c_{n-1}\theta^{n-1}$$

(see Theorem 6.1.3). For $k = 1, 2, \ldots, n$ we set

$$\alpha_k = c_0 + c_1\theta_k + \cdots + c_{n-1}\theta_k^{n-1} \in \mathbb{Q}(\theta_k).$$

Definition 6.3.1 (Complete set of conjugates of α relative to K) *The set of algebraic numbers $\{\alpha_1 = \alpha, \alpha_2, \ldots, \alpha_n\}$ is called a complete set of conjugates of α relative to K. More briefly they are called the "K-conjugates of α" or the "conjugates of α relative to K."*

Example 6.3.1 *Let $K = \mathbb{Q}(\theta)$, where $\theta = \sqrt{2} + \sqrt{3}$. From Example 5.6.1 we see that*

$$\text{irr}_{\mathbb{Q}}(\theta) = x^4 - 10x^2 + 1.$$

As

$$x^4 - 10x^2 + 1 = (x - \sqrt{2} - \sqrt{3})(x - \sqrt{2} + \sqrt{3})(x + \sqrt{2} - \sqrt{3})(x + \sqrt{2} + \sqrt{3})$$

the conjugates of θ are

$$\theta_1 = \sqrt{2} + \sqrt{3}, \ \theta_2 = \sqrt{2} - \sqrt{3}, \ \theta_3 = -\sqrt{2} + \sqrt{3}, \ \theta_4 = -\sqrt{2} - \sqrt{3}.$$

Let $\alpha = 2\sqrt{3}$ *so that* $\alpha \in \mathbb{Q}(\sqrt{3}) \subset \mathbb{Q}(\sqrt{2}, \sqrt{3}) = \mathbb{Q}(\sqrt{2} + \sqrt{3}) = \mathbb{Q}(\theta) = K$
(Example 5.6.1). Hence $\alpha = a + b\theta + c\theta^2 + d\theta^3$ *for some* $a, b, c, d \in \mathbb{Q}$. *Thus*

$$2\sqrt{3} = a + b(\sqrt{2} + \sqrt{3}) + c(\sqrt{2} + \sqrt{3})^2 + d(\sqrt{2} + \sqrt{3})^3$$
$$= (a + 5c) + (b + 11d)\sqrt{2} + (b + 9d)\sqrt{3} + 2c\sqrt{6}.$$

Hence

$$a + 5c = 0, \ b + 11d = 0, \ b + 9d = 2, \ 2c = 0,$$

so that

$$a = 0, \ b = 11, \ c = 0, \ d = -1,$$

giving

$$\alpha = 11\theta - \theta^3.$$

The K-conjugates of α *are*

$$\alpha_1 = \alpha = 11\theta - \theta^3 = 2\sqrt{3},$$
$$\alpha_2 = 11\theta_2 - \theta_2^3 = -2\sqrt{3},$$
$$\alpha_3 = 11\theta_3 - \theta_3^3 = 2\sqrt{3},$$
$$\alpha_4 = 11\theta_4 - \theta_4^3 = -2\sqrt{3}.$$

Thus the complete set of conjugates of α *relative to K is* $\alpha, -\alpha, \alpha, -\alpha$.

The conjugates of α relative to K are obtained from α by applying the monomor-phisms $\sigma_k : K \to \mathbb{C}$ $(k = 1, 2, \ldots, n)$ to α. Clearly $\sigma_k(\alpha) = \alpha_k$ $(k = 1, 2, \ldots, n)$ and $\alpha_k \in \mathbb{Q}(\theta_k)$ $(k = 1, 2, \ldots, n)$. It can be shown that the conjugates of α relative to K do not depend on the choice of θ such that $K = \mathbb{Q}(\theta)$ (Exercise 1 of this Chapter).

Definition 6.3.2 (Field polynomial of α over K) *Let K be an algebraic number field of degree n. Let $\alpha \in K$. Let $\alpha_1 = \alpha, \alpha_2, \ldots, \alpha_n$ be the K-conjugates of α. Then the field polynomial of α over K is the polynomial*

$$\mathrm{fld}_K(\alpha) = \prod_{k=1}^{n} (x - \alpha_k).$$

Clearly $\mathrm{fld}_K(\alpha) \in \mathbb{C}[x]$. However, much more is true as the next theorem shows.

Theorem 6.3.1 *Let K be an algebraic number field of degree n. Let $\alpha \in K$. Then*

$$\mathrm{fld}_K(\alpha) \in \mathbb{Q}[x].$$

Proof: Let $\theta \in K$ be such that $K = \mathbb{Q}(\theta)$. We have $\deg(\mathrm{irr}_{\mathbb{Q}}(\theta)) = [\mathbb{Q}(\theta) : \mathbb{Q}] = [K : \mathbb{Q}] = n$. As $\alpha \in K$, by Theorem 6.1.3 there exist $c_0, c_1, \ldots, c_{n-1} \in \mathbb{Q}$ such that

$$\alpha = c_0 + c_1\theta + \cdots + c_{n-1}\theta^{n-1}.$$

The K-conjugates of α are $\alpha_1 = \alpha, \alpha_2, \ldots, \alpha_n$, where

$$\alpha_k = c_0 + c_1\theta_k + \cdots + c_{n-1}\theta_k^{n-1}, \ \ k = 1, 2, \ldots, n.$$

The field polynomial of α over K is

$$\mathrm{fld}_K(\alpha) = \prod_{k=1}^{n}(x - \alpha_k) = \prod_{k=1}^{n}(x - (c_0 + c_1\theta_k + \cdots + c_{n-1}\theta_k^{n-1})).$$

Clearly $\mathrm{fld}_K(\alpha) \in K_1[x]$, where $K_1 = \mathbb{Q}(\theta_1, \ldots, \theta_n)$. Arguing as in the proof of Theorem 6.2.2, we deduce that the coefficients of $\mathrm{fld}_K(\alpha)$ are polynomials with rational coefficients in the elementary symmetric polynomials in $\theta_1, \ldots, \theta_n$ and so belong to $\mathbb{Q}$. Hence $\mathrm{fld}_K(\alpha) \in \mathbb{Q}[x]$. ∎

Example 6.3.2 *The cubic polynomial $x^3 + 11x + 4 \in \mathbb{Z}[x]$ is irreducible. Let its three roots be $\theta_1 = \theta, \theta_2,$ and θ_3. One of these roots is real and the other two are nonreal and complex conjugates of one another (Exercise 2 of this Chapter). Let $K = \mathbb{Q}(\theta)$ so that $[K : \mathbb{Q}] = \deg(\mathrm{irr}_{\mathbb{Q}}(\theta)) = \deg(x^3 + 11x + 4) = 3$. Let $\alpha = (\theta + \theta^2)/2 \in K$. We determine $\mathrm{fld}_K(\alpha)$. We have*

$$\mathrm{fld}_K(\alpha) = \left(x - \frac{(\theta_1 + \theta_1^2)}{2}\right)\left(x - \frac{(\theta_2 + \theta_2^2)}{2}\right)\left(x - \frac{(\theta_3 + \theta_3^2)}{2}\right)$$
$$= x^3 + a_2x^2 + a_1x + a_0,$$

where

$$a_2 = -\frac{(\theta_1 + \theta_1^2)}{2} - \frac{(\theta_2 + \theta_2^2)}{2} - \frac{(\theta_3 + \theta_3^2)}{2}$$

$$= -\frac{1}{2}(\theta_1 + \theta_2 + \theta_3) - \frac{1}{2}(\theta_1^2 + \theta_2^2 + \theta_3^2),$$

$$a_1 = \frac{1}{4}((\theta_1 + \theta_1^2)(\theta_2 + \theta_2^2) + (\theta_2 + \theta_2^2)(\theta_3 + \theta_3^2) + (\theta_3 + \theta_3^2)(\theta_1 + \theta_1^2))$$

$$= \frac{1}{4}((\theta_1\theta_2 + \theta_2\theta_3 + \theta_3\theta_1) + (\theta_1\theta_2^2 + \theta_1^2\theta_2 + \theta_2\theta_3^2 + \theta_2^2\theta_3 + \theta_3\theta_1^2 + \theta_3^2\theta_1)$$

$$+ (\theta_1^2\theta_2^2 + \theta_2^2\theta_3^2 + \theta_3^2\theta_1^2)),$$

$$a_0 = -\frac{1}{8}(\theta_1 + \theta_1^2)(\theta_2 + \theta_2^2)(\theta_3 + \theta_3^2)$$

$$= -\frac{1}{8}\theta_1\theta_2\theta_3(1 + \theta_1)(1 + \theta_2)(1 + \theta_3)$$

$$= -\frac{1}{8}\theta_1\theta_2\theta_3(1 + (\theta_1 + \theta_2 + \theta_3) + (\theta_1\theta_2 + \theta_2\theta_3 + \theta_3\theta_1) + \theta_1\theta_2\theta_3).$$

Now

$$x^3 + 11x + 4 = (x - \theta_1)(x - \theta_2)(x - \theta_3),$$

so that

$$\theta_1 + \theta_2 + \theta_3 = 0,$$
$$\theta_1\theta_2 + \theta_2\theta_3 + \theta_3\theta_1 = 11,$$
$$\theta_1\theta_2\theta_3 = -4.$$

Hence

$$\theta_1^2 + \theta_2^2 + \theta_3^2 = (\theta_1 + \theta_2 + \theta_3)^2 - 2(\theta_1\theta_2 + \theta_2\theta_3 + \theta_3\theta_1) = -22,$$
$$\theta_1^2\theta_2^2 + \theta_2^2\theta_3^2 + \theta_3^2\theta_1^2 = (\theta_1\theta_2 + \theta_2\theta_3 + \theta_3\theta_1)^2 - 2\theta_1\theta_2\theta_3(\theta_1 + \theta_2 + \theta_3) = 121,$$
$$\theta_1\theta_2^2 + \theta_1^2\theta_2 + \theta_2\theta_3^2 + \theta_2^2\theta_3 + \theta_3\theta_1^2 + \theta_3^2\theta_1$$
$$= \theta_1\theta_2(\theta_2 + \theta_1) + \theta_2\theta_3(\theta_3 + \theta_2) + \theta_3\theta_1(\theta_1 + \theta_3)$$
$$= -3\theta_1\theta_2\theta_3 = 12, \quad as \ \theta_1 + \theta_2 + \theta_3 = 0,$$

so that

$$a_2 = 11, \ a_1 = 36, \ a_0 = 4.$$

Hence

$$\mathrm{fld}_K(\alpha) = x^3 + 11x^2 + 36x + 4,$$

showing that $\alpha \in O_K$.

In the next theorem we relate the field polynomial of α over K to the minimal polynomial of α over $\mathbb{Q}$.

Theorem 6.3.2 *Let K be an algebraic number field of degree n. Let $\alpha \in K$. Then*

$$\mathrm{fld}_K(\alpha) = (\mathrm{irr}_{\mathbb{Q}}(\alpha))^s,$$

where s is the positive integer

$$s = \frac{n}{\deg(\mathrm{irr}_{\mathbb{Q}}(\alpha))}.$$

Proof: Let $\{\alpha_1 = \alpha, \alpha_2, \ldots, \alpha_n\}$ be a complete set of conjugates of α relative to K. Then

$$\mathrm{fld}_K(\alpha) = \prod_{k=1}^{n}(x - \alpha_k) \in \mathbb{Q}[x]$$

by Theorem 6.3.1. As $\mathrm{fld}_K(\alpha)$ has α as a root, we have

$$\mathrm{irr}_{\mathbb{Q}}(\alpha) \mid \mathrm{fld}_K(\alpha)$$

in $\mathbb{Q}[x]$. Hence, as $\mathbb{Q}[x]$ is a unique factorization domain, we have

$$\mathrm{fld}_K(\alpha) = (\mathrm{irr}_{\mathbb{Q}}(\alpha))^s h(x),$$

where $h(x)$ is a monic polynomial of $\mathbb{Q}[x]$, which is not divisible by the irreducible polynomial $\mathrm{irr}_{\mathbb{Q}}(\alpha)$, and s is a positive integer. Suppose that $h(x)$ is a nonconstant polynomial. Then $h(\alpha_k) = 0$ for some $k \in \{1, 2, \ldots, n\}$.

Now choose $\theta \in K$ such that $K = \mathbb{Q}(\theta)$. Let $\theta_1 = \theta, \theta_2, \ldots, \theta_n$ be the conjugates of θ over $\mathbb{Q}$. As $\alpha \in K$ there exists a polynomial

$$r(x) = a_0 + a_1 x + \cdots + a_{n-1}x^{n-1} \in \mathbb{Q}[x]$$

such that $\alpha = r(\theta)$. Thus $\alpha_j = r(\theta_j)$ for $j \in \{1, 2, \ldots, n\}$.

Next let

$$g(x) = h(r(x)) \in \mathbb{Q}[x].$$

Then $g(\theta_k) = h(r(\theta_k)) = h(\alpha_k) = 0$. Thus $g(x)$ is a multiple of $\mathrm{irr}_{\mathbb{Q}}(\theta_k) = \mathrm{irr}_{\mathbb{Q}}(\theta) \in \mathbb{Q}[x]$. Hence $g(\theta_j) = 0$ for $j = 1, 2, \ldots, n$. In particular $g(\theta) = 0$. Thus $h(\alpha) = h(r(\theta)) = g(\theta) = 0$. Hence $h(x)$ is a multiple of $\mathrm{irr}_{\mathbb{Q}}(\alpha)$ in $\mathbb{Q}[x]$, contradicting that $h(x)$ is not divisible by $\mathrm{irr}_{\mathbb{Q}}(\alpha)$.

We have shown that $h(x)$ is a constant polynomial; that is, $h(x) = c, \ c \in \mathbb{Q}$. But $h(x)$ is monic so $c = 1$. Thus

$$\mathrm{fld}_K(\alpha) = (\mathrm{irr}_{\mathbb{Q}}(\alpha))^s$$

as asserted. Comparing degrees of the polynomials in this equation, we see that

$$n = \deg(\mathrm{fld}_K(\alpha)) = s \deg(\mathrm{irr}_{\mathbb{Q}}(\alpha))$$

so that

$$s = \frac{n}{\deg(\mathrm{irr}_{\mathbb{Q}}(\alpha))}.$$ ∎

Theorem 6.3.2 tells us that the conjugates of α with respect to K are the roots of $\mathrm{irr}_{\mathbb{Q}}(\alpha)$ in $\mathbb{C}$ each repeated $s = n/\deg(\mathrm{irr}_{\mathbb{Q}}(\alpha))$ times.

Theorem 6.3.3 *Let K be an algebraic number field. Let $\alpha \in O_K$. Then the K-conjugates of α are algebraic integers.*

Proof: Let $\alpha \in O_K$. Then, by Theorem 5.3.2, we have

$$\mathrm{irr}_{\mathbb{Q}}(\alpha) \in \mathbb{Z}[x],$$

and so by Theorem 6.3.2

$$\mathrm{fld}_K(\alpha) \in \mathbb{Z}[x].$$

Thus the K-conjugates of α being the roots of a monic polynomial with rational integer coefficients are algebraic integers. ∎

Suppose $\alpha \in \mathbb{Q}$, then

$$\alpha_k = \sigma_k(\alpha) = \alpha, \ k = 1, 2, \ldots, n,$$

so all of the K-conjugates of α are equal. Conversely, if all the K-conjugates of α are equal then

$$\mathrm{fld}_K(\alpha) = (x - \alpha)^n.$$

Hence, by Theorem 6.3.2, we have

$$(\mathrm{irr}_{\mathbb{Q}}(\alpha))^s = (x - \alpha)^n.$$

But the roots of $\mathrm{irr}_{\mathbb{Q}}(\alpha)$ are all distinct (Theorem 5.2.1) so that

$$\mathrm{irr}_{\mathbb{Q}}(\alpha) = x - \alpha, \ s = n.$$

As $\mathrm{irr}_{\mathbb{Q}}(\alpha) \in \mathbb{Q}[x]$ we deduce that $\alpha \in \mathbb{Q}$. Hence we have shown the following result.

Theorem 6.3.4 *Let K be an algebraic number field. Let $\alpha \in K$. Then all the K-conjugates of α are equal if and only if $\alpha \in \mathbb{Q}$.*

If the K-conjugates of α are all distinct then $\mathrm{fld}_K(\alpha)$ is a product of distinct linear factors and so by Theorem 6.3.2 we have $s = 1$ and $\mathrm{irr}_{\mathbb{Q}}(\alpha) = \mathrm{fld}_K(\alpha)$. Hence

$$[\mathbb{Q}(\alpha) : \mathbb{Q}] = \deg(\mathrm{irr}_{\mathbb{Q}}(\alpha)) = \deg(\mathrm{fld}_K(\alpha)) = n = [K : \mathbb{Q}].$$

Since $\mathbb{Q}(\alpha) \subseteq K$ we deduce that $K = \mathbb{Q}(\alpha)$. Conversely, if $K = \mathbb{Q}(\alpha)$ then $\deg(\mathrm{irr}_\mathbb{Q}(\alpha)) = [K : \mathbb{Q}] = n$, so that by Theorem 6.3.2 $s = 1$ and $\mathrm{fld}_K(\alpha) = \mathrm{irr}_\mathbb{Q}(\alpha)$. Hence the K-conjugates of α are distinct. We have proved the following theorem.

Theorem 6.3.5 *Let K be an algebraic number field. Let $\alpha \in K$. Then all the K-conjugates of α are distinct if and only if $K = \mathbb{Q}(\alpha)$.*

Let $K = \mathbb{Q}(\theta)$ be an algebraic number field of degree n. Let $\theta_1 = \theta, \theta_2, \ldots, \theta_n$ be the conjugates of θ over $\mathbb{Q}$. Using the preceding ideas it is easy to show that if there are exactly m distinct fields among the conjugate fields $\mathbb{Q}(\theta_1) = K, \mathbb{Q}(\theta_2), \ldots, \mathbb{Q}(\theta_n)$ then m divides n and each distinct field occurs n/m times (Exercise 3 of this Chapter). If $m = 1$ so that $\mathbb{Q}(\theta_1) = \cdots = \mathbb{Q}(\theta_n) = K$, the field K is said to be a normal or Galois extension of $\mathbb{Q}$.

Example 6.3.3 *Let $K = \mathbb{Q}(\sqrt{2}, \sqrt{3}) = \mathbb{Q}(\sqrt{2} + \sqrt{3})$. The conjugates of $\sqrt{2} + \sqrt{3}$ are $\pm\sqrt{2} \pm \sqrt{3}$ and the conjugate fields of K all coincide with K as*

$$\mathbb{Q}(\pm\sqrt{2} \pm \sqrt{3}) = \mathbb{Q}(\sqrt{2} + \sqrt{3}) = K.$$

Thus K is a normal field.

Example 6.3.4 *Let $K = \mathbb{Q}(\sqrt[3]{2})$ so that $K \subseteq \mathbb{R}$. The conjugates of $\sqrt[3]{2}$ are*

$$\sqrt[3]{2}, \ \omega\sqrt[3]{2}, \ \omega^2\sqrt[3]{2},$$

where ω and ω^2 are the two complex cube roots of unity, since

$$\mathrm{irr}_\mathbb{Q}(\sqrt[3]{2}) = x^3 - 2 = (x - \sqrt[3]{2})(x - \omega\sqrt[3]{2})(x - \omega^2\sqrt[3]{2}).$$

The conjugate fields of K are

$$K_1 = \mathbb{Q}(\sqrt[3]{2}) = K, \ K_2 = \mathbb{Q}(\omega\sqrt[3]{2}), \ K_3 = \mathbb{Q}(\omega^2\sqrt[3]{2}).$$

Clearly as K_1 is a real field, and K_2, K_3 are not, we have $K_1 \neq K_2$, $K_1 \neq K_3$. We show that $K_2 \neq K_3$. Suppose $K_2 = K_3$. Then $\omega^2\sqrt[3]{2} \in K_2$ and so there exist $a, b, c \in \mathbb{Q}$ such that

$$\omega^2\sqrt[3]{2} = a + b\omega\sqrt[3]{2} + c(\omega\sqrt[3]{2})^2.$$

Taking complex conjugates, we obtain as $\bar{\omega} = \omega^2$

$$\omega\sqrt[3]{2} = a + b\omega^2\sqrt[3]{2} + c\omega(\sqrt[3]{2})^2.$$

Subtracting we deduce that

$$(\omega^2 - \omega)\sqrt[3]{2} = -b(\omega^2 - \omega)\sqrt[3]{2} + c(\omega^2 - \omega)(\sqrt[3]{2})^2,$$

so that

$$\sqrt[3]{2} = -b\sqrt[3]{2} + c(\sqrt[3]{2})^2.$$

Hence

$$1 + b = c\sqrt[3]{2}.$$

Since $\sqrt[3]{2} \notin \mathbb{Q}$ we must have $1 + b = c = 0$, so that

$$\omega^2\sqrt[3]{2} = a - \omega\sqrt[3]{2}.$$

Thus

$$(\omega^2 + \omega)\sqrt[3]{2} = a;$$

that is (as $\omega^2 + \omega = -1$),

$$\sqrt[3]{2} = -a \in \mathbb{Q},$$

a contradiction.

   Hence all the conjugate fields of $\mathbb{Q}(\sqrt[3]{2})$ are distinct, and $\mathbb{Q}(\sqrt[3]{2})$ is not a normal field.

Example 6.3.5 *Let $K = \mathbb{Q}(\sqrt[4]{2})$ so that $K \subseteq \mathbb{R}$. The conjugates of $\sqrt[4]{2}$ are*

$$\sqrt[4]{2}, \ i\sqrt[4]{2}, \ -\sqrt[4]{2}, \ -i\sqrt[4]{2},$$

as

$$\mathrm{irr}_{\mathbb{Q}}(\sqrt[4]{2}) = x^4 - 2 = (x - \sqrt[4]{2})(x - i\sqrt[4]{2})(x + \sqrt[4]{2})(x + i\sqrt[4]{2}).$$

The conjugate fields of K are

$$\mathbb{Q}(\sqrt[4]{2}) = K,$$
$$\mathbb{Q}(i\sqrt[4]{2}) = L \text{ (say)},$$
$$\mathbb{Q}(-\sqrt[4]{2}) = \mathbb{Q}(\sqrt[4]{2}) = K,$$
$$\mathbb{Q}(-i\sqrt[4]{2}) = \mathbb{Q}(i\sqrt[4]{2}) = L.$$

Clearly $K \neq L$ as K is a real field and L is a nonreal field.

   Hence there are two distinct conjugate fields. $\mathbb{Q}(\sqrt[4]{2})$ is not a normal field.

6.4 The Discriminant of a Set of Elements in an Algebraic Number Field

Let K be an algebraic number field of degree n. Let $\omega_1, \omega_2, \ldots, \omega_n$ be any n elements of K. An important quantity defined in terms of $\omega_1, \omega_2, \ldots, \omega_n$ and their conjugates relative to K is the discriminant $D(\omega_1, \ldots, \omega_n)$. As we shall see the discriminant has some very nice properties. For example, $D(\omega_1, \ldots, \omega_n)$ is always a rational number, which is nonzero if and only if $\omega_1, \ldots, \omega_n$ are linearly independent over $\mathbb{Q}$. Moreover, if $\omega_1, \ldots, \omega_n$ are all algebraic integers then $D(\omega_1, \ldots, \omega_n)$ is a rational integer.

Definition 6.4.1 (Discriminant of n elements in an algebraic number field of degree n) *Let K be an algebraic number field of degree n. Let $\omega_1, \ldots, \omega_n$ be n elements of the field K. Let σ_k $(k = 1, 2, \ldots, n)$ denote the n distinct monomorphisms : $K \longrightarrow \mathbb{C}$. For $i = 1, \ldots, n$ let*

$$\omega_i^{(1)} = \sigma_1(\omega_i) = \omega_i, \quad \omega_i^{(2)} = \sigma_2(\omega_i), \ldots, \omega_i^{(n)} = \sigma_n(\omega_i)$$

denote the conjugates of ω_i relative to K. Then the discriminant of $\{\omega_1, \ldots, \omega_n\}$ is

$$D(\omega_1, \ldots, \omega_n) = \begin{vmatrix} \omega_1^{(1)} & \omega_2^{(1)} & \cdots & \omega_n^{(1)} \\ \omega_1^{(2)} & \omega_2^{(2)} & \cdots & \omega_n^{(2)} \\ \vdots & \vdots & \cdots & \vdots \\ \omega_1^{(n)} & \omega_2^{(n)} & \cdots & \omega_n^{(n)} \end{vmatrix}^2.$$

Example 6.4.1 *Let $K = \mathbb{Q}(\sqrt{2}, \sqrt{3})$ and choose*

$$\omega_1 = 1, \quad \omega_2 = \sqrt{2}, \quad \omega_3 = \sqrt{3}, \quad \omega_4 = \sqrt{2} + \sqrt{3}.$$

By Example 5.6.1 we know that K is a quartic field. The four monomorphisms : $K \longrightarrow \mathbb{C}$ are given by

$$\begin{aligned} \sigma_1(a + b\sqrt{2} + c\sqrt{3} + d\sqrt{6}) &= a + b\sqrt{2} + c\sqrt{3} + d\sqrt{6}, \\ \sigma_2(a + b\sqrt{2} + c\sqrt{3} + d\sqrt{6}) &= a + b\sqrt{2} - c\sqrt{3} - d\sqrt{6}, \\ \sigma_3(a + b\sqrt{2} + c\sqrt{3} + d\sqrt{6}) &= a - b\sqrt{2} + c\sqrt{3} - d\sqrt{6}, \\ \sigma_4(a + b\sqrt{2} + c\sqrt{3} + d\sqrt{6}) &= a - b\sqrt{2} - c\sqrt{3} + d\sqrt{6}, \end{aligned}$$

where $a, b, c, d \in \mathbb{Q}$. Hence

$$D(1, \sqrt{2}, \sqrt{3}, \sqrt{2} + \sqrt{3}) = \begin{vmatrix} 1 & \sqrt{2} & \sqrt{3} & \sqrt{2} + \sqrt{3} \\ 1 & \sqrt{2} & -\sqrt{3} & \sqrt{2} - \sqrt{3} \\ 1 & -\sqrt{2} & \sqrt{3} & -\sqrt{2} + \sqrt{3} \\ 1 & -\sqrt{2} & -\sqrt{3} & -\sqrt{2} - \sqrt{3} \end{vmatrix}^2.$$

As the fourth column of the determinant is the sum of the second and third columns, we deduce that

$$D(1, \sqrt{2}, \sqrt{3}, \sqrt{2} + \sqrt{3}) = 0.$$

We can now define the discriminant $D(\alpha)$ of an element α of an algebraic number field.

Definition 6.4.2 (Discriminant $D(\alpha)$ of an element α) *Let K be an algebraic number field of degree n. Let $\alpha \in K$. Then we define the discriminant $D(\alpha)$ of*

α by

$$D(\alpha) = D(1, \alpha, \alpha^2, \ldots, \alpha^{n-1}).$$

Theorem 6.4.1 *Let K be an algebraic number field of degree n. Let $\alpha \in K$. Then*

$$D(\alpha) = \prod_{1 \le i < j \le n} (\alpha^{(i)} - \alpha^{(j)})^2,$$

where $\alpha^{(1)} = \alpha, \alpha^{(2)}, \ldots, \alpha^{(n)}$ are the conjugates of α with respect to K.

Proof: If $z_1, z_2, \ldots, z_n$ are complex numbers we have the value of the determinant

$$\begin{vmatrix} z_1^{n-1} & z_1^{n-2} & \cdots & z_1 & 1 \\ z_2^{n-1} & z_2^{n-2} & \cdots & z_2 & 1 \\ \vdots & \vdots & \cdots & \vdots & \vdots \\ z_n^{n-1} & z_n^{n-2} & \cdots & z_n & 1 \end{vmatrix} = \prod_{1 \le i < j \le n} (z_i - z_j)$$

(see, for example, [3, pp. 17–18]). Interchanging columns 1 and n, columns 2 and $n-1$, etc., we obtain the evaluation of the Vandermonde determinant

$$\begin{vmatrix} 1 & z_1 & \cdots & z_1^{n-2} & z_1^{n-1} \\ 1 & z_2 & \cdots & z_2^{n-2} & z_2^{n-1} \\ \vdots & \vdots & \cdots & \vdots & \vdots \\ 1 & z_n & \cdots & z_n^{n-2} & z_n^{n-1} \end{vmatrix} = (-1)^{[n/2]} \prod_{1 \le i < j \le n} (z_i - z_j)$$

$$= (-1)^{\frac{n(n-1)}{2}} \prod_{1 \le i < j \le n} (z_i - z_j)$$

$$= \prod_{1 \le i < j \le n} (z_j - z_i),$$

as

$$\left[\frac{n}{2}\right] \equiv \frac{n(n-1)}{2} \pmod 2.$$

Hence for any $\alpha \in K$ we have

$$\begin{vmatrix} 1 & \alpha & \alpha^2 & \cdots & \alpha^{n-1} \\ 1 & \alpha^{(2)} & (\alpha^{(2)})^2 & \cdots & (\alpha^{(2)})^{n-1} \\ \vdots & \vdots & \vdots & \cdots & \vdots \\ 1 & \alpha^{(n)} & (\alpha^{(n)})^2 & \cdots & (\alpha^{(n)})^{n-1} \end{vmatrix} = \prod_{1 \le i < j \le n} (\alpha^{(j)} - \alpha^{(i)})$$

so that

$$D(\alpha) = \left(\prod_{1 \le i < j \le n} (\alpha^{(j)} - \alpha^{(i)}) \right)^2 = \prod_{1 \le i < j \le n} (\alpha^{(i)} - \alpha^{(j)})^2. \qquad \blacksquare$$

Definition 6.4.3 (Discriminant of a polynomial) *Let*

$$f(x) = a_n x^n + a_{n-1} x^{n-1} + \cdots + a_1 x + a_0 \in \mathbb{C}[x],$$

where $n \in \mathbb{N}$ and $a_n \ne 0$. Let $x_1, \ldots, x_n \in \mathbb{C}$ be the roots of $f(x)$. The discriminant of $f(x)$ is the quantity

$$\mathrm{disc}(f(x)) = a_n^{2n-2} \prod_{1 \le i < j \le n} (x_i - x_j)^2 \in \mathbb{C}.$$

Clearly $f(x)$ has a repeated root if and only if $\mathrm{disc}(f(x)) = 0$. The discriminant is a^{2n-2} times a symmetric polynomial in $x_1, \ldots, x_n$. The degree of $\mathrm{disc}(f(x))$ in each x_i is $2(n-1)$. Thus when $\mathrm{disc}(f(x))$ is expressed as a function of $a_0, a_1, \ldots, a_n$, it consists of terms

$$C a_n^{2n-2} \left(\frac{a_{n-1}}{a_n} \right)^{k_1} \left(\frac{a_{n-2}}{a_n} \right)^{k_2} \cdots \left(\frac{a_0}{a_n} \right)^{k_n},$$

where

$$k_1 + k_2 + \cdots + k_n \le 2n - 2,$$

so that $\mathrm{disc}(f(x))$ is a polynomial in the coefficients of $f(x)$.

Theorem 6.4.2 *Let K be an algebraic number field of degree n. Let $\alpha \in K$. Then*

$$D(\alpha) = \mathrm{disc}(\mathrm{fld}_K(\alpha)).$$

Proof: Let $\alpha^{(1)} = \alpha$, $\alpha^{(2)}, \ldots, \alpha^{(n)}$ be the K-conjugates of α. Then the roots of $\mathrm{fld}_K(\alpha)$ are $\alpha^{(1)}, \ldots, \alpha^{(n)}$. Hence, by Definition 6.4.3 and Theorem 6.4.1, we have

$$\mathrm{disc}(\mathrm{fld}_K(\alpha)) = \prod_{1 \le i < j \le n} (\alpha^{(i)} - \alpha^{(j)})^2 = D(\alpha). \qquad \blacksquare$$

Example 6.4.2 *Let $K = \mathbb{Q}(\sqrt[3]{2})$ and choose $\alpha = \sqrt[3]{2}$. Then the conjugates of α are*

$$\alpha_1 = \sqrt[3]{2}, \ \alpha_2 = \omega\sqrt[3]{2}, \ \alpha_3 = \omega^2\sqrt[3]{2},$$

where $\omega = (-1 + \sqrt{-3})/2$. Then

$$\alpha_1 - \alpha_2 = (1 - \omega)\sqrt[3]{2},$$
$$\alpha_2 - \alpha_3 = \omega(1 - \omega)\sqrt[3]{2},$$
$$\alpha_3 - \alpha_1 = \omega^2(1 - \omega)\sqrt[3]{2},$$

so that

$$(\alpha_1 - \alpha_2)(\alpha_2 - \alpha_3)(\alpha_3 - \alpha_1) = (1 - \omega)^3 2.$$

Now

$$(1 - \omega)^3 = 1 - 3\omega + 3\omega^2 - \omega^3 = -3(\omega - \omega^2),$$

so that

$$(1 - \omega)^6 = 3^2(\omega - \omega^2)^2 = 3^2(\omega^2 + \omega - 2) = -3^3.$$

Hence

$$D(\alpha) = ((\alpha_1 - \alpha_2)(\alpha_2 - \alpha_3)(\alpha_3 - \alpha_1))^2 = (1 - \omega)^6 2^2 = -2^2 \cdot 3^3.$$

Theorem 6.4.3 *Let K be an algebraic number field of degree n. Let $\alpha \in K$. Then*

$$K = \mathbb{Q}(\alpha) \text{ if and only if } D(\alpha) \neq 0.$$

Proof: We have by Theorems 6.3.5 and 6.4.1

$$K = \mathbb{Q}(\alpha) \Longleftrightarrow K\text{-conjugates of } \alpha \text{ are distinct} \Longleftrightarrow D(\alpha) \neq 0. \qquad \blacksquare$$

Theorem 6.4.4 *Let K be an algebraic number field of degree n.*

(a) *If $\omega_1, \ldots, \omega_n \in K$ then*

$$D(\omega_1, \ldots, \omega_n) \in \mathbb{Q}.$$

(b) *If $\omega_1, \ldots, \omega_n \in O_K$ then*

$$D(\omega_1, \ldots, \omega_n) \in \mathbb{Z}.$$

(c) *If $\omega_1, \ldots, \omega_n \in K$ then*

$$D(\omega_1, \ldots, \omega_n) \neq 0 \text{ if and only if } \omega_1, \ldots, \omega_n \text{ are linearly independent over } \mathbb{Q}.$$

Proof: **(a)** By Theorem 6.1.1 we have $K = \mathbb{Q}(\theta)$ for some $\theta \in K$. Then, for $i = 1, 2, \ldots, n$, we have

$$\omega_i = c_{0i} + c_{1i}\theta + \cdots + c_{n-1i}\theta^{n-1},$$

where $c_{0i}, \ldots, c_{n-1i} \in \mathbb{Q}$. Hence, for $j = 1, 2, \ldots, n$, we have

$$\omega_i^{(j)} = c_{0i} + c_{1i}\theta_j + \cdots + c_{n-1i}\theta_j^{n-1}, \qquad (6.4.1)$$

where $\theta_1 = \theta,\ \theta_2,\ \ldots,\ \theta_n$ are the conjugates of θ over $\mathbb{Q}$ and $\omega_i^{(1)},\ \ldots,\ \omega_i^{(n)}$ are the K-conjugates of ω_i $(i = 1, 2, \ldots, n)$. Using the expressions (6.4.1) in Definition 6.4.1, we see that any permutation of the conjugates of θ leaves $D(\omega_1, \ldots, \omega_n)$ invariant as it merely causes a permutation of the rows of the matrix of which $D(\omega_1, \ldots, \omega_n)$ is the square of the determinant. Hence $D(\omega_1, \ldots, \omega_n)$ is a symmetric function of the roots of the polynomial

$$(x - \theta_1)(x - \theta_2) \cdots (x - \theta_n) = x^n + a_{n-1}x^{n-1} + \cdots + a_0,$$

where $a_0, a_1, \ldots, a_{n-1} \in \mathbb{Q}$. By the symmetric function theorem, $D(\omega_1, \ldots, \omega_n)$ is a polynomial in the coefficients $a_0, a_1, \ldots, a_{n-1}$ and hence a rational number.

(b) If $\omega_1, \ldots, \omega_n \in O_K$ then $D(\omega_1, \ldots, \omega_n)$, being obtained from them and their conjugates by a series of additions and multiplications, is also in O_K. Since $D(\omega_1, \ldots, \omega_n) \in \mathbb{Q}$, by Theorem 4.2.4 we have $D(\omega_1, \ldots, \omega_n) \in \mathbb{Z}$.

(c) If the set $\{\omega_1, \ldots, \omega_n\}$ is linearly dependent over $\mathbb{Q}$, then there exist rational numbers $c_1, \ldots, c_n$ not all zero such that

$$c_1\omega_1 + \cdots + c_n\omega_n = 0.$$

Applying each monomorphism σ_k $(k = 1, \ldots, n)$ to this equation, we obtain the following homogeneous system of n linear equations in the n quantities $c_1, \ldots, c_n$:

$$\begin{cases} c_1\omega_1^{(1)} + \cdots + c_n\omega_n^{(1)} = 0, \\ c_1\omega_1^{(2)} + \cdots + c_n\omega_n^{(2)} = 0, \\ \qquad\qquad \cdots \\ c_1\omega_1^{(n)} + \cdots + c_n\omega_n^{(n)} = 0. \end{cases}$$

As this system has a nontrivial solution $(c_1, \ldots, c_n) \neq (0, \ldots, 0) \in \mathbb{Q}^n$, its determinant must be zero. Hence

$$D(\omega_1, \ldots, \omega_n) = 0.$$

Now suppose that the set $\{\omega_1, \ldots, \omega_n\}$ is linearly independent over $\mathbb{Q}$. Then $\{\omega_1, \ldots, \omega_n\}$ is a basis for the vector space K over the field $\mathbb{Q}$. In particular as $1, \theta, \ldots, \theta^{n-1} \in K$ there exist rational numbers c_{ij} $(i, j = 1, \ldots, n)$ such that

$$\begin{cases} 1 = c_{11}\omega_1 + \cdots + c_{1n}\omega_n, \\ \theta = c_{21}\omega_1 + \cdots + c_{2n}\omega_n, \\ \qquad\qquad \cdots \\ \theta^{n-1} = c_{n1}\omega_1 + \cdots + c_{nn}\omega_n. \end{cases}$$

Hence

$$D(\theta) = D(1, \theta, \ldots, \theta^{n-1}) = |\det(c_{ij})|^2 D(\omega_1, \ldots, \omega_n).$$

As $K = \mathbb{Q}(\theta)$, by Theorem 6.4.3 we know that $D(\theta) \neq 0$ so that

$$D(\omega_1, \ldots, \omega_n) \neq 0.$$

∎

6.5 Basis of an Ideal

We now use our knowledge of the properties of the discriminant to show that every ideal in the ring O_K of integers of an algebraic number field K has a finite basis considered as an Abelian group, that is, as a $\mathbb{Z}$-module. Thus O_K is Noetherian.

We first prove a preliminary result.

Theorem 6.5.1 *Let K be an algebraic number field with $[K : \mathbb{Q}] = n$. Let I be a nonzero ideal in O_K. Then there exist $\eta_1, \ldots, \eta_n \in I$ such that*

$$D(\eta_1, \ldots, \eta_n) \neq 0.$$

Proof: By Theorem 6.1.2 we have $K = \mathbb{Q}(\theta)$ for some $\theta \in O_K$. By Theorem 6.4.3 $D(\theta) \neq 0$. Further, by Theorem 6.1.7, as I is a nonzero ideal of O_K, there exists $c \in I \cap \mathbb{Z}$ with $c \neq 0$. Hence, as I is an ideal of O_K,

$$\eta_1 = c, \ \eta_2 = c\,\theta, \ldots, \eta_n = c\,\theta^{n-1} \in I$$

and

$$D(\eta_1, \ldots, \eta_n) = D(c, c\theta, \ldots, c\theta^{n-1}) = c^{2n} D(1, \theta, \ldots, \theta^{n-1}) = c^{2n} D(\theta) \neq 0.$$

∎

We are now in a position to prove that every ideal of the ring of integers of an algebraic number field has a finite basis.

Theorem 6.5.2 *Let K be an algebraic number field of degree n. Let I be a nonzero ideal of O_K. Then there exist elements $\eta_1, \ldots, \eta_n$ of I such that every element α of I can be expressed uniquely in the form*

$$\alpha = x_1 \eta_1 + \cdots + x_n \eta_n,$$

where $x_1, \ldots, x_n \in \mathbb{Z}$.

Proof: As I is a nonzero ideal of O_K, by Theorem 6.5.1 there exists a set $\{\eta_1, \ldots, \eta_n\}$ of elements of I such that $D(\eta_1, \ldots, \eta_n) \neq 0$. By Theorem 6.4.4 $D(\eta_1, \ldots, \eta_n) \in \mathbb{Z}$ so that $|D(\eta_1, \ldots, \eta_n)|$ is a positive integer. Let

$$S = \{|D(\eta_1, \ldots, \eta_n)| \ : \ \eta_1, \ldots, \eta_n \in I, \ D(\eta_1, \ldots, \eta_n) \neq 0\}.$$

Clearly S is a nonempty set of positive integers and thus contains a least member, say $|D(\eta_1, \ldots, \eta_n)|$, $\eta_1, \ldots, \eta_n \in I$. As $D(\eta_1, \ldots, \eta_n) \neq 0$, by Theorem 6.4.4(c) $\{\eta_1, \ldots, \eta_n\}$ is a basis for the vector space K over $\mathbb{Q}$. Let $\alpha \in I$. Then there exist unique rational numbers $x_1, \ldots, x_n$ such that

$$\alpha = x_1 \eta_1 + \cdots + x_n \eta_n.$$

Suppose at least one of the x_i is not an integer. By permuting $\eta_1, \ldots, \eta_n$, if necessary, we may suppose that $x_1 \notin \mathbb{Z}$. Then there is a unique integer l such that

$$l < x_1 < l + 1.$$

Set

$$\gamma = \alpha - l\eta_1.$$

As $\alpha \in I$ and $\eta_1 \in I$ we see that $\gamma \in I$. Moreover,

$$\gamma = (x_1 - l)\eta_1 + x_2\eta_2 + \cdots + x_n\eta_n.$$

Applying each monomorphism σ_k $(k = 1, 2, \ldots, n)$ to this equation, we obtain

$$\begin{cases} \gamma^{(1)} = (x_1 - l)\eta_1^{(1)} + x_2\eta_2^{(1)} + \cdots + x_n\eta_n^{(1)}, \\ \gamma^{(2)} = (x_1 - l)\eta_1^{(2)} + x_2\eta_2^{(2)} + \cdots + x_n\eta_n^{(2)}, \\ \qquad\qquad \cdots \\ \gamma^{(n)} = (x_1 - l)\eta_1^{(n)} + x_2\eta_2^{(n)} + \cdots + x_n\eta_n^{(n)}, \end{cases}$$

where $\gamma^{(1)} = \gamma$, $\gamma^{(2)}, \ldots, \gamma^{(n)}$ are the K-conjugates of γ and $\eta_i^{(1)} = \eta_i$, $\eta_i^{(2)}, \ldots, \eta_i^{(n)}$ are the K-conjugates of η_i $(i = 1, 2, \ldots, n)$. By Cramer's rule, we deduce that

$$x_1 - l = \frac{\begin{vmatrix} \gamma^{(1)} & \eta_2^{(1)} & \cdots & \eta_n^{(1)} \\ \gamma^{(2)} & \eta_2^{(2)} & \cdots & \eta_n^{(2)} \\ \vdots & \vdots & \cdots & \vdots \\ \gamma^{(n)} & \eta_2^{(n)} & \cdots & \eta_n^{(n)} \end{vmatrix}}{\begin{vmatrix} \eta_1^{(1)} & \eta_2^{(1)} & \cdots & \eta_n^{(1)} \\ \eta_1^{(2)} & \eta_2^{(2)} & \cdots & \eta_n^{(2)} \\ \vdots & \vdots & \cdots & \vdots \\ \eta_1^{(n)} & \eta_2^{(n)} & \cdots & \eta_n^{(n)} \end{vmatrix}}.$$

Hence

$$(x_1 - l)^2 = \frac{D(\gamma, \eta_2, \ldots, \eta_n)}{D(\eta_1, \eta_2, \ldots, \eta_n)}$$

so that

$$0 < |D(\gamma, \eta_2, \ldots, \eta_n)| = (x_1 - l)^2 |D(\eta_1, \eta_2, \ldots, \eta_n)| < |D(\eta_1, \eta_2, \ldots, \eta_n)|.$$

This contradicts the minimality of $|D(\eta_1, \eta_2, \ldots, \eta_n)|$. Hence all the x_i are integers and each element $\alpha \in I$ can be expressed uniquely in the form $\alpha = x_1\eta_1 + \cdots + x_n\eta_n$. $\blacksquare$

Clearly, as $\eta_1, \ldots, \eta_n \in I$ and I is an ideal, we have

$$\mathbb{Z}\eta_1 + \cdots + \mathbb{Z}\eta_n = \{k_1\eta_1 + \cdots + k_n\eta_n \mid k_1, \ldots, k_n \in \mathbb{Z}\} \subseteq I$$

and Theorem 6.5.2 tells us that

$$I \subseteq \mathbb{Z}\eta_1 + \cdots + \mathbb{Z}\eta_n$$

so that

$$I = \mathbb{Z}\eta_1 + \cdots + \mathbb{Z}\eta_n,$$

showing that I is a finitely generated $\mathbb{Z}$-module.

We now use Theorem 3.5.3 and Theorem 6.5.2 to show that the ring of integers of an algebraic number field is Noetherian.

Theorem 6.5.3 *Let K be an algebraic number field. Then O_K is a Noetherian domain.*

Proof: $\mathbb{Z}$ and O_K are integral domains with $\mathbb{Z} \subseteq O_K$. $\mathbb{Z}$ is a Noetherian domain (Example 3.1.3). $O_K = \langle 1 \rangle$ is a finitely generated $\mathbb{Z}$-module by Theorem 6.5.2. Hence by Theorem 3.5.3 O_K is a Noetherian domain.

Alternatively we can avoid the use of Theorem 3.5.3 by arguing as follows. Let I be an ideal of O_K. If $I = \{0\}$ then $I = \langle 0 \rangle$ is finitely generated. If $I \neq \{0\}$ then I is finitely generated by Theorem 6.5.2. Hence every ideal of O_K is finitely generated and thus the domain O_K is Noetherian. ∎

Definition 6.5.1 (Basis of an ideal) *Let K be an algebraic number field of degree n. Let I be a nonzero ideal of O_K. If $\{\eta_1, \ldots, \eta_n\}$ is a set of elements of I such that every element $\alpha \in I$ can be expressed uniquely in the form*

$$\alpha = x_1\eta_1 + \cdots + x_n\eta_n \ (x_1, \ldots, x_n \in \mathbb{Z})$$

then $\{\eta_1, \ldots, \eta_n\}$ is called a basis for the ideal I.

As the representation of each element α of a nonzero ideal I by a basis $\{\eta_1, \ldots, \eta_n\}$ of I is unique, the basis elements $\eta_1, \ldots, \eta_n$ are linearly independent over $\mathbb{Q}$.

By Theorem 6.5.2 every ideal of the ring of integers of an algebraic number field possesses a basis. Our next result enables us to recognize when a set of elements $\{\lambda_1, \ldots, \lambda_n\}$ of an ideal in the ring of integers of an algebraic number field is a basis for the ideal.

Theorem 6.5.4 *Let K be an algebraic number field of degree n. Let I be a nonzero ideal of O_K.*

(a) *Let $\{\eta_1, \ldots, \eta_n\}$ and $\{\lambda_1, \ldots, \lambda_n\}$ be two bases for I. Then*

$$D(\eta_1, \ldots, \eta_n) = D(\lambda_1, \ldots, \lambda_n)$$

and

$$\eta_i = \sum_{j=1}^{n} c_{ij}\lambda_j, \ i = 1, 2, \ldots, n,$$

where c_{ij} $(i, j = 1, 2, \ldots, n)$ are rational integers such that

$$\det(c_{ij}) = \pm 1.$$

(b) *Let $\{\eta_1, \ldots, \eta_n\}$ be a basis for I and let $\lambda_1, \ldots, \lambda_n \in I$ be such that*

$$D(\lambda_1, \ldots, \lambda_n) = D(\eta_1, \ldots, \eta_n).$$

Then $\{\lambda_1, \ldots, \lambda_n\}$ is a basis for I.

Proof: (a) As $\{\lambda_1, \ldots, \lambda_n\}$ is a basis for I, we have

$$I = \mathbb{Z}\lambda_1 + \cdots + \mathbb{Z}\lambda_n.$$

Since $\eta_1, \ldots, \eta_n \in I$ there exist $c_{ij} \in \mathbb{Z}$ $(i, j = 1, 2, \ldots, n)$ such that

$$\eta_i = \sum_{j=1}^{n} c_{ij}\lambda_j, \ i = 1, 2, \ldots, n. \tag{6.5.1}$$

As $\{\eta_1, \ldots, \eta_n\}$ is a basis for I, we have

$$I = \mathbb{Z}\eta_1 + \cdots + \mathbb{Z}\eta_n.$$

Since $\lambda_1, \ldots, \lambda_n \in I$ there exist $d_{ij} \in \mathbb{Z}$ $(i, j = 1, 2, \ldots, n)$ such that

$$\lambda_j = \sum_{k=1}^{n} d_{jk}\eta_k, \ j = 1, 2, \ldots, n.$$

Thus, for $i = 1, 2, \ldots, n$, we have

$$\eta_i = \sum_{j=1}^{n} c_{ij} \sum_{k=1}^{n} d_{jk}\eta_k = \sum_{k=1}^{n} \left(\sum_{j=1}^{n} c_{ij}d_{jk} \right) \eta_k.$$

As $\{\eta_1, \ldots, \eta_n\}$ is a basis for I, $\eta_1, \ldots, \eta_n$ are linearly independent over $\mathbb{Q}$, so that

$$\sum_{j=1}^{n} c_{ij}d_{jk} = \begin{cases} 1, & \text{if } i = k, \\ 0, & \text{if } i \neq k. \end{cases}$$

We define the $n \times n$ matrices C and D by

$$C = [c_{ij}], \ D = [d_{ij}],$$

so that C and D have rational integer entries. Then

$$CD = I_n,$$

where I_n is the $n \times n$ identity matrix. Thus

$$\det(C)\det(D) = \det(CD) = \det(I_n) = 1.$$

But $\det(C)$, $\det(D) \in \mathbb{Z}$ so

$$\det(C) = \det(D) = \pm 1.$$

From (6.5.1) we have

$$D(\eta_1, \ldots, \eta_n) = (\det(c_{ij}))^2 D(\lambda_1, \ldots, \lambda_n) = (\det(C))^2 D(\lambda_1, \ldots, \lambda_n)$$

so that

$$D(\eta_1, \ldots, \eta_n) = (\pm 1)^2 D(\lambda_1, \ldots, \lambda_n) = D(\lambda_1, \ldots, \lambda_n).$$

This completes the proof of part (a).

(b) As $\{\eta_1, \ldots, \eta_n\}$ is a basis for I and $\lambda_1, \ldots, \lambda_n \in I$, there exist $d_{ij} \in \mathbb{Z}$ ($i, j = 1, 2, \ldots, n$) such that

$$\lambda_i = \sum_{j=1}^{n} d_{ij}\eta_j, \ i = 1, 2, \ldots, n.$$

Hence

$$D(\lambda_1, \ldots, \lambda_n) = (\det(d_{ij}))^2 D(\eta_1, \ldots, \eta_n).$$

As $D(\lambda_1, \ldots, \lambda_n) = D(\eta_1, \ldots, \eta_n)$ we deduce that

$$(\det(d_{ij}))^2 = 1$$

so that

$$\det(d_{ij}) = \pm 1.$$

Thus the matrix $D = (d_{ij})$ has an inverse $D^{-1} = C = (c_{ij})$ all of whose entries are integers, and

$$\eta_j = \sum_{j=1}^{n} c_{ij}\lambda_j \ (i = 1, 2, \ldots, n).$$

Let $\alpha \in I$. Then, as $\{\eta_1, \ldots, \eta_n\}$ is a basis for I, there exist $a_1, \ldots, a_n \in \mathbb{Z}$ such that

$$\alpha = \sum_{i=1}^{n} a_i \eta_i.$$

Hence

$$\alpha = \sum_{i=1}^{n} a_i \sum_{j=1}^{n} c_{ij} \lambda_j = \sum_{j=1}^{n} \left(\sum_{i=1}^{n} a_i c_{ij} \right) \lambda_j,$$

where each $\sum_{i=1}^{n} a_i c_{ij} \in \mathbb{Z}$ $(j = 1, 2, \dots, n)$. This proves that every element α of I can be expressed in the form

$$\alpha = b_1 \lambda_1 + \cdots + b_n \lambda_n$$

for some integers $b_1, \dots, b_n$.

Now suppose that α can be expressed in more than one way in this form, say,

$$\alpha = b_1 \lambda_1 + \cdots + b_n \lambda_n = b_1' \lambda_1 + \cdots + b_n' \lambda_n,$$

where $b_1, \dots, b_n, \ b_1', \dots, b_n' \in \mathbb{Z}$. Hence

$$e_1 \lambda_1 + \cdots + e_n \lambda_n = 0,$$

where $e_i = b_i - b_i' \in \mathbb{Z}$ $(i = 1, 2, \dots, n)$. If at least one of the e_i is nonzero then $\lambda_1, \dots, \lambda_n$ are linearly dependent over $\mathbb{Q}$ and so by Theorem 6.4.4(c) we have

$$D(\lambda_1, \dots, \lambda_n) = 0.$$

Hence

$$D(\eta_1, \dots, \eta_n) = 0,$$

so that $\eta_1, \dots, \eta_n$ are linearly dependent over $\mathbb{Q}$, contradicting that $\{\eta_1, \dots, \eta_n\}$ is a basis for I. Hence $e_i = 0$ $(i = 1, 2, \dots, n)$ and so $b_i = b_i'$ $(i = 1, 2, \dots, n)$, establishing that α is uniquely expressible in the form $b_1 \lambda_1 + \cdots + b_n \lambda_n$ with $b_1, \dots, b_n \in \mathbb{Z}$.

This completes the proof that $\{\lambda_1, \dots, \lambda_n\}$ is a basis for I. ∎

Example 6.5.1 *Let* $K = \mathbb{Q}(\sqrt{7})$ *so that* $O_K = \mathbb{Z} + \mathbb{Z}\sqrt{7} = \{a + b\sqrt{7} \mid a, b \in \mathbb{Z}\}$ *by Theorem 5.4.2. Let* I *be the principal ideal of* O_K *generated by* $2 + \sqrt{7}$*. Then*

$$\begin{aligned} I &= \{(a + b\sqrt{7})(2 + \sqrt{7}) \mid a, b \in \mathbb{Z}\} \\ &= \{a(2 + \sqrt{7}) + b(7 + 2\sqrt{7}) \mid a, b \in \mathbb{Z}\} \\ &= (2 + \sqrt{7})\mathbb{Z} + (7 + 2\sqrt{7})\mathbb{Z}, \end{aligned}$$

so that $\{2 + \sqrt{7}, \ 7 + 2\sqrt{7}\}$ *is a basis for* I*. However, a little more effort yields a "simpler" basis, that is, one having a rational integer as one of the basis elements.*

We have

$$I = \{(a + b\sqrt{7})(2 + \sqrt{7}) \mid a, b \in \mathbb{Z}\}$$
$$= \{(2a + 7b) + (a + 2b)\sqrt{7} \mid a, b \in \mathbb{Z}\}$$
$$= \{(2(c - 2b) + 7b) + c\sqrt{7} \mid b, c \in \mathbb{Z}\}$$
$$= \{3b + c(2 + \sqrt{7}) \mid b, c \in \mathbb{Z}\}$$
$$= 3\mathbb{Z} + (2 + \sqrt{7})\mathbb{Z},$$

showing that $\{3, \ 2 + \sqrt{7}\}$ *is a basis for* I.

If $\{\eta_1, \ldots, \eta_n\}$ and $\{\lambda_1, \ldots, \lambda_n\}$ are two bases for the same nonzero ideal of the ring of integers of an algebraic number field then we know by Theorem 6.5.4 that

$$D(\eta_1, \ldots, \eta_n) = D(\lambda_1, \ldots, \lambda_n).$$

Hence we can make the following definition.

Definition 6.5.2 (Discriminant of an ideal) *Let K be an algebraic number field of degree n. Let I be a nonzero ideal of O_K. Let $\{\eta_1, \ldots, \eta_n\}$ be a basis of I. Then the discriminant $D(I)$ of the ideal I is the nonzero integer given by*

$$D(I) = D(\eta_1, \ldots, \eta_n).$$

Example 6.5.2 *We determine the discriminant of the ideal I in Example 6.5.1. As $\{3, \ 2 + \sqrt{7}\}$ is a basis for I, we have*

$$D(I) = D(3, 2 + \sqrt{7})$$
$$= \begin{vmatrix} 3 & 2 + \sqrt{7} \\ 3 & 2 - \sqrt{7} \end{vmatrix}^2$$
$$= (3(2 - \sqrt{7}) - 3(2 + \sqrt{7}))^2$$
$$= (-6\sqrt{7})^2$$
$$= 252.$$

Next we consider bases of ideals in the ring of integers of a quadratic field.

Theorem 6.5.5 *Let K be a quadratic field. Let m be the unique squarefree integer such that $K = \mathbb{Q}(\sqrt{m})$.*

(a) $m \not\equiv 1 \pmod 4$. *Let $a, b, c \in \mathbb{Z}$ with $a \neq 0$ and $c \neq 0$. Then*

$$\{a, b + c\sqrt{m}\} \text{ is a basis for the ideal } \langle a, b + c\sqrt{m}\rangle$$

if and only if

$$c \mid a, \ c \mid b, \ ac \mid b^2 - mc^2. \tag{6.5.2}$$

(b) $m \equiv 1 \pmod 4$. *Let* $a, b, c \in \mathbb{Z}$ *with* $a \neq 0$, $c \neq 0$, *and* $b \equiv c \pmod 2$. *Then*

$$\{a, \frac{b + c\sqrt{m}}{2}\} \text{ is a basis for the ideal } \langle a, \frac{b + c\sqrt{m}}{2} \rangle$$

if and only if

$$c \mid a, \ c \mid b, \ 4ac \mid b^2 - mc^2. \tag{6.5.3}$$

Proof: (a) Suppose first that (6.5.2) holds. Then there are integers x, y, z such that

$$a = cx, \ b = cy, \ b^2 - mc^2 = acz.$$

Hence

$$bx - ay = 0, \ by - az = mc.$$

Let $\alpha \in I = \langle a, b + c\sqrt{m} \rangle$. Then there exist $\theta \in O_K$ and $\phi \in O_K$ such that

$$\alpha = \theta a + \phi(b + c\sqrt{m}).$$

As $\theta, \phi \in O_K$ and $m \not\equiv 1 \pmod 4$, by Theorem 5.4.2 there exist integers r, s, t, u such that

$$\theta = r + s\sqrt{m}, \ \phi = t + u\sqrt{m}.$$

Hence

$$\begin{aligned}
\alpha &= (r + s\sqrt{m})a + (t + u\sqrt{m})(b + c\sqrt{m}) \\
&= (ra + tb + umc) + (sa + tc + ub)\sqrt{m} \\
&= s(bx - ay) + (ra + tb + u(by - az)) + (scx + tc + ucy)\sqrt{m} \\
&= (r - sy - uz)a + (t + sx + uy)(b + c\sqrt{m}),
\end{aligned}$$

proving that $\{a, b + c\sqrt{m}\}$ is a basis for I.

Conversely, suppose that $\{a, b + c\sqrt{m}\}$ is a basis for the ideal $I = \langle a, b + c\sqrt{m} \rangle$. As $\sqrt{m}a \in I$ and $\sqrt{m}(b + c\sqrt{m}) \in I$ there exist integers x, y, u, v such that

$$\begin{aligned}
\sqrt{m}a &= xa + y(b + c\sqrt{m}), \\
\sqrt{m}(b + c\sqrt{m}) &= ua + v(b + c\sqrt{m}).
\end{aligned}$$

Equating coefficients of 1 and $\sqrt{m}$, we obtain

$$\begin{aligned}
xa + yb &= 0, \\
yc &= a, \\
ua + vb &= cm, \\
vc &= b.
\end{aligned}$$

From the second and fourth equations, we see that $c \mid a$ and $c \mid b$ respectively. From the third and fourth equations, we obtain

$$uac + b^2 = c^2 m$$

so that

$$ac \mid b^2 - mc^2.$$

(b) This case can be treated similarly to part (a).   ■

6.6 Prime Ideals in Rings of Integers

In Theorem 1.5.6 we saw that a maximal ideal of an integral domain D is always a prime ideal. We noted that the converse is not always true but that it is true in a principal ideal domain. In this section we show the important result that a prime ideal is always maximal in the ring of integers of an algebraic number field.

Theorem 6.6.1 *Let P be a prime ideal of the ring O_K of integers of an algebraic number field K. Then P is a maximal ideal of O_K.*

Proof: Suppose that the assertion of the theorem is false. Then there exists a prime ideal P_1 of O_K that is not a maximal ideal. Let S be the set of all proper ideals of O_K that strictly contain P_1. As P_1 is not a maximal ideal, S is a nonempty set. By Theorem 6.5.3 O_K is a Noetherian domain. Hence, by Theorem 3.1.3, S contains a maximal element; that is, there is a maximal ideal P_2 such that

$$P_1 \subset P_2 \subset O_K.$$

By Theorem 1.5.6 P_2 is a prime ideal. Since every nonzero ideal in O_K contains a nonzero rational integer (Theorem 6.1.7) we see that $P_1 \cap \mathbb{Z} \neq \{0\}$. Hence, by Theorem 1.6.2, $P_1 \cap \mathbb{Z}$ is a prime ideal of $\mathbb{Z}$. But $\mathbb{Z}$ is a principal ideal domain (Theorem 1.4.1) so $P_1 \cap \mathbb{Z} = \langle p \rangle$ for some $p \in \mathbb{Z}$. By Theorem 1.5.4 p is a prime. Thus

$$\langle p \rangle = P_1 \cap \mathbb{Z} \subseteq P_2 \cap \mathbb{Z} \subseteq \mathbb{Z}.$$

Now $P_2 \cap \mathbb{Z} \neq \mathbb{Z}$ as $1 \notin P_2$, so as $\langle p \rangle$ is a maximal ideal of $\mathbb{Z}$ (Theorem 1.5.7), we have

$$P_1 \cap \mathbb{Z} = P_2 \cap \mathbb{Z} = \langle p \rangle.$$

As $P_1 \subset P_2$ there exists $\alpha \in P_2$ with $\alpha \notin P_1$. Since $\alpha \in O_K$ there exist a positive integer k and $a_0, \ldots, a_{k-1} \in \mathbb{Z}$ such that

$$\alpha^k + a_{k-1}\alpha^{k-1} + \cdots + a_1\alpha + a_0 = 0,$$

and so

$$\alpha^k + a_{k-1}\alpha^{k-1} + \cdots + a_1\alpha + a_0 \in P_1.$$

Let l be the least positive integer for which there exist $b_0, \ldots, b_{l-1} \in \mathbb{Z}$ such that

$$\alpha^l + b_{l-1}\alpha^{l-1} + \cdots + b_1\alpha + b_0 \in P_1. \tag{6.6.1}$$

Now, as $\alpha \in P_2$, we have

$$\alpha^l + b_{l-1}\alpha^{l-1} + \cdots + b_1\alpha = \alpha(\alpha^{l-1} + b_{l-1}\alpha^{l-2} + \cdots + b_1) \in P_2.$$

Hence, as $P_1 \subset P_2$ and P_2 is an ideal,

$$b_0 = (\alpha^l + \cdots + b_1\alpha + b_0) - (\alpha^l + \cdots + b_1\alpha) \in P_2.$$

But $b_0 \in \mathbb{Z}$ so

$$b_0 \in P_2 \cap \mathbb{Z} = P_1 \cap \mathbb{Z}$$

and thus $b_0 \in P_1$. From (6.6.1) we deduce that

$$\alpha^l + b_{l-1}\alpha^{l-1} + \cdots + b_1\alpha \in P_1.$$

If $l = 1$ then $\alpha \in P_1$, contradicting $\alpha \notin P_1$. Hence $l \geq 2$ and

$$\alpha(\alpha^{l-1} + \cdots + b_1) \in P_1.$$

Since P_1 is a prime ideal and $\alpha \notin P_1$ we deduce that

$$\alpha^{l-1} + \cdots + b_1 \in P_1,$$

contradicting the minimality of l since $l - 1$ is a positive integer as $l \geq 2$. ∎

Exercises

1. Let K be an algebraic number field of degree n. Let $\theta \in K$ be such that $K = \mathbb{Q}(\theta)$. Let $\theta_1 = \theta, \theta_2, \ldots, \theta_n$ be the conjugates of θ over $\mathbb{Q}$. Let $\alpha \in K$ so there exist unique rational numbers $c_0, c_1, \ldots, c_{n-1}$ such that

$$\alpha = c_0 + c_1\theta + \cdots + c_{n-1}\theta^{n-1}.$$

For $k = 1, 2, \ldots, n$ let

$$\alpha_k = c_0 + c_1\theta_k + \cdots + c_{n-1}\theta_k^{n-1}$$

so that $\alpha_1 = \alpha$. Prove that the set of conjugates $\{\alpha_1, \alpha_2, \ldots, \alpha_n\}$ of α relative to K does not depend on the choice of θ.

2. Prove that the cubic equation $x^3 + ax + b = 0$, where $a, b \in \mathbb{R}$, has three distinct real roots if $-4a^3 - 27b^2 > 0$, one real and two nonreal complex conjugate roots if $-4a^3 - 27b^2 < 0$, and at least two equal real roots if $-4a^3 - 27b^2 = 0$.

3. Let $K = \mathbb{Q}(\theta)$ be an algebraic number field of degree n. Let $\theta_1 = \theta, \theta_2, \ldots, \theta_n$ be the conjugates of θ over $\mathbb{Q}$. Suppose that there are exactly m distinct fields among $\mathbb{Q}(\theta_1), \ldots, \mathbb{Q}(\theta_n)$. Prove that $m \mid n$ and each field occurs n/m times.

4. Let m be a squarefree integer. Let $K = \mathbb{Q}(\sqrt{m})$. Prove that

$$\sigma_1(x + y\sqrt{m}) = x + y\sqrt{m} \ (x, y \in \mathbb{Q})$$

and

$$\sigma_2(x + y\sqrt{m}) = x - y\sqrt{m} \ (x, y \in \mathbb{Q})$$

are the only monomorphisms from K to $\mathbb{C}$.

5. Let m be a cubefree integer. Let $K = \mathbb{Q}(\sqrt[3]{m})$. Determine all the monomorphisms from K to $\mathbb{C}$.

6. Let $\theta = \sqrt[3]{1+i} + \sqrt[3]{1-i}$. Determine all the monomorphisms from $\mathbb{Q}(\theta)$ to $\mathbb{C}$.

7. Let θ be a root of the equation $x^6 + 2x^2 + 2 = 0$. Let $K = \mathbb{Q}(\theta)$. How many distinct elements are there in the complete set of conjugates of $\alpha = \theta^2 + \theta^4$ relative to K?

8. Let θ be a root of the equation $x^3 + 2x + 2 = 0$. Let $K = \mathbb{Q}(\theta)$ and $\alpha = \theta - \theta^2$. Determine the field polynomial of α over K.

9. Let $K = \mathbb{Q}(\theta)$, where $\theta^3 - 4\theta + 2 = 0$. Let $\alpha = \theta + \theta^2 \in K$. Determine $D(\alpha)$.

10. Find a basis for the ideal $\langle \sqrt{5} \rangle$ in O_K, where $K = \mathbb{Q}(\sqrt{5})$.

11. Determine the discriminant of the ideal $\langle 5 + \sqrt{2}, 7 + 2\sqrt{2} \rangle$ in O_K, where $K = \mathbb{Q}(\sqrt{2})$.

12. Let m be a squarefree integer $\equiv 2 \pmod 4$. Prove that

$$\langle 2, \sqrt{m} \rangle = 2\mathbb{Z} + \sqrt{m}\,\mathbb{Z}.$$

13. Let m be a squarefree integer $\equiv 3 \pmod 4$. Prove that

$$\langle 2, 1 + \sqrt{m} \rangle = 2\mathbb{Z} + (1 + \sqrt{m})\mathbb{Z}.$$

14. Let m be a squarefree integer $\equiv 1 \pmod 4$. Is

$$\langle 2, \frac{1 + \sqrt{m}}{2} \rangle = 2\mathbb{Z} + \frac{(1 + \sqrt{m})}{2}\mathbb{Z}?$$

15. Prove that the discriminant D of the cubic polynomial $x^3 + ax^2 + bx + c \in \mathbb{Z}[x]$ is

$$D = a^2b^2 - 4b^3 - 4a^3c - 27c^2 + 18abc.$$

Deduce that $D \equiv 0$ or $1 \pmod 4$.

16. Prove that the discriminant of the quartic polynomial $x^4 + ax^2 + bx + c \in \mathbb{Z}[x]$ is

$$D = 16a^4c - 4a^3b^2 - 128a^2c^2 + 144ab^2c - 27b^4 + 256c^3.$$

Deduce that $D \equiv 0$ or $1 \pmod 4$.

17. Let K be an algebraic number field. Let $\alpha \in K$. Let α' be a conjugate of α relative to K. Prove that $D(\alpha) = D(\alpha')$.

18. Let K be an algebraic number field. Let $\alpha \in K$. Let β be a conjugate of α relative to K. Prove that $\mathrm{fld}_K(\alpha) = \mathrm{fld}_K(\beta)$.

19. Let K be an algebraic number field. Let $\alpha, \beta \in K$ be such that $\mathrm{fld}_K(\alpha) = \mathrm{fld}_K(\beta)$. Prove that α and β are conjugates relative to K.

20. Let $K = \mathbb{Q}(\theta)$, where $\theta^3 + 4\theta - 2 = 0$. Is $K = \mathbb{Q}(\theta + \theta^2)$?

21. Let $K = \mathbb{Q}(\theta)$, where $\theta^4 - 4\theta^2 + 8 = 0$. Find a rational number c such that $\mathbb{Q}(\theta + c\theta^3) \neq K$.

22. Prove that the discriminant of the trinomial polynomial $x^n + ax + b \in \mathbb{Z}[x]$, where n is an integer ≥ 2, is

$$(-1)^{(n-1)(n-2)/2}(n-1)^{n-1}a^n + (-1)^{n(n-1)/2}n^n b^{n-1}.$$

23. Prove that the discriminant of the trinomial polynomial $x^n + ax^r + b \in \mathbb{Z}[x]$, where n and r are integers satisfying $n > r \geq 1$ and $(n, r) = 1$, is

$$(-1)^{(n-1)(n-2)/2}(n-r)^{n-r}r^r a^n b^{r-1} + (-1)^{n(n-1)/2}n^n b^{n-1}.$$

Suggested Reading

1. E. T. Bell, *Gauss and the early development of algebraic numbers*, National Mathematics Magazine 18 (1944), 188–204, 219–233.

   Bell provides a very readable account of the early development of algebraic numbers.

2. R. L. Goodstein, *The discriminant of a certain polynomial*, Mathematical Gazette 53 (1969), 60–61.

   The formula for the discriminant of $x^n + ax^r + b$ is derived

3. L. Mirsky, *An Introduction to Linear Algebra*, Oxford University Press, London 1972.

   The evaluation of the Vandermonde determinant is carried out on pages 17 and 18.

4. D. W. Masser, *The discriminants of special equations*, Mathematical Gazette 50 (1966), 158–160.

   The formula for the discriminant of $x^n + ax + b$ is derived.

Biographies

1. The website

   http://www-groups.dcs.st-and.ac.uk/~history/

   has a biography of A.-T. Vandermonde (1735–1796). Nowhere in his four mathematical papers does the so-called Vandermonde determinant appear!

7

Integral Bases

7.1 Integral Basis of an Algebraic Number Field

A basis of the principal ideal of the ring O_K of integers of an algebraic number field K generated by 1, that is, O_K itself, is called an integral basis for K.

Definition 7.1.1 (Integral basis of an algebraic number field) *Let K be an algebraic number field. A basis for O_K is called an integral basis for K.*

In view of this definition the following theorem, which gives an integral basis for a quadratic field, is just a restatement of Theorem 5.4.2.

Theorem 7.1.1 *Let K be a quadratic field. Let m be the unique squarefree integer such that $K = \mathbb{Q}(\sqrt{m})$. Then $\{1, \sqrt{m}\}$ is an integral basis for K if $m \not\equiv 1 \pmod 4$ and $\left\{1, \frac{1+\sqrt{m}}{2}\right\}$ is an integral basis for K if $m \equiv 1 \pmod 4$.*

If $\{\eta_1, \ldots, \eta_n\}$ and $\{\lambda_1, \ldots, \lambda_n\}$ are two integral bases for an algebraic number field K then Theorem 6.5.4 shows that $D(\eta_1, \ldots, \eta_n) = D(\lambda_1, \ldots, \lambda_n)$, and that if $\{\eta_1, \ldots, \eta_n\}$ is an integral basis for K and $\lambda_1, \ldots, \lambda_n \in O_K$ are such that $D(\lambda_1, \ldots, \lambda_n) = D(\eta_1, \ldots, \eta_n)$ then $\{\lambda_1, \ldots, \lambda_n\}$ is also an integral basis for K. We can therefore make the following definition.

Definition 7.1.2 (Discriminant of an algebraic number field) *Let K be an algebraic number field of degree n. Let $\{\eta_1, \ldots, \eta_n\}$ be an integral basis for K. Then $D(\eta_1, \ldots, \eta_n)$ is called the discriminant of K and is denoted by $d(K)$.*

Clearly if K is an algebraic number field of degree n and $\lambda_1, \ldots, \lambda_n \in O_K$ are such that $D(\lambda_1, \ldots, \lambda_n) = d(K)$, then $\{\lambda_1, \ldots, \lambda_n\}$ is an integral basis for K.

We determine the discriminant of a quadratic field.

Theorem 7.1.2 *Let K be a quadratic field. Let m be the unique squarefree integer such that $K = \mathbb{Q}(\sqrt{m})$. Then the discriminant $d(K)$ of K is given by*

$$d(K) = \begin{cases} 4m, & \text{if } m \not\equiv 1 \ (\text{mod } 4), \\ m, & \text{if } m \equiv 1 \ (\text{mod } 4). \end{cases}$$

Proof: We appeal to Theorem 7.1.1. If $m \not\equiv 1$ (mod 4), an integral basis for K is $\{1, \sqrt{m}\}$ so that

$$d(K) = \begin{vmatrix} 1 & \sqrt{m} \\ 1 & -\sqrt{m} \end{vmatrix}^2 = (-2\sqrt{m})^2 = 4m.$$

If $m \equiv 1$ (mod 4), an integral basis for K is $\left\{1, \frac{1+\sqrt{m}}{2}\right\}$ so that

$$d(K) = \begin{vmatrix} 1 & \dfrac{1 + \sqrt{m}}{2} \\ 1 & \dfrac{1 - \sqrt{m}}{2} \end{vmatrix}^2 = (-\sqrt{m})^2 = m. \qquad \blacksquare$$

Since $\sqrt{d(K)} = \sqrt{m}$ or $2\sqrt{m}$ the next theorem follows immediately from Theorem 7.1.2.

Theorem 7.1.3 *Let K be a quadratic field. Then $K = \mathbb{Q}(\sqrt{d(K)})$.*

We note that the quadratic field K is a real field if and only if $d(K) > 0$.

Next we define the norm of an ideal in the ring of integers of an algebraic number field.

Definition 7.1.3 (Norm of an ideal) *Let K be an algebraic number field of degree n. Let I be a nonzero ideal of O_K. Then the norm of the ideal I, written $N(I)$, is the positive integer defined by*

$$N(I) = \sqrt{\frac{D(I)}{d(K)}}.$$

We now justify that $N(I)$ is indeed a positive integer.

Theorem 7.1.4 *Let K be an algebraic number field of degree n. Let I be a nonzero ideal of O_K. Then the norm $N(I)$ of the ideal I is a positive integer.*

Proof: Let $\{\eta_1, \ldots, \eta_n\}$ be a basis for I and let $\{\omega_1, \ldots, \omega_n\}$ be an integral basis for K. As $\eta_1, \ldots, \eta_n \in O_K$ there exist c_{ij} $(i, j = 1, \ldots, n) \in \mathbb{Z}$ such that

$$\eta_i = \sum_{j=1}^{n} c_{ij}\omega_j, \ i = 1, \ldots, n.$$

Hence

$$D(\eta_1, \ldots, \eta_n) = (\det(c_{ij}))^2 D(\omega_1, \ldots, \omega_n)$$

so that

$$D(I) = \det(c_{ij}))^2 d(K).$$

Since $D(I) \neq 0$ we have $\det(c_{ij}) \neq 0$ so that

$$N(I) = \sqrt{\frac{D(I)}{d(K)}} = |\det(c_{ij})|$$

is a positive integer.   ∎

Example 7.1.1 *Let* $K = \mathbb{Q}(\sqrt{-5})$. *Let* I *be the ideal of* O_K *generated by 2 and* $1 + \sqrt{-5}$; *that is,* $I = \langle 2, \ 1 + \sqrt{-5} \rangle$. *We determine the norm* $N(I)$ *of the ideal* I. *First we find a basis for* I. *As* $O_K = \{x + y\sqrt{-5} \mid x, y \in \mathbb{Z}\}$ *we have*

$$
\begin{aligned}
I &= \{2(a + b\sqrt{-5}) + (1 + \sqrt{-5})(c + d\sqrt{-5}) \mid a, b, c, d \in \mathbb{Z}\} \\
&= \{(2a + c - 5d) + (2b + c + d)\sqrt{-5} \mid a, b, c, d \in \mathbb{Z}\} \\
&= \{(2a + (y - 2b - d) - 5d) + y\sqrt{-5} \mid a, b, d, y \in \mathbb{Z}\} \\
&= \{2(a - b - 3d) + y + y\sqrt{-5} \mid a, b, d, y \in \mathbb{Z}\} \\
&= \{2x + y + y\sqrt{-5} \mid x, y \in \mathbb{Z}\} \\
&= \{2x + (1 + \sqrt{-5})y \mid x, y \in \mathbb{Z}\},
\end{aligned}
$$

so that $\{2, \ 1 + \sqrt{-5}\}$ *is a basis for* I. *Hence*

$$D(I) = D(2, 1 + \sqrt{-5}) = \begin{vmatrix} 2 & 1 + \sqrt{-5} \\ 2 & 1 - \sqrt{-5} \end{vmatrix}^2 = (-4\sqrt{-5})^2 = -80.$$

By Theorem 7.1.2 we have

$$d(K) = 4(-5) = -20.$$

Hence

$$N(I) = \sqrt{\frac{D(I)}{d(K)}} = \sqrt{\frac{-80}{-20}} = \sqrt{4} = 2.$$

Example 7.1.2 *Let* $K = \mathbb{Q}(\sqrt{m})$, *where* m *is a squarefree integer with* $m \not\equiv 1 \pmod 4$, *so that* $O_K = \{x + y\sqrt{m} \mid x, y \in \mathbb{Z}\}$. *Let* $\alpha = a + b\sqrt{m} \in O_K$. *We* *determine the norm of the principal ideal $\langle \alpha \rangle$. We have*

$$
\begin{aligned}
\langle \alpha \rangle &= \langle a + b\sqrt{m} \rangle \\
&= \{(x + y\sqrt{m})(a + b\sqrt{m}) \mid x, y \in \mathbb{Z}\} \\
&= \{x(a + b\sqrt{m}) + y(bm + a\sqrt{m}) \mid x, y \in \mathbb{Z}\}
\end{aligned}
$$

so that $\{a + b\sqrt{m}, \ bm + a\sqrt{m}\}$ is a basis for $\langle \alpha \rangle$. Hence

$$
\begin{aligned}
D(\langle \alpha \rangle) &= \begin{vmatrix} a + b\sqrt{m} & bm + a\sqrt{m} \\ a - b\sqrt{m} & bm - a\sqrt{m} \end{vmatrix}^2 \\
&= ((a + b\sqrt{m})(bm - a\sqrt{m}) - (a - b\sqrt{m})(bm + a\sqrt{m}))^2.
\end{aligned}
$$

Recalling the identity

$$
(A + B)(C - D) - (A - B)(C + D) = 2(BC - AD),
$$

we see that

$$
D(\langle \alpha \rangle) = 2^2 m (a^2 - mb^2)^2.
$$

Now $d(K) = 4m$ (Theorem 7.1.2) so that

$$
N(\langle \alpha \rangle) = \sqrt{\frac{2^2 m (a^2 - mb^2)^2}{4m}} = |a^2 - mb^2|.
$$

We now use Theorem 6.5.5 to determine the norms of a wide class of ideals in a quadratic field.

Theorem 7.1.5 *Let K be a quadratic field. Let m be the unique squarefree integer such that $K = \mathbb{Q}(\sqrt{m})$.*

(a) *$m \not\equiv 1 \pmod 4$. Let $a, b, c \in \mathbb{Z}$ be such that*

$$
a \neq 0, \ c \neq 0, \ c \mid a, \ c \mid b, \ ac \mid b^2 - mc^2.
$$

Then

$$
N(\langle a, b + c\sqrt{m} \rangle) = |ac|.
$$

(b) *$m \equiv 1 \pmod 4$. Let $a, b, c \in \mathbb{Z}$ be such that*

$$
a \neq 0, \ c \neq 0, \ b \equiv c \pmod 2, \ c \mid a, \ c \mid b, \ 4ac \mid b^2 - mc^2.
$$

Then

$$
N(\langle a, \frac{b + c\sqrt{m}}{2} \rangle) = |ac|.
$$

Proof: **(a)** By Theorem 6.5.5(a) $\{a, \ b + c\sqrt{m}\}$ is a basis for the ideal $\langle a, b + c\sqrt{m}\rangle$. Hence

$$D(\langle a, b + c\sqrt{m}\rangle) = D(a, b + c\sqrt{m})$$

$$= \begin{vmatrix} a & b + c\sqrt{m} \\ a & b - c\sqrt{m} \end{vmatrix}^2$$

$$= 4a^2c^2m.$$

As $m \not\equiv 1 \pmod 4$ we have

$$d(K) = 4m,$$

by Theorem 7.1.2. Thus

$$N(\langle a, b + c\sqrt{m}\rangle) = \sqrt{\frac{4a^2c^2m}{4m}} = |ac|.$$

(b) By Theorem 6.5.5(b) we see that $\left\{a, \frac{b+c\sqrt{m}}{2}\right\}$ is a basis for the ideal $\langle a, \frac{b+c\sqrt{m}}{2}\rangle$. Hence

$$D\left(\langle a, \frac{b + c\sqrt{m}}{2}\rangle\right) = D\left(a, \frac{b + c\sqrt{m}}{2}\right)$$

$$= \begin{vmatrix} a & \dfrac{b + c\sqrt{m}}{2} \\ a & \dfrac{b - c\sqrt{m}}{2} \end{vmatrix}^2 = a^2c^2m.$$

As $m \equiv 1 \pmod 4$ we have

$$d(K) = m,$$

by Theorem 7.1.2. Thus

$$N\left(\langle a, \frac{b + c\sqrt{m}}{2}\rangle\right) = \sqrt{\frac{a^2c^2m}{m}} = |ac|. \qquad \blacksquare$$

We remark that Example 7.1.1 is the special case $a = 2$, $b = c = 1$, $m = -5$ of Theorem 7.1.5(a).

The next theorem determines the norm of a principal ideal in the ring of integers of an arbitrary algebraic number field, which is generated by a rational integer.

Theorem 7.1.6 *Let K be an algebraic number field of degree n. Let c be a nonzero rational integer. Then the norm of the principal ideal $\langle c \rangle$ of O_K generated by c is*

$$N(\langle c \rangle) = |c|^n.$$

Proof: Let $\{\omega_1, \ldots, \omega_n\}$ be an integral basis for K. Then

$$\alpha \in \langle c \rangle \Longleftrightarrow \alpha = c\beta \text{ for a unique } \beta \in O_K$$
$$\Longleftrightarrow \alpha = c(x_1\omega_1 + \cdots + x_n\omega_n) \text{ for unique } x_1, \ldots, x_n \in \mathbb{Z}$$
$$\Longleftrightarrow \alpha = x_1(c\omega_1) + \cdots + x_n(c\omega_n) \text{ for unique } x_1, \ldots, x_n \in \mathbb{Z}.$$

This shows that $\{c\omega_1, \ldots, c\omega_n\}$ is a basis for the principal ideal $\langle c \rangle$. Hence

$$D(\langle c \rangle) = D(c\omega_1, \ldots, c\omega_n) = c^{2n} D(\omega_1, \ldots, \omega_n) = c^{2n} d(K)$$

so that

$$N(\langle c \rangle) = \sqrt{\frac{D(\langle c \rangle)}{d(K)}} = \sqrt{c^{2n}} = |c|^n. \qquad \blacksquare$$

If K is an algebraic number field of degree n then by Theorem 6.1.2 there exists $\theta \in O_K$ such that $K = \mathbb{Q}(\theta)$. By Theorem 6.4.3 we have $D(\theta) \neq 0$. Let $\omega_1, \ldots, \omega_n$ be an integral basis for K. Then there exist c_{ij} $(i, j = 1, \ldots, n) \in \mathbb{Z}$ such that

$$\begin{cases} 1 = c_{11}\omega_1 + \cdots + c_{1n}\omega_n, \\ \theta = c_{21}\omega_1 + \cdots + c_{2n}\omega_n, \\ \quad \cdots \\ \theta^{n-1} = c_{n1}\omega_1 + \cdots + c_{nn}\omega_n. \end{cases}$$

Hence

$$D(\theta) = D(1, \theta, \ldots, \theta^{n-1}) = (\det(c_{ij}))^2 D(\omega_1, \ldots, \omega_n) = |\det(c_{ij})|^2 d(K),$$

showing that

$$D(\theta) = m^2 d(K)$$

for some positive integer $m (= |\det(c_{ij})|)$. The positive integer m is called the index of θ.

Definition 7.1.4 (Index of θ) *Let K be an algebraic number field. Let $\theta \in O_K$ be such that $K = \mathbb{Q}(\theta)$. Then the index of θ, written* ind θ, *is the positive integer given by*

$$D(\theta) = (\text{ind } \theta)^2 d(K).$$

Theorem 7.1.7 *Let K be an algebraic number field of degree n. Let $\theta \in O_K$ be such that $K = \mathbb{Q}(\theta)$. Then $\{1, \theta, \theta^2, \ldots, \theta^{n-1}\}$ is an integral basis for K if and only if* ind $\theta = 1$.

Proof: We have

$\{1, \theta, \theta^2, \dots, \theta^{n-1}\}$ is an integral basis for K

$$\Longleftrightarrow D(1, \theta, \dots, \theta^{n-1}) = d(K) \text{ (Theorem 6.5.4)}$$

$$\Longleftrightarrow D(\theta) = d(K)$$

$$\Longleftrightarrow \text{ind } \theta = 1.$$

∎

Clearly if $D(\theta)$ is squarefree then ind $\theta = 1$ so that, by Theorem 7.1.7, $\{1, \theta, \dots, \theta^{n-1}\}$ is an integral basis.

Theorem 7.1.8 *Let K be an algebraic number field of degree n. Let $\theta \in O_K$ be such that $K = \mathbb{Q}(\theta)$. If $D(\theta)$ is squarefree then $\{1, \theta, \dots, \theta^{n-1}\}$ is an integral basis for K.*

To apply Theorem 7.1.8 in a particular example we need to calculate $D(\theta)$. The following result is often useful in this connection.

Theorem 7.1.9 *Let θ be an algebraic number of degree n. Let $\theta_1 = \theta, \theta_2, \dots, \theta_n$ be the conjugates of θ over $\mathbb{Q}$, that is, the roots of $f(x) = \mathrm{irr}_{\mathbb{Q}}(\theta)$. Then*

$$D(\theta) = (-1)^{n(n-1)/2} \prod_{i=1}^{n} f'(\theta_i).$$

Proof: We have

$$f(x) = \mathrm{irr}_{\mathbb{Q}}(\theta) = \prod_{i=1}^{n} (x - \theta_i).$$

Differentiating $f(x)$ using the product rule, we obtain

$$f'(x) = \sum_{i=1}^{n} \prod_{\substack{j=1 \\ j \neq i}}^{n} (x - \theta_j)$$

so that

$$f'(\theta_i) = \prod_{\substack{j=1 \\ j \neq i}}^{n} (\theta_i - \theta_j), \quad i = 1, 2, \dots, n.$$

Hence

$$\prod_{i=1}^{n} f'(\theta_i) = \prod_{i=1}^{n} \prod_{\substack{j=1 \\ j \neq i}}^{n} (\theta_i - \theta_j)$$

$$= \prod_{1 \leq i < j \leq n} (\theta_i - \theta_j) \prod_{1 \leq j < i \leq n} (\theta_i - \theta_j)$$

$$= (-1)^{n(n-1)/2} \prod_{1 \leq i < j \leq n} (\theta_i - \theta_j) \prod_{1 \leq j < i \leq n} (\theta_j - \theta_i)$$

$$= (-1)^{n(n-1)/2} \prod_{1 \leq i < j \leq n} (\theta_i - \theta_j) \prod_{1 \leq i < j \leq n} (\theta_i - \theta_j)$$

$$= (-1)^{n(n-1)/2} \prod_{1 \leq i < j \leq n} (\theta_i - \theta_j)^2$$

$$= (-1)^{n(n-1)/2} D(\theta),$$

by Theorem 6.4.1. ∎

Clearly with the notation of Theorem 7.1.9, we have

$$\text{disc}(\text{irr}_{\mathbb{Q}}\theta) = \prod_{1 \leq i < j \leq n} (\theta_i - \theta_j)^2 = (-1)^{n(n-1)/2} \prod_{i=1}^{n} f'(\theta_i) = D(\theta),$$

in agreement with Theorem 6.4.2.

We apply Theorem 7.1.9 in the case when θ is an algebraic integer of degree 3.

Theorem 7.1.10 *Let a, b be integers such that $x^3 + ax + b \in \mathbb{Z}[x]$ is irreducible. Let $\theta \in \mathbb{C}$ be a root of $x^3 + ax + b$ so that $K = \mathbb{Q}(\theta)$ is a cubic field and $\theta \in O_K$. Then*

$$D(\theta) = -4a^3 - 27b^2.$$

Proof: Let $f(x) = \text{irr}_{\mathbb{Q}}(\theta) = x^3 + ax + b$. Let $\theta_1 = \theta, \theta_2, \theta_3$ be the conjugates of θ over $\mathbb{Q}$ so that

$$(x - \theta_1)(x - \theta_2)(x - \theta_3) = x^3 + ax + b.$$

Equating coefficients we obtain

$$\theta_1 + \theta_2 + \theta_3 = 0,$$
$$\theta_1\theta_2 + \theta_2\theta_3 + \theta_3\theta_1 = a,$$
$$\theta_1\theta_2\theta_3 = -b.$$

Now

$$f'(x) = 3x^2 + a$$

so that

$$f'(\theta_1)f'(\theta_2)f'(\theta_3) = (3\theta_1^2 + a)(3\theta_2^2 + a)(3\theta_3^2 + a)$$
$$= a^3 + 3a^2(\theta_1^2 + \theta_2^2 + \theta_3^2) + 9a(\theta_1^2\theta_2^2 + \theta_2^2\theta_3^2 + \theta_3^2\theta_1^2)$$
$$+ 27\theta_1^2\theta_2^2\theta_3^2.$$

Next we observe that

$$\theta_1^2 + \theta_2^2 + \theta_3^2 = (\theta_1 + \theta_2 + \theta_3)^2 - 2(\theta_1\theta_2 + \theta_2\theta_3 + \theta_3\theta_1) = -2a,$$
$$\theta_1^2\theta_2^2 + \theta_2^2\theta_3^2 + \theta_3^2\theta_1^2 = (\theta_1\theta_2 + \theta_2\theta_3 + \theta_3\theta_1)^2 - 2\theta_1\theta_2\theta_3(\theta_1 + \theta_2 + \theta_3) = a^2,$$
$$\theta_1^2\theta_2^2\theta_3^2 = (\theta_1\theta_2\theta_3)^2 = b^2,$$

so

$$f'(\theta_1)f'(\theta_2)f'(\theta_3) = a^3 + 3a^2(-2a) + 9a(a^2) + 27b^2 = 4a^3 + 27b^2.$$

Hence by Theorem 7.1.9 we obtain

$$D(\theta) = (-1)^{\frac{3\cdot2}{2}} f'(\theta_1)f'(\theta_2)f'(\theta_3) = -4a^3 - 27b^2,$$

as asserted. ∎

In the next example we find an integral basis for a particular cubic field using Theorems 7.1.8 and 7.1.10.

Example 7.1.3 *Let* $K = \mathbb{Q}(\theta)$, *where* θ *is a root of* $x^3 + x + 1$. *The cubic polynomial* $x^3 + x + 1$ *is irreducible in* $\mathbb{Z}[x]$ *so that* $[K : \mathbb{Q}] = 3$. *Also* $\theta \in O_K$. *By Theorem 7.1.10 we have*

$$D(\theta) = -4(1)^3 - 27(1)^2 = -31.$$

As -31 *is squarefree, by Theorem 7.1.8* $\{1, \theta, \theta^2\}$ *is an integral basis for* K.

More generally we have the following result.

Theorem 7.1.11 *Let* a, b *be integers such that* $x^3 + ax + b \in \mathbb{Z}[x]$ *is irreducible and* $-4a^3 - 27b^2$ *is squarefree. Let* θ *be a root of* $x^3 + ax + b$. *Then* $\{1, \theta, \theta^2\}$ *is an integral basis for the cubic field* $\mathbb{Q}(\theta)$.

Other values of a and b satisfying the conditions of Theorem 7.1.11 are

$$(a, b) = (-1, -1), (2, 1), (4, 1), (-1, 3), \text{ and } (5, 3).$$

Similarly to Theorem 7.1.10 we can use Theorem 7.1.9 to prove the following result.

Theorem 7.1.12 *Let* a, b *be integers such that* $x^4 + ax + b \in \mathbb{Z}[x]$ *is irreducible. Let* θ *be a root of* $x^4 + ax + b$ *so that* $K = \mathbb{Q}(\theta)$ *is a quartic field and* $\theta \in O_K$. *Then*

$$D(\theta) = -27a^4 + 256b^3.$$

Appealing to Theorems 7.1.8 and 7.1.12 we obtain

Theorem 7.1.13 *Let* a, b *be integers such that* $x^4 + ax + b \in \mathbb{Z}[x]$ *is irreducible and* $-27a^4 + 256b^3$ *is squarefree. Let* θ *be a root of* $x^4 + ax + b$. *Then* $\{1, \theta, \theta^2, \theta^3\}$ *is an integral basis for the quartic field* $\mathbb{Q}(\theta)$.

The quartic polynomial $x^4 + x + 1$ is irreducible and has discriminant $-27(1)^4 + 256(1)^3 = 229$, which is prime. Hence, by Theorem 7.1.13, the quartic field $\mathbb{Q}(\theta)$, where $\theta^4 + \theta + 1 = 0$, has $\{1, \theta, \theta^2, \theta^3\}$ as an integral basis.

If we take K to be the cubic field $\mathbb{Q}(\theta)$, where $\theta^3 - 2 = 0$, then $D(\theta) = -108 = -3 \cdot 6^2$ and Theorem 7.1.11 is not applicable. As $D(\theta)/d(K)$ is a perfect square $(= (\text{ind } \theta)^2)$, we have

$$\frac{D(\theta)}{d(K)} = 1, 4, 9, \text{ or } 36$$

so that

$$d(K) = -108, -27, -12, \text{ or } -3,$$

and further information is required to determine which case actually occurs.

In some cases the following result first proved by Ludwig Stickelberger (1850–1936) in 1897 is useful (see [18]).

Theorem 7.1.14 *Let* K *be an algebraic number field. Then*

$$d(K) \equiv 0 \text{ or } 1 \pmod 4.$$

Proof: Let $\{\omega_1, \omega_2, \ldots, \omega_n\}$ be an integral basis for K. Let $\omega_i^{(1)} = \omega_i, \omega_i^{(2)}, \ldots, \omega_i^{(n)}$ be the K-conjugates of ω_i $(i = 1, 2, \ldots, n)$. In the expansion of the determinant

$$\begin{vmatrix} \omega_1^{(1)} & \omega_2^{(1)} & \cdots & \omega_n^{(1)} \\ \omega_1^{(2)} & \omega_2^{(2)} & \cdots & \omega_n^{(2)} \\ \vdots & \vdots & \cdots & \vdots \\ \omega_1^{(n)} & \omega_2^{(n)} & \cdots & \omega_n^{(n)} \end{vmatrix}$$

there are $n!$ terms, half of which occur with positive signs and half with negative signs. Let the sum of those with positive signs be λ and those with negative signs

μ so that

$$\det(\omega_i^{(j)}) = \lambda - \mu.$$

Set

$$A = \lambda + \mu, \quad B = \lambda\mu.$$

Then

$$d(K) = (\det(\omega_i^{(j)}))^2 = (\lambda - \mu)^2 = (\lambda + \mu)^2 - 4\lambda\mu = A^2 - 4B.$$

As $\omega_i \in O_K$ ($i = 1, 2, \ldots, n$), by Theorem 6.3.3 each $\omega_i^{(j)}$ ($i, j = 1, 2, \ldots, n) \in \Omega$. Hence $\lambda, \mu \in \Omega$ so

$$A \in \Omega, \quad B \in \Omega.$$

Let $\theta \in O_K$ be such that $K = \mathbb{Q}(\theta)$. Let $\theta_1 = \theta, \theta_2, \ldots, \theta_n$ be the conjugates of θ over $\mathbb{Q}$. If we express each ω_j as a polynomial in θ with rational coefficients, A becomes a symmetric function of $\theta_1, \ldots, \theta_n$ with rational coefficients, and so

$$A \in \mathbb{Q}.$$

Hence $A \in \Omega \cap \mathbb{Q} = \mathbb{Z}$. Then

$$B = \frac{A^2 - d(K)}{4} \in \mathbb{Q},$$

so that $B \in \Omega \cap \mathbb{Q} = \mathbb{Z}$. Finally, as $A, B \in \mathbb{Z}$, we have

$$d(K) = A^2 - 4B \equiv 0 \text{ or } 1 \pmod{4}. \qquad \blacksquare$$

The next example illustrates the use of Theorem 7.1.14 to determine the discriminant of an algebraic number field $K = \mathbb{Q}(\theta)$ when $D(\theta)$ is not squarefree.

Example 7.1.4 *The cubic polynomial $x^3 - x - 2 \in \mathbb{Z}[x]$ is irreducible. Let θ be a root of $x^3 - x - 2$ and set $K = \mathbb{Q}(\theta)$ so that $[K : \mathbb{Q}] = 3$. By Theorem 7.1.10 we have*

$$D(\theta) = -4(-1)^3 - 27(-2)^2 = -104 = -26 \cdot 2^2.$$

Since $D(\theta)/d(K)$ must be a square in $\mathbb{Z}$, we have

$$\frac{D(\theta)}{d(K)} = 1 \text{ or } 4$$

so that

$$d(K) = -104 \text{ or } -26.$$

But by Theorem 7.1.14, $d(K) \equiv 0$ or $1 \pmod 4$ so that $d(K) \neq -26$. Hence $d(K) = -104$, and $\{1, \theta, \theta^2\}$ is an integral basis for K.

The following result is a straightforward generalization of Example 7.1.4.

Theorem 7.1.15 *Let a, b be integers such that $x^3 + ax + b \in \mathbb{Z}$ is irreducible and*

$$-4a^3 - 27b^2 = 4m,$$

where m is a squarefree integer $\equiv 2$ or $3 \pmod 4$. Let θ be a root of $x^3 + ax + b$. Then $\{1, \theta, \theta^2\}$ is an integral basis for the cubic field $\mathbb{Q}(\theta)$.

In the next example we give a cubic field $\mathbb{Q}(\theta)$ for which $\{1, \theta, \theta^2\}$ is not an integral basis.

Example 7.1.5 *The cubic polynomial $x^3 + 11x + 4 \in \mathbb{Z}[x]$ is irreducible. Let θ be a root of $x^3 + 11x + 4$. Set $K = \mathbb{Q}(\theta)$ so that K is a cubic field. By Theorem 7.1.10 we have*

$$D(\theta) = -4(11)^3 - 27(4)^2 = -5756 = -1439 \cdot 2^2,$$

where 1439 is prime. As $D(\theta)/d(K)$ is the square of an integer, we have

$$\frac{D(\theta)}{d(K)} = 1 \text{ or } 4$$

so that

$$d(K) = -4 \cdot 1439 \text{ or } -1439.$$

The first of these is $\equiv 0 \pmod 4$ and the second is $\equiv 1 \pmod 4$, so we cannot use Theorem 7.1.14 to distinguish between them. We recall from Example 6.3.2 that $(\theta + \theta^2)/2$ is an integer of K as it is a root of the polynomial $x^3 + 11x^2 + 36x + 4 \in \mathbb{Z}[x]$. Hence $\{1, \theta, \theta^2\}$ is not an integral basis for K. Thus

$$d(K) \neq D(1, \theta, \theta^2) = D(\theta) = -1439 \cdot 2^2,$$

so $d(K) = -1439$. Since

$$D\left(1, \theta, \frac{\theta + \theta^2}{2}\right) = \begin{vmatrix} 1 & 0 & 0 \\ 0 & 1 & 0 \\ 0 & \frac{1}{2} & \frac{1}{2} \end{vmatrix}^2 D(1, \theta, \theta^2)$$

$$= \frac{1}{4}D(\theta) = \frac{1}{4}(-5756) = -1439,$$

$\{1, \theta, (\theta + \theta^2)/2\}$ is an integral basis for $\mathbb{Q}(\theta)$.

In the next example Theorem 7.1.14 is not sufficient to distinguish between the possible values of the discriminant and we have to carry out a more detailed analysis.

Example 7.1.6 *Let* $\theta = \sqrt[3]{2}$ *and set* $K = \mathbb{Q}(\theta) = \mathbb{Q}(\sqrt[3]{2})$. *Since* θ *is a root of the irreducible polynomial* $x^3 - 2 \in \mathbb{Z}[x]$, *we have* $\mathrm{irr}_{\mathbb{Q}}(\theta) = x^3 - 2$ *and* $[K : \mathbb{Q}] = \deg(\mathrm{irr}_{\mathbb{Q}}(\theta)) = 3$. *By Theorem 7.1.10 we have*

$$D(\theta) = -4(0)^3 - 27(-2)^2 = -108 = -2^2 \cdot 3^3.$$

As $D(\theta)/d(K)$ *is a perfect square, we must have*

$$\frac{D(\theta)}{d(K)} = 1^2, 2^2, 3^2, \text{ or } 6^2$$

so that

$$d(K) = -108, -27, -12, \text{ or } -3.$$

Each of these possibilities is congruent to 0 or 1 modulo 4, so we cannot use Theorem 7.1.14 to distinguish among them. We proceed instead by showing that if $x_1 + x_2\theta + x_3\theta^2 \in O_K$, *where* $x_1, x_2, x_3 \in \mathbb{Q}$, *then* $x_1, x_2, x_3 \in \mathbb{Z}$ *so that* $\{1, \theta, \theta^2\}$ *is an integral basis for K and* $d(K) = -108$. *Clearly* $\mathbb{Z} + \mathbb{Z}\theta + \mathbb{Z}\theta^2 \subseteq O_K$, *so we wish to show that* $O_K \subseteq \mathbb{Z} + \mathbb{Z}\theta + \mathbb{Z}\theta^2$.

Let $\alpha \in O_K$. *Then* $\alpha \in K$ *and thus there exist* $x_1, x_2, x_3 \in \mathbb{Q}$ *such that*

$$\alpha = x_1 + x_2\theta + x_3\theta^2.$$

The K-conjugates of α *are*

$$\begin{cases} \alpha = x_1 + x_2\theta + x_3\theta^2, \\ \alpha' = x_1 + x_2\omega\theta + x_3\omega^2\theta^2, \\ \alpha'' = x_1 + x_2\omega^2\theta + x_3\omega\theta^2, \end{cases}$$

where ω *is a complex cube root of unity. Hence, as* $1 + \omega + \omega^2 = 0$, *we have*

$$\begin{cases} \alpha + \alpha' + \alpha'' = 3x_1, \\ \theta^2(\alpha + \omega^2\alpha' + \omega\alpha'') = 6x_2, \\ \theta(\alpha + \omega\alpha' + \omega^2\alpha'') = 6x_3. \end{cases}$$

As $\alpha \in O_K$, *by Theorem 6.3.3 we have* $\alpha, \alpha', \alpha'' \in \Omega$; *as* $\theta \in O_K$ *we have* $\theta, \theta^2 \in \Omega$; *and as* $\omega, \omega^2 \in \mathbb{Z} + \mathbb{Z}\left(\frac{1+\sqrt{-3}}{2}\right) = O_{\mathbb{Q}(\sqrt{-3})}$ *we have* $\omega, \omega^2 \in \Omega$. *Thus*

$$\alpha + \alpha' + \alpha'', \ \theta^2(\alpha + \omega^2\alpha' + \omega\alpha''), \ \theta(\alpha + \omega\alpha' + \omega^2\alpha'') \in \Omega$$

so that

$$3x_1 \in \Omega \cap \mathbb{Q} = \mathbb{Z}, \ 6x_2 \in \Omega \cap \mathbb{Q} = \mathbb{Z}, \ 6x_3 \in \Omega \cap \mathbb{Q} = \mathbb{Z}.$$

Set

$$y_i = 6x_i \in \mathbb{Z}, \ i = 1, 2, 3,$$

so that

$$6\alpha = y_1 + y_2\theta + y_3\theta^2. \tag{7.1.1}$$

Before proceeding we note the following simple result. Suppose $\theta \mid n$ in O_K, where $n \in \mathbb{Z}$. Then $n = \theta w$ for $w \in O_K$. Thus $n^3 = \theta^3 w^3 = 2w^3$. Now $w^3 = n^3/2 \in \mathbb{Q}$ and $w^3 \in O_K \subseteq \Omega$ so that $w^3 \in \mathbb{Q} \cap \Omega = \mathbb{Z}$. Thus $2 \mid n^3$ in $\mathbb{Z}$. But 2 is a prime, so $2 \mid n$ in $\mathbb{Z}$. We have shown that

$$\theta \mid n \ in \ O_K \implies 2 \mid n \ in \ \mathbb{Z}.$$

We also note that θ, θ^2, and θ^3 divide 2 in O_K as $2 = \theta^3$. Using these results in (7.1.1), we see that $\theta \mid y_1$ so that $2 \mid y_1$. Then $\theta^2 \mid y_2\theta$ so $\theta \mid y_2$ and thus $2 \mid y_2$. Finally, $\theta^3 \mid y_3\theta^2$, so $\theta \mid y_3$ and thus $2 \mid y_3$. Set

$$y_i = 2z_i, \ i = 1, 2, 3,$$

so that

$$3\alpha = z_1 + z_2\theta + z_3\theta^2, \ z_1, z_2, z_3 \in \mathbb{Z}.$$

If $z_2 = z_3 = 0$ then $3\alpha = z_1$ so that $\alpha = z_1/3 \in \mathbb{Q}$. But $\alpha \in O_K \subseteq \Omega$ so that $\alpha \in \mathbb{Q} \cap \Omega = \mathbb{Z}$. Hence $\alpha \in \mathbb{Z} + \mathbb{Z}\theta + \mathbb{Z}\theta^2$ as required. If $(z_2, z_3) \neq (0, 0)$ then $\alpha = \frac{1}{3}(z_1 + z_2\theta + z_3\theta^2) \notin \mathbb{Q}$ since $\deg(\mathrm{irr}_\mathbb{Q}(\theta)) = 3$. Hence

$$\mathbb{Q}(\alpha) = \mathbb{Q}(\frac{1}{3}(z_1 + z_2\theta + z_3\theta^2)) \neq \mathbb{Q}.$$

Now $\alpha \in \mathbb{Q}(\theta)$ so $\mathbb{Q}(\alpha) \subseteq \mathbb{Q}(\theta)$ and thus $[\mathbb{Q}(\alpha) : \mathbb{Q}] \mid [\mathbb{Q}(\theta) : \mathbb{Q}] = 3$; that is, $[\mathbb{Q}(\alpha) : \mathbb{Q}] = 1$ or 3. But $\mathbb{Q}(\alpha) \neq \mathbb{Q}$ so $[\mathbb{Q}(\alpha) : \mathbb{Q}] \neq 1$. Hence $[\mathbb{Q}(\alpha) : \mathbb{Q}] = 3$. Thus the minimal polynomial of α over $\mathbb{Q}$ is of degree 3. Now α is a root of

$$x^3 + c_1x^2 + c_2x + c_3 \in \mathbb{Q}[x],$$

where

$$c_1 = -(\alpha + \alpha' + \alpha'') = -z_1,$$
$$c_2 = \alpha\alpha' + \alpha'\alpha'' + \alpha''\alpha = \frac{1}{3}(z_1^2 - 2z_2z_3),$$
$$c_3 = -\alpha\alpha'\alpha'' = \frac{-1}{27}(z_1^3 + 2z_2^3 + 4z_3^3 - 6z_1z_2z_3).$$

Hence

$$\mathrm{irr}_\mathbb{Q}(\alpha) = x^3 + c_1x^2 + c_2x + c_3.$$

Since $\alpha \in \Omega$ we must have $x^3 + c_1 x^2 + c_2 x + c_3 \in \mathbb{Z}[x]$ so that $c_1, c_2, c_3 \in \mathbb{Z}$; that is,

$$z_1^2 - 2z_2z_3 \equiv 0 \,(\mathrm{mod}\ 3), \qquad (7.1.2)$$
$$z_1^3 + 2z_2^3 + 4z_3^3 - 6z_1z_2z_3 \equiv 0 \,(\mathrm{mod}\ 27). \qquad (7.1.3)$$

Suppose that at least one of z_1, z_2, z_3 is not divisible by 3. Then (7.1.2) and (7.1.3) show that 3 does not divide any of z_1, z_2, z_3. From (7.1.3) we have

$$z_1^3 + 2z_2^3 + 4z_3^3 \equiv 0 \,(\mathrm{mod}\ 3).$$

As $z^3 \equiv z \,(\mathrm{mod}\ 3)$ for any integer z, we have

$$z_1 + 2z_2 + z_3 \equiv 0 \,(\mathrm{mod}\ 3).$$

Thus, as $3 \nmid z_1, z_2, z_3$, we must have

$$(z_1, z_2, z_3) \equiv (1, 2, 1) \ or \ (2, 1, 2) \,(\mathrm{mod}\ 3)$$

so that

$$z_2 \equiv 2z_1 \,(\mathrm{mod}\ 3), \quad z_3 \equiv z_1 \,(\mathrm{mod}\ 3).$$

Define integers t and u by

$$z_2 = 2z_1 + 3t, \quad z_3 = z_1 + 3u.$$

Then

$$
\begin{aligned}
z_1^3 &+ 2z_2^3 + 4z_3^3 - 6z_1z_2z_3 \\
&= z_1^3 + 2(2z_1 + 3t)^3 + 4(z_1 + 3u)^3 - 6z_1(2z_1 + 3t)(z_1 + 3u) \\
&= 9z_1^3 + 54(tz_1^2 + 2t^2z_1 + t^3 + 2u^2z_1 + 2u^3 - tuz_1) \\
&\equiv 9z_1^3 \,(\mathrm{mod}\ 27) \\
&\not\equiv 0 \,(\mathrm{mod}\ 27),
\end{aligned}
$$

as $3 \nmid z_1$, contradicting (7.1.3). Hence $z_1 \equiv z_2 \equiv z_3 \equiv 0 \,(\mathrm{mod}\ 3)$. Thus we can define integers w_1, w_2, w_3 by

$$z_i = 3w_i, \quad i = 1, 2, 3.$$

Then

$$\alpha = w_1 + w_2\theta + w_3\theta^2 \in \mathbb{Z} + \mathbb{Z}\theta + \mathbb{Z}\theta^2,$$

proving $O_K \subseteq \mathbb{Z} + \mathbb{Z}\theta + \mathbb{Z}\theta^2$ as required.

We have shown that $\{1, \theta, \theta^2\}$ is an integral basis for $K = \mathbb{Q}(\theta) = \mathbb{Q}(\sqrt[3]{2})$, so that

$$d(\mathbb{Q}(\sqrt[3]{2})) = \begin{vmatrix} 1 & \sqrt[3]{2} & (\sqrt[3]{2})^2 \\ 1 & \omega\sqrt[3]{2} & \omega^2(\sqrt[3]{2})^2 \\ 1 & \omega^2\sqrt[3]{2} & \omega(\sqrt[3]{2})^2 \end{vmatrix}^2 = -108.$$

Richard Dedekind (1831–1916) [5] determined an integral basis for the cubic field $\mathbb{Q}(\sqrt[3]{m})$ (with m a cubefree integer) in 1900 (see Theorem 7.3.2).

We conclude this section with the determination of an integral basis for the quartic field $\mathbb{Q}(\sqrt{-1} + \sqrt{2})$.

Example 7.1.7 *Let K be the quartic field $\mathbb{Q}(\sqrt{-1} + \sqrt{2}) = \mathbb{Q}(\sqrt{-1}, \sqrt{2})$. We show that*

$$O_K = \mathbb{Z} + \mathbb{Z}\sqrt{-1} + \mathbb{Z}\sqrt{2} + \mathbb{Z}\left(\frac{1}{2}(\sqrt{2} + \sqrt{-2})\right)$$

and

$$d(K) = 256.$$

It is easy to check that

$$\mathbb{Z} + \mathbb{Z}\sqrt{-1} + \mathbb{Z}\sqrt{2} + \mathbb{Z}\left(\frac{1}{2}(\sqrt{2} + \sqrt{-2})\right) \subseteq O_K,$$

so we have to prove that

$$O_K \subseteq \mathbb{Z} + \mathbb{Z}\sqrt{-1} + \mathbb{Z}\sqrt{2} + \mathbb{Z}\left(\frac{1}{2}(\sqrt{2} + \sqrt{-2})\right).$$

Let $\theta \in O_K$. The subfields of K are $\mathbb{Q}$, $\mathbb{Q}(\sqrt{-1})$, $\mathbb{Q}(\sqrt{2})$, and $\mathbb{Q}(\sqrt{-2})$. If $\theta \in \mathbb{Q}$ then $\theta \in \mathbb{Z} \subset \mathbb{Z} + \mathbb{Z}\sqrt{-1} + \mathbb{Z}\sqrt{2} + \mathbb{Z}\left(\frac{1}{2}(\sqrt{2} + \sqrt{-2})\right)$. If $\theta \in \mathbb{Q}(\sqrt{-1})$ then $\theta \in O_{\mathbb{Q}(\sqrt{-1})} = \mathbb{Z} + \mathbb{Z}\sqrt{-1} \subset \mathbb{Z} + \mathbb{Z}\sqrt{-1} + \mathbb{Z}\sqrt{2} + \mathbb{Z}\left(\frac{1}{2}(\sqrt{2} + \sqrt{-2})\right)$. If $\theta \in \mathbb{Q}(\sqrt{2})$ then $\theta \in O_{\mathbb{Q}(\sqrt{2})} = \mathbb{Z} + \mathbb{Z}\sqrt{2} \subset \mathbb{Z} + \mathbb{Z}\sqrt{-1} + \mathbb{Z}\sqrt{2} + \mathbb{Z}\left(\frac{1}{2}(\sqrt{2} + \sqrt{-2})\right)$. If $\theta \in \mathbb{Q}(\sqrt{-2})$ then $\theta \in O_{\mathbb{Q}(\sqrt{-2})} = \mathbb{Z} + \mathbb{Z}\sqrt{-2} \subset \mathbb{Z} + \mathbb{Z}\sqrt{-1} + \mathbb{Z}\sqrt{2} + \mathbb{Z}\left(\frac{1}{2}(\sqrt{2} + \sqrt{-2})\right)$. Hence we may suppose that θ does not belong to any of the subfields of K. As $\theta \in K$ we have

$$\theta = a_0 + a_1\sqrt{-1} + a_2\sqrt{2} + a_3\sqrt{-2}, \ a_0, a_1, a_2, a_3 \in \mathbb{Q}.$$

The conjugates of θ over $\mathbb{Q}$ are

$$\theta = a_0 + a_1\sqrt{-1} + a_2\sqrt{2} + a_3\sqrt{-2},$$
$$\theta' = a_0 - a_1\sqrt{-1} + a_2\sqrt{2} - a_3\sqrt{-2},$$
$$\theta'' = a_0 + a_1\sqrt{-1} - a_2\sqrt{2} - a_3\sqrt{-2},$$
$$\theta''' = a_0 - a_1\sqrt{-1} - a_2\sqrt{2} + a_3\sqrt{-2}.$$

Then $\theta + \theta' = 2a_0 + 2a_2\sqrt{2}$, $\theta + \theta'' = 2a_0 + 2a_1\sqrt{-1}$, $\theta + \theta''' = 2a_0 + 2a_3\sqrt{-2}$ *must be integers of* $\mathbb{Q}(\sqrt{2})$, $\mathbb{Q}(\sqrt{-1})$, $\mathbb{Q}(\sqrt{-2})$ *respectively. Hence* $2a_0, 2a_1, 2a_2, 2a_3 \in \mathbb{Z}$. *Define integers b_i by* $b_i = 2a_i$ $(i = 0, 1, 2, 3)$ *so that*

$$\theta = \frac{1}{2}(b_0 + b_1\sqrt{-1} + b_2\sqrt{2} + b_3\sqrt{-2}).$$

Set

$$c = b_0^2 + 2b_3^2 \in \mathbb{Z},$$
$$d = b_0^2 + b_1^2 - 2b_2^2 - 2b_3^2 \in \mathbb{Z},$$
$$e = b_0 b_3 - b_1 b_2 \in \mathbb{Z},$$

so that θ is a root of

$$f(x) = x^4 - 2b_0 x^3 + \left(c + \frac{d}{2}\right)x^2 + \left(-2b_3 e - \frac{b_0 d}{2}\right)x + \left(\frac{d^2 + 8e^2}{16}\right) \in \mathbb{Q}[x].$$

As θ is of degree 4 over $\mathbb{Q}$ (since it does not belong to any of the subfields of K) the polynomial $f(x)$ must be the minimal polynomial of θ over $\mathbb{Q}$. Hence, as $\theta \in O_K$, we have $f(x) \in \mathbb{Z}[x]$, and so $d/2 \in \mathbb{Z}$ and $(d^2 + 8e^2)/16 \in \mathbb{Z}$. Hence

$$d \equiv 0 \,(\text{mod } 2), \quad d^2 + 8e^2 \equiv 0 \,(\text{mod } 16).$$

From these congruences we deduce that

$$d \equiv 0 \,(\text{mod } 4), \quad e \equiv 0 \,(\text{mod } 2).$$

Hence

$$b_0^2 + b_1^2 - 2b_2^2 - 2b_3^2 \equiv 0 \,(\text{mod } 4) \qquad (7.1.4)$$

and

$$b_0 b_3 - b_1 b_2 \equiv 0 \,(\text{mod } 2). \qquad (7.1.5)$$

If b_0 or b_1 is odd from (7.1.4) we see that the other is odd as well. Then from (7.1.5) we deduce that $b_2 \equiv b_3 \,(\text{mod } 2)$, and (7.1.4) gives the contradiction $2 \equiv 0 \,(\text{mod } 4)$. Thus

$$b_0 \equiv b_1 \equiv 0 \,(\text{mod } 2) \text{ and } b_2 \equiv b_3 \,(\text{mod } 2).$$

Hence we can define integers c_0, c_1, c_2, c_3 by

$$b_0 = 2c_0, \quad b_1 = 2c_1, \quad b_2 = 2c_2 + c_3, \quad b_3 = c_3.$$

Then

$$\theta = \frac{1}{2}(b_0 + b_1\sqrt{-1} + b_2\sqrt{2} + b_3\sqrt{-2})$$

$$= c_0 + c_1\sqrt{-1} + c_2\sqrt{2} + c_3\left(\frac{1}{2}(\sqrt{2} + \sqrt{-2})\right)$$

$$\in \mathbb{Z} + \mathbb{Z}\sqrt{-1} + \mathbb{Z}\sqrt{2} + \mathbb{Z}\left(\frac{1}{2}(\sqrt{2} + \sqrt{-2})\right)$$

as required.

Thus $\{1, \sqrt{-1}, \sqrt{2}, \frac{1}{2}(\sqrt{2} + \sqrt{-2})\}$ is an integral basis for K and

$$d(K) = \begin{vmatrix} 1 & \sqrt{-1} & \sqrt{2} & \frac{1}{2}(\sqrt{2} + \sqrt{-2}) \\ 1 & -\sqrt{-1} & \sqrt{2} & \frac{1}{2}(\sqrt{2} - \sqrt{-2}) \\ 1 & \sqrt{-1} & -\sqrt{2} & \frac{1}{2}(-\sqrt{2} - \sqrt{-2}) \\ 1 & -\sqrt{-1} & -\sqrt{2} & \frac{1}{2}(-\sqrt{2} + \sqrt{-2}) \end{vmatrix}^2 = 256.$$

An integral basis for the quartic field $\mathbb{Q}(\sqrt{m} + \sqrt{n}) = \mathbb{Q}(\sqrt{m}, \sqrt{n})$, where m and n are distinct squarefree integers, was determined by K. S. Williams [19] in 1970.

Definition 7.1.5 (Monogenic number field) *Let K be an algebraic number field of degree n. If there exists an element $\theta \in O_K$ such that $\{1, \theta, \ldots, \theta^{n-1}\}$ is an integral basis for K then K is said to be monogenic and the integral basis $\{1, \theta, \ldots, \theta^{n-1}\}$ is called a power basis for K.*

Clearly every quadratic field is monogenic. The cubic fields in Examples 7.1.4 and 7.1.6 are monogenic. Dedekind showed in 1878 that not every algebraic number field is monogenic by proving that the cubic field

$$K = \mathbb{Q}(\theta), \quad \theta^3 - \theta^2 - 2\theta - 8 = 0,$$

is not monogenic (see [4]).

Example 7.1.8 *We show that the quartic field $K = \mathbb{Q}(\sqrt{-1}, \sqrt{2})$ considered in Example 7.1.7 is monogenic. Let*

$$\theta = \frac{\sqrt{2} + i\sqrt{2}}{2},$$

so that

$$\theta^2 = i, \ \theta^3 = \frac{1}{2}(-\sqrt{2} + i\sqrt{2}), \ \theta^4 = -1.$$

Then, by Example 7.1.7, we have

$$O_K = \mathbb{Z} + \mathbb{Z}i + \mathbb{Z}\sqrt{2} + \mathbb{Z}\left(\frac{1}{2}(\sqrt{2} + i\sqrt{2})\right)$$

$$= \mathbb{Z} + \mathbb{Z}\theta^2 + \mathbb{Z}(\theta - \theta^3) + \mathbb{Z}\theta$$

$$= \mathbb{Z} + \mathbb{Z}\theta + \mathbb{Z}\theta^2 + \mathbb{Z}\theta^3,$$

so that K is monogenic with power basis $\{1, \theta, \theta^2, \theta^3\}$.

We conclude this section with a simple upper bound for the absolute value of the discriminant of an algebraic number field as well as a theorem giving the sign of the discriminant.

Theorem 7.1.16 *Let K be an algebraic number field of degree n. Let $\lambda_1, \ldots, \lambda_n \in O_K$ be such that $D(\lambda_1, \ldots, \lambda_n) \neq 0$. Then*

$$|d(K)| \leq |D(\lambda_1, \ldots, \lambda_n)|.$$

Moreover, if $D(\lambda_1, \ldots, \lambda_n)$ is squarefree then $\{\lambda_1, \ldots, \lambda_n\}$ is an integral basis for O_K.

Proof: Let $\{\eta_1, \ldots, \eta_n\}$ be an integral basis for K. Then there exist c_{ij} $(i, j = 1, 2, \ldots, n) \in \mathbb{Z}$ such that

$$\begin{cases} \lambda_1 = c_{11}\eta_1 + \cdots + c_{1n}\eta_n, \\ \lambda_2 = c_{21}\eta_1 + \cdots + c_{2n}\eta_n, \\ \quad \cdots \\ \lambda_n = c_{n1}\eta_1 + \cdots + c_{nn}\eta_n. \end{cases}$$

Hence

$$D(\lambda_1, \ldots, \lambda_n) = (\det c_{ij})^2 D(\eta_1, \ldots, \eta_n) = (\det c_{ij})^2 d(K).$$

As $D(\lambda_1, \ldots, \lambda_n) \neq 0$ we see that $\det (c_{ij}) \neq 0$. Thus, as $\det (c_{ij}) \in \mathbb{Z}$, we have $(\det c_{ij})^2 \geq 1$ and so

$$|D(\lambda_1, \ldots, \lambda_n)| \geq |d(K)|.$$

If $D(\lambda_1, \ldots, \lambda_n)$ is squarefree then from

$$D(\lambda_1, \ldots, \lambda_n) = (\det c_{ij})^2 d(K),$$

we deduce that $\det c_{ij} = \pm 1$. Hence $D(\lambda_1, \ldots, \lambda_n) = d(K)$, proving that $\{\lambda_1, \ldots, \lambda_n\}$ is an integral basis for K. ∎

The next theorem is due to Alexander Brill (1842–1935) (see [3]).

Theorem 7.1.17 *Let K be an algebraic number field of degree n. Let $\theta \in O_K$ be such that $K = \mathbb{Q}(\theta)$. Let $\theta_1 = \theta, \theta_2, \ldots, \theta_n$ be the conjugates of θ. Let r be the number of $\theta_1, \ldots, \theta_n$ that are real. Then*

$$\operatorname{sgn}(d(K)) = (-1)^{(n-r)/2}.$$

Proof: As r is the number of $\theta_1, \ldots, \theta_n$ that are real, the number of $\theta_1, \ldots, \theta_n$ that are nonreal is $n - r$. Since the nonreal conjugates occur in complex conjugate pairs, $n - r$ is even, say $n - r = 2s$, so $n = r + 2s$. Now let $\{\omega_1, \ldots, \omega_n\}$ be an integral basis for K. Let $\omega_k^{(j)}$ ($j = 1, 2, \ldots, n$) be the conjugates of ω_k ($k = 1, 2, \ldots, n$). Then $d(K) = \det(\omega_k^{(j)})^2$. Set $\det(\omega_k^{(j)}) = A + iB$ with $A, B \in \mathbb{R}$. Since the change of i into $-i$ in this determinant is equivalent to the interchange of s pairs of rows, we have $A - iB = (-1)^s \det(\omega_k^{(j)})$. Hence $A - iB = (-1)^s(A + iB)$. If s is even then $A - iB = A + iB$ so $B = 0$ and $d(K) = \det(\omega_k^{(j)})^2 = A^2$ is positive. If s is odd then $A - iB = -(A + iB)$ so $A = 0$ and $d(K) = \det(\omega_k^{(j)})^2 = (iB)^2 = -B^2$ is negative. Hence $\operatorname{sgn}(d(K)) = (-1)^s = (-1)^{(n-r)/2}$. ∎

7.2 Minimal Integers

Let K be an algebraic number field of degree n. Let $\theta \in O_K$ be such that $K = \mathbb{Q}(\theta)$. Then every $\alpha \in O_K$ can be expressed in the form

$$\alpha = a_0 + a_1\theta + \cdots + a_{n-1}\theta^{n-1}, \tag{7.2.1}$$

where $a_0, a_1, \ldots, a_{n-1}$ are rational numbers uniquely determined by α and θ. If $k \in \{1, 2, \ldots, n - 1\}$ is such that

$$a_k \neq 0, \quad a_{k+1} = \cdots = a_{n-1} = 0$$

so that

$$\alpha = a_0 + a_1\theta + \cdots + a_k\theta^k$$

then α is called an integer of degree k in θ. If $a_1 = a_2 = \cdots = a_{n-1} = 0$ so that $\alpha = a_0$ then α is called an integer of degree 0 in θ. The integers of degree 0 in θ are precisely the rational integers.

We are going to show that the denominators of the a_j are bounded. Of course if $\{1, \theta, \ldots, \theta^{n-1}\}$ is a power basis for K then all the denominators are equal to 1.

First we prove the following result.

Theorem 7.2.1 *Let K be an algebraic number field of degree n. Let $\omega_1, \ldots, \omega_n \in O_K$ be such that*

$$D(\omega_1, \ldots, \omega_n) \neq 0.$$

Then for each $\alpha \in O_K$ there exist unique rational integers $x_1, \ldots, x_n$ such that

$$\alpha = \sum_{j=1}^{n} \frac{x_j}{D(\omega_1, \ldots, \omega_n)} \omega_j$$

and

$$D(\omega_1, \ldots, \omega_n) \mid x_j^2, \quad j = 1, 2, \ldots, n.$$

Proof: As $\omega_1, \ldots, \omega_n \in O_K$, by Theorem 6.4.4(b) $D(\omega_1, \ldots, \omega_n)$ is a rational integer, which is nonzero by assumption. Further, as $D(\omega_1, \ldots, \omega_n) \neq 0$, by Theorem 6.4.4(c) $\omega_1, \ldots, \omega_n$ are linearly independent over $\mathbb{Q}$ and thus form a basis for K over $\mathbb{Q}$. Hence there exist unique rational numbers $y_1, \ldots, y_n$ such that

$$\alpha = \sum_{j=1}^{n} y_j \omega_j. \tag{7.2.2}$$

Let $\sigma_1 (= 1), \sigma_2, \ldots, \sigma_n$ be the n monomorphisms: $K \longrightarrow \mathbb{C}$. Applying these to (7.2.2), we obtain

$$\sigma_k(\alpha) = \sum_{j=1}^{n} y_j \sigma_k(\omega_j), \quad k = 1, 2, \ldots, n. \tag{7.2.3}$$

Regarding (7.2.3) as a system of n linear equations in the n unknowns $y_1, \ldots, y_n$, we obtain by Cramer's rule

$$y_j = \frac{\begin{vmatrix} \sigma_1(\omega_1) \cdots \sigma_1(\omega_{j-1}) & \sigma_1(\alpha) & \sigma_1(\omega_{j+1}) \cdots \sigma_1(\omega_n) \\ \cdots & & \cdots \\ \sigma_n(\omega_1) \cdots \sigma_n(\omega_{j-1}) & \sigma_n(\alpha) & \sigma_n(\omega_{j+1}) \cdots \sigma_n(\omega_n) \end{vmatrix}}{\det(\sigma_i(\omega_j))}$$

for $j = 1, 2, \ldots, n$. Hence

$$y_j^2 D(\omega_1, \ldots, \omega_n) = \begin{vmatrix} \sigma_1(\omega_1) \cdots \sigma_1(\omega_{j-1}) & \sigma_1(\alpha) & \sigma_1(\omega_{j+1}) \cdots \sigma_1(\omega_n) \\ \cdots & & \cdots \\ \sigma_n(\omega_1) \cdots \sigma_n(\omega_{j-1}) & \sigma_n(\alpha) & \sigma_n(\omega_{j+1}) \cdots \sigma_n(\omega_n) \end{vmatrix}^2$$

is an algebraic integer for $j = 1, 2, \ldots, n$ and since $y_j^2 D(\omega_1, \ldots, \omega_n) \in \mathbb{Q}$ we deduce that $y_j^2 D(\omega_1, \ldots, \omega_n) \in \mathbb{Z}$. Set $y_j = r_j/s_j$, where $r_j \in \mathbb{Z}$, $s_j \in \mathbb{N}$, and $(r_j, s_j) = 1$. Then

$$\frac{r_j^2}{s_j^2} D(\omega_1, \ldots, \omega_n) \in \mathbb{Z}, \quad j = 1, 2, \ldots, n.$$

As $(r_j, s_j) = 1$ we deduce that

$$s_j^2 \mid D(\omega_1, \ldots, \omega_n), \quad j = 1, 2, \ldots, n.$$

Let

$$x_j = y_j D(\omega_1, \ldots, \omega_n) = \frac{r_j}{s_j} D(\omega_1, \ldots, \omega_n) \in \mathbb{Z}, \ j = 1, 2, \ldots, n.$$

Then, from (7.2.2), we obtain

$$\alpha = \sum_{j=1}^{n} \frac{x_j}{D(\omega_1, \ldots, \omega_n)} \omega_j.$$

Finally, we observe that

$$\frac{x_j^2}{D(\omega_1, \ldots, \omega_n)} = r_j^2 \frac{D(\omega_1, \ldots, \omega_n)}{s_j^2} \in \mathbb{Z},$$

so that

$$D(\omega_1, \ldots, \omega_n) \mid x_j^2, \ j = 1, 2, \ldots, n.$$

This completes the proof of the theorem.   ■

We note that if $D(\omega_1, \ldots, \omega_n)$ is squarefree then by Theorem 7.2.1 we have

$$D(\omega_1, \ldots, \omega_n) \mid x_j, \ j = 1, 2, \ldots, n.$$

Hence, by Theorem 7.2.1, for each $\alpha \in O_K$ there exist unique rational integers $a_j = x_j / D(\omega_1, \ldots, \omega_n)$ such that

$$\alpha = \sum_{j=1}^{n} a_j \omega_j,$$

proving that $\{\omega_1, \ldots, \omega_n\}$ is an integral basis for K, a result that we have seen before in Theorem 7.1.16.

We now use Theorem 7.2.1 to bound the denominators of the a_j in (7.2.1).

Theorem 7.2.2 *Let K be an algebraic number field of degree n. Let $\theta \in O_K$ be such that $K = \mathbb{Q}(\theta)$. Let $\alpha \in O_K$. Then there exist unique rational numbers r_j/s_j ($j = 1, 2, \ldots, n$) with $(r_j, s_j) = 1$ and $s_j > 0$ such that*

$$\alpha = \sum_{j=1}^{n} \frac{r_j}{s_j} \theta^{j-1}$$

and

$$1 \le s_j \le |D(\theta)|, \ s_j^2 \mid D(\theta).$$

Proof: As $\theta \in O_K$ we have

$$1, \theta, \theta^2, \ldots, \theta^{n-1} \in O_K,$$

so by Theorem 6.4.4(b)

$$D(\theta) = D(1, \theta, \theta^2, \dots, \theta^{n-1}) \in \mathbb{Z}.$$

Further, as $K = \mathbb{Q}(\theta)$, by Theorem 6.4.3 we have

$$D(\theta) \neq 0.$$

Then by Theorem 7.2.1 there exist unique rational integers $x_1, \dots, x_n$ such that

$$\alpha = \sum_{j=1}^{n} \frac{x_j}{D(\theta)} \theta^{j-1}$$

and

$$D(\theta) \mid x_j^2, \quad j = 1, 2, \dots, n.$$

For $j = 1, 2, \dots, n$ we define coprime integers r_j and s_j (> 0) by

$$r_j = \frac{\operatorname{sgn}(D(\theta)) x_j}{(x_j, D(\theta))}, \quad s_j = \frac{|D(\theta)|}{(x_j, D(\theta))},$$

so that

$$\frac{r_j}{s_j} = \frac{x_j}{D(\theta)}, \quad \alpha = \sum_{j=1}^{n} \frac{r_j}{s_j} \theta^{j-1},$$

and

$$1 \leq s_j \leq |D(\theta)|.$$

Finally, for $j = 1, 2, \dots, n$ we have

$$\frac{r_j^2}{s_j^2} D(\theta) = \frac{x_j^2}{D(\theta)} \in \mathbb{Z},$$

so that as $(r_j, s_j) = 1$

$$s_j^2 \mid D(\theta). \qquad \blacksquare$$

Theorem 7.2.2 enables us to define the concept of a "minimal integer of degree k in θ." Let K be an algebraic number field of degree n. Fix $\theta \in O_K$ such that $K = \mathbb{Q}(\theta)$. For $k \in \{0, 1, 2, \dots, n-1\}$ define the set S_k by

$$S_k = \{a_k \in \mathbb{Q} \mid a_0 + a_1\theta + \dots + a_k\theta^k \in O_K$$
$$\text{for some } a_0, a_1, \dots, a_{k-1} \in \mathbb{Q}\}. \qquad (7.2.4)$$

Clearly

$$S_0 = \mathbb{Z}$$

and

$$S_k \supseteq \mathbb{Z}, \ k = 1, 2, \ldots, n - 1.$$

By Theorem 7.2.2 the denominators of the rational numbers in S_k are bounded. Hence S_k has a least positive element a_k^*. Clearly $a_0^* = 1$.

Definition 7.2.1 (Minimal integer of degree k in θ) *With the preceding notation any integer of K that is of the form*

$$a_0 + a_1\theta + \cdots + a_{k-1}\theta^{k-1} + a_k^*\theta^k,$$

where $a_0, a_1, \ldots, a_{k-1} \in \mathbb{Q}$, is called a minimal integer of degree k in θ.

The next theorem gives the structure of the set S_k.

Theorem 7.2.3 *With the preceding notation*

$$S_k = a_k^* \mathbb{Z}.$$

Proof: Let $a \in S_k$. Let m be the least positive integer such that

$$ma \in \mathbb{Z}, \ ma_k^* \in \mathbb{N}.$$

By the division algorithm there exist $q \in \mathbb{Z}$ and $r \in \mathbb{Z}$ such that

$$ma = qma_k^* + r, \ 0 \le r < ma_k^*.$$

Hence

$$a = qa_k^* + \frac{r}{m}, \ 0 \le \frac{r}{m} < a_k^*.$$

As $a \in S_k$ there exist $b_0, b_1, \ldots, b_{k-1} \in \mathbb{Q}$ such that

$$b_0 + b_1\theta + \cdots + b_{k-1}\theta^{k-1} + a\theta^k \in O_K.$$

Similarly, as $a_k^* \in S_k$, there exist $c_0, c_1, \ldots, c_{k-1} \in \mathbb{Q}$ such that

$$c_0 + c_1\theta + \cdots + c_{k-1}\theta^{k-1} + a_k^*\theta^k \in O_K.$$

Then

$$(b_0 - qc_0) + (b_1 - qc_1)\theta + \cdots + (b_{k-1} - qc_{k-1})\theta^{k-1} + \frac{r}{m}\theta^k \in O_K,$$

so that

$$\frac{r}{m} \in S_k.$$

If $0 < r/m < a_k^*$ this contradicts the minimality of a_k^*. Hence $r/m = 0$ and $a = qa_k^*$, proving $S_k = a_k^*\mathbb{Z}$. ∎

The next result gives the form of a_k^*.

Theorem 7.2.4 *For $k = 0, 1, 2, \ldots, n - 1$*

$$a_k^* = \frac{1}{d_k}$$

for some $d_k \in \mathbb{N}$.

Proof: We prove the assertion by induction on $k \in \{0, 1, 2, \ldots, n - 1\}$. The result is true for $k = 0$ as

$$a_0^* = 1 = \frac{1}{d_0}$$

with $d_0 = 1$. Assume now that

$$a_k^* = \frac{1}{d_k}, \ d_k \in \mathbb{N}, \ k = 0, 1, \ldots, l - 1,$$

where $1 \leq l \leq n - 1$. By the definition of a_{l-1}^* and the inductive hypothesis there exist rational numbers $a_0, a_1, \ldots, a_{l-2}$ such that

$$a_0 + a_1\theta + \cdots + a_{l-2}\theta^{l-2} + \frac{1}{d_{l-1}}\theta^{l-1} \in O_K.$$

Then

$$a_0\theta + a_1\theta^2 + \cdots + a_{l-2}\theta^{l-1} + \frac{1}{d_{l-1}}\theta^l \in O_K.$$

Hence

$$\frac{1}{d_{l-1}} \in S_l.$$

Thus, by Theorem 7.2.3, there exists $m \in \mathbb{Z}$ such that

$$\frac{1}{d_{l-1}} = a_l^* m.$$

This proves that

$$a_l^* = \frac{1}{d_l}$$

with $d_l = md_{l-1}$. This completes the inductive step and the result follows by the principle of mathematical induction. ∎

The next theorem shows that each d_{k-1} ($k = 1, 2, \ldots, n - 1$) divides its successor d_k.

Theorem 7.2.5 *For $k = 1, 2, \ldots, n - 1$*

$$d_{k-1} \mid d_k.$$

Proof: Let $k \in \{1, 2, \ldots, n-1\}$. Exactly as in the proof of Theorem 7.2.4 we deduce that

$$a_{k-1}^* \in S_k = a_k^* \mathbb{Z}$$

so that

$$a_{k-1}^* = m a_k^*$$

for some $m \in \mathbb{Z}$. Hence, by Theorem 7.2.4, we have

$$\frac{1}{d_{k-1}} = m \frac{1}{d_k}$$

so that

$$d_k = m d_{k-1},$$

proving

$$d_{k-1} \mid d_k, \quad k = 1, 2, \ldots, n-1. \qquad \blacksquare$$

The next theorem gives the form of an integer of degree k ($k = 0, 1, 2, \ldots, n-1$) in θ. As an immediate consequence we obtain the form of a minimal integer of degree k in θ.

Theorem 7.2.6 *If α is an integer of degree k in θ then there exist $a_0, a_1, \ldots, a_k \in \mathbb{Z}$ such that*

$$\alpha = \frac{a_0 + a_1 \theta + \cdots + a_{k-1} \theta^{k-1} + a_k \theta^k}{d_k}.$$

In particular if α is a minimal integer of degree k in θ then there exist $a_0, a_1, \ldots, a_{k-1} \in \mathbb{Z}$ such that

$$\alpha = \frac{a_0 + a_1 \theta + \cdots + a_{k-1} \theta^{k-1} + \theta^k}{d_k}.$$

Proof: We prove the assertion by induction on $k \in \{0, 1, 2, \ldots, n-1\}$. Let α be an integer of degree 0 in θ. Then $\alpha = a_0$ for some $a_0 \in \mathbb{Z}$. But $d_0 = 1$ so that $\alpha = a_0/d_0$ is of the asserted form and the result is true for $k = 0$.

Assume now that all integers of degree up to $l - 1$ in θ are of the specified form, where $l \in \{1, 2, \ldots, n-1\}$. Let α be any integer of degree l in θ. By Theorems 7.2.3 and 7.2.4 there exist $r_0, r_1, \ldots, r_{l-1} \in \mathbb{Q}$ and $a_l \in \mathbb{Z}$ such that

$$\alpha = r_0 + r_1 \theta + \cdots + r_{l-1} \theta^{l-1} + \frac{a_l}{d_l} \theta^l.$$

Let β be a minimal integer in θ of degree $l - 1$. By the minimality of β and the inductive hypothesis, there exist $s_0, s_1, \ldots, s_{l-2} \in \mathbb{Z}$ such that

$$\beta = \frac{s_0 + s_1\theta + \cdots + s_{l-2}\theta^{l-2} + \theta^{l-1}}{d_{l-1}}.$$

By Theorem 7.2.5 we have $d_{l-1} \mid d_l$ so that

$$\frac{d_l}{d_{l-1}}\alpha - a_l\theta\beta \in O_K.$$

Thus

$$\frac{d_l}{d_{l-1}}r_0 + \sum_{j=1}^{l-1} \frac{d_l r_j - a_l s_{j-1}}{d_{l-1}}\theta^j$$

is an integer of degree $l - 1$ in θ. Hence, by the inductive hypothesis, there exist $c_0, c_1, \ldots, c_{l-1} \in \mathbb{Z}$ such that

$$\frac{d_l}{d_{l-1}}r_0 + \sum_{j=1}^{l-1} \frac{d_l r_j - a_l s_{j-1}}{d_{l-1}}\theta^j = \sum_{j=0}^{l-1} \frac{c_j}{d_{l-1}}\theta^j.$$

Equating coefficients we obtain

$$r_0 = \frac{c_0}{d_l},$$
$$r_j = \frac{a_l s_{j-1} + c_j}{d_l}, \quad j = 1, 2, \ldots, l - 1.$$

Define integers $a_0, a_1, \ldots, a_{l-1}$ by

$$a_0 = c_0, \quad a_j = a_l s_{j-1} + c_j, \quad j = 1, 2, \ldots, l - 1.$$

Then

$$r_j = \frac{a_j}{d_l}, \quad j = 0, 1, \ldots, l - 1,$$

and

$$\alpha = \frac{a_0 + a_1\theta + \cdots + a_{l-1}\theta^{l-1} + a_l\theta^l}{d_l}.$$

This completes the inductive step and the theorem follows by the principle of mathematical induction and (for the second part) Theorem 7.2.4. ∎

We now come to the main theorem of this section. We show that if α_k ($k = 0, 1, \ldots, n - 1$) is a minimal integer in θ of degree k then $\{\alpha_0, \alpha_1, \ldots, \alpha_{n-1}\}$ is an integral basis for $K = \mathbb{Q}(\theta)$.

Theorem 7.2.7 *Let K be an algebraic number field of degree n. Let $\theta \in O_K$ be such that $K = \mathbb{Q}(\theta)$. For $k = 0, 1, 2, \ldots, n - 1$ let α_k be a minimal integer in θ of degree k. Then $\{\alpha_0, \alpha_1, \ldots, \alpha_{n-1}\}$ is an integral basis for K.*

Proof: In any integral basis for $K = \mathbb{Q}(\theta)$ at least one of the basis elements must be of the form $a_0 + a_1\theta + \cdots + a_{n-1}\theta^{n-1}$ $(a_0, a_1, \ldots, a_{n-1} \in \mathbb{Q})$ with $a_{n-1} \neq 0$; otherwise the integral basis could not represent θ^{n-1}. Replacing the basis element by its negative, if necessary, we may suppose that $a_{n-1} > 0$. We choose an integral basis $\{\omega_1, \ldots, \omega_n\}$ for K with

$$\omega_n = a_0 + a_1\theta + \cdots + a_{n-1}\theta^{n-1}, \quad a_{n-1} > 0, \quad a_{n-1} \text{ least.}$$

Let $k \in \{1, 2, \ldots, n - 1\}$ and suppose that

$$\omega_k = b_0 + b_1\theta + \cdots + b_{n-1}\theta^{n-1} \quad (b_0, \ldots, b_{n-1} \in \mathbb{Q}).$$

Replacing ω_k by $-\omega_k$ if necessary we may suppose that $b_{n-1} \geq 0$. Let m be the unique nonnegative integer such that

$$\frac{b_{n-1}}{a_{n-1}} - 1 < m \leq \frac{b_{n-1}}{a_{n-1}}.$$

Then

$$0 \leq b_{n-1} - ma_{n-1} < a_{n-1}.$$

If $b_{n-1} - ma_{n-1} \neq 0$ we set

$$\omega_k' = \omega_k - m\omega_{n-1}.$$

Then $\{\omega_1, \ldots, \omega_{k-1}, \omega_k', \omega_{k+1}, \ldots, \omega_n\}$ is an integral basis for K. This contradicts the minimality of a_{n-1} as the coefficient of θ^{n-1} in ω_k' is $b_{n-1} - ma_{n-1}$, which is positive and strictly less than a_{n-1}. Hence $b_{n-1} - ma_{n-1} = 0$, so that b_{n-1} is a rational integral multiple of a_{n-1}. Thus there exist rational integers $m_1, \ldots, m_{n-2}$ such that $\omega_1' = \omega_1 - m_1\omega_n$, $\omega_2' = \omega_1 - m_2\omega_n, \ldots, \omega_{n-1}' = \omega_{n-1} - m_{n-1}\omega_n$ are integers of degrees at most $n - 2$ in θ. Moreover, $\{\omega_1', \ldots, \omega_{n-1}', \omega_n\}$ is an integral basis for K. Among all integral bases $\{\omega_1, \ldots, \omega_n\}$ for which $\omega_1, \ldots, \omega_{n-1}$ are integers of degree at most $n - 2$ in θ, we choose one for which the coefficient of θ^{n-2} is positive and minimal, and we continue our construction until we arrive at an integral basis $\alpha_0, \alpha_1, \alpha_2, \ldots, \alpha_{n-1}$, where each α_i is of degree i in θ. Let

$$\alpha_i = \sum_{k=0}^{i} a_{ik}\theta^k, \quad a_{i\,k} \in \mathbb{Q}.$$

Then

$$d(K) = D(\alpha_1, \ldots, \alpha_n) = (a_{00}a_{11} \cdots a_{n-1\,n-1})^2 D(\theta).$$

For $i = 0, 1, \ldots, n - 1$ let β_i be a minimal integer of degree i. Then

$$D(\beta_0, \ldots, \beta_{n-1}) = (a_0^* a_1^* \cdots a_{n-1}^*)^2 D(\theta).$$

By Theorem 7.1.16 we have

$$|d(K)| \leq |D(\beta_0, \ldots, \beta_{n-1})|$$

so that

$$a_{00} a_{11} \cdots a_{n-1\,n-1} \leq a_0^* a_1^* \cdots a_{n-1}^*$$

and thus

$$\frac{a_{00}}{a_0^*} \cdot \frac{a_{11}}{a_1^*} \cdots \frac{a_{n-1\,n-1}}{a_{n-1}^*} \leq 1.$$

As $a_{ii} \in S_i$ $(i = 0, 1, \ldots, n - 1)$, by Theorem 7.2.3 each a_{ii}/a_i^* $(i = 0, 1, \ldots, n - 1)$ is a positive integer and so

$$a_{ii} = a_i^*, \ i = 0, 1, \ldots, n - 1.$$

Thus each α_i $(i = 0, 1, \ldots, n - 1)$ is a minimal integer of degree i in θ. ∎

Theorem 7.2.7 gives a method of finding an integral basis for an algebraic number field of degree n. We have only to find a minimal integer of each degree up to $n - 1$. This is illustrated for some cubic fields in the next section.

Our final theorem of this section gives some further useful information about the denominators of minimal integers.

Theorem 7.2.8 *Let K be an algebraic number field of degree n. Let $\theta \in O_K$ be such that $K = \mathbb{Q}(\theta)$. For $k = 0, 1, 2, \ldots, n - 1$ let*

$$\alpha_k = \frac{a_{k0} + a_{k1}\theta + \cdots + a_{kk-1}\theta^{k-1} + \theta^k}{d_k} \quad (a_{k0}, \ldots, a_{kk-1} \in \mathbb{Z})$$

be a minimal integer in θ of degree k, so that $\alpha_0 = d_0 = 1$. Then

$$d_0 d_1 \cdots d_{n-1} = \mathrm{ind}\,\theta$$

and

$$d_i^{2(n-i)} \mid D(\theta), \ i = 0, 1, \ldots, n - 1.$$

Proof: By Theorem 7.2.7 $\{\alpha_0, \alpha_1, \ldots, \alpha_{n-1}\}$ is an integral basis for K. Hence

$$D(\alpha_0, \alpha_1, \ldots, \alpha_{n-1}) = d(K).$$

However,

$$D(\alpha_0, \alpha_1, \ldots, \alpha_{n-1}) = \frac{D(\theta)}{(d_0 \cdots d_{n-1})^2} = \frac{(\mathrm{ind}\,\theta)^2 d(K)}{(d_0 \cdots d_{n-1})^2}.$$

Hence, as $d_0, \ldots, d_{n-1}$, ind θ are positive integers, we have

$$\text{ind } \theta = d_0 d_1 \cdots d_{n-1}.$$

Further

$$\frac{D(\theta)}{(d_0 \cdots d_{n-1})^2} = d(K) \in \mathbb{Z}$$

so that

$$(d_0 \cdots d_{n-1})^2 \mid D(\theta).$$

For $i = 0, 1, \ldots, n-1$ we have by Theorem 7.2.5

$$d_i \mid d_{i+1} \mid \cdots \mid d_{n-1}$$

so that

$$d_i^{2(n-i)} \mid D(\theta). \qquad \blacksquare$$

7.3 Some Integral Bases in Cubic Fields

In this section we use Theorems 7.2.7 and 7.2.8 to find integral bases for some cubic fields. The following elementary theorem will be very helpful in connection with the calculations.

Theorem 7.3.1 *Let θ be a root of the cubic equation $x^3 + ax + b = 0$ $(a, b \in \mathbb{Q})$. Then $y_0 + y_1\theta + y_2\theta^2$ $(y_0, y_1, y_2 \in \mathbb{Q})$ is a root of the cubic equation $x^3 + Ax^2 + Bx + C = 0$, where*

$A = -3y_0 + 2ay_2,$
$B = 3y_0^2 + ay_1^2 + a^2 y_2^2 - 4ay_0 y_2 + 3by_1 y_2,$
$C = -y_0^3 + by_1^3 - b^2 y_2^3 - ay_0 y_1^2 - a^2 y_0 y_2^2 + 2ay_0^2 y_2 + aby_1 y_2^2 - 3by_0 y_1 y_2.$

Proof: Let $\theta, \theta', \theta'' \in \mathbb{C}$ be the three roots of the cubic equation $x^3 + ax + b = 0$ so that

$$\theta + \theta' + \theta'' = 0,$$
$$\theta\theta' + \theta'\theta'' + \theta''\theta = a,$$
$$\theta\theta'\theta'' = -b.$$

Then

$$\theta^2 + \theta'^2 + \theta''^2 = (\theta + \theta' + \theta'')^2 - 2(\theta\theta' + \theta'\theta'' + \theta''\theta) = -2a,$$

$$\theta^2\theta'^2 + \theta'^2\theta''^2 + \theta''^2\theta^2 = (\theta\theta' + \theta'\theta'' + \theta''\theta)^2 - 2\theta\theta'\theta''(\theta + \theta' + \theta'') = a^2,$$

$$\theta\theta'^2 + \theta^2\theta' + \theta\theta''^2 + \theta^2\theta'' + \theta'\theta''^2 + \theta'^2\theta'' = \theta\theta'(\theta + \theta') + \theta\theta''(\theta + \theta'')$$
$$+ \theta'\theta''(\theta' + \theta'') = -3\theta\theta'\theta'' = 3b.$$

Now set

$$\alpha = y_0 + y_1\theta + y_2\theta^2,$$
$$\alpha' = y_0 + y_1\theta' + y_2\theta'^2,$$
$$\alpha'' = y_0 + y_1\theta'' + y_2\theta''^2.$$

Then

$$\alpha + \alpha' + \alpha'' = 3y_0 + y_1(\theta + \theta' + \theta'') + y_2(\theta^2 + \theta'^2 + \theta''^2) = 3y_0 - 2ay_2,$$

$$\alpha\alpha' + \alpha'\alpha'' + \alpha''\alpha = 3y_0^2 + y_1^2(\theta\theta' + \theta'\theta'' + \theta''\theta) + y_2^2(\theta^2\theta'^2 + \theta'^2\theta''^2 + \theta''^2\theta^2)$$
$$+ 2y_0y_1(\theta + \theta' + \theta'') + 2y_0y_2(\theta^2 + \theta'^2 + \theta''^2)$$
$$+ y_1y_2(\theta\theta'^2 + \theta^2\theta' + \theta\theta''^2 + \theta^2\theta'' + \theta'\theta''^2 + \theta'^2\theta'')$$
$$= 3y_0^2 + ay_1^2 + a^2y_2^2 - 4ay_0y_2 + 3by_1y_2$$

and

$$\alpha\alpha'\alpha'' = y_0^3 + y_1^3\theta\theta'\theta'' + y_2^3(\theta\theta'\theta'')^2 + y_0y_1^2(\theta\theta' + \theta'\theta'' + \theta''\theta)$$
$$+ y_0y_2^2(\theta^2\theta'^2 + \theta'^2\theta''^2 + \theta''^2\theta^2) + y_0^2y_1(\theta + \theta' + \theta'')$$
$$+ y_0^2y_2(\theta^2 + \theta'^2 + \theta''^2) + y_1^2y_2\theta\theta'\theta''(\theta + \theta' + \theta'')$$
$$+ y_1y_2^2\theta\theta'\theta''(\theta\theta' + \theta'\theta'' + \theta''\theta)$$
$$+ y_0y_1y_2(\theta\theta'^2 + \theta^2\theta' + \theta\theta''^2 + \theta^2\theta'' + \theta'\theta''^2 + \theta'^2\theta'')$$
$$= y_0^3 - by_1^3 + b^2y_2^3 + ay_0y_1^2 + a^2y_0y_2^2 - 2ay_0^2y_2 - aby_1y_2^2 + 3by_0y_1y_2.$$

The result now follows as $\alpha = y_0 + y_1\theta + y_2\theta^2$ is a root of

$$(x - \alpha)(x - \alpha')(x - \alpha'') = x^3 - (\alpha + \alpha' + \alpha'')x^2$$
$$+ (\alpha\alpha' + \alpha\alpha'' + \alpha'\alpha'')x - \alpha\alpha'\alpha''. \qquad \blacksquare$$

Example 7.3.1 *Let θ be a root of $\theta^3 - 3\theta + 9 = 0$. In the notation of Theorem 7.3.1 we have $a = -3$, $b = 9$. We determine the polynomial of which $\theta^2/3$ is a root. We have $y_0 = y_1 = 0$, $y_2 = 1/3$. Then, by Theorem 7.3.1, we obtain*

$$A = 2(-3)\frac{1}{3} = -2,$$

$$B = (-3)^2\frac{1}{3^2} = 1,$$

$$C = -9^2\frac{1}{3^3} = -3,$$

and $\theta^2/3$ is a root of $x^3 - 2x^2 + x - 3 = 0$. This shows that $\theta^2/3$ is an algebraic integer of $\mathbb{Q}(\theta)$. In this case it is easy to check that $\theta^2/3$ is a root of $x^3 - 2x^2 +$

$x - 3 = 0$ *directly. Let* $\alpha = \theta^2/3$. *Then* $\theta^2 = 3\alpha$. *Hence*

$$81 = (-9)^2 = (\theta^3 - 3\theta)^2 = \theta^6 - 6\theta^4 + 9\theta^2 = 27\alpha^3 - 54\alpha^2 + 27\alpha,$$

so that $\alpha^3 - 2\alpha^2 + \alpha - 3 = 0$.

Example 7.3.2 *Let* θ *be a root of the cubic equation* $x^3 - x + 4 = 0$. *Here* $a = -1$, $b = 4$. *We consider* $\alpha = \frac{1}{2}\theta + \frac{1}{2}\theta^2$, *so that* $y_0 = 0$, $y_1 = 1/2$, $y_2 = 1/2$. *Then, by Theorem 7.3.1, we obtain*

$$A = 2(-1)\frac{1}{2} = -1,$$

$$B = \frac{1}{4}(-1 + 1 + 12) = 3,$$

$$C = \frac{1}{8}(4 - 16 - 4) = -2,$$

so that α *is a root of* $x^3 - x^2 + 3x - 2 = 0$. *This proves that* $(\theta + \theta^2)/2$ *is an integer of* $\mathbb{Q}(\theta)$.

Example 7.3.3 *Let* θ *be a root of the cubic equation* $x^3 + 11x + 4 = 0$. *Here* $a = 11$, $b = 4$. *We consider* $\alpha = \frac{1}{2}\theta + \frac{1}{2}\theta^2$, *so that* $y_0 = 0$, $y_1 = 1/2$, $y_2 = 1/2$. *Then, by Theorem 7.3.1, we obtain*

$$A = 2 \cdot 11 \cdot \frac{1}{2} = 11,$$

$$B = 11 \cdot \frac{1}{2^2} + 11^2 \cdot \frac{1}{2^2} + 3 \cdot 4 \cdot \frac{1}{2} \cdot \frac{1}{2} = 36,$$

$$C = 4 \cdot \frac{1}{2^3} - 4^2 \cdot \frac{1}{2^3} + 4 \cdot 11 \cdot \frac{1}{2^3} = 4,$$

so that α *is a root of the cubic equation* $x^3 + 11x^2 + 36x + 4 = 0$ *and thus an integer of* $\mathbb{Q}(\theta)$ *(see Example 6.3.2).*

Example 7.3.4 *Let* θ *be a root of the cubic equation* $x^3 - 21x - 236 = 0$. *Here* $a = -21$, $b = -236$. *We consider* $\alpha = (1 + \theta)/3$, *so that* $y_0 = y_1 = 1/3$, $y_2 = 0$. *By Theorem 7.3.1 we obtain*

$$A = -3\left(\frac{1}{3}\right) = -1,$$

$$B = 3 \cdot \frac{1}{3^2} - 21 \cdot \frac{1}{3^2} = -2,$$

$$C = -\frac{1}{3^3} - 236\frac{1}{3^3} + 21\frac{1}{3^3} = \frac{-216}{27} = -8,$$

so that α *is a root of the equation* $x^3 - x^2 - 2x - 8 = 0$. *Hence* $(1 + \theta)/3$ *is an integer of the cubic field* $\mathbb{Q}(\theta)$.

Example 7.3.5 *Let θ be a root of $x^3 - 21x - 236 = 0$. Here $a = -21$, $b = -236$. We consider $\alpha = (-2 - \theta + \theta^2)/18$, so that*

$$y_0 = \frac{-2}{18}, \ y_1 = \frac{-1}{18}, \ y_2 = \frac{1}{18}.$$

By Theorem 7.3.1 we obtain

$$A = \frac{1}{18}(6 - 42) = \frac{-36}{18} = -2,$$

$$B = \frac{1}{18^2}(12 - 21 + 441 - 168 + 708) = \frac{972}{324} = 3,$$

$$C = \frac{1}{18^3}(8 + 236 - 55696 - 42 + 882 - 168 - 4956 + 1416)$$

$$= \frac{-58320}{5832} = -10,$$

so that α is a root of the cubic equation $x^3 - 2x^2 + 3x - 10 = 0$. Hence $(-2 - \theta + \theta^2)/18$ is an integer of the cubic field $\mathbb{Q}(\theta)$.

In the next four examples we use Theorems 7.2.7 and 7.2.8 to give integral bases for the following cubic fields:

$$\mathbb{Q}(\theta), \ \theta^3 - 3\theta + 9 = 0 \text{ (Example 7.3.6)},$$
$$\mathbb{Q}(\theta), \ \theta^3 - \theta + 4 = 0 \text{ (Example 7.3.7)},$$
$$\mathbb{Q}(\theta), \ \theta^3 + 11\theta + 4 = 0 \text{ (Example 7.3.8)},$$
$$\mathbb{Q}(\theta), \ \theta^3 - 21\theta - 236 = 0 \text{ (Example 7.3.9)}.$$

Example 7.3.6 *Let $K = \mathbb{Q}(\theta)$, $\theta^3 - 3\theta + 9 = 0$. The polynomial $x^3 - 3x + 9 \in \mathbb{Z}[x]$ is irreducible, so K is a cubic field ($n = 3$). By Theorem 7.1.10 we have*

$$D(\theta) = -4(-3)^3 - 27 \cdot 9^2 = -2079 = -3^3 \cdot 7 \cdot 11.$$

Let d_1 be the denominator of a minimal integer in θ of degree 1. By Theorem 7.2.8 we see that $d_1^{2(3-1)} \mid D(\theta)$, that is, $d_1^4 \mid -3^3 \cdot 7 \cdot 11$, so that $d_1 = 1$. Hence θ is a minimal integer of degree 1. Let d_2 be the denominator of a minimal integer of degree 2. By Theorem 7.2.8 we have $d_2^{2(3-2)} \mid D(\theta)$, that is, $d_2^2 \mid -3^3 \cdot 7 \cdot 11$, so that $d_2 = 1$ or 3. But it was shown in Example 7.3.1 that $\theta^2/3$ is an integer of K. Hence $d_2 = 3$ and $\theta^2/3$ is a minimal integer in θ of degree 2. Then, by Theorem 7.2.7, we deduce that $\{1, \theta, \theta^2/3\}$ is an integral basis for K. By Theorem 7.2.8 we have $\mathrm{ind}\,\theta = d_0 d_1 d_2 = 3$. Thus

$$d(K) = \frac{D(\theta)}{(\mathrm{ind}\,\theta)^2} = \frac{-3^3 \cdot 7 \cdot 11}{3^2} = -3 \cdot 7 \cdot 11 = -231.$$

Example 7.3.7 Let $K = \mathbb{Q}(\theta)$, $\theta^3 - \theta + 4 = 0$. The polynomial $x^3 - x + 4 \in \mathbb{Z}[x]$ is irreducible, so K is a cubic field $(n = 3)$. By Theorem 7.1.10 we have

$$D(\theta) = -4(-1)^3 - 27 \cdot 4^2 = -2^2 \cdot 107.$$

Let d_1 be the denominator of a minimal integer in θ of degree 1. By Theorem 7.2.8 we see that $d_1^{2(3-1)} \mid D(\theta)$, that is, $d_1^4 \mid -2^2 \cdot 107$. But 107 is a prime, so $d_1 = 1$. Hence θ is a minimal integer of degree 1. Let d_2 be the denominator of a minimal integer in θ of degree 2. By Theorem 7.2.8 we have $d_2^{2(3-2)} \mid D(\theta)$, that is, $d_2^2 \mid -2^2 \cdot 107$, showing that $d_2 = 1$ or 2. Now $\frac{1}{2}\theta + \frac{1}{2}\theta^2$ is an integer of K (Example 7.3.2) so it is a minimal integer in θ of degree 2. Hence by Theorem 7.2.7 $\{1, \theta, (\theta + \theta^2)/2\}$ is an integral basis for K. By Theorem 7.2.8 we have ind $\theta = d_0 d_1 d_2 = 2$. Thus

$$d(K) = \frac{D(\theta)}{(\text{ind } \theta)^2} = \frac{-2^2 \cdot 107}{2^2} = -107.$$

Example 7.3.8 Let $K = \mathbb{Q}(\theta)$, $\theta^3 + 11\theta + 4 = 0$. By Theorem 7.1.10 $D(\theta) = -4 \cdot 11^3 - 27 \cdot 4^2 = -5756 = -2^2 \cdot 1439$, where 1439 is a prime. By Theorem 7.2.8 we have $d_1^4 \mid -2^2 \cdot 1439$ and $d_2^2 \mid -2^2 \cdot 1439$, so that $d_1 = 1$ and $d_2 = 1$ or 2. By Example 7.3.3 $(\theta + \theta^2)/2$ is an integer of K, so that $d_2 = 2$. Thus by Theorem 7.2.7 $\{1, \theta, (\theta + \theta^2)/2\}$ is an integral basis for K. By Theorem 7.2.8 we have ind $\theta = d_0 d_1 d_2 = 2$. Finally,

$$d(K) = \frac{D(\theta)}{(\text{ind } \theta)^2} = \frac{-2^2 \cdot 1439}{2^2} = -1439.$$

Example 7.3.9 Let $K = \mathbb{Q}(\theta)$, $\theta^3 - 21\theta - 236 = 0$. By Theorem 7.1.10 $D(\theta) = -4(-21)^3 - 27(-236)^2 = 37044 - 1503792 = -1466748 = -2^2 \cdot 3^6 \cdot 503$, where 503 is a prime. By Theorem 7.2.8 we have $d_1^4 \mid -2^2 \cdot 3^6 \cdot 503$, so that $d_1 = 1$ or 3. In Example 7.3.4 it was shown that $(1 + \theta)/3$ is an integer of K, so we must have $d_1 = 3$. By Example 7.3.5 $(-2 - \theta + \theta^2)/18$ is an integer of K. Thus $18 \mid d_2$, say $d_2 = 2 \cdot 3^2 \cdot m$, where $m \in \mathbb{N}$. By Theorem 7.2.8 we have

$$-2^2 \cdot 3^6 \cdot 503 = D(\theta) = d(K)(d_0 d_1 d_2)^2 = 2^2 \cdot 3^6 \cdot m^2 d(K),$$

so that $m = 1$, $d(K) = -503$, and

$$\left\{ 1, \frac{1 + \theta}{3}, \frac{-2 - \theta + \theta^2}{18} \right\}$$

is an integral basis for K.

Definition 7.3.1 (Pure cubic field) A cubic field K is said to be pure if there exists a rational integer m, which is not a perfect cube, such that $K = \mathbb{Q}(\sqrt[3]{m})$.

In Example 7.1.6 we found an integral basis for the pure cubic field $\mathbb{Q}(\sqrt[3]{2})$.

Example 7.3.10 *We show that the cubic field K given by*

$$K = \mathbb{Q}(\theta), \ \theta^3 + 6\theta + 2 = 0, \ \theta \in \mathbb{R},$$

is the pure cubic field $\mathbb{Q}(\sqrt[3]{2})$. Clearly $-1 < \theta < 0$. Because the function $x - 2/x$ increases monotonically from -1 to 0 as x varies from -2 to $-\sqrt{2}$, there exists a unique real number a with $-2 < a < -\sqrt{2}$ such that

$$\theta = a - \frac{2}{a}.$$

Then

$$\left(a - \frac{2}{a}\right)^3 + 6\left(a - \frac{2}{a}\right) + 2 = 0$$

so that

$$a^3 - \frac{8}{a^3} + 2 = 0.$$

Hence

$$a^3 = 2 \text{ or } -4.$$

As $a < 0$ we must have $a^3 = -4$; that is,

$$a = -2^{2/3}.$$

Thus

$$\theta = 2^{1/3} - 2^{2/3}.$$

This shows that

$$K = \mathbb{Q}(\theta) \subseteq \mathbb{Q}(2^{1/3}).$$

Further,

$$\theta^2 = -4 + 2 \cdot 2^{1/3} + 2^{2/3}$$

so that

$$\theta + \theta^2 = -4 + 3 \cdot 2^{1/3}.$$

Hence

$$2^{1/3} = \frac{4}{3} + \frac{1}{3}\theta + \frac{1}{3}\theta^2,$$

proving that

$$\mathbb{Q}(2^{1/3}) \subseteq \mathbb{Q}(\theta) = K.$$

This completes the proof that $K = \mathbb{Q}(2^{1/3})$.

As

$$2^{1/3} = \frac{4 + \theta + \theta^2}{3}, \quad 2^{2/3} = \frac{4 - 2\theta + \theta^2}{3},$$

and $\{1, 2^{1/3}, 2^{2/3}\}$ *is an integral basis for* $K = \mathbb{Q}(2^{1/3})$*, we see that*

$$\left\{ 1, \frac{4 + \theta + \theta^2}{3}, \frac{4 - 2\theta + \theta^2}{3} \right\}$$

is an integral basis for $K = \mathbb{Q}(\theta)$*. Since*

$$\frac{4 + \theta + \theta^2}{3} - \frac{4 - 2\theta + \theta^2}{3} = \theta$$

and

$$\frac{4 + \theta + \theta^2}{3} - 1 = \frac{1 + \theta + \theta^2}{3},$$

a simpler integral basis is

$$\left\{ 1, \theta, \frac{1 + \theta + \theta^2}{3} \right\}.$$

We now give an integral basis for the pure cubic field $\mathbb{Q}(\sqrt[3]{m})$. As we have already mentioned this basis was first given by Dedekind [5] in 1900.

Theorem 7.3.2 *Let m be a cubefree integer. Set $m = hk^2$, where h is squarefree, so that k is squarefree and $(h, k) = 1$. Set $\theta = m^{1/3}$ and $K = \mathbb{Q}(\theta)$. Then an integral basis for K is*

$$\left\{ 1, \theta, \frac{\theta^2}{k} \right\}, \quad \text{if } m^2 \not\equiv 1 \ (\text{mod } 9),$$

$$\left\{ 1, \theta, \frac{k^2 \pm k^2\theta + \theta^2}{3k} \right\}, \quad \text{if } m \equiv \pm 1 \ (\text{mod } 9).$$

The discriminant $d(K)$ of K is given by

$$d(K) = \begin{cases} -27h^2k^2, & \text{if } m^2 \not\equiv 1 \ (\text{mod } 9), \\ -3h^2k^2, & \text{if } m \equiv \pm 1 \ (\text{mod } 9). \end{cases}$$

We leave the proof of Theorem 7.3.2 as an exercise (Exercise 6). From Theorem 7.3.2 we obtain Table 1.

If K is a pure cubic field given in the form $K = \mathbb{Q}(\theta)$, $\theta^3 + a\theta + b = 0$, $a, b \in \mathbb{Z}$, it is known that $-4a^3 - 27b^2 = -3c^2$ for some positive integer c (in Example 7.3.10 we have $a = 6$, $b = 2$, $c = 18$), and an integral basis for K has been given by Spearman and Williams [15].

Table 1. *Integral bases and discriminants for*
$$\mathbb{Q}(\sqrt[3]{k}), \ 2 \le k \le 20, \ k \ cubefree$$

k	Integral basis ($\theta = \sqrt[3]{k}$)	Discriminant
2	$\{1, \theta, \theta^2\}$	$-108 = -2^2 \cdot 3^3$
3	$\{1, \theta, \theta^2\}$	$-243 = -3^5$
5	$\{1, \theta, \theta^2\}$	$-675 = -3^3 \cdot 5^2$
6	$\{1, \theta, \theta^2\}$	$-972 = -2^2 \cdot 3^5$
7	$\{1, \theta, \theta^2\}$	$-1323 = -3^3 \cdot 7^2$
10	$\{1, \theta, (1 + \theta + \theta^2)/3\}$	$-300 = -2^2 \cdot 3 \cdot 5^2$
11	$\{1, \theta, \theta^2\}$	$-3267 = -3^3 \cdot 11^2$
12	$\{1, \theta, \theta^2/2\}$	$-972 = -2^2 \cdot 3^5$
13	$\{1, \theta, \theta^2\}$	$-4563 = -3^3 \cdot 13^2$
14	$\{1, \theta, \theta^2\}$	$-5292 = -2^2 \cdot 3^3 \cdot 7^2$
15	$\{1, \theta, \theta^2\}$	$-6075 = -3^5 \cdot 5^2$
17	$\{1, \theta, (1 - \theta + \theta^2)/3\}$	$-867 = -3 \cdot 17^2$
19	$\{1, \theta, (1 + \theta + \theta^2)/3\}$	$-1083 = -3 \cdot 19^2$
20	$\{1, \theta, \theta^2/2\}$	$-2700 = -2^2 \cdot 3^3 \cdot 5^2$

Note: $\mathbb{Q}(\sqrt[3]{4}) = \mathbb{Q}(\sqrt[3]{2})$, $\mathbb{Q}(\sqrt[3]{9}) = \mathbb{Q}(\sqrt[3]{3})$, $\mathbb{Q}(\sqrt[3]{16}) = \mathbb{Q}(\sqrt[3]{2})$, and $\mathbb{Q}(\sqrt[3]{18}) = \mathbb{Q}(\sqrt[3]{12})$.

The discriminant of an arbitrary cubic field $K = \mathbb{Q}(\theta)$, $\theta^3 + a\theta + b = 0$, was obtained by Llorente and Nart [12] in 1983, and an integral basis was first given by Alaca [1].

We conclude this section by mentioning that Funakura [8] has given an integral basis for a pure quartic field $\mathbb{Q}(\sqrt[4]{k})$, where $k \in \mathbb{Z}$ is such that $x^4 - k$ is irreducible over $\mathbb{Q}$. Appealing to his results, we obtain Tables 2 and 3.

Table 2. *Integral bases and discriminants for* $\mathbb{Q}(\sqrt[4]{k})$,
$x^4 - k$ *irreducible in* $\mathbb{Q}[x]$, $2 \le k \le 10$

k	Integral basis ($\theta = \sqrt[4]{k}$)	Discriminant
2	$\{1, \theta, \theta^2, \theta^3\}$	$-2048 = -2^{11}$
3	$\{1, \theta, \theta^2, \theta^3\}$	$-6912 = -2^8 \cdot 3^3$
5	$\{1, \theta, (1 + \theta^2)/2, (\theta + \theta^3)/2\}$	$-2000 = -2^4 \cdot 5^3$
6	$\{1, \theta, \theta^2, \theta^3\}$	$-55296 = -2^{11} \cdot 3^3$
7	$\{1, \theta, \theta^2, \theta^3\}$	$-87808 = -2^8 \cdot 7^3$
10	$\{1, \theta, \theta^2, \theta^3\}$	$-256000 = -2^{11} \cdot 5^3$

Note: $\mathbb{Q}(\sqrt[4]{8}) = \mathbb{Q}(\sqrt[4]{2})$.

Table 3.   *Integral bases and discriminants for* $\mathbb{Q}(\sqrt[4]{-k})$,
$x^4 + k$ *irreducible in* $\mathbb{Q}[x]$, $1 \leq k \leq 10$

k	Integral basis ($\theta = \sqrt[4]{-k}$, $\arg \theta = \pi/4$)	Discriminant
1	$\{1, \theta, \theta^2, \theta^3\}$	$256 = 2^8$
2	$\{1, \theta, \theta^2, \theta^3\}$	$2048 = 2^{11}$
3	$\{1, \theta, (1+\theta^2)/2, (\theta+\theta^3)/2\}$	$432 = 2^4 \cdot 3^3$
5	$\{1, \theta, \theta^2, \theta^3\}$	$32000 = 2^8 \cdot 5^3$
6	$\{1, \theta, \theta^2, \theta^3\}$	$55296 = 2^{11} \cdot 3^3$
7	$\{1, \theta, (1+\theta^2)/2, (1+\theta+\theta^2+\theta^3)/4\}$	$1372 = 2^2 \cdot 7^3$
9	$\{1, \theta, \theta^2/3, \theta^3/3\}$	$2304 = 2^8 \cdot 3^2$
10	$\{1, \theta, \theta^2, \theta^3\}$	$256000 = 2^{11} \cdot 5^3$

Note: $\mathbb{Q}(\sqrt[4]{-4}) = \mathbb{Q}(i)$ and $\mathbb{Q}(\sqrt[4]{-8}) = \mathbb{Q}(\sqrt[4]{-2})$.

7.4 Index and Minimal Index of an Algebraic Number Field

Let K be an algebraic number field of degree n over $\mathbb{Q}$. An element $\alpha \in O_K$ is called a generator of K if $K = \mathbb{Q}(\alpha)$. By Theorem 6.4.3 α is a generator of K if and only if $D(\alpha) \neq 0$. For a generator α of K, the index of α is the positive integer ind α given by

$$D(\alpha) = (\text{ind } \alpha)^2 d(K)$$

(see Definition 7.1.4). We now define the index $i(K)$ and minimal index $m(K)$ of the field K.

Definition 7.4.1 (Index of a field) *The index of K is*

$$i(K) = \gcd \{\text{ind } \alpha \mid \alpha \ \ a \ generator \ of \ K\}.$$

Definition 7.4.2 (Minimal index of a field) *The minimal index of K is*

$$m(K) = \min \{\text{ind } \alpha \mid \alpha \ \ a \ generator \ of \ K\}.$$

Clearly

$$i(K) \mid m(K). \tag{7.4.1}$$

Theorem 7.4.1 *Let K be an algebraic number field. Then $m(K) = 1$ if and only if K possesses a power basis.*

Proof: Suppose $m(K) = 1$. Then there exists a generator α of K such that ind $\alpha = 1$. Hence $D(1, \alpha, \ldots, \alpha^{n-1}) = D(\alpha) = (\text{ind } \alpha)^2 d(K) = d(K)$ so that

$\{1, \alpha, \ldots, \alpha^{n-1}\}$ is an integral basis for K. Hence K possesses a power basis.

Conversely, suppose K possesses a power basis, say $\{1, \alpha, \ldots, \alpha^{n-1}\}$. Then $\{1, \alpha, \ldots, \alpha^{n-1}\}$ is an integral basis for K and so

$$D(1, \alpha, \ldots, \alpha^{n-1}) = d(K).$$

But

$$D(1, \alpha, \ldots, \alpha^{n-1}) = D(\alpha) = (\text{ind } \alpha)^2 d(K),$$

so ind $\alpha = 1$ and hence $m(K) = 1$. ∎

From (7.4.1) and Theorem 7.4.1 we obtain

Theorem 7.4.2 *Let K be an algebraic number field such that K possesses a power basis. Then $i(K) = 1$.*

In Example 7.4.4 we give an algebraic number field K for which $i(K) = 1$ but K does not possess a power basis. This shows that the converse of Theorem 7.4.2 is not true. Theorem 7.4.2 gives a convenient way of establishing that an algebraic number field does not have a power basis; all we have to do is to show that $i(K) \geq 2$.

In the next theorem we determine the index and minimal index of a quadratic field directly from their definitions.

Theorem 7.4.3 *Let K be a quadratic field. Then $i(K) = m(K) = 1$.*

Proof: As K is a quadratic field, by Theorem 5.4.1 there exists a unique squarefree integer m such that $K = \mathbb{Q}(\sqrt{m})$.

First we suppose that $m \equiv 1 \pmod 4$ so that $\left\{1, \frac{1+\sqrt{m}}{2}\right\}$ is an integral basis for K (Theorem 5.4.2) and $d(K) = m$ (Theorem 7.1.2). Let $\alpha \in O_K$. Then $\alpha = a + b\left(\frac{1+\sqrt{m}}{2}\right)$ for some $a, b \in \mathbb{Z}$. Now

$$D(\alpha) = \begin{vmatrix} 1 & a + b\left(\dfrac{1+\sqrt{m}}{2}\right) \\ 1 & a + b\left(\dfrac{1-\sqrt{m}}{2}\right) \end{vmatrix}^2 = (-b\sqrt{m})^2 = b^2 m,$$

so that

$$D(\alpha) \neq 0 \text{ if and only if } b \neq 0.$$

Thus α is a generator of K if and only if $b \neq 0$. Further,

$$\text{ind } \alpha = \sqrt{\frac{D(\alpha)}{d(K)}} = \sqrt{\frac{b^2 m}{m}} = |b|,$$

so that

$$i(K) = \gcd \{|b| \mid b \in \mathbb{Z}, \ b \neq 0\} = 1$$

and

$$m(K) = \min \{|b| \mid b \in \mathbb{Z}, \ b \neq 0\} = 1.$$

Next we suppose that $m \equiv 2$ or $3 \pmod 4$, so that $\{1, \sqrt{m}\}$ is an integral basis for K (Theorem 5.4.2) and $d(K) = 4m$ (Theorem 7.1.2). Let $\alpha \in O_K$. Then $\alpha = a + b\sqrt{m}$ for some $a, b \in \mathbb{Z}$. Now

$$D(\alpha) = \begin{vmatrix} 1 & a + b\sqrt{m} \\ 1 & a - b\sqrt{m} \end{vmatrix}^2 = (-2b\sqrt{m})^2 = 4b^2 m,$$

so that

$$D(\alpha) \neq 0 \text{ if and only if } b \neq 0.$$

Thus α is a generator of K if and only if $b \neq 0$. Further,

$$\text{ind } \alpha = \sqrt{\frac{D(\alpha)}{d(K)}} = \sqrt{\frac{4mb^2}{4m}} = |b|,$$

so that

$$i(K) = \gcd \{|b| \mid b \in \mathbb{Z}, \ b \neq 0\} = 1$$

and

$$m(K) = \min \{|b| \mid b \in \mathbb{Z}, \ b \neq 0\} = 1. \qquad \blacksquare$$

Of course we could have argued that a quadratic field clearly has a power basis so that by Theorem 7.4.1 $m(K) = 1$ and then by (7.4.1) $i(K) = 1$.

In the next four examples we determine $i(K)$ and $m(K)$ for some cubic fields K.

Example 7.4.1 *We determine the index $i(K)$ and the minimal index $m(K)$ of the cubic field $K = \mathbb{Q}(\theta)$, where θ is a root of $f(x) = x^3 - 3x + 9$. Let θ' and θ'' be the other two roots of $f(x)$, so that $x^3 - 3x + 9 = (x - \theta)(x - \theta')(x - \theta'')$. By Example 7.3.6 we know that $\{1, \theta, \theta^2/3\}$ is an integral basis for K, $D(\theta) = -3^3 \cdot 7 \cdot 11$, and $d(K) = -3 \cdot 7 \cdot 11$. Let $\alpha \in O_K$. Then $\alpha = a + b\theta + c\theta^2/3$ for* *some $a, b, c \in \mathbb{Z}$. The conjugates of α are*

$$\alpha = a + b\theta + c\frac{\theta^2}{3},$$

$$\alpha' = a + b\theta' + c\frac{\theta'^2}{3},$$

$$\alpha'' = a + b\theta'' + c\frac{\theta''^2}{3}.$$

Hence, as $\theta + \theta' + \theta'' = 0$, we have

$$\alpha - \alpha' = (\theta - \theta')\left(b + \frac{c}{3}(\theta + \theta')\right) = (\theta - \theta')\left(b - \frac{c}{3}\theta''\right),$$

$$\alpha - \alpha'' = (\theta - \theta'')\left(b + \frac{c}{3}(\theta + \theta'')\right) = (\theta - \theta'')\left(b - \frac{c}{3}\theta'\right),$$

$$\alpha' - \alpha'' = (\theta' - \theta'')\left(b + \frac{c}{3}(\theta' + \theta'')\right) = (\theta' - \theta'')\left(b - \frac{c}{3}\theta\right).$$

Thus, by Theorem 6.4.1,

$$
\begin{aligned}
D(\alpha) &= (\alpha - \alpha')^2(\alpha - \alpha'')^2(\alpha' - \alpha'')^2 \\
&= (\theta - \theta')^2(\theta - \theta'')^2(\theta' - \theta'')^2(b - \frac{c}{3}\theta)^2(b - \frac{c}{3}\theta')^2(b - \frac{c}{3}\theta'')^2 \\
&= D(\theta)\left\{\left(\frac{c}{3}\right)^3 f\left(\frac{3b}{c}\right)\right\}^2 \\
&= -3^3 \cdot 7 \cdot 11 \left(b^3 - \frac{bc^2}{3} + \frac{c^3}{3}\right)^2 \\
&= -3 \cdot 7 \cdot 11 (3b^3 - bc^2 + c^3)^2.
\end{aligned}
$$

Then

$$\text{ind } \alpha = \sqrt{\frac{D(\alpha)}{d(K)}} = \sqrt{\frac{-3 \cdot 7 \cdot 11(3b^3 - bc^2 + c^3)^2}{-3 \cdot 7 \cdot 11}} = |3b^3 - bc^2 + c^3|.$$

Hence

$$m(K) = \min\{|3b^3 - bc^2 + c^3| \mid b, c \in \mathbb{Z}, \ 3b^3 - bc^2 + c^3 \neq 0\} = 1$$

as

$$3b^3 - bc^2 + c^3 = 1 \text{ for } (b, c) = (1, -1).$$

By (7.4.1) $i(K) = 1$. As $m(K) = 1$, K has a power basis by Theorem 7.4.1. Now

$$D\left(1, \frac{\theta^2}{3}, \left(\frac{\theta^2}{3}\right)^2\right) = D\left(1, \frac{\theta^2}{3}, \frac{\theta^4}{9}\right) = D\left(1, \frac{\theta^2}{3}, \frac{\theta^2}{3} - \theta\right) = D\left(1, \frac{\theta^2}{3}, -\theta\right)$$

$$= D\left(1, \theta, \frac{\theta^2}{3}\right) = d(K),$$

so

$$\left\{ 1, \frac{\theta^2}{3}, \left(\frac{\theta^2}{3}\right)^2 \right\}$$

is a power basis for K. This is easily seen directly as

$$a + b\theta + c\frac{\theta^2}{3} = a + (b+c)\frac{\theta^2}{3} - b\left(\frac{\theta^2}{3}\right)^2$$

for all $a, b, c \in \mathbb{Z}$.

Example 7.4.2 *We show that $m(K) = i(K) = 1$ for the cubic field $K = \mathbb{Q}(\theta)$, where θ is a root of $f(x) = x^3 - x + 4$.*

Let θ' and θ'' be the other two roots of $f(x)$ so that

$$f(x) = x^3 - x + 4 = (x - \theta)(x - \theta')(x - \theta'').$$

By Example 7.3.7 we know that $\left\{1, \theta, \frac{\theta+\theta^2}{2}\right\}$ is an integral basis for K and $d(K) = -107$. Let $\alpha \in O_K$. Then $\alpha = a + b\theta + c\left(\frac{\theta+\theta^2}{2}\right)$ for some $a, b, c \in \mathbb{Z}$. Exactly as in Example 7.4.1 we find that

$$D(\alpha) = -107(2b^3 + 3b^2c + bc^2 + c^3)^2.$$

Then

$$\text{ind } \alpha = \sqrt{\frac{D(\alpha)}{d(K)}} = \sqrt{\frac{-107(2b^3 + 3b^2c + bc^2 + c^3)^2}{-107}} = |2b^3 + 3b^2c + bc^2 + c^3|.$$

Hence

$$m(K) = \min \{|2b^3 + 3b^2c + bc^2 + c^3| \mid b, c \in \mathbb{Z},\ 2b^3 + 3b^2c + bc^2 + c^3 \neq 0\} = 1$$

as

$$2b^3 + 3b^2c + bc^2 + c^3 = 1 \text{ for } (b, c) = (-1, 1).$$

Then, by (7.4.1), $i(K) = 1$.

As $m(K) = 1$, K has a power basis by Theorem 7.4.1. Now, as $\theta^3 = \theta - 4$, $\theta^4 = \theta^2 - 4\theta$, we obtain

$$D\left(1, \frac{\theta+\theta^2}{2}, \left(\frac{\theta+\theta^2}{2}\right)^2\right) = D\left(1, \frac{\theta+\theta^2}{2}, \frac{\theta^2-\theta}{2} - 2\right)$$

$$= D\left(1, \frac{\theta+\theta^2}{2}, \frac{\theta^2-\theta}{2}\right) = D\left(1, \frac{\theta+\theta^2}{2}, \theta\right)$$

$$= D(1, \theta, \frac{\theta+\theta^2}{2}) = d(K),$$

so

$$\left\{ 1, \frac{\theta + \theta^2}{2}, \left(\frac{\theta + \theta^2}{2} \right)^2 \right\}$$

is a power basis for K.

Example 7.4.3 *We show that*

$$m(K) = i(K) = 2 \text{ for the cubic field } K = \mathbb{Q}(\theta), \ \theta^3 - 21\theta - 236 = 0.$$

Let θ' and θ'' be the other two roots of $f(x) = x^3 - 21x - 236$ so that

$$f(x) = x^3 - 21x - 236 = (x - \theta)(x - \theta')(x - \theta''). \tag{7.4.2}$$

By Example 7.3.9 we know that $\left\{ 1, \frac{1+\theta}{3}, \frac{-2-\theta+\theta^2}{18} \right\}$ is an integral basis for K and $d(K) = -503$. Let $\alpha \in O_K$. Then

$$\alpha = a + b\frac{(1 + \theta)}{3} + c\left(\frac{-2 - \theta + \theta^2}{18} \right)$$

for some $a, b, c \in \mathbb{Z}$. The other conjugates of α are

$$\alpha' = a + b\frac{(1 + \theta')}{3} + c\left(\frac{-2 - \theta' + \theta'^2}{18} \right)$$

and

$$\alpha'' = a + b\frac{(1 + \theta'')}{3} + c\left(\frac{-2 - \theta'' + \theta''^2}{18} \right).$$

From (7.4.2) we deduce that $\theta + \theta' + \theta'' = 0$. Thus,

$$\theta^2 - \theta'^2 = (\theta - \theta')(\theta + \theta') = -(\theta - \theta')\theta'',$$

and we obtain

$$\alpha - \alpha' = (\theta - \theta')\left(\frac{b}{3} - \frac{c}{18} - \frac{c}{18}\theta'' \right),$$

and similarly

$$\alpha' - \alpha'' = (\theta' - \theta'')\left(\frac{b}{3} - \frac{c}{18} - \frac{c}{18}\theta \right), \quad \alpha - \alpha'' = (\theta - \theta'')\left(\frac{b}{3} - \frac{c}{18} - \frac{c}{18}\theta' \right).$$

Hence, appealing to (7.4.2), we obtain, as $D(\theta) = -2^2 \cdot 3^6 \cdot 503$ *from Example 7.3.9,*

$$D(\alpha) = \left\{ (\alpha - \alpha')(\alpha' - \alpha'')(\alpha - \alpha'') \right\}^2$$

$$= \left\{ (\theta - \theta')(\theta' - \theta'')(\theta - \theta'') \right\}^2 \frac{c^6}{18^6} \left\{ f\left(\frac{6b}{c} - 1\right) \right\}^2$$

$$= D(\theta) \frac{c^6}{2^6 3^{12}} \left\{ \left(\frac{6b}{c} - 1\right)^3 - 21\left(\frac{6b}{c} - 1\right) - 236 \right\}^2$$

$$= \frac{-2^2 \cdot 3^6 \cdot 503}{2^6 3^{12}} \left\{ (6b - c)^3 - 21c^2(6b - c) - 236c^3 \right\}^2$$

$$= -\frac{503}{2^4 \cdot 3^6} \{216b^3 - 108b^2c - 108bc^2 - 216c^3\}^2$$

$$= -503\{2b^3 - b^2c - bc^2 - 2c^3\}^2.$$

Then

$$\text{ind } \alpha = \sqrt{\frac{D(\alpha)}{d(K)}} = \sqrt{\frac{-503(2b^3 - b^2c - bc^2 - 2c^3)^2}{-503}} = |2b^3 - b^2c - bc^2 - 2c^3|.$$

Now

$$2b^3 - b^2c - bc^2 - 2c^3 \equiv 0 \,(\text{mod } 2)$$

for all $b, c \in \mathbb{Z}$, *so that*

$$\text{ind } \alpha \equiv 0 \,(\text{mod } 2)$$

for all $\alpha \in O_K$. *But*

$$2b^3 - b^2c - bc^2 - 2c^3 = 2$$

for $(b, c) = (1, 0)$, *so that*

$$m(K) = \min \left\{ |2b^3 - b^2c - bc^2 - 2c^3| \,|\, b, c \in \mathbb{Z}, \ 2b^3 - b^2c - bc^2 - 2c^3 \neq 0 \right\} = 2$$

and

$$i(K) = \gcd \left\{ |2b^3 - b^2c - bc^2 - 2c^3| \,|\, b, c \in \mathbb{Z}, \ 2b^3 - b^2c - bc^2 - 2c^3 \neq 0 \right\} = 2.$$

As $m(K) = 2$, K *does not possess a power basis by Theorem 7.4.1.*

Dedekind [4] gave in 1878 the first example of an algebraic number field without a power basis, namely, the cubic field L given by

$$L = \mathbb{Q}(\phi), \ \ \phi^3 - \phi^2 - 2\phi - 8 = 0.$$

The field L is in fact the same field as the field $K = \mathbb{Q}(\theta)$, $\theta^3 - 21\theta - 236 = 0$, in Example 7.4.3, as θ and ϕ are related by $\theta = 3\phi - 1$.

The next example gives a cubic field K for which

$$i(K) = 1, \; m(K) = 2, \; K \text{ does not possess a power basis,}$$

which shows that the converse of Theorem 7.4.2 does not hold.

Example 7.4.4 *Let $K = \mathbb{Q}(\sqrt[3]{175})$. An integral basis for K is given by $\{1, 175^{1/3}, 245^{1/3}\}$ and $d(K) = -3^3 \cdot 5^2 \cdot 7^2$ (see Theorem 7.3.2). Let $\alpha \in O_K$. Then there exist $a, b, c \in \mathbb{Z}$ such that*

$$\alpha = a + b175^{1/3} + c245^{1/3}.$$

The other conjugates of α are

$$\alpha' = a + b\omega 175^{1/3} + c\omega^2 245^{1/3},$$
$$\alpha'' = a + b\omega^2 175^{1/3} + c\omega 245^{1/3},$$

where ω is a complex cube root of unity. As

$$1 + \omega + \omega^2 = 0, \; \omega^3 = 1,$$

we obtain

$$\alpha - \alpha' = (1 - \omega)(b175^{1/3} - c\omega^2 245^{1/3}),$$
$$\alpha - \alpha'' = (1 - \omega^2)(b175^{1/3} - c\omega 245^{1/3}),$$
$$\alpha' - \alpha'' = (\omega - \omega^2)(b175^{1/3} - c245^{1/3}).$$

Hence

$$\begin{aligned}
D(\alpha) &= (\alpha - \alpha')^2(\alpha - \alpha'')^2(\alpha' - \alpha'')^2 \\
&= \{(1 - \omega)(1 - \omega^2)(\omega - \omega^2)\}^2(175b^3 - 245c^3)^2 \\
&= -27(175b^3 - 245c^3)^2 = -3^3 \cdot 5^2 \cdot 7^2(5b^3 - 7c^3)^2.
\end{aligned}$$

Then

$$\text{ind } \alpha = \sqrt{\frac{D(\alpha)}{d(K)}} = \sqrt{\frac{-3^3 \cdot 5^2 \cdot 7^2(5b^3 - 7c^3)^2}{-3^3 \cdot 5^2 \cdot 7^2}} = |5b^3 - 7c^3|.$$

Thus

$$i(K) = \gcd\{|5b^3 - 7c^3| \mid b, c \in \mathbb{Z}, \; 5b^3 - 7c^3 \neq 0\}$$

and

$$m(K) = \min\{|5b^3 - 7c^3| \mid b, c \in \mathbb{Z}, \; 5b^3 - 7c^3 \neq 0\}.$$

Since $|5 \cdot 1^3 - 7 \cdot 1^3| = 2$ and $|5 \cdot 1^3 - 7 \cdot 0^3| = 5$ we see that $i(K) = 1$ and $m(K) = 1$ or 2. Suppose $m(K) = 1$. Then there exist integers B and C such that

$5B^3 - 7C^3 = \pm 1$. *Thus* $5B^3 \equiv \pm 1 \pmod{7}$, *so* $B^3 \equiv \pm 3 \pmod{7}$. *But this is impossible as the only cubes modulo* 7 *are* $0, \pm 1$. *Hence* $m(K) = 2$. *By Theorem* 7.4.1 K *does not possess a power basis.*

Llorente and Nart [12, Theorem 4, p. 585] have given a necessary and sufficient condition for a cubic field to have index 2.

Definition 7.4.3 (Inessential discriminant divisor) *Let* K *be an algebraic number field of degree* n *over* $\mathbb{Q}$. *A prime* p *is called an inessential discriminant divisor or common index divisor if* $p \mid$ ind α *for every generator* α *of* K.

The inessential discriminant divisors of an algebraic number field K are precisely the prime factors of the index $i(K)$. Example 7.4.3 shows that the only inessential discriminant divisor of the cubic field $\mathbb{Q}(\theta)$, $\theta^3 - 21\theta - 236 = 0$, is the prime 2. Indeed the set of inessential discriminant divisors of a cubic field is either the empty set ϕ or $\{2\}$. This is a special case of the general result due to Zyliński [20] that a prime p can be an inessential discriminant divisor of an algebraic number field of degree n only if $p < n$. Thus when K is a quartic field the set of inessential discriminant divisors is ϕ, $\{2\}$, $\{3\}$, or $\{2, 3\}$.

7.5 Integral Basis of a Cyclotomic Field

Let m be a positive integer. The number of positive integers less than or equal to m that are coprime with m is denoted by $\phi(m)$. The arithmetic function $\phi(m)$ is called Euler's phi function. Let ζ_m be any primitive mth root of unity. There are $\phi(m)$ primitive mth roots of unity, namely ζ_m^r, $r = 1, 2, \ldots, m$, $(r, m) = 1$. Let $K_m = \mathbb{Q}(\zeta_m)$. It is easy to show that $K_m = \mathbb{Q}(\zeta_m^r)$ for any $r \in \{1, 2, \ldots, m\}$ with $(r, m) = 1$, so that K_m is independent of the primitive mth root of unity chosen. The field K_m is called the mth cyclotomic field. For odd m the fields K_m and K_{2m} coincide as $-\zeta_m$ is a primitive $2m$th root of unity. Clearly ζ_m is a root of the polynomial

$$f_m(x) = \prod_{\substack{r = 1 \\ (r, m) = 1}}^{m} (x - \zeta_m^r).$$

It is known that $f_m(x) \in \mathbb{Z}[x]$ and that $f_m(x)$ is irreducible, so that

$$\mathrm{irr}_{\mathbb{Q}}(\zeta_m) = f_m(x).$$

Moreover, the degree of $f_m(x)$ is $\phi(m)$ so that

$$[K_m : \mathbb{Q}] = \phi(m).$$

The smallest field containing both K_m and K_n is $K_{[m,n]}$, where $[m, n]$ denotes the least common multiple of m and n. Also, $K_m \cap K_n = K_{(m,n)}$. If $m \not\equiv 2 \pmod 4$ then $K_m \subseteq K_n$ holds if and only if $m \mid n$. Thus if m and n are distinct and not congruent to 2 (mod 4) the cyclotomic fields K_m and K_n are distinct.

The next theorem gives an integral basis for K_m as well as a formula for the discriminant $d(K_m)$.

Theorem 7.5.1 *Let m be a positive integer. Let ζ_m be a primitive mth root of unity. Let K_m denote the cyclotomic field $\mathbb{Q}(\zeta_m)$. Then $\{1, \zeta_m, \zeta_m^2, \ldots, \zeta_m^{\phi(m)-1}\}$ is an integral basis for K_m. Further,*

$$d(K_m) = (-1)^{\frac{\phi(m)}{2}} \frac{m^{\phi(m)}}{\prod_{p \mid m} p^{\frac{\phi(m)}{p-1}}},$$

where the product is over all primes p dividing m.

We refer the reader to Narkiewicz [13, Theorem 4.10, p. 169] for a proof of this theorem.

Taking $m = 3, 4, 5, 8$ in Theorem 7.5.1, we obtain $d(K_3) = -3$, $d(K_4) = -4$, $d(K_5) = 125$, $d(K_8) = 256$. The first two of these are familar to us as $K_3 = \mathbb{Q}(\sqrt{-3})$ and $K_4 = \mathbb{Q}(\sqrt{-1})$ are quadratic fields. The fourth equality is also known to us as $\frac{1}{2}(\sqrt{2} + \sqrt{-2})$ is a primitive eighth root of unity, so $K_8 = \mathbb{Q}(\sqrt{2} + \sqrt{-2}) = \mathbb{Q}(\sqrt{2}, \sqrt{-2}) = \mathbb{Q}(\sqrt{2}, i)$ and we showed that $d(\mathbb{Q}(\sqrt{2}, i)) = 256$ in Example 7.1.7.

Example 7.5.1 *We show that*

$$K_5 = \mathbb{Q}\left(i\sqrt{10 + 2\sqrt{5}} \right).$$

Let β be the primitive fifth root of unity, $e^{2\pi i/5}$, so that

$$\beta = e^{2\pi i/5} = \cos\frac{2\pi}{5} + i \sin\frac{2\pi}{5} = c + is,$$

where

$$c = \cos\frac{2\pi}{5}, \quad s = \sin\frac{2\pi}{5}.$$

Then

$$1 = \beta^5 = (c + is)^5 = (c^5 - 10c^3 s^2 + 5cs^4) + i(5c^4 s - 10c^2 s^3 + s^5).$$

Hence

$$5c^4 s - 10c^2 s^3 + s^5 = 0.$$

As $s \neq 0$ and $c^2 + s^2 = 1$ we obtain

$$5c^4 - 10c^2(1 - c^2) + (1 - c^2)^2 = 0,$$

that is,

$$16c^4 - 12c^2 + 1 = 0,$$

so that

$$c^2 = \frac{3 \pm \sqrt{5}}{8}.$$

Now $c \approx 0.3$, $c^2 \approx 0.09$, $(3 - \sqrt{5})/8 \approx 0.09$, so

$$c^2 = \frac{3 - \sqrt{5}}{8}$$

and

$$c = \sqrt{\frac{3 - \sqrt{5}}{8}} = \sqrt{\frac{6 - 2\sqrt{5}}{16}} = \frac{\sqrt{5} - 1}{4}.$$

Hence

$$s^2 = 1 - c^2 = 1 - \left(\frac{\sqrt{5} - 1}{4} \right)^2 = \frac{10 + 2\sqrt{5}}{16},$$

so

$$s = \frac{\sqrt{10 + 2\sqrt{5}}}{4}.$$

We have shown that

$$\beta = e^{\frac{2\pi i}{5}} = \frac{1}{4} \left(\sqrt{5} - 1 + i\sqrt{10 + 2\sqrt{5}} \right).$$

Squaring we obtain

$$\beta^2 = e^{\frac{4\pi i}{5}} = \frac{1}{4} \left(-\sqrt{5} - 1 + i\sqrt{10 - 2\sqrt{5}} \right),$$

as

$$\sqrt{10 - 2\sqrt{5}} = \frac{\sqrt{5} - 1}{2} \sqrt{10 + 2\sqrt{5}}.$$

Further,

$$\beta^3 = \bar{\beta}^2 = \frac{1}{4} \left(-\sqrt{5} - 1 - i\sqrt{10 - 2\sqrt{5}} \right),$$

$$\beta^4 = \bar{\beta} = \frac{1}{4} \left(\sqrt{5} - 1 - i\sqrt{10 + 2\sqrt{5}} \right).$$

Hence

$$\beta - \beta^4 = \frac{1}{2}i\sqrt{10 + 2\sqrt{5}},$$

so

$$\mathbb{Q}\left(i\sqrt{10 + 2\sqrt{5}}\right) = \mathbb{Q}(2\beta - 2\beta^4) \subseteq \mathbb{Q}(\beta).$$

Also,

$$\beta = \frac{1}{8}\left(-12 + 2\left(i\sqrt{10 + 2\sqrt{5}}\right) - \left(i\sqrt{10 + 2\sqrt{5}}\right)^2\right),$$

so

$$\mathbb{Q}(\beta) \subseteq \mathbb{Q}\left(i\sqrt{10 + 2\sqrt{5}}\right).$$

This shows that

$$K_5 = \mathbb{Q}\left(i\sqrt{10 + 2\sqrt{5}}\right)$$

and

$$d\left(\mathbb{Q}\left(i\sqrt{10 + 2\sqrt{5}}\right)\right) = d(K_5) = 125.$$

Integral bases for quartic fields like $\mathbb{Q}(i\sqrt{10 + 2\sqrt{5}})$, which contain a quadratic subfield, have been given by Huard, Spearman, and Williams [10].

The final theorem of this chapter is immediate from Definition 7.1.5 and Theorem 7.5.1.

Theorem 7.5.2 *The cyclotomic field $K_m = \mathbb{Q}(\zeta_m)$ is monogenic for every positive integer m.*

Exercises

1. Let D denote the discriminant of

$$f(x) = x^n + a_{n-1}x^{n-1} + \cdots + a_1 x + a_0 \in \mathbb{Z}[x].$$

   Prove that

$$D \equiv 0 \text{ or } 1 \pmod{4}.$$

2. Using the method of Example 7.1.6, prove that $\{1, \sqrt[3]{3}, (\sqrt[3]{3})^2\}$ is an integral basis for $\mathbb{Q}(\sqrt[3]{3})$. What is the discriminant of $\mathbb{Q}(\sqrt[3]{3})$?

3. Prove that

$$\left\{1, \ \frac{1+\sqrt{5}}{2}, \ \frac{1+\sqrt{13}}{2}, \ \frac{1+\sqrt{5}+\sqrt{13}+\sqrt{65}}{4}\right\}$$

is an integral basis for $K = \mathbb{Q}(\sqrt{5}, \sqrt{13})$. What is $d(K)$?

4. Let $K = \mathbb{Q}(\sqrt{5}, \sqrt{13})$. Use Exercise 3 to prove that

$$O_K = \{\frac{1}{4}(x + y\sqrt{5} + z\sqrt{13} + w\sqrt{65}) \mid x, y, z, w \in \mathbb{Z},$$
$$x \equiv y \equiv z \equiv w \,(\mathrm{mod}\ 2), \ x - y - z + w \equiv 0\,(\mathrm{mod}\ 4)\}.$$

5. Let $K = \mathbb{Q}(\theta)$, where $\theta^3 - 9\theta - 6 = 0$. Prove that $\{1, \ \theta, \ \theta^2\}$ is an integral basis for K and that $d(K) = 2^3 \cdot 3^5$.

6. Prove Theorem 7.3.2.

7. Let $K = \mathbb{Q}(\theta)$, where $\theta^3 - 6\theta + 36 = 0$. Prove that $\theta^2/6 \in O_K$. Show that $\{1, \theta, \theta^2/6\}$ is an integral basis for K and that $d(K) = -2^2 \cdot 3 \cdot 79$.

8. Let $K = \mathbb{Q}(\theta)$, where $\theta^3 - 3\theta + 56 = 0$. Prove that $(\theta - 1)/3 \in O_K$ and $(\theta^2 + \theta - 2)/9 \in O_K$. Show that $\left\{1, \frac{\theta-1}{3}, \frac{\theta^2+\theta-2}{9}\right\}$ is an integral basis for K and that $d(K) = -2^2 \cdot 29$.

9. Let $K_1 = \mathbb{Q}(\theta_1)$, where $\theta_1^3 + 27\theta_1 + 240 = 0$, and $K_2 = \mathbb{Q}(\theta_2)$, where $\theta_2^3 + 27\theta_2 + 72 = 0$. Prove that $d(K_1) = d(K_2) = -3^5$. Is $K_1 = K_2$?

10. Let $K = \mathbb{Q}(\theta)$, where $\theta^4 - 17\theta^2 - 34\theta - 17 = 0$. Prove that $d(K) = 17^3$.

11. Let $K = \mathbb{Q}(\sqrt[4]{2})$. Prove that $d(K) = -2^{11}$.

12. Prove from first principles that $K = \mathbb{Q}(\theta)$, $\theta^3 + 30\theta + 90 = 0, \theta \in \mathbb{R}$, is a pure cubic field, and express K in the form $K = \mathbb{Q}(m^{1/3})$ for some cubefree integer m.

13. Let $K = \mathbb{Q}(\theta)$, where $\theta^3 - 4\theta + 2 = 0$. Prove that $\{1, \theta, \theta^2\}$ is an integral basis for K and that $d(K) = 2^2 \cdot 37$.

14. Prove that $K_5 = \mathbb{Q}(i\sqrt{5 + 2\sqrt{5}})$.

15. If p is an odd prime prove that

$$[\mathbb{Q}(e^{2\pi i/p} + e^{-2\pi i/p}) : \mathbb{Q}] = \frac{p-1}{2}.$$

16. Suppose that $x^3 + ax + b \in \mathbb{Z}[x]$ is irreducible. Prove that $K = \mathbb{Q}(\theta)$, $\theta^3 + a\theta + b = 0$, $\theta \in \mathbb{R}$, is a pure cubic field if and only if $-4a^3 - 27b^2 = -3c^2$ for some positive integer c.

17. Let K be an algebraic number field. Let L be a conjugate field of K. Prove that $d(K) = d(L)$.

18. Let K be an algebraic number field. Let σ be a monomorphism : $K \longrightarrow \mathbb{C}$. Let L be the conjugate field $\sigma(K)$. Let $\{\omega_1, \ldots, \omega_n\}$ be an integral basis for K. Is $\{\sigma(\omega_1), \ldots, \sigma(\omega_n)\}$ an integral basis for L?

19. Determine an integral basis for

$$K = \mathbb{Q}(\theta), \ \theta^3 + 30\theta + 15 = 0.$$

20. If K and L are algebraic number fields with $K \subseteq L$ prove that $d(K) \mid d(L)$.

21. Let K be an algebraic number field of degree n over $\mathbb{Q}$. Let $\theta \in O_K$ be such that $K = \mathbb{Q}(\theta)$. Let $\alpha \in O_K$. Express α in the form

$$\alpha = \sum_{j=0}^{n-1} y_j \theta^j,$$

where $y_0, y_1, \ldots, y_{n-1} \in \mathbb{Q}$. Prove that

$$y_j D(\theta) \in \mathbb{Z}, \quad j = 0, 1, \ldots, n - 1.$$

22. Let K be an algebraic number field of degree n. Is it possible to find $\lambda_1, \ldots, \lambda_n \in O_K$ such that $D(\lambda_1, \ldots, \lambda_n) = -d(K)$?

23. Let K be an algebraic number field. Prove from first principles that an integral basis for K can always be chosen to include 1.

24. Prove that

$$\left\{ 1, \sqrt[3]{10}, \frac{1 + \sqrt[3]{10} + (\sqrt[3]{10})^2}{3} \right\}$$

is an integral basis for $\mathbb{Q}(\sqrt[3]{10})$ using Theorem 7.2.7.

25. Prove that $\{1, \sqrt[4]{2}, (\sqrt[4]{2})^2, (\sqrt[4]{2})^3\}$ is an integral basis for $\mathbb{Q}(\sqrt[4]{2})$ using the ideas of Example 7.1.6.

26. Prove that

$$\left\{ 1, \sqrt[4]{5}, \frac{1 + (\sqrt[4]{5})^2}{2}, \frac{\sqrt[4]{5} + (\sqrt[4]{5})^3}{2} \right\}$$

is an integral basis for $\mathbb{Q}(\sqrt[4]{5})$ using Theorem 7.2.7.

27. Use Brill's theorem to show that

$$\text{sgn}(d(K_m)) = (-1)^{\frac{\phi(m)}{2}}.$$

28. Let m be a positive integer. Let ζ_m be a primitive mth root of unity. What is $\text{sgn}(d(\mathbb{Q}(\zeta_m + \zeta_m^{-1})))$?

Suggested Reading

1. Ş. Alaca, *p-integral bases of a cubic field*, Proceedings of the American Mathematical Society 126 (1998), 1949–1953.

   A p-integral basis of a cubic field K is determined for each rational prime p, and then an integral basis of K and the discriminant $d(K)$ of K are obtained from its p-integral bases.

2. Ş. Alaca, *p-integral bases of algebraic number fields*, Utilitas Mathematica 56 (1999), 97–106.

   The properties of p-integral bases of an algebraic number field K are developed and used to show how an integral basis of K can be obtained from its p-integral bases.

3. A. Brill, *Ueber die Discriminante*, Mathematische Annalen 12 (1877), 87–89.

   This is the original paper of Brill giving the sign of the discriminant of an algebraic number field.

4. R. Dedekind, *Über den Zusammenhang zwischen der Theorie der Ideale und der Theorie der höheren Kongruenzen,* Abh. Kgl. Ges. Wiss. Göttingen 23 (1878), 1–23. (*Gesammelte Mathematische Werke I*, pp. 202–232, Vieweg, Wiesbaden, 1930.)

It is shown that the cubic field $K = \mathbb{Q}(\theta)$, $\theta^3 - \theta^2 - 2\theta - 8 = 0$, does not have a power basis.

5. R. Dedekind, *Über die Anzahl der Idealklassen in reinen kubischen Zahlkörpern,* Journal für die reine und angewandte Mathematik 121 (1900), 40–123. (*Gesammelte Mathematische Werke II*, pp. 148–233, Vieweg, Wiesbaden, 1931.)

An integral basis is given for the pure cubic field $\mathbb{Q}(\sqrt[3]{m})$.

6. D. S. Dummit and H. Kisilevsky, *Indices in cyclic cubic fields,* in Zassenhaus, H. (ed.), *Number Theory and Algebra, Collected Papers Dedicated to Henry B. Mann, Arnold E. Ross and Olga Taussky-Todd,* pp. 29–42, Academic Press, New York, 1979.

It is shown that infinitely many cyclic cubic fields have a power basis.

7. H. T. Engstrom, *On the common index divisors of an algebraic field,* Transactions of the American Mathematical Society 32 (1930), 223–237.

The basic properties of the index of an algebraic number field are given.

8. T. Funakura, *On integral bases of pure quartic fields,* Mathematical Journal of Okayama University 26 (1984), 27–41.

An explicit integral basis is given for a pure quartic field.

9. M.-N. Gras, *Sur les corps cubiques cycliques dont l'anneau des entiers est monogène,* Annales Scientifiques de l'Université de Besançon, Mathematics, 1973, 26 pp.

Necessary and sufficient conditions are given for a cyclic cubic field to have a power basis. It can be deduced from these that infinitely many cyclic cubic fields do not have a power basis.

10. J. G. Huard, B. K. Spearman, and K. S. Williams, *Integral bases for quartic fields with quadratic subfields,* Journal of Number Theory 51 (1995), 87–102.

Let L be quartic field with quadratic subfield $\mathbb{Q}(\sqrt{c})$, where c is a squarefree integer. Then $L = \mathbb{Q}(\sqrt{c}, \sqrt{a + b\sqrt{c}})$, where $a + b\sqrt{c}$ is not a square in $\mathbb{Q}(\sqrt{c})$. The discriminant of L and an integral basis for L are determined explicitly.

11. R. H. Hudson and K. S. Williams, *The integers of a cyclic quartic field,* Rocky Mountain Journal of Mathematics 20 (1990), 145–150.

An explicit integral basis is given for a quartic field with Galois group $\mathbb{Z}_4$.

12. P. Llorente and E. Nart, *Effective determination of the rational primes in a cubic field,* Proceedings of the American Mathematical Society 87 (1983), 579–585.

A necessary and sufficient condition is given for a cubic field to have index 2.

13. W. Narkiewicz, *Elementary and Analytic Theory of Algebraic Numbers,* Springer-Verlag, Berlin, 1990.

The principal properties of cyclotomic fields are summarized in Theorem 4.10, p. 169.

14. B. K. Spearman and K. S. Williams, *The conductor of a cyclic quartic field,* Publicationes Mathematicae 48 (1996), 13–43.

An explicit formula is given for the discriminant of a cyclic quartic field $\mathbb{Q}(\theta)$, where $\theta^4 + A\theta^2 + B\theta + C = 0$.

15. B. K. Spearman and K. S. Williams, *An explicit integral basis for a pure cubic field,* Far East Journal of Mathematical Sciences 6 (1998), 1–14.

An explicit integral basis is given for a pure cubic field $K = \mathbb{Q}(\theta)$, $\theta^3 + a\theta + b = 0$.

16. B. K. Spearman and K. S. Williams, *Cubic fields with a power basis,* Rocky Mountain Journal of Mathematics 31 (2001), 1103–1109.

   It is shown that there exist infinitely many cubic fields L with a power basis such that the splitting field M of L contains a given quadratic field K.

17. B. K. Spearman and K. S. Williams, *Cubic fields with index* 2, Monatshefte für Mathematik 134 (2002), 331–336.

   Let d be a squarefree integer with $d = 1$ allowed. If $d \not\equiv 1 \pmod 8$ it is shown that there do not exist any cubic fields with index 2 whose splitting field contains $\mathbb{Q}(\sqrt{d})$. If $d \equiv 1 \pmod 8$ it is shown that there exist infinitely many cubic fields K with $i(K) = m(K) = 2$ whose splitting field contains $\mathbb{Q}(\sqrt{d})$.

18. L. Stickelberger, *Über eine neue Eigenschaft der Diskriminanten algebraischer Zahlkörper,* International Mathematische Kongress, Zürich, 1897, 182–193.

   It is shown that $d(K) \equiv 0$ or $1 \pmod 4$ for an algebraic number field K.

19. K. S. Williams, *Integers of biquadratic fields,* Canadian Mathematical Bulletin 13 (1970), 519–526.

   This paper gives an explicit integral basis for the quartic field $\mathbb{Q}(\sqrt{m} + \sqrt{n}) = \mathbb{Q}(\sqrt{m}, \sqrt{n})$, where m and n are distinct squarefree integers.

20. E. Zyliński, *Zur Theorie der ausserwesentlichen Discriminantenteiler algebraischer Körper,* Mathematische Annalen 73 (1913), 273–274.

   It is shown that a prime p can only be an inessential discriminant divisor of an algebraic number field of degree n if $p < n$.

Biographies

1. E. T. Bell, *Men of Mathematics,* Simon and Schuster, New York, 1937.

   Chapter 27 is devoted to Ernst Kummer (1810–1893) and Richard Dedekind (1831–1916).

2. R. A. Mollin, *Algebraic Number Theory,* Chapman and Hall/CRC Press, London/Boca Raton, Florida, 1999.

   A brief biography of Ludwig Stickelberger (1850–1936) is given on page 43.

3. The website

   http://www-groups.dcs.st-and.ac.uk/~history/

   has biographies of Alexander Brill (1842–1935) and Richard Dedekind (1831–1916).

8

Dedekind Domains

8.1 Dedekind Domains

In Chapter 6 it was shown that the ring of algebraic integers O_K of an algebraic number field K has the following three properties:

O_K is a Noetherian domain (Theorem 6.5.3),
O_K is integrally closed (Theorem 6.1.6), and
each prime ideal P of O_K is a maximal ideal (Theorem 6.6.1).

An integral domain with these properties is called a Dedekind domain after Richard Dedekind, the creator of the modern theory of ideals.

Definition 8.1.1 (Dedekind domain) *An integral domain D that satisfies the following three properties:*

$$D \text{ is a Noetherian domain,} \qquad (8.1.1)$$
$$D \text{ is integrally closed, and} \qquad (8.1.2)$$
$$\text{each prime ideal of } D \text{ is a maximal ideal,} \qquad (8.1.3)$$

is called a Dedekind domain.

In view of the remarks before Definition 8.1.1, we have

Theorem 8.1.1 *Let K be an algebraic number field. Let O_K be the ring of integers of K. Then O_K is a Dedekind domain.*

The next theorem gives another class of integral domains that are Dedekind domains.

Theorem 8.1.2 *Let D be a principal ideal domain. Then D is a Dedekind domain.*

Proof: Let D be a principal ideal domain. By Theorem 3.1.2 D is a Noetherian domain, so (8.1.1) holds. By Theorem 3.3.1 D is a unique factorization domain and thus, by Theorem 4.2.5, D is integrally closed, so (8.1.2) holds. By Theorem 1.5.7 each prime ideal of D is maximal so that (8.1.3) holds. Hence D is a Dedekind domain. ∎

Our main objective in this chapter is to show that every ideal $I(\neq \langle 0 \rangle, \langle 1 \rangle)$ of a Dedekind domain can be expressed uniquely as a product of prime ideals. We also show that every ideal of a Dedekind domain is generated by at most two elements.

8.2 Ideals in a Dedekind Domain

The first step toward our objective of proving that in a Dedekind domain every proper ideal is a product of prime ideals is to show that every such ideal contains a product of prime ideals. This is actually true in a Noetherian domain.

Theorem 8.2.1 *In a Noetherian domain every nonzero ideal contains a product of one or more prime ideals.*

Proof: Suppose that D is a Noetherian domain that possesses at least one nonzero ideal that does not contain a product of one or more prime ideals. Let S be the set of all such ideals. By assumption S is not empty. As D is Noetherian, by Theorem 3.1.3 S contains a (nonzero) ideal A maximal with respect to the property of not containing a product of one or more prime ideals. Clearly A itself is not a prime ideal. Hence, by Theorem 1.6.1, there exist ideals B and C such that

$$BC \subseteq A, \ B \not\subseteq A, \ C \not\subseteq A.$$

Define the ideals B_1 and C_1 of D by

$$B_1 = A + B, \ C_1 = A + C.$$

Clearly

$$A \subset B_1, \ A \subset C_1,$$

so that $B_1 \notin S, C_1 \notin S$. Hence there exist prime ideals $P_1, \ldots, P_k$ such that

$$B_1 \supseteq P_1 \cdots P_h, \ C_1 \supseteq P_{h+1} \cdots P_k.$$

But

$$B_1 C_1 = (A + B)(A + C) \subseteq A,$$

so

$$A \supseteq P_1 \cdots P_k,$$

contradicting that $A \in S$. ∎

As a Dedekind domain is a Noetherian domain, the next theorem is an immediate consequence of Theorem 8.2.1.

Theorem 8.2.2 *In a Dedekind domain every nonzero ideal contains a product of one or more prime ideals.*

Our next step is to obtain an inverse of a prime ideal P in a Dedekind domain. To do this, we extend the notion of an "ideal" to that of a "fractional ideal."

Definition 8.2.1 (Fractional ideal) *Let D be an integral domain. Let K be the quotient field of D. A nonempty subset A of K with the following three properties:*

(i) $\alpha \in A,\ \beta \in A \Longrightarrow \alpha + \beta \in A$,
(ii) $\alpha \in A,\ r \in D \Longrightarrow r\alpha \in A$, *and*
(iii) *there exists $\gamma \in D$ with $\gamma \neq 0$ such that $\gamma A \subseteq D$*

is called a fractional ideal of D.

Condition (iii) means that the elements of a fractional ideal have γ as a "common denominator."

Example 8.2.1 *Let*

$$A = \left\{ \frac{n}{25} \mid n \in \mathbb{Z} \right\}$$

so that A is a nonempty subset of $\mathbb{Q}$. Clearly A has properties (i) and (ii). Also, $25A = \mathbb{Z}$ so that (iii) holds. Hence A is a fractional ideal of $\mathbb{Z}$.

Example 8.2.2 *Let*

$$A = \left\{ \frac{n}{5^m} \mid n \in \mathbb{Z},\ m \in \mathbb{N} \cup \{0\} \right\}.$$

Clearly A is a nonempty subset of $\mathbb{Q}$ having properties (i) and (ii). However, there is no nonzero integer k such that $kA \subseteq \mathbb{Z}$, so (iii) does not hold. Thus A is not a fractional ideal of $\mathbb{Z}$.

A fractional ideal of D that is a subset of D is clearly an ideal of D in the ordinary sense. Moreover, an ideal of D is a fractional ideal of D that is a subset of D. We often refer to the ideals of D in the ordinary sense as integral ideals. If A is a fractional ideal of D and γ is a common denominator for A then γA is an integral ideal of D.

It follows immediately from Definition 8.2.1 that if A is a fractional ideal of D then

$$A = \frac{1}{\gamma} I,$$

where $\gamma \in D \setminus \{0\}$ and I is an integral ideal of D. This representation is not unique as

$$A = \frac{1}{\gamma \delta}(\delta I)$$

for any $\delta \in D \setminus \{0\}$.

If D is a Noetherian domain each integral ideal I of D is finitely generated. Hence

$$I = \langle \alpha_1, \ldots, \alpha_k \rangle,$$

for some $\alpha_1, \ldots, \alpha_k \in D$, and thus

$$A = \frac{1}{\gamma} I = \frac{1}{\gamma} \langle \alpha_1, \ldots, \alpha_k \rangle = \langle \frac{\alpha_1}{\gamma}, \ldots, \frac{\alpha_k}{\gamma} \rangle;$$

that is, every fractional ideal A of D is also finitely generated.

It is easily verified that if A and B are fractional ideals of D so are $A + B$ and AB. We note that if γ and δ are common denominators for A and B respectively, then $\gamma \delta$ is a common denominator for both $A + B$ and AB.

Definition 8.2.2 (The set $\tilde{P}$ for a prime ideal P) *Let D be an integral domain and let K be the quotient field of D. For each prime ideal P of D we define the set $\tilde{P}$ by*

$$\tilde{P} = \{\alpha \in K : \alpha P \subseteq D\}.$$

Theorem 8.2.3 *Let D be an integral domain and let P be a prime ideal of D. Then $\tilde{P}$ is a fractional ideal of D.*

Proof: If $\alpha \in \tilde{P}$ and $\beta \in \tilde{P}$ then $\alpha P \subseteq D$ and $\beta P \subseteq D$. Hence $(\alpha + \beta)P \subseteq \alpha P + \beta P \subseteq D$, so that $\alpha + \beta \in \tilde{P}$.

If $\alpha \in \tilde{P}$ and $r \in D$ then $\alpha P \subseteq D$ and thus $r\alpha P \subseteq D$, so that $r\alpha \in \tilde{P}$.

Take $\pi \in P \setminus \{0\}$. For any $\alpha \in \tilde{P}$ we have $\alpha P \subseteq D$ so that in particular $\alpha \pi \in D$. Hence $\pi \tilde{P} \subseteq D$.

Thus the three properties in Definition 8.2.1 hold, showing that $\tilde{P}$ is a fractional ideal of D. ∎

Theorem 8.2.4 *Let D be a Dedekind domain. Let P be a prime ideal of D. Then $P\tilde{P} = D$.*

Proof: We first show that

$$P\tilde{P} = D \text{ or } P\tilde{P} = P.$$

As $\tilde{P}$ and P are both fractional ideals of D so is $P\tilde{P}$. Clearly $P\tilde{P} \subseteq D$ so that $P\tilde{P}$ is an integral ideal of D. As $1 \in \tilde{P}$ we have $P \subseteq P\tilde{P}$. Since P is a prime ideal and D is a Dedekind domain, P is a maximal ideal. Thus $P\tilde{P} = P$ or $P\tilde{P} = D$.

Next we show that $D \subset \tilde{P}$. If $\alpha \in D$ then $\alpha P \subseteq D$ so that $\alpha \in \tilde{P}$. Hence $D \subseteq \tilde{P}$. To prove that $D \subset \tilde{P}$ we show that $\tilde{P}$ contains an element γ of K that does not lie in D. Let $\beta \in P \setminus \{0\}$. By Theorem 8.2.2 there exist prime ideals $P_1, \ldots, P_k$ $(k \geq 1)$ with

$$\langle \beta \rangle \supseteq P_1 \cdots P_k.$$

We choose k to be the least positive integer for which such an inclusion holds. Since

$$P_1 \cdots P_k \subseteq \langle \beta \rangle \subseteq P$$

and P is a prime ideal, we have

$$P_i \subseteq P, \text{ for some } i \in \{1, 2, \ldots, k\}.$$

Relabeling P_1 as P_i and P_i as P_1, if necessary, we may suppose that

$$P_1 \subseteq P.$$

But as D is a Dedekind domain, P_1 is a maximal ideal, and so

$$P_1 = P.$$

We now consider two cases according as $k = 1$ or $k \geq 2$. If $k = 1$ then

$$P = P_1 = \langle \beta \rangle.$$

As $\beta \neq 0$ we can define $\gamma = 1/\beta \in K$. Suppose $\gamma \in D$. Then β is a unit of D and $P = \langle \beta \rangle = D$, contradicting that P is a prime ideal. Hence $\gamma \notin D$. Also,

$$\gamma P = \frac{1}{\beta} \langle \beta \rangle = \langle 1 \rangle = D,$$

so that $\gamma \in \tilde{P}$. Hence $\gamma \in \tilde{P} \setminus D$ in this case. If $k \geq 2$ then by the minimality of k we have

$$P_2 \cdots P_k \not\subseteq \langle \beta \rangle.$$

Hence there exists $\delta \in P_2 \cdots P_k$ but $\delta \notin \langle \beta \rangle$. As $\beta \neq 0$ we can define $\gamma = \delta/\beta \in K$. As $\delta \notin \langle \beta \rangle$, we see that $\gamma = \delta/\beta \notin D$. However,

$$P\langle \delta \rangle = P_1 \langle \delta \rangle \subseteq P_1 \cdots P_k \subseteq \langle \beta \rangle,$$

so

$$P\gamma = P\delta/\beta \subseteq D$$

and thus $\gamma \in \tilde{P}$. Hence $\gamma \in \tilde{P} \setminus D$ in this case. This completes the proof that

$$D \subset \tilde{P}.$$

Finally, we show that

$$P\tilde{P} = D.$$

Recall that we have shown that $P\tilde{P} = P$ or $P\tilde{P} = D$. We show that $P\tilde{P} \neq P$, so that we must have $P\tilde{P} = D$. Suppose that $P\tilde{P} = P$. We show that $\tilde{P}$ is closed under multiplication. Let $\alpha \in \tilde{P}$ and $\beta \in \tilde{P}$. Then $\alpha P \subseteq P\tilde{P} = P$ and $\beta P \subseteq P\tilde{P} = P$. Thus

$$\alpha\beta P \subseteq \alpha P \subseteq P,$$

showing that $\alpha\beta \in \tilde{P}$. Hence $\tilde{P}$ is closed under multiplication. This proves that $\tilde{P}$ is an integral domain, which strictly contains D. As D is a Noetherian domain, all its ideals (integral or fractional) are finitely generated. Hence $\tilde{P}$ is a finitely generated fractional ideal of D. Thus $\tilde{P}$ is a finitely generated D-module. Hence, by the remark following Theorem 4.1.4, $\tilde{P}$ is integral over D. However, D is integrally closed in its quotient field (since D is a Dedekind domain) so that $D = \tilde{P}$. This contradicts that $\tilde{P} \supset D$. Hence $P\tilde{P} = D$.   ∎

Example 8.2.3 *Let*

$$D = \mathbb{Z} + \mathbb{Z}\sqrt{6}.$$

As D is the ring of integers of $K = \mathbb{Q}(\sqrt{6})$, D is a Dedekind domain with quotient field K. Let

$$P = \langle 2, \sqrt{6} \rangle.$$

It is easily checked that P is a prime ideal of D with $P = 2\mathbb{Z} + \mathbb{Z}\sqrt{6}$. Then

$$
\begin{aligned}
\tilde{P} &= \{\alpha \in K \mid \alpha P \subseteq D\} \\
&= \{x + y\sqrt{6} \mid x, y \in \mathbb{Q}, \ (x + y\sqrt{6})\langle 2, \sqrt{6} \rangle \subseteq \mathbb{Z} + \mathbb{Z}\sqrt{6}\} \\
&= \{x + y\sqrt{6} \mid x, y \in \mathbb{Q}, \ 2(x + y\sqrt{6}) \in \mathbb{Z} + \mathbb{Z}\sqrt{6}, \ (x + y\sqrt{6})\sqrt{6} \in \mathbb{Z} + \mathbb{Z}\sqrt{6}\} \\
&= \{x + y\sqrt{6} \mid 2x \in \mathbb{Z}, \ 2y \in \mathbb{Z}, \ x \in \mathbb{Z}, \ 6y \in \mathbb{Z}\} \\
&= \{x + y\sqrt{6} \mid x \in \mathbb{Z}, \ 2y \in \mathbb{Z}\} \\
&= \{m + \frac{n}{2}\sqrt{6} \mid m, n \in \mathbb{Z}\} \\
&= \{\frac{2m + n\sqrt{6}}{2} \mid m, n \in \mathbb{Z}\} \\
&= \frac{1}{2}\{2m + n\sqrt{6} \mid m, n \in \mathbb{Z}\} \\
&= \frac{1}{2}(2\mathbb{Z} + \mathbb{Z}\sqrt{6}) \\
&= \frac{1}{2}P.
\end{aligned}
$$

8.3 Factorization into Prime Ideals

We now use Theorem 8.2.4 to prove the fundamental property of a Dedekind domain D, namely, that every proper integral ideal of D can be expressed uniquely (up to order) as a product of prime ideals.

Theorem 8.3.1 *If D is a Dedekind domain every integral ideal ($\neq \langle 0 \rangle$, D) is a product of prime ideals and this factorization is unique in the sense that if*

$$P_1 P_2 \cdots P_k = Q_1 Q_2 \cdots Q_l,$$

where the P_i and Q_j are prime ideals, then $k = l$ and after relabeling (if necessary)

$$P_i = Q_i, \quad i = 1, 2, \ldots, k.$$

Proof: Suppose there exist integral ideals ($\neq \langle 0 \rangle$, D) of D that are not products of prime ideals. As D is a Dedekind domain, it is Noetherian, and so by the maximal principle (Theorem 3.1.3) there is an ideal $A(\neq \langle 0 \rangle$, D) of D maximal with respect to the property of not being a product of prime ideals. By Theorem 8.2.1 there exist prime ideals $P_1, \ldots, P_k$ ($k \geq 1$) of D such that

$$P_1 \cdots P_k \subseteq A.$$

Let k be the smallest positive integer for which such a product exists. If $k = 1$ then $P_1 \subseteq A \subset D$. As P_1 is a prime ideal, it is a maximal ideal since D is a Dedekind domain. Thus $A = P_1$. This is impossible as A is not a product of prime ideals. Hence $k \geq 2$. By Theorem 8.2.4 we have $\tilde{P}_1 P_1 = D$ so that

$$\tilde{P}_1 P_1 P_2 \cdots P_k = D P_2 \cdots P_k.$$

Hence

$$\tilde{P}_1 A \supseteq \tilde{P}_1 P_1 \cdots P_k = P_2 \cdots P_k.$$

From the proof of Theorem 8.2.4 we have $D \subset \tilde{P}_1$ so that $A \subseteq \tilde{P}_1 A$. If $A = \tilde{P}_1 A$ then

$$A \supseteq P_2 \cdots P_k,$$

which contradicts the minimality of k as $k - 1 \geq 1$. Hence $A \subset \tilde{P}_1 A$. Since $\tilde{P}_1 A$ is an ideal of D, by the maximality property of A, we have

$$\tilde{P}_1 A = Q_2 \cdots Q_h$$

for prime ideals $Q_2, \ldots, Q_h$. Then

$$A = AD = A\tilde{P}_1 P_1 = P_1 Q_2 \cdots Q_h$$

is also a product of prime ideals, which contradicts the way A was chosen. Hence every ideal ($\neq \langle 0 \rangle$, D) of D is a product of prime ideals.

Suppose now that factorization of ideals as products of prime ideals is not always unique. By the maximal principle we may choose B to be an ideal ($\neq \langle 0 \rangle$, D) maximal with respect to the property of having at least two distinct factorizations as the product of prime ideals, say,

$$B = P_1 \cdots P_k = Q_1 \cdots Q_l,$$

where $P_1, \ldots, P_k, Q_1, \ldots Q_l$ are prime ideals. Then, as

$$P_1 \cdots P_k \subseteq Q_1,$$

and Q_1 is a prime ideal, by Theorem 1.6.1 we have

$$P_i \subseteq Q_1$$

for some $i \in \{1, 2, \ldots, k\}$. Relabeling P_1 as P_i and vice versa, we may suppose that

$$P_1 \subseteq Q_1.$$

Since P_1 is a prime ideal, it is a maximal ideal as D is a Dedekind domain, and thus

$$P_1 = Q_1.$$

Therefore

$$B \tilde{P}_1 = \tilde{P}_1 P_1 P_2 \cdots P_k = P_2 \cdots P_k$$

and

$$B \tilde{P}_1 = B \tilde{Q}_1 = \tilde{Q}_1 Q_1 \cdots Q_h = Q_2 \cdots Q_h.$$

If $B \tilde{P}_1 = B$ then $B \tilde{P}_1 P_1 = B P_1$, so $B = B P_1$. Define the fractional ideal $\tilde{B}$ of D by

$$\tilde{B} = \tilde{P}_1 \cdots \tilde{P}_k.$$

Then

$$B \tilde{B} = P_1 \cdots P_k \tilde{P}_1 \cdots \tilde{P}_k = P_1 \tilde{P}_1 \cdots P_k \tilde{P}_k = D$$

so that

$$D = B \tilde{B} = B \tilde{P}_1 \tilde{B} = P_1,$$

which is false as P_1 (being a prime ideal) is a proper ideal of D. Hence $B \tilde{P}_1 \neq B$. As $D \subset \tilde{P}_1$ we have $B \subseteq B \tilde{P}_1$. But $B \tilde{P}_1 \neq B$, so we must have

$$B \subset B \tilde{P}_1.$$

Since $B \tilde{P}_1$ is an ideal of D strictly containing B, by the maximality of B, $B \tilde{P}_1$ has exactly one factorization as a product of prime ideals. Thus from $B \tilde{P}_1 = P_2 \cdots P_k = Q_2 \cdots Q_h$ we deduce that $k - 1 = h - 1$ (that is, $k = h$) and after relabeling, we obtain $P_i = Q_i$ $(i = 2, \ldots, k)$. This implies that the two factorizations of B into prime ideals are the same, which is a contradiction.

This completes the proof of the theorem. ∎

Theorem 8.3.2 *Let K be an algebraic number field. Then every proper integral ideal of O_K can be expressed uniquely up to order as a product of prime ideals.*

Proof: This follows immediately from Theorems 8.1.1 and 8.3.1. ∎

Example 8.3.1 *Let*

$$D = \{a + b\sqrt{-5} \mid a, b \in \mathbb{Z}\}.$$

As $D = O_K$, where $K = \mathbb{Q}(\sqrt{-5})$, D is a Dedekind domain. D is not a unique factorization domain as

$$6 = 2 \cdot 3 = (1 + \sqrt{-5})(1 - \sqrt{-5}),$$

where $2, 3, 1 + \sqrt{-5}$, and $1 - \sqrt{-5}$ are nonassociated irreducibles of D. We show how the use of prime ideals restores unique factorization. We let

$$P = \langle 2, 1 + \sqrt{-5} \rangle,$$
$$P_1 = \langle 3, 1 + \sqrt{-5} \rangle,$$
$$P_2 = \langle 3, 1 - \sqrt{-5} \rangle.$$

Then

$$\begin{aligned} P &= \langle 2, 1 + \sqrt{-5} \rangle = \langle 2, 1 + \sqrt{-5}, 2 - (1 + \sqrt{-5}) \rangle \\ &= \langle 2, 1 + \sqrt{-5}, 1 - \sqrt{-5} \rangle = \langle 2, 2 - (1 - \sqrt{-5}), 1 - \sqrt{-5} \rangle \\ &= \langle 2, 1 - \sqrt{-5} \rangle, \end{aligned}$$

$$\begin{aligned} P^2 &= \langle 2, 1 + \sqrt{-5} \rangle^2 = \langle 2, 1 + \sqrt{-5} \rangle \langle 2, 1 - \sqrt{-5} \rangle \\ &= \langle 4, 2(1 + \sqrt{-5}), 2(1 - \sqrt{-5}), 6 \rangle \\ &= \langle 2 \rangle \langle 2, 1 + \sqrt{-5}, 1 - \sqrt{-5}, 3 \rangle \\ &= \langle 2 \rangle \langle 1 \rangle = \langle 2 \rangle, \end{aligned}$$

$$\begin{aligned} P_1 P_2 &= \langle 3, 1 + \sqrt{-5} \rangle \langle 3, 1 - \sqrt{-5} \rangle \\ &= \langle 9, 3(1 + \sqrt{-5}), 3(1 - \sqrt{-5}), 6 \rangle \\ &= \langle 3 \rangle \langle 3, 1 + \sqrt{-5}, 1 - \sqrt{-5}, 2 \rangle \\ &= \langle 3 \rangle \langle 1 \rangle = \langle 3 \rangle, \end{aligned}$$

$$\begin{aligned} P P_1 &= \langle 2, 1 + \sqrt{-5} \rangle \langle 3, 1 + \sqrt{-5} \rangle \\ &= \langle 6, 2(1 + \sqrt{-5}), 3(1 + \sqrt{-5}), (1 + \sqrt{-5})^2 \rangle \\ &= \langle 1 + \sqrt{-5} \rangle \langle 1 - \sqrt{-5}, 2, 3, 1 + \sqrt{-5} \rangle \\ &= \langle 1 + \sqrt{-5} \rangle \langle 1 \rangle = \langle 1 + \sqrt{-5} \rangle, \end{aligned}$$

$$P P_2 = \langle 2, 1 + \sqrt{-5} \rangle \langle 3, 1 - \sqrt{-5} \rangle$$
$$= \langle 2, 1 - \sqrt{-5} \rangle \langle 3, 1 - \sqrt{-5} \rangle$$
$$= \langle 6, 2(1 - \sqrt{-5}), 3(1 - \sqrt{-5}), (1 - \sqrt{-5})^2 \rangle$$
$$= \langle 1 - \sqrt{-5} \rangle \langle 1 + \sqrt{-5}, 2, 3, 1 - \sqrt{-5} \rangle$$
$$= \langle 1 - \sqrt{-5} \rangle \langle 1 \rangle$$
$$= \langle 1 - \sqrt{-5} \rangle.$$

Hence

$$\langle 2 \rangle = P^2, \ \langle 3 \rangle = P_1 P_2, \ \langle 1 + \sqrt{-5} \rangle = P P_1, \ \langle 1 - \sqrt{-5} \rangle = P P_2,$$

and

$$\langle 6 \rangle = \langle 2 \rangle \langle 3 \rangle = \langle 1 + \sqrt{-5} \rangle \langle 1 - \sqrt{-5} \rangle = P^2 P_1 P_2.$$

It is known from Exercises 20 and 21 of Chapter 1 that P, P_1, and P_2 are distinct prime ideals.

If A is a proper integral ideal of a Dedekind domain D then Theorem 8.3.1 tells us that we can express A uniquely (apart from order) in the form

$$A = Q_1 \cdots Q_h,$$

where $Q_1, \ldots, Q_h$ are prime ideals. Let $P_1, \ldots, P_n$ denote the distinct prime ideals among $Q_1, \ldots, Q_h$. Suppose that P_i $(i = 1, 2, \ldots, n)$ occurs a_i times among $Q_1, \ldots, Q_h$ so that each $a_i \geq 1$ $(i = 1, 2, \ldots, n)$ and $a_1 + a_2 + \cdots + a_n = h$. Then

$$A = P_1^{a_1} \cdots P_n^{a_n},$$

where $a_1, \ldots, a_n$ are positive integers. Clearly this representation of A is unique. We extend the factorization $A = P_1^{a_1} \cdots P_n^{a_n}$ to allow the possibility $A = \langle 1 \rangle = D$ by taking $a_1 = \cdots = a_n = 0$ in this case; that is, $D = \langle 1 \rangle$ is regarded as the unique empty product of prime ideals. With this convention every nonzero integral ideal of a Dedekind domain can be expressed uniquely as a product of powers of prime ideals.

Let A and B be nonzero integral ideals of a Dedekind domain D. Then AB is a nonzero integral ideal of D. Let $P_1, \ldots, P_n$ denote the distinct prime ideals of D that occur in the prime ideal factorizartions of at least one of A, B, and AB. Then

$$A = \prod_{i=1}^{n} P_i^{a_i}, \ B = \prod_{i=1}^{n} P_i^{b_i}, \ AB = \prod_{i=1}^{n} P_i^{c_i},$$

where we have grouped together all equal prime ideal factors so that a_i, b_i, c_i ($i = 1, 2, \ldots, n$) are nonnegative integers. Hence

$$\prod_{i=1}^{n} P_i^{c_i} = AB = \prod_{i=1}^{n} P_i^{a_i} \prod_{i=1}^{n} P_i^{b_i} = \prod_{i=1}^{n} P_i^{a_i + b_i},$$

so that by Theorem 8.3.1

$$c_i = a_i + b_i, \quad i = 1, 2, \ldots, n.$$

Hence, if $A = \prod_{i=1}^{n} P_i^{a_i}$ and $B = \prod_{i=1}^{n} P_i^{b_i}$, then $AB = \prod_{i=1}^{n} P_i^{a_i + b_i}$.

Definition 8.3.1 (Divisibility of integral ideals) *Let D be a Dedekind domain. Let A and B be nonzero integral ideals of D. We say that A divides B, written $A \mid B$, if there exists an integral ideal C of D such that $B = AC$.*

If $A = \prod_{i=1}^{n} P_i^{a_i}$ and $B = \prod_{i=1}^{n} P_i^{b_i}$, where $P_1, \ldots, P_n$ are distinct prime ideals and $a_1, \ldots, a_n, b_1, \ldots, b_n$ are nonnegative integers, then

$$A \mid B \iff a_i \leq b_i, \quad i = 1, 2, \ldots, n.$$

We now wish to extend the representation of integral ideals as products of prime ideals to fractional ideals; in this case negative as well as zero and positive exponents of the prime ideals will occur.

Let A be a nonzero fractional ideal of the Dedekind domain D. Let $\alpha \in D \setminus \{0\}$ and $\beta \in D \setminus \{0\}$ be any two common denominators for A. Then

$$\langle \alpha \rangle A = B, \quad \langle \beta \rangle A = C,$$

where B and C are nonzero integral ideals of D. Suppose that

$$\langle \alpha \rangle = \prod_{i=1}^{n} P_i^{r_i}, \quad B = \prod_{i=1}^{n} P_i^{s_i},$$

$$\langle \beta \rangle = \prod_{i=1}^{n} P_i^{t_i}, \quad C = \prod_{i=1}^{n} P_i^{u_i},$$

where $P_1, \ldots, P_n$ are distinct prime ideals and r_i, s_i, t_i, u_i ($i = 1, 2, \ldots, n$) are nonnegative integers. Then as

$$\langle \alpha \rangle C = \langle \alpha \rangle (\langle \beta \rangle A) = \langle \beta \rangle (\langle \alpha \rangle A) = \langle \beta \rangle B$$

we have

$$\prod_{i=1}^{n} P_i^{r_i + u_i} = \prod_{i=1}^{n} P_i^{s_i + t_i},$$

so that by Theorem 8.3.1

$$r_i + u_i = s_i + t_i, \quad i = 1, 2, \ldots, n.$$

Hence we can define the prime ideal factorization of the fractional ideal A to be

$$A = \prod_{i=1}^{n} P_i^{s_i - r_i}$$

and this definition is a valid one since it is independent of the choice of common denominator of A. With this notation, as $P\tilde{P} = \langle 1 \rangle$ for any prime ideal P of D, we have

$$\tilde{P} = P^{-1}.$$

If $P_1, \ldots, P_n$ are prime ideals such that

$$\prod_{i=1}^{n} P_i^{a_i} = \prod_{i=1}^{n} P_i^{b_i},$$

where a_i, b_i $(i = 1, 2, \ldots, n)$ are integers (positive, negative, or zero), then multiplying both sides by $\prod_{i=1}^{n} P_i^{M}$, where M is an integer such that $a_i + M > 0$ and $b_i + M > 0$ for all i, and appealing to Theorem 8.3.1, we deduce that $a_i + M = b_i + M$ $(i = 1, 2, \ldots, n)$, that is, $a_i = b_i$ $(i = 1, 2, \ldots, n)$. Hence the representation of a nonzero fractional ideal as a product of prime ideals is unique.

The following result is now clear.

Theorem 8.3.3 *The set of all nonzero integral and fractional ideals of a Dedekind domain D forms an Abelian group with respect to multiplication. The identity element of the group is $\langle 1 \rangle = D$ and the inverse of $A = \prod_{i=1}^{n} P_i^{a_i}$, where $P_1, \ldots, P_n$ are distinct prime ideals and $a_1, \ldots, a_n$ are integers (positive, negative, or zero), is*

$$A^{-1} = \prod_{i=1}^{n} P_i^{-a_i}.$$

Theorem 8.3.4 *Let K be an algebraic number field. Let O_K be the ring of integers of K. Then the set of all nonzero integral and fractional ideals of O_K forms an Abelian group $I(K)$ with respect to multiplication.*

Proof: This follows immediately from Theorems 8.1.1 and 8.3.3. ∎

Example 8.3.2 *With the notation of Example 8.2.3 we have*

$$P = \langle 2, \sqrt{6} \rangle$$

and

$$\tilde{P} = \frac{1}{2}P.$$

Thus

$$P^{-1} = \frac{1}{2}P.$$

We check this directly. We have

$$P\left(\frac{1}{2}P\right) = \frac{1}{2}P^2 = \frac{1}{2}\langle 2, \sqrt{6}\rangle^2$$

$$= \frac{1}{2}\langle 4, 2\sqrt{6}, 6\rangle = \langle 2, \sqrt{6}, 3\rangle = \langle 1\rangle,$$

as $1 = 3 - 2 \in \langle 2, \sqrt{6}, 3\rangle$. *This shows that* $P^{-1} = \frac{1}{2}P$.

Example 8.3.3   *Let* $D = \mathbb{Z} + \mathbb{Z}\sqrt{6}$. *We determine the inverse* A^{-1} *of the ideal* $A = \langle\sqrt{6}\rangle$ *of D, illustrating the ideas of this section. Let*

$$P = \langle 2, \sqrt{6}\rangle, \quad Q = \langle 3, \sqrt{6}\rangle.$$

P and Q are distinct prime ideals of D such that

$$PQ = \langle 6, 2\sqrt{6}, 3\sqrt{6}, 6\rangle = \langle\sqrt{6}\rangle\langle\sqrt{6}, 2, 3\rangle = \langle\sqrt{6}\rangle = A.$$

Thus $A = PQ$ *is the prime ideal factorization of A and so*

$$A^{-1} = P^{-1}Q^{-1},$$

where

$$P^{-1} = \tilde{P} = \frac{1}{2}P \quad (\textit{Example 8.3.2})$$

and

$$Q^{-1} = \tilde{Q} = \frac{1}{3}Q.$$

Thus

$$A^{-1} = \left(\frac{1}{2}P\right)\left(\frac{1}{3}Q\right) = \frac{1}{6}PQ = \frac{1}{6}A.$$

This is clear as $A(\frac{1}{6}A) = \frac{1}{6}A^2 = \frac{1}{6}\langle 6\rangle = \langle 1\rangle$.

8.4 Order of an Ideal with Respect to a Prime Ideal

Let A be a nonzero fractional or integral ideal of a Dedekind domain D. Then A can be written uniquely in the form

$$A = \prod_{i=1}^{n} P_i^{a_i},$$

where the P_i are distinct prime ideals and the a_i are integers (positive, negative, or zero).

Definition 8.4.1 (Order of an ideal with respect to a prime ideal) *With the preceding notation, the order of the nonzero ideal A of the Dedekind domain D with respect to the prime ideal P_i ($i = 1, 2, \ldots, n$), written $\mathrm{ord}_{P_i}(A)$, is defined by*

$$\mathrm{ord}_{P_i}(A) = a_i.$$

For any prime ideal $P \neq P_1, \ldots, P_n$ we define

$$\mathrm{ord}_P(A) = 0.$$

Clearly $\mathrm{ord}_P(\langle 1 \rangle) = 0$ and $\mathrm{ord}_P(P^k) = k$ for all prime ideals P.

Example 8.4.1 *Let $D = \mathbb{Z} + \mathbb{Z}\sqrt{6}$. Let B be the ideal $\langle 12, 6\sqrt{6} \rangle$. Then, with the notation of Example 8.3.3, we have*

$$B = \langle 12, 6\sqrt{6} \rangle = \langle 6 \rangle \langle 2, \sqrt{6} \rangle = \langle \sqrt{6} \rangle^2 \langle 2, \sqrt{6} \rangle$$
$$= A^2 P = (PQ)^2 P = P^3 Q^2,$$

so that

$$\mathrm{ord}_P(B) = 3, \quad \mathrm{ord}_Q(B) = 2.$$

We now extend the concept of divisibility from integral ideals (Definition 8.3.1) to fractional ideals.

Definition 8.4.2 (Divisibility of fractional ideals) *Let D be a Dedekind domain. Let A and B be nonzero fractional ideals of D. We say that A divides B, written $A \mid B$, if there exists an integral ideal C of D such that $B = AC$.*

Clearly if A and B are nonzero fractional or integral ideals of a Dedekind domain, we have

$$A \mid B \iff \mathrm{ord}_P(A) \leq \mathrm{ord}_P(B) \text{ for all prime ideals } P.$$

Example 8.4.2 *Let $D = \mathbb{Z} + \mathbb{Z}\sqrt{6}$. Let A and B be the fractional ideals of D given by*

$$A = \langle \frac{3}{2}, \frac{\sqrt{6}}{2} \rangle, \quad B = \langle 3, \frac{3}{2}\sqrt{6} \rangle.$$

We show that $A \mid B$. Let $P = \langle 2, \sqrt{6} \rangle$ and $Q = \langle 3, \sqrt{6} \rangle$. P and Q are distinct prime ideals of D (Example 8.3.3). We have, as $P^2 = \langle 2 \rangle$ and $Q^2 = \langle 3 \rangle$,

$$A = \frac{1}{2} \langle 3, \sqrt{6} \rangle = \frac{1}{2} Q = P^{-2} Q$$

and

$$B = \frac{3}{2} \langle 2, \sqrt{6} \rangle = P^{-1} Q^2.$$

Since $\operatorname{ord}_P A = -2 < -1 = \operatorname{ord}_P B$ *and* $\operatorname{ord}_Q A = 1 < 2 = \operatorname{ord}_Q B$, *we see that* $A \mid B$. *Indeed* $B = AC$ *with* $C = PQ$.

The next theorem gives a necessary and sufficient condition for an ideal A to divide an ideal B. It is usually remembered as "To contain is to divide."

Theorem 8.4.1 *Let D be a Dedekind domain. Let A and B be nonzero integral or fractional ideals of D. Then*

$$A \mid B \text{ if and only if } A \supseteq B.$$

Proof: As A is a nonzero integral or fractional ideal of D, A^{-1} is a nonzero integral or fractional ideal of D. Thus BA^{-1} is a nonzero integral or fractional ideal of D. Then

$$
\begin{aligned}
A \supseteq B &\Longleftrightarrow AA^{-1} \supseteq BA^{-1} \\
&\Longleftrightarrow D \supseteq BA^{-1} \\
&\Longleftrightarrow BA^{-1} \text{ is an integral ideal of } D \\
&\Longleftrightarrow BA^{-1} = C \text{ for some integral ideal } C \text{ of } D \\
&\Longleftrightarrow B = AC \text{ for some integral ideal } C \text{ of } D \\
&\Longleftrightarrow A \mid B.
\end{aligned}
$$

∎

The two basic properties of the function $\operatorname{ord}_P(A)$ are given in the next theorem.

Theorem 8.4.2 *Let D be a Dedekind domain. Let P be a prime ideal of D. Let A and B be nonzero integral or fractional ideals of D. Then*

(a) $\operatorname{ord}_P(AB) = \operatorname{ord}_P(A) + \operatorname{ord}_P(B)$,
(b) $\operatorname{ord}_P(A + B) = \min(\operatorname{ord}_P(A), \operatorname{ord}_P(B))$.

Proof: (a) We have

$$A = \prod_P P^{\operatorname{ord}_P(A)}, \quad B = \prod_P P^{\operatorname{ord}_P(B)},$$

where the products are taken over all prime ideals P of D, so that

$$\prod_P P^{\operatorname{ord}_P(AB)} = AB = \prod_P P^{\operatorname{ord}_P(A)+\operatorname{ord}_P(B)}.$$

Of course only finitely many of the exponents in the products are nonzero. Hence, by the uniqueness property, we have

$$\operatorname{ord}_P(AB) = \operatorname{ord}_P(A) + \operatorname{ord}_P(B)$$

for all prime ideals P of D.

(b) Set $C = A + B$. As A and B are nonzero ideals, so is C. Then

$$AC^{-1} + BC^{-1} = (A+B)C^{-1} = CC^{-1} = D.$$

Hence $AC^{-1} \subseteq AC^{-1} + BC^{-1} = D$ and $BC^{-1} \subseteq AC^{-1} + BC^{-1} = D$. So AC^{-1} and BC^{-1} are both integral ideals of D. Suppose $AC^{-1} \subseteq P$ and $BC^{-1} \subseteq P$. Then

$$D = AC^{-1} + BC^{-1} \subseteq P + P = P,$$

which is impossible. Thus either $AC^{-1} \not\subseteq P$ or $BC^{-1} \not\subseteq P$; that is, by Theorem 8.4.1, $P \nmid AC^{-1}$ or $P \nmid BC^{-1}$, so

$$\min(\operatorname{ord}_P(AC^{-1}), \operatorname{ord}_P(BC^{-1})) = 0.$$

Finally, by part (a), we obtain

$$\min(\operatorname{ord}_P(A), \operatorname{ord}_P(B)) = \operatorname{ord}_P(C) = \operatorname{ord}_P(A+B). \qquad \blacksquare$$

We next define the order of a nonzero element with respect to a prime ideal.

Definition 8.4.3 (Order of a nonzero element with respect to a prime ideal) *Let D be a Dedekind domain with quotient field K. For $\alpha \in K$, $\alpha \neq 0$, we define*

$$\operatorname{ord}_P(\alpha) = \operatorname{ord}_P(\langle \alpha \rangle)$$

for any prime ideal P of D.

The next theorem allows us to recognize when an element α belongs to an ideal A of D in terms of the orders of α and A with respect to prime ideals P.

Theorem 8.4.3 *Let D be a Dedekind domain with quotient field K. Let A be a nonzero ideal of D. Let $\alpha \in K, \alpha \neq 0$. Then*

$$\alpha \in A \text{ if and only if } \operatorname{ord}_P(\alpha) \geq \operatorname{ord}_P(A) \text{ for all prime ideals } P \text{ of } D.$$

Proof: We have

$$\alpha \in A \Longleftrightarrow \langle \alpha \rangle \subseteq A$$
$$\Longleftrightarrow A \mid \langle \alpha \rangle \text{ (by Theorem 8.4.1)}$$
$$\Longleftrightarrow \text{ord}_P(A) \leq \text{ord}_P(\langle \alpha \rangle) \text{ for all prime ideals } P \text{ of } D$$
$$\Longleftrightarrow \text{ord}_P(\alpha) \geq \text{ord}_P(A) \text{ for all prime ideals } P \text{ of } D.$$

∎

The next theorem gives the basic properties of the order of an element with respect to a prime ideal.

Theorem 8.4.4 *Let D be a Dedekind domain with quotient field K. Let P be a prime ideal of D.*

(a) *For $\alpha \in K^*$ and $\beta \in K^*$*

$$\text{ord}_P(\alpha\beta) = \text{ord}_P(\alpha) + \text{ord}_P(\beta).$$

(b) *For $\alpha, \beta, \alpha + \beta \in K^*$,*

$$\text{ord}_P(\alpha + \beta) \geq \min(\text{ord}_P(\alpha), \text{ord}_P(\beta)).$$

(c) *If $\alpha, \beta, \alpha + \beta \in K^*$ are such that $\text{ord}_P(\alpha) \neq \text{ord}_P(\beta)$ then*

$$\text{ord}_P(\alpha + \beta) = \min(\text{ord}_P(\alpha), \text{ord}_P(\beta)).$$

Proof: (a) We have for any prime ideal P of D

$$\text{ord}_P(\alpha\beta) = \text{ord}_P(\langle \alpha\beta \rangle)$$
$$= \text{ord}_P(\langle \alpha \rangle \langle \beta \rangle)$$
$$= \text{ord}_P(\langle \alpha \rangle) + \text{ord}_P(\langle \beta \rangle)$$
$$= \text{ord}_P(\alpha) + \text{ord}_P(\beta).$$

(b) As $\alpha + \beta \in \langle \alpha \rangle + \langle \beta \rangle$ we have by Theorems 8.4.2(b) and 8.4.3

$$\text{ord}_P(\alpha + \beta) \geq \text{ord}_P(\langle \alpha \rangle + \langle \beta \rangle)$$
$$= \min(\text{ord}_P(\langle \alpha \rangle), \text{ord}_P(\langle \beta \rangle))$$
$$= \min(\text{ord}_P(\alpha), \text{ord}_P(\beta)).$$

(c) Without loss of generality we may suppose that

$$\text{ord}_P(\alpha) > \text{ord}_P(\beta).$$

Then by part (b) we have

$$\text{ord}_P(\alpha + \beta) \geq \text{ord}_P(\beta).$$

Thus

$$
\begin{aligned}
\mathrm{ord}_P(\beta) &= \mathrm{ord}_P((\alpha + \beta) - \alpha) \\
&\geq \min(\mathrm{ord}_P(\alpha + \beta), \mathrm{ord}_P(\alpha)) \\
&= \mathrm{ord}_P(\alpha + \beta) \quad (\text{as } \mathrm{ord}_P(\alpha) > \mathrm{ord}_P(\beta)) \\
&\geq \mathrm{ord}_P(\beta).
\end{aligned}
$$

Hence

$$
\mathrm{ord}_P(\alpha + \beta) = \mathrm{ord}_P(\beta) = \min(\mathrm{ord}_P(\alpha), \mathrm{ord}_P(\beta)).
$$ ∎

Example 8.4.3 *We give a simple example to show that if* $\mathrm{ord}_P(\alpha) = \mathrm{ord}_P(\beta)$ *then* $\mathrm{ord}_P(\alpha + \beta)$ *may actually be larger than* $\mathrm{ord}_P(\alpha)$. *Take* $D = \mathbb{Z}$, $\alpha = 1$, $\beta = 4$, $P = \langle 5 \rangle$. *Then*

$$
\mathrm{ord}_P(\alpha) = 0, \;\; \mathrm{ord}_P(\beta) = 0, \;\; \mathrm{ord}_P(\alpha + \beta) = 1.
$$

Theorem 8.4.5 *Let D be a Dedekind domain with quotient field K. Given any finite set of prime ideals $P_1, \ldots, P_k$ of D and a corresponding set of integers $a_1, \ldots, a_k$ then there exists $\alpha \in K$ such that*

$$
\mathrm{ord}_{P_i}(\alpha) = a_i, \;\; i = 1, 2, \ldots, k,
$$

and

$$
\mathrm{ord}_P(\alpha) \geq 0, \;\; \text{for any prime ideal } P \neq P_1, \ldots, P_k.
$$

Proof: As

$$
P_1^{a_1} \prod_{i=2}^{k} P_i^{a_i+1} \mid P_1^{a_1+1} \prod_{i=2}^{k} P_i^{a_i+1},
$$

by Theorem 8.4.1 we have

$$
P_1^{a_1} \prod_{i=2}^{k} P_i^{a_i+1} \supseteq P_1^{a_1+1} \prod_{i=2}^{k} P_i^{a_i+1}.
$$

By the uniqueness property we have

$$
P_1^{a_1} \prod_{i=2}^{k} P_i^{a_i+1} \neq P_1^{a_1+1} \prod_{i=2}^{k} P_i^{a_i+1},
$$

so that

$$
P_1^{a_1} \prod_{i=2}^{k} P_i^{a_i+1} \supset P_1^{a_1+1} \prod_{i=2}^{k} P_i^{a_i+1}.
$$

Hence there exists

$$\alpha_1 \in P_1^{a_1} \prod_{i=2}^{k} P_i^{a_i+1}, \quad \alpha_1 \notin P_1^{a_1+1} \prod_{i=2}^{k} P_i^{a_i+1}.$$

Thus

$$\mathrm{ord}_{P_1}(\alpha_1) = a_1$$

and

$$\mathrm{ord}_{P_i}(\alpha_1) \geq a_i + 1 \text{ for } i \neq 1.$$

Similarly we can define $\alpha_j \in K$ for $j = 2, \ldots, k$ such that

$$\mathrm{ord}_{P_j}(\alpha_j) = a_j$$

and

$$\mathrm{ord}_{P_i}(\alpha_j) \geq a_i + 1 \text{ for } i \neq j.$$

Now set

$$\alpha = \alpha_1 + \alpha_2 + \cdots + \alpha_k \in K.$$

Then, by Theorem 8.4.4(b), we have

$$\mathrm{ord}_{P_1}(\alpha_2 + \cdots + \alpha_k) \geq \min(\mathrm{ord}_{P_1}(\alpha_2), \ldots, \mathrm{ord}_{P_1}(\alpha_k)) \geq a_1 + 1 > \mathrm{ord}_{P_1}(\alpha_1),$$

so that

$$\begin{aligned}
\mathrm{ord}_{P_1}(\alpha) &= \mathrm{ord}_{P_1}(\alpha_1 + (\alpha_2 + \cdots + \alpha_k)) \\
&= \min(\mathrm{ord}_{P_1}(\alpha_1), \mathrm{ord}_{P_1}(\alpha_2 + \cdots + \alpha_k)) \\
&= \mathrm{ord}_{P_1}(\alpha_1),
\end{aligned}$$

that is,

$$\mathrm{ord}_{P_1}(\alpha) = a_1.$$

Similarly,

$$\mathrm{ord}_{P_j}(\alpha) = a_j, \quad j = 2, \ldots, k.$$

Finally, for $P \neq P_1, \ldots, P_k$ we have

$$\mathrm{ord}_P(\alpha_i) \geq 0, \quad i = 1, 2, \ldots, k,$$

so that

$$\mathrm{ord}_P(\alpha) \geq 0. \qquad \blacksquare$$

If D is a Dedekind domain with quotient field K, A is a nonzero fractional or integral ideal of D, and $a, b, c \in A$, then we write

$$a \equiv b \pmod{A} \text{ if and only if } A \mid \langle a - b \rangle.$$

We observe that

$$A \mid \langle a - b \rangle \iff \langle a - b \rangle \subseteq A \iff a - b \in A \iff a + A = b + A.$$

The properties

$$a \equiv a \ (\text{mod } A),$$
$$a \equiv b \ (\text{mod } A) \implies b \equiv a \ (\text{mod } A),$$
$$a \equiv b \ (\text{mod } A), \ b \equiv c \ (\text{mod } A) \implies a \equiv c \ (\text{mod } A),$$
$$a \equiv b \ (\text{mod } A) \implies ac \equiv bc \ (\text{mod } A)$$

are easily proved.

Theorem 8.4.6 (Chinese remainder theorem) *Let D be a Dedekind domain.*

(a) *Let $P_1, \ldots, P_k$ be distinct prime ideals in D. Let $a_1, \ldots, a_k$ be positive integers. Let $\alpha_1, \ldots, \alpha_k$ be elements of D. Then there exists $\alpha \in D$ such that*

$$\alpha \equiv \alpha_i \ (\text{mod } P_i^{a_i}), \ i = 1, 2, \ldots, k.$$

(b) *Let $I_1, \ldots, I_k$ be pairwise relatively prime ideals of D. Let $\alpha_1, \ldots, \alpha_k$ be elements of D. Then there exists $\alpha \in D$ such that*

$$\alpha \equiv \alpha_i \ (\text{mod } I_i), \ i = 1, 2, \ldots, k.$$

Proof: **(a)** Consider the ideal

$$Q_1 = P_1^{a_1} + P_2^{a_2} \cdots P_k^{a_k}$$

of D. Suppose P is a prime ideal such that

$$P \mid Q_1.$$

Now

$$P_1^{a_1} \subseteq Q_1, \ P_2^{a_2} \cdots P_k^{a_k} \subseteq Q_1,$$

so, by Theorem 8.4.1, we have

$$Q_1 \mid P_1^{a_1}, \ Q_1 \mid P_2^{a_2} \cdots P_k^{a_k}.$$

Hence

$$P \mid P_1^{a_1}, \ P \mid P_2^{a_2} \cdots P_k^{a_k}.$$

From the first of these we deduce that $P \mid P_1$ so that $P = P_1$. Hence

$$P_1 \mid P_2^{a_2} \cdots P_k^{a_k},$$

which contradicts that P_1 is a prime ideal distinct from $P_2, \ldots, P_k$. Thus there is no prime ideal dividing Q_1. Hence $Q_1 = D$; that is

$$P_1^{a_1} + P_2^{a_2} \cdots P_k^{a_k} = \langle 1 \rangle.$$

Hence there exist $x_1 \in P_1^{a_1}$ and $y_1 \in P_2^{a_2} \cdots P_k^{a_k}$ such that

$$x_1 + y_1 = 1.$$

Thus

$$y_1 \equiv 1 \,(\text{mod } P_1^{a_1}), \ \ y_1 \equiv 0 \,(\text{mod } P_i^{a_i}), \ \ i = 2, \ldots, k.$$

Similarly, for $j = 2, \ldots, k$ we can find y_j such that

$$y_j \equiv 1 \,(\text{mod } P_j^{a_j}), \ \ y_j \equiv 0 \,(\text{mod } P_i^{a_i}), \ \ i \neq j.$$

Now let

$$\alpha = \alpha_1 y_1 + \cdots + \alpha_k y_k \in D.$$

Then as $\alpha_2 y_2 + \cdots + \alpha_k y_k \in P_1^{a_1}$ we have

$$\alpha \equiv \alpha_1 y_1 \equiv \alpha_1 \,(\text{mod } P_1^{a_1}).$$

Similarly,

$$\alpha \equiv \alpha_j \,(\text{mod } P_j^{a_j}), \ \ j = 2, \ldots, k.$$

This completes the proof of part (a).

   (b) Part (b) follows from part (a) by observing that any congruence of the form $x \equiv \alpha \,(\text{mod } I)$ is equivalent to the system of congruences $x \equiv \alpha \,(\text{mod } P_i^{a_i})$, $i = 1, 2, \ldots, r$, where $I = P_1^{a_1} \cdots P_r^{a_r}$. ∎

Example 8.4.4 *Let $D = \mathbb{Z}[x]$. D is not a Dedekind domain as the prime ideal $\langle x \rangle$ is not a maximal ideal (Example 1.5.6). Consider the pair of congruences*

$$\alpha \equiv 0 \,(\text{mod } \langle 2 \rangle),$$
$$\alpha \equiv 1 \,(\text{mod } \langle x \rangle).$$

The moduli $\langle 2 \rangle$ and $\langle x \rangle$ are distinct prime ideals. However, the congruences are not simultaneously solvable in D, since any solution of $\alpha \equiv 0 \,(\text{mod } \langle 2 \rangle)$ has an even constant term, whereas any solution of $\alpha \equiv 1 \,(\text{mod } \langle x \rangle)$ has a constant term equal to 1. This shows that the Chinese remainder theorem does not necessarily hold in an integral domain that is not a Dedekind domain.

8.5 Generators of Ideals in a Dedekind Domain

In this section we show that every fractional or integral ideal of a Dedekind domain is generated by at most two elements.

Theorem 8.5.1 *Let D be a Dedekind domain. Let A be a fractional or integral ideal of D. Then A is generated by at most two elements.*

Proof: If $A = \{0\}$ then $A = \langle 0 \rangle$, and if $A = D$ then $A = \langle 1 \rangle$, so that we may suppose that $A \neq \{0\}$, D. Let $\beta \in A$, $\beta \neq 0$, and $\beta \neq$ unit. Then $\langle \beta \rangle \subseteq A$ so that $A \mid \langle \beta \rangle$. Hence there exists a nonzero integral ideal B of D such that

$$\langle \beta \rangle = AB.$$

Let $P_1, \ldots, P_n$ be the set of distinct prime ideals for which either

$$\mathrm{ord}_{P_i}(A) \neq 0 \text{ or } \mathrm{ord}_{P_i}(AB) \neq 0 \text{ (or both)}.$$

This set is nonempty as $A \neq D$. By Theorem 8.4.5 there exists $\alpha \in K$ (the quotient field of D) such that

$$\mathrm{ord}_{P_i}(\alpha) = \mathrm{ord}_{P_i}(A), \quad i = 1, 2, \ldots, n,$$
$$\mathrm{ord}_P(\alpha) \geq 0, \quad P \neq P_1, \ldots, P_n.$$

For $P \neq P_1, \ldots, P_n$ we have $\mathrm{ord}_P(A) = 0$ so that

$$\mathrm{ord}_P(\alpha) \geq \mathrm{ord}_P(A) \text{ for all prime ideals } P.$$

Hence

$$\alpha \in A.$$

For $i = 1, 2, \ldots, n$ we have

$$
\begin{aligned}
\mathrm{ord}_{P_i}(A) &= \min(\mathrm{ord}_{P_i}(A), \mathrm{ord}_{P_i}(AB)) \text{ (as } B \text{ is an integral ideal)} \\
&= \min(\mathrm{ord}_{P_i}(\alpha), \mathrm{ord}_{P_i}(AB)) \\
&= \min(\mathrm{ord}_{P_i}(\langle \alpha \rangle), \mathrm{ord}_{P_i}(AB)) \\
&= \mathrm{ord}_{P_i}(\langle \alpha \rangle + AB),
\end{aligned}
$$

by Theorem 8.4.2(b). For $P \neq P_1, \ldots, P_n$ we have $\mathrm{ord}_P(A) = \mathrm{ord}_P(AB) = 0$, so that

$$
\begin{aligned}
\mathrm{ord}_P(A) &= \min(\mathrm{ord}_P(\alpha), \mathrm{ord}_P(AB)) \\
&= \min(\mathrm{ord}_P(\langle \alpha \rangle), \mathrm{ord}_P(AB)) \\
&= \mathrm{ord}_P(\langle \alpha \rangle + AB)
\end{aligned}
$$

by Theorem 8.4.2(b). Hence

$$\mathrm{ord}_P(A) = \mathrm{ord}_P(\langle \alpha \rangle + AB) \text{ for all prime ideals } P.$$

Hence

$$A = \langle \alpha \rangle + AB.$$

Finally,

$$A = \langle \alpha \rangle + \langle \beta \rangle = \langle \alpha, \beta \rangle. \qquad \blacksquare$$

Exercises

1. Let D be a Dedekind domain. Let A and B be integral ideals of D with $A \neq D$, $B \neq D$. Prove from first principles that $AB \neq D$.
2. Let D be a Dedekind domain. Let A be a nonzero integral ideal of D. Let P be a prime ideal of D. If P does not divide A prove that $\text{ord}_P(A) = 0$.
3. Determine all fractional ideals of $\mathbb{Z} + \mathbb{Z}\sqrt{-1}$.
4. Find all ideals in $\mathbb{Z} + \mathbb{Z}\sqrt{-6}$ that contain 6.
5. Let D be a principal ideal domain with quotient field K. Prove that every fractional ideal of D is of the form $\{d\alpha \mid d \in D\}$ for some $\alpha \in K$.
6. Let K be an algebraic number field of degree n. Let a be a nonzero rational integer. Prove that a belongs to at most a^n integral ideals of O_K.
7. Show that $\langle 3, 1 + 2\sqrt{-5} \rangle \mid \langle 1 + 2\sqrt{-5} \rangle$ in O_K, where $K = \mathbb{Q}(\sqrt{-5})$. Determine an integral ideal A such that

$$\langle 1 + 2\sqrt{-5} \rangle = \langle 3, 1 + 2\sqrt{-5} \rangle A.$$

8. Determine the fractional ideal $\langle 3, 1 + 2\sqrt{-5} \rangle^{-1}$ of O_K, where $K = \mathbb{Q}(\sqrt{-5})$.
9. Let K be an algebraic number field. Let I be an integral ideal of O_K. Let $a \in I$. Prove that there exists an integral ideal I' of O_K such that $\langle a \rangle = II'$.
10. Let K be an algebraic number field. Let I be a nonzero integral ideal of O_K. Let $\alpha \in K$ have the following property:

$$a \in I \implies a\alpha \in I.$$

Prove that $\alpha \in O_K$.
11. Let I be the ideal of $\mathbb{Z} + \mathbb{Z}\sqrt{-5}$ generated by $1 + \sqrt{-5}$, $3 + \sqrt{-5}$, and $19 + 9\sqrt{-5}$. Determine $\alpha, \beta \in \mathbb{Z} + \mathbb{Z}\sqrt{-5}$ such that $I = \langle \alpha, \beta \rangle$.
12. Let I and J be nonzero integral ideals of a Dedekind domain D. Let $P_1, \ldots, P_k$ be the distinct prime ideals dividing either I or J (or both) so that

$$I = P_1^{a_1} \cdots P_k^{a_k}, \quad J = P_1^{b_1} \cdots P_k^{b_k},$$

for nonnegative integers $a_1, \ldots, a_k, b_1, \ldots, b_k$. The greatest common divisor $\gcd(I, J)$ and the least common multiple $\text{lcm}(I, J)$ of I and J are defined by

$$\gcd(I, J) = P_1^{\min(a_1, b_1)} \cdots P_k^{\min(a_k, b_k)},$$
$$\text{lcm}(I, J) = P_1^{\max(a_1, b_1)} \cdots P_k^{\max(a_k, b_k)}.$$

Prove that

$$\gcd(I, J) = I + J$$

and

$$\text{lcm}\,(I, J) = I \cap J.$$

13. Prove that a Dedekind domain is a unique factorization domain if and only if it is a principal ideal domain.

14. Let D be a Dedekind domain. Let A, B, C be ideals of D with $A \neq \langle 0 \rangle$ and $AB = AC$. Prove that $B = C$.

15. Let D be a Dedekind domain. Let A and B be nonzero integral ideals of D. Prove that there exists $a \in A$ such that $\gcd\,(AB, \langle a \rangle) = A$.

16. Let D be a Dedekind domain. Let A and B be nonzero integral ideals of D. Prove that there is an integral ideal C of D such that AC is a principal ideal and $\gcd\,(B, C) = D$.

17. Let D be a Dedekind domain. Let A be a nonzero integral ideal of D. Prove that there exist only finitely many integral ideals of D that divide A.

18. Let D be a Dedekind domain. A nonzero integral ideal I of D is said to be primary if the following condition holds:

$$a, b \in D, \ ab \in I, \ a \notin I \implies b^m \in I \text{ for some } m \in \mathbb{N}.$$

Prove that a primary ideal must be a power of a prime ideal.

19. Let K be an algebraic number field. Prove that O_K contains infinitely many prime ideals.

20. Determine the prime ideal factorization of $\langle 54 \rangle$ in $\mathbb{Z} + \mathbb{Z}\sqrt{-6}$.

21. Let D be a Dedekind domain. Let I be an ideal of D with $I \neq \langle 0 \rangle$, $\langle 1 \rangle$. Prove that

$$D/I \simeq D/P_1^{a_1} \times \cdots \times D/P_r^{a_r},$$

where

$$I = P_1^{a_1} \cdots P_r^{a_r}$$

is the factorization of I into distinct prime ideals $P_1, \ldots, P_r$.

Suggested Reading

1. F. T. Howard, *A generalized Chinese remainder theorem*, The College Mathematics Journal 33 (2002), 279–282.

   An extension of the Chinese remainder theorem that allows the moduli of the linear congruences to have common factors is proved.

2. O. Zariski and P. Samuel, *Commutative Algebra*, Volume 1, van Nostrand, Princeton, New Jersey, 1958.

   Chapter 5 of this classic book on algebra is devoted to Dedekind domains and the classical theory of ideals.

9

Norms of Ideals

9.1 Norm of an Integral Ideal

We have already defined (Definition 7.1.3) the norm $N(A)$ of a nonzero integral ideal A in the ring O_K of integers of an algebraic number field K by

$$N(A) = \sqrt{\frac{D(A)}{d(K)}}, \qquad (9.1.1)$$

where $D(A)$ is the discriminant of the ideal A (Definition 6.5.2) and $d(K)$ is the discriminant of the field K (Definition 7.1.2). Two main results of this chapter are the following:

$$N(A) = \text{card}(O_K/A), \qquad (9.1.2)$$

where O_K/A is the factor ring of O_K by A, and

$$N(AB) = N(A)N(B) \qquad (9.1.3)$$

for any two nonzero integral ideals A and B of O_K.

We require a couple of preliminary results to establish (9.1.2).

Theorem 9.1.1 *Let C be an $n \times n$ matrix with rational integer entries. Then, by applying to C a finite sequence of elementary operations of the types*

(1) *interchange of two rows or two columns,*
(2) *addition of an integral multiple of one row (or column) to another row (or column),*
    we can transform C into a diagonal matrix

$$\begin{bmatrix} d_1 & 0 & \cdots & 0 \\ 0 & d_2 & \cdots & 0 \\ \vdots & \vdots & \cdots & \vdots \\ 0 & 0 & \cdots & d_n \end{bmatrix},$$

where the integers $d_1, d_2, \ldots, d_n$ are such that

$$|\det C| = |d_1| \cdots |d_n|.$$

Proof: The proof is by induction on the size n of the matrix C. If $n = 1$ the result is clearly true. Now suppose that it is true for all $(n - 1) \times (n - 1)$ matrices with rational integer entries. Let C be a given $n \times n$ matrix with rational integer entries. If C is the zero matrix there is nothing to prove, so we may suppose that $C \neq O_n$. Let k denote any one of the nonzero entries in C. By means of elementary operations of type 1 we can transform C into a matrix $B = (b_{ij})$ in which $b_{11} = k$. If k does not divide all the remaining entries in the first row and first column we can find an integer j $(2 \leq j \leq n)$ such that b_{1j} or $b_{j1} = qk + r$ with $0 < r < |k|$; then by an elementary operation of type 2, subtracting q times the first row or column from the jth row or column, we obtain a matrix with an entry $r < |k|$. Applying elementary operations of type 1 we can move r to the $(1, 1)$ position and repeat the process. After a finite number of operations we obtain a matrix in which the $(1, 1)$ entry divides all the entries in the first row and column. Thus by means of a finite number of operations of types 1 and 2 we can transform the matrix into one of the form

$$
\begin{bmatrix}
d_1 & 0 & \cdots & 0 \\
0 & & & \\
& (n-1) \times (n-1) & \\
\vdots & \text{submatrix with} & \\
& \text{integer entries} & \\
0 & & &
\end{bmatrix}.
$$

Applying the inductive hypothesis to the $(n - 1) \times (n - 1)$ submatrix we finally get a matrix of the required diagonal type.

Clearly, elementary operations of types 1 and 2 at most change the sign of the determinant so that

$$|\det C| = |d_1| \cdots |d_n|. \qquad \blacksquare$$

It can be shown that $d_1, d_2, \ldots, d_n$ in Theorem 9.1.1 can be arranged to satisfy $d_1 \mid d_2 \mid \cdots \mid d_n$, in which case the matrix is said to be in Smith normal form.

Theorem 9.1.2 *Let G be a free Abelian group with n generators $\omega_1, \ldots, \omega_n$, so that each element of G is uniquely expressible as*

$$x_1\omega_1 + \cdots + x_n\omega_n, \quad x_1, \ldots, x_n \in \mathbb{Z}.$$

Let H be the subgroup of G generated by the n elements $\eta_1, \ldots, \eta_n$ so that

$$H = \{y_1\eta_1 + \cdots + y_n\eta_n \mid y_1, \ldots, y_n \in \mathbb{Z}\}.$$

As each $\eta_i \in H \subseteq G$ we have

$$\eta_i = \sum_{j=1}^{n} c_{ij}\omega_j, \quad i = 1, 2, \ldots, n,$$

where each $c_{ij} \in \mathbb{Z}$. Let C be the $n \times n$ matrix whose (i, j) entry is c_{ij}. Then

$$[G : H] = \begin{cases} |\det C|, & \text{if } \det C \neq 0, \\ \infty, & \text{if } \det C = 0. \end{cases}$$

Proof: We wish to transform the matrix C into the form given in Theorem 9.1.1 by means of elementary operations of types 1 and 2.

An elementary operation of type 1, that is, interchanging rows or columns of C, corresponds to rearranging the order of the generators $\eta_1, \ldots, \eta_n$ of H or $\omega_1, \ldots, \omega_n$ of G, and it so leaves $[G : H]$ unchanged.

The elementary operation of type 2, which adds k times the ith row to the lth row, corresponds to replacing c_{ij} by $c_{ij} + kc_{lj}$ ($j = 1, 2, \ldots, n$) and hence replaces

$$\eta_i = \sum_{j=1}^{n} c_{ij}\omega_j$$

by

$$\eta_i + k\eta_l = \sum_{j=1}^{n} (c_{ij} + kc_{lj})\omega_j.$$

But it is clear that $\{\eta_1, \ldots, \eta_n\}$ and $\{\eta_1, \ldots, \eta_i + k\eta_l, \ldots, \eta_l, \ldots, \eta_n\}$ generate the same subgroup so that again $[G : H]$ is unaltered.

Finally, the elementary operation of type 2, which adds k times the lth column to the jth column, corresponds to replacing c_{ij} by $c_{ij} + kc_{il}$ ($i = 1, 2, \ldots, n$) and thus is equivalent to replacing the generators $\omega_1, \ldots, \omega_n$ of G by the equivalent set $\{\omega_1, \ldots, \omega_l - k\omega_j, \ldots, \omega_n\}$ since

$$\eta_i = \sum_{h=1}^{n} c_{ih}\omega_h$$

$$= \sum_{\substack{h=1 \\ h \neq j,l}}^{n} c_{ih}\omega_h + c_{ij}\omega_j + c_{il}\omega_l$$

$$= \sum_{\substack{h=1 \\ h \neq j,l}}^{n} c_{ih}\omega_h + (c_{ij} + kc_{il})\omega_j + c_{il}(\omega_l - k\omega_j)$$

$$= \sum_{h=1}^{n} c'_{ih}\omega'_h,$$

where

$$c'_{ih} = \begin{cases} c_{ih}, & h \neq j, \\ c_{ij} + kc_{il}, & h = j, \end{cases}$$

and

$$\omega_h' = \begin{cases} \omega_h, & h \neq l, \\ \omega_l - k\omega_j, & h = l. \end{cases}$$

Thus $[G : H]$ remains unchanged.

Hence transforming C into $\operatorname{diag}(d_1, \ldots, d_n)$, where $|d_1| \cdots |d_n| = |\det C|$, by elementary operations as in Theorem 9.1.1, we obtain a set of generators for G, namely,

$$G = \langle \overline{\omega_1}, \ldots, \overline{\omega_n} \rangle,$$

such that

$$H = \langle d_1\overline{\omega_1}, \ldots, d_n\overline{\omega_n} \rangle.$$

Clearly

$$x_1\overline{\omega_1} + \cdots + x_n\overline{\omega_n} \in H$$
$$\Longleftrightarrow x_1\overline{\omega_1} + \cdots + x_n\overline{\omega_n} = y_1d_1\overline{\omega_1} + \cdots + y_nd_n\overline{\omega_n}$$
$$\text{for some } y_i \in \mathbb{Z} \ (i = 1, 2, \ldots, n)$$
$$\Longleftrightarrow x_i = y_id_i \ (i = 1, 2, \ldots, n)$$
$$\Longleftrightarrow d_i \mid x_i \ (i = 1, 2, \ldots, n).$$

Suppose now that $\det C \neq 0$. Hence $d_1 \cdots d_n \neq 0$ so that each $d_i \neq 0$ ($i = 1, 2, \ldots, n$). Then a complete set of coset representatives for G modulo H is

$$\{x_1\overline{\omega_1} + \cdots + x_n\overline{\omega_n} \mid x_1 = 0, 1, \ldots, |d_1| - 1; \ldots; x_n = 0, 1, \ldots, |d_n| - 1\},$$

and thus

$$[G : H] = |d_1| \cdots |d_n| = |\det C|.$$

Finally, suppose that $\det C = 0$. Hence $d_1 \cdots d_n = 0$ so that $d_i = 0$ for some $i \in \{1, 2, \ldots, n\}$. Then $k\overline{\omega_i} + H$ ($k = 0, 1, 2, \ldots$) are distinct cosets of H in G so that $[G : H] = \infty$. ∎

We can now prove (9.1.2).

Theorem 9.1.3 *Let K be an algebraic number field with $[K : \mathbb{Q}] = n$. Let O_K be the ring of integers of K. Let A be a nonzero integral ideal of O_K. Then*

$$N(A) = \operatorname{card}(O_K/A).$$

Proof: Let $\{\eta_1, \ldots, \eta_n\}$ be a basis for A and $\{\omega_1, \ldots, \omega_n\}$ an integral basis for K. Then

$$\eta_i = \sum_{j=1}^{n} c_{ij}\omega_j, \ i = 1, 2, \ldots, n,$$

for $c_{ij} \in \mathbb{Z}$ $(i, j = 1, 2, \ldots, n)$. Thus

$$N(A) = \sqrt{\frac{D(A)}{d(K)}} = \sqrt{\frac{D(\eta_1, \ldots, \eta_n)}{D(\omega_1, \ldots, \omega_n)}} = |\det(c_{ij})| = [O_K : A] = \mathrm{card}(O_K/A),$$

by Theorem 9.1.2.   ■

9.2 Norm and Trace of an Element

If K is an algebraic number field of degree n and α is an element of K then there are two very important quantities associated with α, namely,

$$\sigma_1(\alpha) + \sigma_2(\alpha) + \cdots + \sigma_n(\alpha) \text{ and } \sigma_1(\alpha)\sigma_2(\alpha) \cdots \sigma_n(\alpha),$$

where

$$\sigma_k : K \longrightarrow \mathbb{C}, \ k = 1, 2, \ldots, n,$$

are the n distinct monomorphisms from K to $\mathbb{C}$. These quantities are called the trace and norm of α respectively.

Definition 9.2.1 (Norm and trace of an element) *Let K be an algebraic number field of degree n. Let $\alpha \in K$. Let $\alpha_1 = \alpha, \alpha_2, \ldots, \alpha_n$ be the K-conjugates of α. Then the trace of α is denoted by $\mathrm{tr}(\alpha)$ and is defined by*

$$\mathrm{tr}(\alpha) = \alpha_1 + \alpha_2 + \cdots + \alpha_n,$$

and the norm of α is denoted by $N(\alpha)$ and is defined by

$$N(\alpha) = \alpha_1 \alpha_2 \cdots \alpha_n.$$

If $\alpha \in \mathbb{Q}$ then by Theorem 6.3.4 we know that all the K-conjugates of α are all equal to α. Hence for $\alpha \in \mathbb{Q}$ we have

$$\mathrm{tr}(\alpha) = \alpha + \cdots + \alpha = n\alpha$$

and

$$N(\alpha) = \alpha \cdots \alpha = \alpha^n.$$

If K is a quadratic field then $K = \mathbb{Q}(\sqrt{m})$ for some squarefree integer m. Let $\alpha \in K$. Then $\alpha = r + s\sqrt{m}$ for some $r, s \in \mathbb{Q}$. The K-conjugates of α are $\alpha = r + s\sqrt{m}$ and $\alpha' = r - s\sqrt{m}$. The trace of α is

$$\mathrm{tr}(\alpha) = \alpha + \alpha' = 2r$$

and the norm of α is

$$N(\alpha) = \alpha\alpha' = r^2 - s^2 m.$$

Recalling Definition 2.2.1 we observe that $\phi_m(\alpha) = |N(\alpha)|$.

From Definitions 6.3.2 and 9.2.1, we see that

$$\text{fld}_K(\alpha) = x^n - \text{tr}(\alpha)x^{n-1} + \cdots + (-1)^n N(\alpha).$$

In particular when K is a quadratic field (so that $n = 2$) we have

$$\text{fld}_K(\alpha) = x^2 - \text{tr}(\alpha)x + N(\alpha).$$

From Theorem 6.3.1 we deduce that for an arbitrary algebraic number field K

$$\text{tr}(\alpha) \in \mathbb{Q} \text{ and } N(\alpha) \in \mathbb{Q}.$$

Further, if $\alpha \in O_K$ then, by Theorem 6.3.3, $\alpha_1 = \alpha, \alpha_2, \ldots, \alpha_n$ are algebraic integers so that $\text{fld}(\alpha) \in \mathbb{Z}[x]$ and thus

$$\text{tr}(\alpha) \in \mathbb{Z} \text{ and } N(\alpha) \in \mathbb{Z}.$$

In the next theorem we show that the trace is additive and the norm is multiplicative.

Theorem 9.2.1 *Let K be an algebraic number field of degree n. Let $\alpha, \beta \in K$. Then*

$$\text{tr}(\alpha + \beta) = \text{tr}(\alpha) + \text{tr}(\beta)$$

and

$$N(\alpha\beta) = N(\alpha)N(\beta).$$

Proof: Let $\sigma_k : K \longrightarrow \mathbb{C} \ (k = 1, 2, \ldots, n)$ be the n distinct monomorphisms from K to $\mathbb{C}$. Then

$$\text{tr}(\alpha + \beta) = \sum_{k=1}^{n} \sigma_k(\alpha + \beta) = \sum_{k=1}^{n} (\sigma_k(\alpha) + \sigma_k(\beta))$$

$$= \sum_{k=1}^{n} \sigma_k(\alpha) + \sum_{k=1}^{n} \sigma_k(\beta) = \text{tr}(\alpha) + \text{tr}(\beta)$$

and

$$N(\alpha\beta) = \prod_{k=1}^{n} \sigma_k(\alpha\beta) = \prod_{k=1}^{n} (\sigma_k(\alpha)\sigma_k(\beta))$$

$$= \left(\prod_{k=1}^{n} \sigma_k(\alpha) \right) \left(\prod_{k=1}^{n} \sigma_k(\beta) \right) = N(\alpha)N(\beta). \qquad \blacksquare$$

The next theorem tells us about the norm of a unit.

Theorem 9.2.2 *Let K be an algebraic number field of degree n.*

(a) *If α is a unit of O_K then $N(\alpha) = \pm 1$.*
(b) *If $\alpha \in O_K$ and $N(\alpha) = \pm 1$ then α is a unit of O_K.*

Proof: (a) Let $\alpha \in O_K$ be a unit. Then there exists $\beta \in O_K$ such that $\alpha\beta = 1$. Taking norms we obtain by Theorem 9.2.1

$$N(\alpha)N(\beta) = N(\alpha\beta) = N(1) = 1.$$

As $N(\alpha) \in \mathbb{Z}$ and $N(\beta) \in \mathbb{Z}$ we deduce that $N(\alpha) = \pm 1$.

(b) Let $\alpha \in O_K$ be such that $N(\alpha) = \pm 1$. Let $\sigma_k : K \longrightarrow \mathbb{C}$ $(k = 1, 2, \ldots, n)$ be the n distinct monomorphisms from K to $\mathbb{C}$ with $\sigma_1 = 1$. Then

$$\prod_{k=1}^{n} \sigma_k(\alpha) = N(\alpha) = \pm 1.$$

Set

$$\beta = \pm \prod_{k=2}^{n} \sigma_k(\alpha),$$

so that

$$\alpha\beta = 1.$$

As $\alpha \in O_K$, each $\sigma_k(\alpha) \in O_K$ $(k = 1, 2, \ldots, n)$ by Theorem 6.3.3, so that $\beta \in O_K$. Hence α is a unit of O_K. ∎

The next theorem is often useful in showing that an algebraic integer is an irreducible.

Theorem 9.2.3 *Let K be an algebraic number field. If $\alpha \in O_K$ is such that*

$$N(\alpha) = \pm p,$$

where p is a rational prime, then α is an irreducible.

Proof: Suppose that $\alpha \in O_K$ is such that $N(\alpha) = \pm p$, where p is a prime. Clearly $\alpha \neq 0$ as $N(0) = 0$. Moreover, α is not a unit as the norm of a unit is ± 1 by Theorem 9.2.2. Thus if α is not irreducible then there exist nonzero, nonunit elements β and γ of O_K such that

$$\alpha = \beta\gamma.$$

Then, by Theorem 9.2.1, we have

$$\pm p = N(\alpha) = N(\beta\gamma) = N(\beta)N(\gamma).$$

As $N(\beta) \in \mathbb{Z}$, $N(\gamma) \in \mathbb{Z}$, and p is a prime, we must have

$$N(\beta) \text{ or } N(\gamma) = \pm 1.$$

Hence, by Theorem 9.2.2(b), β or γ is a unit, which is a contradiction. This proves that α is irreducible. ∎

We emphasize that the converse of Theorem 9.2.3 is not true. To see this take $K = \mathbb{Q}(\sqrt{-5})$ and $\alpha = 1 + \sqrt{-5}$. Then α is an irreducible in O_K but $N(\alpha) = (1 + \sqrt{-5})(1 - \sqrt{-5}) = 6$ is not prime.

In the next theorem we use Theorems 9.2.2 and 9.2.3 to give a condition that, when satisfied by an element α of a cubic field, guarantees that α is a unit, and we give a similar condition for α to be an irreducible.

Theorem 9.2.4 *Let a and b be integers such that the cubic polynomial $x^3 + ax + b$ is irreducible in $\mathbb{Z}[x]$. Let*

$$K = \mathbb{Q}(\theta), \ \text{where} \ \theta^3 + a\theta + b = 0,$$

so that K is a cubic field.

(a) *If r, s, t are integers such that*

$$r^3 - bs^3 + b^2t^3 + ars^2 + a^2rt^2 - 2ar^2t - abst^2 + 3brst = \pm 1$$

then $r + s\theta + t\theta^2$ is a unit in O_K.

(b) *If r, s, t are integers such that*

$$r^3 - bs^3 + b^2t^3 + ars^2 + a^2rt^2 - 2ar^2t - abst^2 + 3brst = \pm p,$$

where p is a prime then $r + s\theta + t\theta^2$ is an irreducible in O_K.

Proof: Let $\theta, \ \theta', \ \theta''$ be the roots of $x^3 + ax + b = 0$ so that

$$\theta + \theta' + \theta'' = 0,$$
$$\theta\theta' + \theta'\theta'' + \theta''\theta = a,$$
$$\theta\theta'\theta'' = -b.$$

Then

$$\theta^2 + {\theta'}^2 + {\theta''}^2 = (\theta + \theta' + \theta'')^2 - 2(\theta\theta' + \theta'\theta'' + \theta''\theta) = -2a,$$
$$\theta^2{\theta'}^2 + {\theta'}^2{\theta''}^2 + {\theta''}^2\theta^2 = (\theta\theta' + \theta'\theta'' + \theta''\theta)^2 - 2\theta\theta'\theta''(\theta + \theta' + \theta'') = a^2,$$
$$\theta{\theta'}^2 + \theta^2\theta' + \theta{\theta''}^2 + \theta^2\theta'' + \theta'{\theta''}^2 + {\theta'}^2\theta''$$
$$= (\theta + \theta')\theta\theta' + (\theta + \theta'')\theta\theta'' + (\theta' + \theta'')\theta'\theta''$$
$$= -\theta\theta'\theta'' - \theta\theta'\theta'' - \theta\theta'\theta'' = -3\theta\theta'\theta'' = 3b.$$

Hence

$$
\begin{aligned}
N(r + s\theta + t\theta^2) &= (r + s\theta + t\theta^2)(r + s\theta' + t\theta'^2)(r + s\theta'' + t\theta''^2) \\
&= r^3 + s^3\theta\theta'\theta'' + t^3(\theta\theta'\theta'')^2 + rs^2(\theta\theta' + \theta'\theta'' + \theta''\theta) \\
&\quad + rt^2(\theta^2\theta'^2 + \theta'^2\theta''^2 + \theta''^2\theta^2) + r^2s(\theta + \theta' + \theta'') \\
&\quad + r^2t(\theta^2 + \theta'^2 + \theta''^2) + s^2t\theta\theta'\theta''(\theta + \theta' + \theta'') \\
&\quad + st^2\theta\theta'\theta''(\theta\theta' + \theta'\theta'' + \theta''\theta) \\
&\quad + rst(\theta\theta'^2 + \theta^2\theta' + \theta\theta''^2 + \theta^2\theta'' + \theta'\theta''^2 + \theta'^2\theta'') \\
&= r^3 - bs^3 + b^2t^3 + ars^2 + a^2rt^2 - 2ar^2t - abst^2 + 3brst.
\end{aligned}
$$

The assertions of the theorem now follow from Theorems 9.2.2 and 9.2.3.   ∎

Example 9.2.1   *Let K be the cubic field given by $K = \mathbb{Q}(\theta)$, where $\theta^3 - 4\theta + 2 = 0$. We show that $\theta - 1$ is a unit of O_K. This is the special case*

$$
a = -4, \; b = 2, \; r = -1, \; s = 1, \; t = 0
$$

of Theorem 9.2.4(a) as

$$
\begin{aligned}
r^3 - bs^3 &+ b^2t^3 + ars^2 + a^2rt^2 - 2ar^2t - abst^2 + 3brst \\
&= r^3 - bs^3 + ars^2 \\
&= -1 - 2 + 4 = 1.
\end{aligned}
$$

Similarly, we can show that $2\theta - 1$ is a unit of O_K since in this case

$$
a = -4, \; b = 2, \; r = -1, \; s = 2, \; t = 0
$$

and

$$
\begin{aligned}
r^3 - bs^3 &+ b^2t^3 + ars^2 + a^2rt^2 - 2ar^2t - abst^2 + 3brst \\
&= r^3 - bs^3 + ars^2 \\
&= -1 - 16 + 16 = -1.
\end{aligned}
$$

In the next theorem we relate $N(\langle \alpha \rangle)$ and $N(\alpha)$.

Theorem 9.2.5 *Let K be an algebraic number field of degree n. Let O_K be the ring of integers of K. Let $\alpha \in O_K$. Then*

$$
N(\langle \alpha \rangle) = |N(\alpha)|.
$$

Proof: Let $\{\omega_1, \ldots, \omega_n\}$ be an integral basis for K. Then $\{\alpha\omega_1, \ldots, \alpha\omega_n\}$ is a minimal basis for the principal ideal $\langle \alpha \rangle$. Hence

$$D(\langle\alpha\rangle) = \begin{vmatrix} \sigma_1(\alpha\omega_1) & \cdots & \sigma_1(\alpha\omega_n) \\ \sigma_2(\alpha\omega_1) & \cdots & \sigma_2(\alpha\omega_n) \\ \vdots & \cdots & \vdots \\ \sigma_n(\alpha\omega_1) & \cdots & \sigma_n(\alpha\omega_n) \end{vmatrix}^2,$$

where $\sigma_k : K \longrightarrow \mathbb{C}$ $(k = 1, 2, \ldots, n)$ are the n distinct monomorphisms from K to $\mathbb{C}$. Thus

$$D(\langle\alpha\rangle) = \begin{vmatrix} \sigma_1(\alpha)\sigma_1(\omega_1) & \cdots & \sigma_1(\alpha)\sigma_1(\omega_n) \\ \sigma_2(\alpha)\sigma_2(\omega_1) & \cdots & \sigma_2(\alpha)\sigma_2(\omega_n) \\ \vdots & \cdots & \vdots \\ \sigma_n(\alpha)\sigma_n(\omega_1) & \cdots & \sigma_n(\alpha)\sigma_n(\omega_n) \end{vmatrix}^2$$

$$= (\sigma_1(\alpha)\sigma_2(\alpha) \cdots \sigma_n(\alpha))^2 \begin{vmatrix} \sigma_1(\omega_1) & \cdots & \sigma_1(\omega_n) \\ \sigma_2(\omega_1) & \cdots & \sigma_2(\omega_n) \\ \vdots & \cdots & \vdots \\ \sigma_n(\omega_1) & \cdots & \sigma_n(\omega_n) \end{vmatrix}^2$$

$$= N(\alpha)^2 d(K),$$

so that

$$N(\langle\alpha\rangle) = \sqrt{\frac{D(\langle\alpha\rangle)}{d(K)}} = \sqrt{N(\alpha)^2} = |N(\alpha)|. \qquad \blacksquare$$

We see that Example 7.1.2 and Theorem 7.1.6 are special cases of Theorem 9.2.5. We also observe that if $\alpha \in O_K$, where K is an algebraic number field, then

$$\text{card}(O_K / \langle\alpha\rangle) = N(\langle\alpha\rangle) = |N(\alpha)|.$$

Next we determine the norm of the principal ideal $\langle\alpha\rangle$ in terms of the constant term of the minimal polynomial of α over $\mathbb{Q}$.

Theorem 9.2.6 *Let K be an algebraic number field of degree n. Let $\alpha \in K$. Let*

$$\text{irr}_\mathbb{Q}(\alpha) = x^m + b_{m-1}x^{m-1} + \cdots + b_0 \in \mathbb{Q}[x].$$

Then

$$N(\langle\alpha\rangle) = |b_0|^{n/m}.$$

Proof: By Theorem 6.3.2 we know that $m \mid n$. Let $\alpha_1 = \alpha, \alpha_2, \ldots, \alpha_m$ be the roots of $\text{irr}_\mathbb{Q}(\alpha)$ so that

$$x^m + b_{m-1}x^{m-1} + \cdots + b_0 = (x - \alpha_1)(x - \alpha_2) \cdots (x - \alpha_m)$$

and thus

$$b_0 = (-1)^m \alpha_1 \alpha_2 \cdots \alpha_m.$$

Again by Theorem 6.3.2 a complete set of conjugates of α is

$$\alpha_1, \ldots, \alpha_1, \alpha_2, \ldots, \alpha_2, \ldots, \alpha_m, \ldots, \alpha_m,$$

where each α_i is repeated n/m times. Hence

$$N(\alpha) = \alpha_1^{n/m} \alpha_2^{n/m} \cdots \alpha_m^{n/m} = (\alpha_1 \alpha_2 \cdots \alpha_m)^{n/m}$$
$$= ((-1)^m b_0)^{n/m} = (-1)^n b_0^{n/m}$$

and thus by Theorem 9.2.5

$$N(\langle \alpha \rangle) = |N(\alpha)| = |(-1)^n b_0^{n/m}| = |b_0|^{n/m}. \qquad \blacksquare$$

Example 9.2.2 *Let* $K = \mathbb{Q}(\sqrt{2} + \sqrt{3})$. *We determine* $N(\langle \sqrt{2} \rangle)$. *The minimal polynomial of* $\sqrt{2}$ *over* $\mathbb{Q}$ *is* $x^2 - 2$. *Hence in the notation of Theorem 9.2.6 we have* $n = 4$, $m = 2$, $b_1 = 0$, $b_0 = -2$. *Thus by Theorem 9.2.6 we obtain*

$$N(\langle \sqrt{2} \rangle) = |-2|^{4/2} = 2^2 = 4.$$

9.3 Norm of a Product of Ideals

In this section we prove the multiplicative property (9.1.3) of norms of ideals. We will need the following result, the proof of which closely resembles that of Theorem 8.5.1.

Theorem 9.3.1 *Let* D *be a Dedekind domain. Let* A *be a fractional or integral ideal of* D *with* $A \neq \langle 0 \rangle$, $\langle 1 \rangle$. *Let* B *be an integral ideal of* D *with* $B \neq \langle 0 \rangle$, $\langle 1 \rangle$. *Then there exists* $\gamma \in A$ *such that*

$$A = \langle \gamma \rangle + AB.$$

Proof: Let $P_1, \ldots, P_n$ be the set of distinct prime ideals for which either

$$\text{ord}_{P_i}(A) \neq 0 \text{ or } \text{ord}_{P_i}(AB) \neq 0 \text{ (or both)}.$$

This set is nonempty as $A \neq D$. By Theorem 8.4.5 we can find an element γ of the quotient field of D such that

$$\text{ord}_{P_i}(\gamma) = \text{ord}_{P_i}(A), \quad i = 1, 2, \ldots, n,$$
$$\text{ord}_P(\gamma) \geq 0, \quad P \neq P_1, \ldots, P_n.$$

Thus

$$\text{ord}_P(\gamma) \geq \text{ord}_P(A) \text{ for all prime ideals } P,$$

and so

$$\gamma \in A.$$

Now for $i = 1, 2, \ldots, n$ we have

$$\begin{aligned}
\mathrm{ord}_{P_i}(\langle \gamma \rangle + AB) &= \min(\mathrm{ord}_{P_i}(\langle \gamma \rangle), \mathrm{ord}_{P_i}(AB)) \\
&= \min(\mathrm{ord}_{P_i}(\gamma), \mathrm{ord}_{P_i}(AB)) \\
&= \min(\mathrm{ord}_{P_i}(A), \mathrm{ord}_{P_i}(AB)) \\
&= \mathrm{ord}_{P_i}(A),
\end{aligned}$$

as B is an integral ideal. For a prime ideal $P \neq P_1, \ldots, P_n$ we have $\mathrm{ord}_P(A) = \mathrm{ord}_P(AB) = 0$ so that

$$\begin{aligned}
\mathrm{ord}_P(\langle \gamma \rangle + AB) &= \min(\mathrm{ord}_P(\langle \gamma \rangle), \mathrm{ord}_P(AB)) \\
&= \min(\mathrm{ord}_P(\gamma), 0) \\
&= 0 \\
&= \mathrm{ord}_P(A).
\end{aligned}$$

Hence

$$\mathrm{ord}_P(\langle \gamma \rangle + AB) = \mathrm{ord}_P(A)$$

for all prime ideals P, and so

$$A = \langle \gamma \rangle + AB. \qquad \blacksquare$$

Theorem 9.3.2 *Let K be an algebraic number field. Let A and B be nonzero integral ideals in $D = O_K$. Then*

$$N(AB) = N(A)N(B).$$

Proof: If A or $B = D$ then the result is trivially true as $N(D) = \mathrm{card}(D/D) = 1$. Hence we may assume that $A \neq D$ and $B \neq D$. Let $k = N(A)$ and $l = N(B)$. Then, by Theorem 9.1.3, the ring D/A has k elements, say,

$$\alpha_1 + A, \ldots, \alpha_k + A.$$

Also, D/B has l elements, say,

$$\beta_1 + B, \ldots, \beta_l + B.$$

By Theorem 9.3.1 there is an element γ of A such that

$$A = \langle \gamma \rangle + AB.$$

If $\gamma = 0$ then $A = AB$ so that $B = D$, contradicting $B \neq D$. Hence $\gamma \neq 0$. Let $\delta \in D$. Then there is a unique integer i $(1 \leq i \leq k)$ such that

$$\delta \equiv \alpha_i \pmod{A}.$$

Clearly,

$$\delta - \alpha_i \in A = \langle \gamma \rangle + AB$$

so there exist $\sigma \in D$ and $\tau \in AB$ such that

$$\delta - \alpha_i = \sigma \gamma + \tau.$$

Similarly, there is a unique integer j $(1 \le j \le l)$ such that

$$\sigma \equiv \beta_j \pmod{B},$$

that is,

$$\sigma - \beta_j \in B.$$

As $\gamma \in A$ we have

$$(\sigma - \beta_j)\gamma \in AB.$$

Hence

$$\delta = \alpha_i + \sigma\gamma + \tau = \alpha_i + \beta_j\gamma + (\sigma - \beta_j)\gamma + \tau \equiv \alpha_i + \beta_j\gamma \pmod{AB}.$$

This shows that the set of kl elements $\alpha_i + \beta_j\gamma + AB$ $(i = 1, \dots, k; j = 1, \dots, l)$ is a complete set of representatives of D/AB. We must still show that they are distinct. Suppose

$$\alpha_i + \beta_j\gamma + AB = \alpha_p + \beta_q\gamma + AB.$$

Then

$$\alpha_i + \beta_j\gamma \equiv \alpha_p + \beta_q\gamma \pmod{AB}$$

and thus

$$\alpha_i - \alpha_p \equiv (\beta_q - \beta_j)\gamma \pmod{AB}.$$

But $\gamma \in A$ so

$$\alpha_i - \alpha_p \in A.$$

Thus $i = p$ and

$$\beta_j\gamma \equiv \beta_q\gamma \pmod{AB}.$$

Hence

$$(\beta_j - \beta_q)\gamma \in AB.$$

Now let $B = \prod_{i=1}^{h} P_i^{b_i}$ $(b_i > 0)$ be the prime ideal decomposition of B. Then

$$\operatorname{ord}_{P_i}(A) = \operatorname{ord}_{P_i}(\langle \gamma \rangle + AB)$$
$$= \min(\operatorname{ord}_{P_i}(\langle \gamma \rangle), \operatorname{ord}_{P_i}(AB))$$
$$= \min(\operatorname{ord}_{P_i}(\gamma), \operatorname{ord}_{P_i}(A) + \operatorname{ord}_{P_i}(B))$$

and it follows that

$$\mathrm{ord}_{P_i}(A) = \mathrm{ord}_{P_i}(\gamma), \quad i = 1, \ldots, h. \tag{9.3.1}$$

If $\beta_j - \beta_q \neq 0$ then $(\beta_j - \beta_q)\gamma$ is a nonzero element of AB so that for $i = 1, \ldots, h$ we have

$$\mathrm{ord}_{P_i}((\beta_j - \beta_q)\gamma) \geq \mathrm{ord}_{P_i}(AB)$$

and thus

$$\mathrm{ord}_{P_i}(\beta_j - \beta_q) + \mathrm{ord}_{P_i}(\gamma) \geq \mathrm{ord}_{P_i}(A) + \mathrm{ord}_{P_i}(B).$$

Then, appealing to (9.3.1), we deduce that

$$\mathrm{ord}_{P_i}(\beta_j - \beta_q) \geq \mathrm{ord}_{P_i}(B), \quad i = 1, \ldots, h,$$

which shows that

$$\beta_j - \beta_q \in B,$$

and hence $j = q$, contradicting $\beta_j - \beta_q \neq 0$. This proves that $\beta_j - \beta_q = 0$ so that $j = q$. Hence $\{\alpha_i + \beta_j\gamma + AB \mid i = 1, \ldots, k; \ j = 1, \ldots, l\}$ is a complete set of distinct representatives of D/AB and so

$$N(AB) = \mathrm{card}(D/AB) = kl = N(A)N(B). \qquad \blacksquare$$

9.4 Norm of a Fractional Ideal

The multiplicative property of the norm (Theorem 9.3.2) allows us to extend the definition of the norm of an integral ideal (of the ring of integers of an algebraic number field) to the norm of a fractional ideal.

Definition 9.4.1 (Norm of a fractional ideal) *Let K be an algebraic number field. Let O_K be its ring of integers. Let A be a nonzero fractional ideal of O_K. Then there exists a nonzero integral ideal I of O_K and a nonzero element α of O_K such that*

$$A = \frac{1}{\alpha} I.$$

We define the norm $N(A)$ of the fractional ideal A by

$$N(A) = \frac{N(I)}{N(\langle\alpha\rangle)},$$

where $N(I)$, $N(\langle\alpha\rangle)$ are the norms of the integral ideals I and $\langle\alpha\rangle$.

Definition 9.4.1 is valid for if I and J are nonzero integral ideals of O_K and α and β are nonzero elements of O_K such that

$$A = \frac{1}{\alpha} I = \frac{1}{\beta} J$$

then

$$\beta I = \alpha J,$$

so that we have the equal products of integral ideals

$$\langle \beta \rangle I = \langle \alpha \rangle J,$$

and thus by Theorem 9.3.2

$$N(\langle \beta \rangle)N(I) = N(\langle \beta \rangle I) = N(\langle \alpha \rangle J) = N(\langle \alpha \rangle)N(J),$$

so that

$$\frac{N(I)}{N(\langle \alpha \rangle)} = \frac{N(J)}{N(\langle \beta \rangle)}.$$

When the fractional ideal A of Definition 9.4.1 is actually an integral ideal the two definitions of the norm coincide.

The next theorem shows that the multiplicative property of the norm carries over to fractional ideals.

Theorem 9.4.1 *Let K be an algebraic number field. Let O_K be its ring of integers. Let A and B be nonzero fractional ideals of O_K. Then*

$$N(AB) = N(A)N(B).$$

Proof: As A and B are nonzero fractional ideals of O_K there exist nonzero integral ideals I and J of O_K and nonzero elements α and β of O_K such that

$$A = \frac{1}{\alpha}I, \quad B = \frac{1}{\beta}J.$$

Then

$$AB = \frac{1}{\alpha\beta}IJ$$

so that

$$
\begin{aligned}
N(AB) &= \frac{N(IJ)}{N(\langle \alpha\beta \rangle)} \quad \text{(Definition 9.4.1)} \\
&= \frac{N(IJ)}{N(\langle \alpha \rangle \langle \beta \rangle)} \quad \text{(Definition 1.6.2)} \\
&= \frac{N(I)N(J)}{N(\langle \alpha \rangle)N(\langle \beta \rangle)} \quad \text{(Theorem 9.3.2)} \\
&= \frac{N(I)}{N(\langle \alpha \rangle)} \cdot \frac{N(J)}{N(\langle \beta \rangle)} \\
&= N(A)N(B).
\end{aligned}
$$

∎

Example 9.4.1 *Let $K = \mathbb{Q}(\sqrt{6})$ so that $O_K = \{a + b\sqrt{6} \mid a, b \in \mathbb{Z}\}$. Let A be the fractional ideal of O_K given by*

$$A = \langle 1, \frac{1}{2}\sqrt{6} \rangle.$$

Then

$$A = \frac{1}{2}I,$$

where I is the integral ideal $\langle 2, \sqrt{6} \rangle$. Now

$$I^2 = \langle 2, \sqrt{6} \rangle^2 = \langle 4, 2\sqrt{6}, 6 \rangle = \langle 2 \rangle \langle 2, \sqrt{6}, 3 \rangle = \langle 2 \rangle$$

(as $1 = 3 - 2 \in \langle 2, \sqrt{6}, 3 \rangle$) so that

$$N(I)^2 = N(I^2) = N(\langle 2 \rangle) = 2^2$$

and thus

$$N(I) = 2.$$

Hence

$$N(A) = N(\tfrac{1}{2}I) = \frac{N(I)}{N(\langle 2 \rangle)} = \frac{2}{2^2} = \frac{1}{2}.$$

Exercises

1. Let p be a prime such that $p \equiv 3$ or $5 \pmod 8$. Prove that there does not exist an element $\alpha \in O_{\mathbb{Q}(\sqrt{p})}$ such that $N(\alpha) = 2$.
2. Let p be a prime such that $p \equiv 5$ or $7 \pmod 8$. Prove that there does not exist an element $\alpha \in O_{\mathbb{Q}(\sqrt{p})}$ such that $N(\alpha) = -2$.
3. Let K be an algebraic number field and O_K its ring of integers. If I is an integral ideal of O_K such that $N(I)$ is a prime, then I is a prime ideal.
4. Let K be an algebraic number field and O_K its ring of integers. If I is a nonzero integral ideal of O_K, prove that $I \mid \langle N(I) \rangle$.
5. Let K be an algebraic number field. Let n be a given positive integer. Prove that there are only finitely many integral ideals I of O_K such that $N(I) = n$.
6. Let $K = \mathbb{Q}(\theta)$, where $\theta^3 - \theta - 1 = 0$. Prove that $\langle 23, 3 - \theta \rangle$ is a prime ideal in O_K.
7. Let $K = \mathbb{Q}(\sqrt{-23})$. Let $I = \langle 2, \frac{1}{2}(1 + \sqrt{-23}) \rangle$.
   (a) Prove that $N(I) = 2$.
   (b) Prove that $I^3 = \langle (-3 + \sqrt{-23})/2 \rangle$.
   (c) Use each of (a) and (b) to prove that I is not a principal ideal.
8. Let K be an algebraic number field and O_K its ring of integers. Let I be an integral ideal of O_K such that $N(I) = |N(a)|$ for some $a \in I$. Prove that $I = \langle a \rangle$.
9. Let K be an algebraic number field and O_K its ring of integers. Let P be a prime ideal of O_K. Prove that $G = \{a + P \mid a \in O_K, \ a \notin P\}$ is a cyclic group with respect to multiplication. What is the order of G?
10. Let K be an algebraic number field and O_K its ring of integers. Let P be a prime ideal of O_K. Prove that $P \cap \mathbb{Z} = \langle p \rangle$ for some prime p.

11. Let K be an algebraic number field and O_K its ring of integers. Show that

$$a^p \equiv a \,(\mathrm{mod}\ P) \Longleftrightarrow a \equiv m \,(\mathrm{mod}\ P) \text{ for some } m \in \mathbb{Z},$$

where $P \cap \mathbb{Z} = \langle p \rangle$ (see Exercise 10).

12. Determine the fractional ideals of $\mathbb{Z} + \mathbb{Z}i$.

13. Let K be an algebraic number field and O_K its ring of integers. Let $m \in \mathbb{Z} \setminus \{0\}$. Prove that there exist only finitely many integral ideals of O_K to which m belongs.

14. Find all the ideals of $\mathbb{Z} + \mathbb{Z}\sqrt{-5}$ that contain 6.

15. Find all the ideals of $\mathbb{Z} + \mathbb{Z}\sqrt{2}$ with norm 12.

16. Determine the set of positive integers that are not norms of ideals of $\mathbb{Z} + \mathbb{Z}i$.

17. Let K be an algebraic number field. Let O_K be its ring of integers. Let P be a prime ideal of O_K. Let $a \in O_K$ be such that $P \nmid \langle a \rangle$. Prove that

$$a^{N(P)} - 1 \equiv 0 \,(\mathrm{mod}\ P).$$

18. Give an example of an algebraic number field K and an integral ideal $I = \langle a, b, c \rangle$ of O_K such that $I \neq \langle a, b \rangle,\ \langle a, c \rangle,\ \langle b, c \rangle$.

19. Determine all complex quadratic fields K for which O_K possesses elements of norm 38 and trace 11.

20. Let K be an algebraic number field. Let $\alpha, \beta \in O_K \setminus \{0\}$. Prove that $N(\alpha)\mathrm{tr}(\beta/\alpha) \in \mathbb{Z}$.

21. Solve $3x \equiv 5 \,(\mathrm{mod}\ A)$ in $\mathbb{Z} + \mathbb{Z}\sqrt{-5}$, where $A = \langle 3\sqrt{-5}, 10 + 10\sqrt{-5} \rangle$.

22. Let K be an algebraic number field. Let I be an integral ideal of O_K. Let m be the least positive integer in I. Prove that $m \mid N(I)$.

23. Let K be an algebraic number field. Let I be an integral ideal of O_K such that $pq \mid N(I)$, where p and q are distinct primes. Prove that I is not a prime ideal.

24. Let K be a quadratic field. Let $\alpha \in O_K$ be such that $|N(\alpha)| = ab$, where a and b are coprime positive integers. Prove that

$$\langle a, \alpha \rangle \langle b, \alpha \rangle = \langle \alpha \rangle.$$

25. Prove that $d_1, d_2, \ldots, d_n$ in Theorem 9.1.1 can be arranged to satisfy $d_1 \mid d_2 \mid \cdots \mid d_n$.

26. Let K be an algebraic number field. Prove that O_K is a principal ideal domain if and only if for every pair $(\alpha, \beta) \in O_K \times O_K$ such that

$$\alpha \neq 0,\ \beta \neq 0,\ \beta \nmid \alpha,\ |N(\alpha)| \geq |N(\beta)|,$$

there exist $\gamma \in O_K$ and $\delta \in O_K$ such that

$$0 < |N(\alpha\gamma - \beta\delta)| < |N(\beta)|.$$

Suggested Reading

1. P. B. Bhattacharya, S. K. Jain, and S. R. Nagpaul, *Basic Abstract Algebra,* second edition, Cambridge University Press, Cambridge, United Kingdom, 1994.

   Chapter 20 is devoted to the Smith normal form of a matrix over a principal ideal domain.

2. C. C. MacDuffee, *An Introduction to Abstract Algebra,* Wiley, New York, 1956.

   Section 105 discusses the Smith normal form of a matrix and mentions that this form is named for H. J. S. Smith.

3. I. N. Stewart and D. O. Tall, *Algebraic Number Theory,* second edition, Chapman and Hall, London, 1987.

    Section 1.6 contains a very readable discussion of free Abelian groups.

Biographies

1. H. J. S. Smith, *Report on the Theory of Numbers*, Chelsea, New York, 1964.

    This book contains a biographical sketch of the Irish mathematician Henry John Stephen Smith (1826–1883).

2. The website

    http://www-groups.dcs.st-and.ac.uk/~history/

    contains a biography of H. J. S. Smith.

10

Factoring Primes in a Number Field

10.1 Norm of a Prime Ideal

We begin by showing that each prime ideal in the ring of integers of an algebraic number field is associated with a unique rational prime.

Theorem 10.1.1 *Let K be an algebraic number field. Let P be a prime ideal of O_K. Then there exists a unique rational prime p such that*

$$P \mid \langle p \rangle.$$

Proof: As P is a prime ideal in O_K, $P \cap \mathbb{Z}$ is a prime ideal in $\mathbb{Z}$ (Theorem 1.6.2). Hence, by Theorems 1.4.1 and 1.5.4, we have

$$P \cap \mathbb{Z} = \langle p \rangle$$

for some rational prime p. Thus

$$P \supseteq \langle p \rangle$$

and so by Theorem 8.4.1 we have

$$P \mid \langle p \rangle.$$

Suppose q is another rational prime such that

$$P \mid \langle q \rangle.$$

Then $P \supseteq \langle p \rangle$ and $P \supseteq \langle q \rangle$ so that

$$P \supseteq \langle p, q \rangle.$$

As p and q are distinct primes we have $\gcd(p, q) = 1$ so that there are integers a and b such that $ap + bq = 1$. Hence $1 \in \langle p, q \rangle \subseteq P$. Thus $O_K \subseteq P$, which is impossible.

Hence the prime p is uniquely determined by $P \mid \langle p \rangle$. ∎

236

The rational prime p in Theorem 10.1.1 is called the prime lying below P as $P \supseteq \langle p \rangle$. Given a rational prime p, any prime ideal P such that $P \mid \langle p \rangle$ is said to be a prime ideal lying over p.

Next we relate the norm of the prime ideal P to the prime p lying below P.

Theorem 10.1.2 *Let K be an algebraic number field with $[K : \mathbb{Q}] = n$. Let P be a prime ideal of O_K. Let p be the rational prime lying below P. Then*

$$N(P) = p^f$$

for some integer $f \in \{1, 2, \ldots, n\}$.

Proof: As p lies below P we have $P \mid \langle p \rangle$. Hence $\langle p \rangle = PQ$ for some integral ideal Q of O_K. By Theorem 9.3.2 we have

$$N(\langle p \rangle) = N(PQ) = N(P)N(Q).$$

As the K-conjugates of p comprise p repeated n times (Theorem 6.3.4), we have

$$N(p) = p^n,$$

so that, by Theorem 9.2.5,

$$N(\langle p \rangle) = |N(p)| = p^n.$$

Hence we have

$$p^n = N(P)N(Q)$$

so that

$$N(P) = p^f$$

for some $f \in \{1, 2, \ldots, n\}$. ∎

Definition 10.1.1 (Inertial degree) *Let K be an algebraic number field with $[K : \mathbb{Q}] = n$. Let p be the rational prime lying below P. Then the positive integer f such that*

$$N(P) = p^f$$

is called the inertial degree of P in O_K and is denoted by $f_K(P)$.

From Theorems 9.1.3 and 10.1.2 we see that

$$\text{card}(O_K/P) = p^f$$

so that O_K/P is a finite field with p^f elements. Consider the elements

$$a + P \ (a \in \mathbb{Z})$$

of O_K/P. If $a, a' \in \mathbb{Z}$ are such that $a \equiv a' \pmod{p}$ then $p \mid a - a'$ so that $\langle p \rangle \mid \langle a - a' \rangle$. But $P \mid \langle p \rangle$ so $P \mid \langle a - a' \rangle$ and thus $a \equiv a' \pmod{P}$, that is, $a + P = a' + P$. Conversely suppose that $a + P = a' + P$ $(a, a' \in \mathbb{Z})$. Then $a - a' \in P$, $\langle a - a' \rangle \subseteq P$, and $P \mid \langle a - a' \rangle$, so

$$\langle a - a' \rangle = PQ$$

for some integral ideal Q of O_K. Taking norms we obtain

$$|a - a'|^n = N(\langle a - a' \rangle) = N(PQ) = N(P)N(Q) = p^f N(Q),$$

so that $p \mid (a - a')^n$ and thus, as p is a prime, $p \mid a - a'$ and so $a \equiv a' \pmod{p}$. We have shown that for $a, a' \in \mathbb{Z}$

$$a + P = a' + P \iff a \equiv a' \pmod{p}.$$

Hence the cosets $a + P$ $(a \in \{0, 1, \ldots, p - 1\})$ are distinct and the prime field of O_K/P is

$$F = \{a + P \mid a = 0, 1, \ldots, p - 1\} \simeq \mathbb{Z}/\langle p \rangle.$$

The inertial degree $f = f_K(P)$ is given by

$$f_K(P) = [O_K/P : F] = [O_K/P : \mathbb{Z}/\langle p \rangle].$$

If $f_K(P) = 1$ then $[O_K/P : F] = 1$ so that $O_K/P = F$; that is,

$$O_K/P = \{a + P \mid a = 0, 1, \ldots, p - 1\}.$$

Let $\alpha \in O_K$. Then $\alpha + P \in O_K/P$ so that $\alpha + P = a + P$ for some $a \in \{0, 1, \ldots, p - 1\}$. Hence $\alpha \equiv a \pmod{P}$ for some $a \in \mathbb{Z}$, when $f_K(P) = 1$.

Theorem 10.1.3 *Let K be an algebraic number field with $[K : Q] = n$. Let p be a rational prime. Suppose that the principal ideal $\langle p \rangle$ factors in O_K in the form*

$$\langle p \rangle = P_1^{e_1} \cdots P_g^{e_g}, \tag{10.1.1}$$

where $P_1, \ldots, P_g$ are distinct prime ideals of O_K and $e_1, \ldots, e_g$ are positive integers. Suppose that f_i is the inertial degree of P_i $(i = 1, 2, \ldots, g)$ in K, that is, $f_i = f_K(P_i)$. Then

$$e_1 f_1 + \cdots + e_g f_g = n.$$

Proof: As f_i is the inertial degree of P_i in K, we have $N(P_i) = p^{f_i}$ $(i = 1, 2, \ldots, g)$ so that

$$p^n = N(\langle p \rangle) = N(P_1^{e_1} \cdots P_g^{e_g})$$
$$= N(P_1)^{e_1} \cdots N(P_g)^{e_g}$$
$$= (p^{f_1})^{e_1} \cdots (p^{f_g})^{e_g}$$
$$= p^{e_1 f_1 + \cdots + e_g f_g}.$$

Hence

$$e_1 f_1 + \cdots + e_g f_g = n. \qquad \blacksquare$$

It follows immediately from Theorem 10.1.3 that

$$e_i \in \{1, 2, \ldots, n\}, \quad f_i \in \{1, 2, \ldots, n\}, \quad i = 1, 2, \ldots, g, \qquad (10.1.2)$$

and that

$$P^n \mid \langle p \rangle \text{ (for some prime ideal } P) \Longrightarrow \langle p \rangle = P^n. \qquad (10.1.3)$$

Definition 10.1.2 (Decomposition number) *With the notation of Theorem 10.1.3, the positive integer g defined in (10.1.1) is called the decomposition number of p in K and is written $g_K(p)$.*

Theorem 10.1.4 *Let K be an algebraic number field of degree n. Let p be a rational prime. Then*

$$g_K(p) \le n.$$

Proof: With the notation of Theorem 10.1.3, as $e_1, \ldots, e_g, f_1, \ldots, f_g$ are positive integers, we deduce that

$$n = e_1 f_1 + \cdots + e_g f_g \ge 1 + \cdots + 1 = g. \qquad \blacksquare$$

Thus in an algebraic number field of degree n, the principal ideal $\langle p \rangle$ (with p a rational prime) cannot split into a product of more than n distinct prime powers.

Definition 10.1.3 (Ramification index) *Let K be an algebraic number field of degree n. Let P be a prime ideal of O_K. Let p be a rational prime lying below P. Then the unique positive integer e such that*

$$P^e \mid \langle p \rangle, \quad P^{e+1} \nmid \langle p \rangle$$

is called the ramification index of P in K and is written $e_K(P)$. From (10.1.2) we see that $e_K(P) \le n$.

In the notation of Theorem 10.1.3 we have

$$e_K(P_i) = e_i, \quad f_K(P_i) = f_i, \quad g_K(p) = g.$$

Definition 10.1.4 (Ramification) *Let K be an algebraic number field of degree n. Let p be a rational prime. Let $P_1, \ldots, P_g$ be the prime ideals of O_K lying above p. Then*

$$\langle p \rangle = P_1^{e_1} \cdots P_g^{e_g},$$

where

$$e_i = e_K(P_i), \quad i = 1, 2, \ldots, g.$$

If $e_i > 1$ for some $i \in \{1, 2, \ldots, g\}$ then p is said to ramify in K. If $e_i = 1$ for $i = 1, 2, \ldots, g$ then p is said to be unramified in K.

The following theorem of Dedekind, which we shall not prove here, enables us to recognize when a rational prime p ramifies in an algebraic number field K.

Theorem 10.1.5 (Dedekind) *Let K be an algebraic number field. Then the rational prime p ramifies in K if and only if $p \mid d(K)$.*

In the next section we examine the factorization of a rational prime p into prime ideals in O_K, when K is a quadratic field.

We conclude this section by proving the following simple but useful result.

Theorem 10.1.6 *Let K be an algebraic number field. Let $I (\neq \langle 0 \rangle)$ be an ideal of O_K.*

(a) *If $N(I) = p$, where p is a prime, then I is a prime ideal.*
(b) *$N(I) \in I$.*

Proof: **(a)** Clearly $I \neq \langle 1 \rangle$ as $N(I) \neq 1$. By Theorem 8.3.1 I is a product of prime ideals. If there exist prime ideals P and Q such that PQ divides I (where P and Q may or may not be distinct) then $I = PQA$ for some integral ideal A of O_K. Hence, by Theorem 9.4.1, we have

$$p = N(I) = N(PQA) = N(P)N(Q)N(A),$$

which contradicts that p is a rational prime as $N(P)$, $N(Q)$, and $N(A)$ are positive integers with $N(P) > 1$ and $N(Q) > 1$. Thus I is a prime ideal.

(b) By Theorem 9.1.3 we have

$$N(I) = \text{card } (O_K/I).$$

Hence

$$N(I)(x + I) = 0 + I, \quad \text{for all } x \in O_K;$$

that is,

$$N(I)x \in I, \text{ for all } x \in O_K.$$

Taking $x = 1 \in O_K$ we obtain

$$N(I) \in I$$

as asserted. ∎

From Theorems 10.1.6(b) and 8.4.1 we deduce that

$$\langle N(I) \rangle \subseteq I$$

and

$$I \mid \langle N(I) \rangle$$

for all ideals I of the ring of integers of an algebraic number field.

10.2 Factoring Primes in a Quadratic Field

Let p be a rational prime and let K be a quadratic field. By Theorem 10.1.4 we have

$$g = g_K(p) \leq 2$$

so that

$$g = 1 \text{ or } 2.$$

If $g = 2$ by Theorem 10.1.3 we have

$$e_1 f_1 + e_2 f_2 = 2$$

so that

$$e_1 = f_1 = e_2 = f_2 = 1.$$

If $g = 1$ we have

$$e_1 f_1 = 2$$

so that

$$(e_1, f_1) = (2, 1) \text{ or } (1, 2).$$

Thus in the case of a quadratic field there are just three possibilities:

(i) $g = 2$, $e_1 = f_1 = e_2 = f_2 = 1$,
(ii) $g = 1$, $e_1 = 2$, $f_1 = 1$,
(iii) $g = 1$, $e_1 = 1$, $f_1 = 2$.

In other words,

$$\text{(i) } \langle p \rangle = P_1 P_2, \ N(P_1) = N(P_2) = p, \ P_1 \neq P_2,$$
$$\text{(ii) } \langle p \rangle = P^2, \ N(P) = p,$$
$$\text{(iii) } \langle p \rangle = P, \ N(P) = p^2,$$

where P_1, P_2, P denote prime ideals of O_K. In case (i) we say that p splits in K, in case (ii) that p ramifies in K, and in case (iii) that p is inert (or remains prime) in K. In cases (i) and (iii) p is unramified in K.

Our next theorem gives necessary and sufficient conditions for each of (i), (ii), (iii) to occur. As usual $\left(\frac{m}{p} \right)$ denotes the Legendre symbol of the integer m modulo the odd prime p.

Theorem 10.2.1 *Let K be a quadratic field so that there exists a squarefree integer m such that $K = \mathbb{Q}(\sqrt{m})$. Let p be a rational prime.*

(a) *If $p > 2$, $\left(\frac{m}{p} \right) = 1$ or $p = 2$, $m \equiv 1 \,(\mathrm{mod}\ 8)$ then*

$$\langle p \rangle = P_1 P_2,$$

where P_1 and P_2 are distinct prime ideals with $N(P_1) = N(P_2) = p$.
(b) *If $p > 2$, $p \mid m$ or $p = 2$, $m \equiv 2$ or $3 \,(\mathrm{mod}\ 4)$ then*

$$\langle p \rangle = P^2,$$

where P is a prime ideal with $N(P) = p$.
(c) *If $p > 2$, $\left(\frac{m}{p} \right) = -1$ or $p = 2$, $m \equiv 5 \,(\mathrm{mod}\ 8)$ then*

$$\langle p \rangle \text{ is a prime ideal of } O_K.$$

Proof: As m is squarefree we have $m \not\equiv 0 \,(\mathrm{mod}\ 4)$ so that $m \equiv 1, 2,$ or $3 \,(\mathrm{mod}\ 4)$. We consider seven cases.

(i): $p > 2$, $\left(\frac{m}{p} \right) = 1$. As $\left(\frac{m}{p} \right) = 1$ there exists $a \in \mathbb{Z}$ such that $a^2 \equiv m \,(\mathrm{mod}\ p)$. As $p \nmid m$ we have $p \nmid a$. Let $P_1 = \langle p, a + \sqrt{m} \rangle$ and $P_2 = \langle p, a - \sqrt{m} \rangle$.

We show first that $P_1 \neq P_2$. Suppose on the contrary that $P_1 = P_2$. Then

$$2a = (a + \sqrt{m}) + (a - \sqrt{m}) \in P_1.$$

But $2a \in \mathbb{Z}$ so

$$2a \in P_1 \cap \mathbb{Z} = \langle p \rangle.$$

Hence $p \mid 2a$. This is impossible as p is odd and $p \nmid a$.

Next we show that $\langle p \rangle = P_1 P_2$. We have

$$
\begin{aligned}
P_1 P_2 &= \langle p, a + \sqrt{m} \rangle \langle p, a - \sqrt{m} \rangle \\
&= \langle p^2, \ p(a + \sqrt{m}), \ p(a - \sqrt{m}), \ a^2 - m \rangle \\
&= \langle p \rangle \langle p, \ a + \sqrt{m}, \ a - \sqrt{m}, \ (a^2 - m)/p \rangle \\
&= \langle p \rangle I,
\end{aligned}
$$

where I is the ideal

$$I = \langle p, a + \sqrt{m}, a - \sqrt{m}, (a^2 - m)/p \rangle.$$

As $\gcd(2a, p) = 1$ there exist integers x and y such that

$$xp + y(2a) = 1.$$

Thus

$$1 = xp + y(a + \sqrt{m}) + y(a - \sqrt{m}) \in I$$

and so $I = \langle 1 \rangle$. This proves that $\langle p \rangle = P_1 P_2$, showing that this case falls under (a).

(ii): $p > 2$, $\left(\frac{m}{p} \right) = -1$. Suppose that

$$\langle p \rangle = P_1 P_2 \ (P_1 \neq P_2) \text{ or } P_1^2.$$

In each case we have $N(P_1) = p$ so that $f_K(P_1) = 1$. Hence, as $\sqrt{m} \in O_K$, there exists $a \in \mathbb{Z}$ such that

$$\sqrt{m} \equiv a \ (\text{mod } P_1)$$

and so

$$m \equiv a^2 \ (\text{mod } P_1).$$

But, as $m \in \mathbb{Z}$, $a^2 \in \mathbb{Z}$ and $P_1 \mid \langle p \rangle$, we must have

$$m \equiv a^2 \ (\text{mod } p),$$

contradicting $\left(\frac{m}{p} \right) = -1$. Thus $\langle p \rangle$ is a prime ideal in O_K, and this case falls under (c).

(iii): $p > 2$, $p \mid m$. Set $P = \langle p, \sqrt{m} \rangle$. Then

$$P^2 = \langle p, \sqrt{m} \rangle \langle p, \sqrt{m} \rangle = \langle p^2, p\sqrt{m}, m \rangle = \langle p \rangle I,$$

where I is the ideal

$$I = \langle p, \sqrt{m}, m/p \rangle.$$

As m is squarefree, we have $\gcd(p, m/p) = 1$, so that there exist integers x and y such that

$$xp + y(m/p) = 1.$$

Hence

$$1 = xp + y(m/p) \in I,$$

so $I = \langle 1 \rangle$, that is, $\langle p \rangle = P^2$, and this case falls under (b).

(iv): $p = 2$, $m \equiv 2 \ (\text{mod } 4)$. Set $P = \langle 2, \sqrt{m} \rangle$. Then

$$P^2 = \langle 2, \sqrt{m} \rangle \langle 2, \sqrt{m} \rangle = \langle 4, 2\sqrt{m}, m \rangle = \langle 2 \rangle I,$$

where I is the ideal

$$I = \langle 2, \sqrt{m}, m/2 \rangle.$$

As $m/2$ is odd, say $m/2 = 2k + 1$, then

$$1 = (-k)2 + (1)m/2 \in I,$$

so $I = \langle 1 \rangle$, and $P^2 = \langle 2 \rangle$. This case falls under (b).

(v): $p = 2$, $m \equiv 3 \pmod 4$. Let $P = \langle 2, 1 + \sqrt{m} \rangle$. As $1 - \sqrt{m} = 2 - (1 + \sqrt{m})$ we see that $P = \langle 2, 1 - \sqrt{m} \rangle$. Then

$$
\begin{aligned}
P^2 &= \langle 2, 1 + \sqrt{m} \rangle \langle 2, 1 - \sqrt{m} \rangle \\
&= \langle 4, 2(1 + \sqrt{m}), 2(1 - \sqrt{m}), 1 - m \rangle \\
&= \langle 2 \rangle I,
\end{aligned}
$$

where I is the ideal

$$I = \langle 2, 1 + \sqrt{m}, 1 - \sqrt{m}, (1 - m)/2 \rangle.$$

As $m \equiv 3 \pmod 4$, $(1 - m)/2$ is an odd integer, and $I = \langle 1 \rangle$. Hence $\langle 2 \rangle = P^2$ and this case falls under (b).

(vi): $p = 2$, $m \equiv 1 \pmod 8$. Let

$$P_1 = \langle 2, \frac{1 + \sqrt{m}}{2} \rangle \text{ and } P_2 = \langle 2, \frac{1 - \sqrt{m}}{2} \rangle.$$

Thus

$$P_1 P_2 = \langle 4, 2\left(\frac{1 + \sqrt{m}}{2}\right), 2\left(\frac{1 - \sqrt{m}}{2}\right), \frac{1 - m}{4} \rangle = \langle 2 \rangle I,$$

where I is the ideal

$$I = \langle 2, \frac{1 + \sqrt{m}}{2}, \frac{1 - \sqrt{m}}{2}, \frac{1 - m}{8} \rangle.$$

Now

$$1 = \frac{1 + \sqrt{m}}{2} + \frac{1 - \sqrt{m}}{2} \in I$$

so that $I = \langle 1 \rangle$ and thus $\langle 2 \rangle = P_1 P_2$. If $P_1 = P_2$ then

$$1 = \frac{1 + \sqrt{m}}{2} + \frac{1 - \sqrt{m}}{2} \in P_1,$$

which is impossible, so $P_1 \neq P_2$, and this case falls under (a).

(vii): $p = 2$, $m \equiv 5 \pmod 8$. Suppose that $\langle 2 \rangle = P_1 P_2$ ($P_1 \neq P_2$) or P_1^2. Then in both cases $N(P_1) = 2$ so that $f_K(P_1) = 1$. Hence, as $(1 + \sqrt{m})/2 \in O_K$, there exists a rational integer a such that

$$\frac{1 + \sqrt{m}}{2} \equiv a \ (\text{mod } P_1).$$

Thus

$$\frac{1 - \sqrt{m}}{2} = 1 - \left(\frac{1 + \sqrt{m}}{2} \right) \equiv 1 - a \ (\text{mod } P_1)$$

and so

$$\frac{1 - m}{4} = \left(\frac{1 + \sqrt{m}}{2} \right) \left(\frac{1 - \sqrt{m}}{2} \right) \equiv a(1 - a) \ (\text{mod } P_1).$$

Since $(1 - m)/4 \in \mathbb{Z}$, $a(1 - a) \in \mathbb{Z}$, and $P_1 \mid \langle 2 \rangle$, we have

$$\frac{1 - m}{4} \equiv a(1 - a) \ (\text{mod } 2).$$

Thus

$$\frac{1 - m}{4} \equiv 0 \ (\text{mod } 2)$$

so that $m \equiv 1 \ (\text{mod } 8)$, contradicting $m \equiv 5 \ (\text{mod } 8)$. Hence $\langle 2 \rangle$ is inert in O_K and this case falls under (c). ∎

From the proof of Theorem 10.2.1 we see that we can express the factorizations of the principal ideal $\langle p \rangle$ (with p prime) into prime ideals of O_K, where K is the quadratic field $\mathbb{Q}(\sqrt{m})$ (m squarefree), as follows:

$$\langle 2 \rangle = \begin{cases} \langle 2 \rangle, & \text{if } m \equiv 5 \ (\text{mod } 8), \\ \langle 2, \frac{1}{2}(1 + \sqrt{m}) \rangle \langle 2, \frac{1}{2}(1 - \sqrt{m}) \rangle, & \text{if } m \equiv 1 \ (\text{mod } 8), \\ \langle 2, 1 + \sqrt{m} \rangle^2, & \text{if } m \equiv 3 \ (\text{mod } 4), \\ \langle 2, \sqrt{m} \rangle^2, & \text{if } m \equiv 2 \ (\text{mod } 4), \end{cases}$$

and for $p > 2$

$$\langle p \rangle = \begin{cases} \langle p \rangle, & \text{if } p \nmid m \text{ and } x^2 \equiv m \ (\text{mod } p) \\ & \quad \text{is insolvable,} \\ \langle p, x + \sqrt{m} \rangle \langle p, x - \sqrt{m} \rangle & \text{if } p \nmid m \text{ and } x^2 \equiv m \ (\text{mod } p) \\ & \quad \text{is solvable,} \\ \langle p, \sqrt{m} \rangle^2, & \text{if } p \mid m. \end{cases}$$

Recalling that for a squarefree integer m

$$d(\mathbb{Q}(\sqrt{m})) = \begin{cases} m, & \text{if } m \equiv 1 \ (\text{mod } 4), \\ 4m, & \text{if } m \equiv 2, 3 \ (\text{mod } 4), \end{cases}$$

we see from Theorem 10.2.1 that

$$p \text{ ramifies in } \mathbb{Q}(\sqrt{m}) \iff p \mid d(\mathbb{Q}(\sqrt{m})),$$

in agreement with Theorem 10.1.5.

We next simplify the statement of Theorem 10.2.1 by introducing the Kronecker symbol, which is an extension of the Legendre symbol from an odd prime to the prime 2.

Definition 10.2.1 (Kronecker symbol) *Let d be a nonsquare integer with $d \equiv 0$ or 1 (mod 4). The Kronecker symbol $\left(\frac{d}{2}\right)$ is defined by*

$$\left(\frac{d}{2}\right) = \begin{cases} 0, & \text{if } d \equiv 0 \text{ (mod 4)}, \\ 1, & \text{if } d \equiv 1 \text{ (mod 8)}, \\ -1, & \text{if } d \equiv 5 \text{ (mod 8)}. \end{cases}$$

Thus $\left(\frac{-4}{2}\right) = 0$, $\left(\frac{17}{2}\right) = 1$, $\left(\frac{-3}{2}\right) = -1$, and $\left(\frac{-6}{2}\right)$ is not defined. Making use of the Kronecker symbol, Theorem 10.2.1 can be reformulated as follows:

Theorem 10.2.2 *Let K be a quadratic field. Let $d = d(K)$. Let p be a rational prime. Then*

$$(i) \quad \langle p \rangle \text{ splits} \iff \left(\frac{d}{p}\right) = 1,$$

$$(ii) \quad \langle p \rangle \text{ ramifies} \iff \left(\frac{d}{p}\right) = 0,$$

$$(iii) \quad \langle p \rangle \text{ is inert} \iff \left(\frac{d}{p}\right) = -1,$$

where $\left(\frac{d}{p}\right)$ is the Legendre symbol for $p > 2$ and the Kronecker symbol for $p = 2$.

Example 10.2.1

(a) $\langle 11 \rangle$ *is inert in* $\mathbb{Q}(\sqrt{-163})$ *as*

$$\left(\frac{-163}{11}\right) = \left(\frac{2}{11}\right) = -1.$$

(b) $\langle 23 \rangle$ *is inert in* $\mathbb{Q}(\sqrt{37})$ *as*

$$\left(\frac{37}{23}\right) = \left(\frac{14}{23}\right) = \left(\frac{2}{23}\right)\left(\frac{7}{23}\right) = (+1)(-1) = -1.$$

(c) $\langle 2 \rangle$ *ramifies in* $\mathbb{Q}(\sqrt{7})$ *as*

$$\left(\frac{28}{2}\right) = 0.$$

Indeed $\langle 2 \rangle = \langle 2, 1 + \sqrt{7} \rangle^2$.

(d) $\langle 3 \rangle$ *splits in* $\mathbb{Q}(\sqrt{7})$ *as*

$$\left(\frac{28}{3}\right) = \left(\frac{1}{3}\right) = 1.$$

Indeed as $x^2 \equiv 7 \,(\mathrm{mod}\ 3)$ *has the solution* $x = 1$, *we have*

$$\langle 3 \rangle = \langle 3, 1 + \sqrt{7} \rangle \langle 3, 1 - \sqrt{7} \rangle.$$

(e) $\langle 7 \rangle$ *ramifies in* $\mathbb{Q}(\sqrt{7})$ *as* $\left(\frac{28}{7}\right) = 0$. *Indeed* $\langle 7 \rangle = \langle \sqrt{7} \rangle^2$.

Let K be the quadratic field $\mathbb{Q}(\sqrt{m})$, where m is a squarefree integer. There are exactly two monomorphisms $: K \longrightarrow \mathbb{C}$, namely, 1 and σ given by

$$1(a + b\sqrt{m}) = a + b\sqrt{m},$$
$$\sigma(a + b\sqrt{m}) = a - b\sqrt{m},$$

for all $a, b \in \mathbb{Q}$. Let I be an ideal of O_K. By Theorem 8.5.1 we know that I is generated by at most two elements. Hence $I = \langle \alpha \rangle$ or $I = \langle \alpha, \beta \rangle$ and we define the conjugate ideal $\sigma(I)$ of I by

$$\sigma(I) = \langle \sigma(\alpha) \rangle \text{ or } \langle \sigma(\alpha), \sigma(\beta) \rangle$$

respectively. It is customary to write α' for $\sigma(\alpha)$ and I' for $\sigma(I)$. Clearly $(\alpha')' = \sigma^2(\alpha) = \alpha$ so that $(I')' = I$. Thus if $I = \langle 2 + \sqrt{3}, 1 - 2\sqrt{3} \rangle$ then $I' = \langle 2 - \sqrt{3}, 1 + 2\sqrt{3} \rangle$. We note that

$$(\alpha + \beta)' = \sigma(\alpha + \beta) = \sigma(\alpha) + \sigma(\beta) = \alpha' + \beta'$$

and

$$(\alpha\beta)' = \sigma(\alpha\beta) = \sigma(\alpha)\sigma(\beta) = \alpha'\beta'$$

for all α and β in K. From these properties it is easy to show that if $I = \langle \alpha_1, \ldots, \alpha_n \rangle$ then $I' = \langle \alpha_1', \ldots, \alpha_n' \rangle$.

The basic property of conjugate ideals is given in the next theorem.

Theorem 10.2.3 *Let I and J be ideals of the ring of integers of a quadratic field K. Then*

$$(IJ)' = I'J'.$$

Proof: As $\langle \alpha \rangle = \langle \alpha, \alpha \rangle$ we may suppose that both I and J are generated by two elements, say,

$$I = \langle \alpha, \beta \rangle, \quad J = \langle \gamma, \delta \rangle.$$

Then

$$IJ = \langle \alpha, \beta \rangle \langle \gamma, \delta \rangle = \langle \alpha\gamma, \beta\gamma, \alpha\delta, \beta\delta \rangle$$

so that

$$(IJ)' = \langle(\alpha\gamma)', (\beta\gamma)', (\alpha\delta)', (\beta\delta)'\rangle = \langle\alpha'\gamma', \beta'\gamma', \alpha'\delta', \beta'\delta'\rangle$$
$$= \langle\alpha', \beta'\rangle\langle\gamma', \delta'\rangle = I'J'$$

as asserted. ∎

If I is an ideal of the ring of integers of an arbitrary algebraic number field K, we have already observed that $I \mid \langle N(I)\rangle$. In the case of quadratic fields, we can say something stronger.

Theorem 10.2.4 *Let K be a quadratic field. Let I be an ideal of O_K. Then*

$$\langle N(I)\rangle = II'.$$

Proof: The assertion of the theorem is trivial if $I = \langle 0\rangle$ or $\langle 1\rangle$, so we may suppose that $I \neq \langle 0\rangle, \langle 1\rangle$. Let

$$I = P_1^{a_1} P_1'^{b_1} \cdots P_r^{a_r} P_r'^{b_r} Q_1^{c_1} \cdots Q_s^{c_s} R_1^{d_1} \cdots R_t^{d_t}$$

be the prime ideal decomposition of I, where $P_1, \ldots, P_r$ are distinct prime ideals such that

$$PP' = \langle p\rangle, \quad N(P) = N(P') = p, \quad P \neq P',$$

$Q_1, \ldots, Q_s$ are distinct prime ideals such that

$$Q = Q' = \langle q\rangle, \quad N(Q) = q^2,$$

$R_1, \ldots, R_t$ are distinct prime ideals such that

$$R = R', \quad R^2 = \langle r\rangle, \quad N(R) = r,$$

and p, q, r denote rational primes. Then, by Theorem 10.2.3, we have

$$I' = P_1'^{a_1} P_1^{b_1} \cdots P_r'^{a_r} P_r^{b_r} Q_1^{c_1} \cdots Q_s^{c_s} R_1^{d_1} \cdots R_t^{d_t}.$$

Hence

$$II' = P_1^{a_1+b_1} P_1'^{a_1+b_1} \cdots P_r^{a_r+b_r} P_r'^{a_r+b_r} Q_1^{2c_1} \cdots Q_s^{2c_s} R_1^{2d_1} \cdots R_t^{2d_t}$$
$$= \langle p_1\rangle^{a_1+b_1} \cdots \langle p_r\rangle^{a_r+b_r} \langle q_1\rangle^{2c_1} \cdots \langle q_s\rangle^{2c_s} \langle r_1\rangle^{d_1} \cdots \langle r_t\rangle^{d_t}$$
$$= \langle p_1^{a_1+b_1} \cdots p_r^{a_r+b_r} q_1^{2a} \cdots q_s^{2c_s} r_1^{d_1} \cdots r_t^{d_t}\rangle.$$

Further, by Theorem 9.3.2, we have

$$N(I) = N(P_1)^{a_1} N(P_1')^{b_1} \cdots N(P_r)^{a_r} N(P_r')^{b_r} N(Q_1)^{c_1} \cdots$$
$$N(Q_s)^{c_s} N(R_1)^{d_1} \cdots N(R_t)^{d_t}$$
$$= p_1^{a_1} p_1^{b_1} \cdots p_r^{a_r} p_r^{b_r} q_1^{2c_1} \cdots q_s^{2c_s} r_1^{d_1} \cdots r_t^{d_t}.$$

Thus

$$II' = \langle N(I) \rangle$$

as asserted. ∎

By Theorem 10.2.4 we have for a nonzero ideal I of O_K

$$\langle N(I) \rangle = II' = I'(I')' = \langle N(I') \rangle$$

so that

$$N(I) = \epsilon N(I')$$

for some unit $\epsilon \in O_K$. But $N(I)$ and $N(I')$ are both positive integers so that $\epsilon = +1$ and thus

$$N(I) = N(I').$$

This result is trivially true if $I = \langle 0 \rangle$.

We close this section by noting that (in the notation of Theorem 10.2.2): If $\left(\frac{d}{p}\right)$ $= 1$ then $\langle p \rangle = PP'$ for some prime ideal P with $P \neq P'$, $N(P) = N(P') = p$; if $\left(\frac{d}{p}\right) = 0$ then $\langle p \rangle = P^2$ for some prime ideal P with $P = P'$, $N(P) = p$; and if $\left(\frac{d}{p}\right) = -1$ then $\langle p \rangle = P$ for some prime ideal P with $P = P'$, $N(P) = p^2$.

An ideal I such that $I = I'$ is called self-conjugate.

10.3 Factoring Primes in a Monogenic Number Field

Let K be an algebraic number field. Recall that K is said to be monogenic (Definition 7.1.5) if there exists $\theta \in O_K$ such that

$$O_K = \mathbb{Z} + \mathbb{Z}\theta + \cdots + \mathbb{Z}\theta^{n-1},$$

where $[K : \mathbb{Q}] = n$. The next theorem shows how to factor $\langle p \rangle$ (with p a rational prime) into prime ideals in a monogenic number field. It was originally proved by Dedekind [3] in 1878.

Theorem 10.3.1 *Let $K = \mathbb{Q}(\theta)$ be an algebraic number field of degree n such that*

$$O_K = \mathbb{Z} + \mathbb{Z}\theta + \cdots + \mathbb{Z}\theta^{n-1}.$$

Let p be a rational prime. Let

$$f(x) = \mathrm{irr}_{\mathbb{Q}}\theta \in \mathbb{Z}[x].$$

Let $\bar{}$ denote the natural map $: \mathbb{Z}[x] \to \mathbb{Z}_p[x]$, where $\mathbb{Z}_p = \mathbb{Z}/p\mathbb{Z}$. Let

$$\bar{f}(x) = g_1(x)^{e_1} \cdots g_r(x)^{e_r},$$

where $g_1(x), \ldots, g_r(x)$ are distinct monic irreducible polynomials in $\mathbb{Z}_p[x]$ and $e_1, \ldots, e_r$ are positive integers. For $i = 1, 2, \ldots, r$ let $f_i(x)$ be any monic polynomial of $\mathbb{Z}[x]$ such that $\bar{f}_i = g_i$. Set

$$P_i = \langle p, f_i(\theta) \rangle, \quad i = 1, 2, \ldots, r.$$

Then $P_1, \ldots, P_r$ are distinct prime ideals of O_K with

$$\langle p \rangle = P_1^{e_1} \cdots P_r^{e_r}$$

and

$$N(P_i) = p^{\deg f_i}, \quad i = 1, 2, \ldots, r.$$

Proof: For $i = 1, 2, \ldots, r$ let θ_i be a root of g_i in a suitable extension field of $\mathbb{Z}_p$. This extension field is the finite field $\mathbb{Z}_p[\theta_i] \simeq \mathbb{Z}_p[x]/\langle g_i(x) \rangle$. Let $v_i : \mathbb{Z}[\theta] \to \mathbb{Z}_p[\theta_i]$ be the surjective homomorphism given by

$$v_i(h(\theta)) = \bar{h}(\theta_i).$$

Then

$$\mathbb{Z}[\theta]/\ker v_i \simeq v_i(\mathbb{Z}[\theta]) = \mathbb{Z}_p[\theta_i]$$

is a field, so that $\ker v_i$ is a prime ideal of $\mathbb{Z}[\theta] = O_K$.

Clearly

$$v_i(p) = 0, \quad v_i(f_i(\theta)) = \bar{f}_i(\theta_i) = g_i(\theta_i) = 0,$$

so that

$$p \in \ker v_i, \quad f_i(\theta) \in \ker v_i,$$

and thus

$$\langle p, f_i(\theta) \rangle \subseteq \ker v_i.$$

If $g(\theta) \in \ker v_i$ then

$$\bar{g}(\theta_i) = v_i(g(\theta)) = 0$$

so that $g_i(x) \mid \bar{g}(x)$ in $\mathbb{Z}_p[x]$. Thus

$$\bar{g}(x) = \bar{f}_i(x)\bar{h}(x) \text{ for some } \bar{h} \in \mathbb{Z}_p[x].$$

Hence $(g - f_i h)(x) \in \mathbb{Z}[x]$ has coefficients that are divisible by p so that

$$\begin{aligned}
g(\theta) &= (g(\theta) - f_i(\theta)h(\theta)) + f_i(\theta)h(\theta) \\
&\in \langle p \rangle + \langle f_i(\theta) \rangle \\
&= \langle p, f_i(\theta) \rangle,
\end{aligned}$$

proving

$$\ker v_i \subseteq \langle p, f_i(\theta) \rangle.$$

We have shown that

$$P_i = \langle p, f_i(\theta) \rangle = \ker v_i, \quad i = 1, 2, \ldots, r.$$

Thus each P_i $(i = 1, 2, \ldots, r)$ is a prime ideal of O_K.

Next we show that the prime ideals P_i $(i = 1, 2, \ldots, r)$ are distinct. Suppose that $P_i = P_j$ for some $i, j \in \{1, 2, \ldots, r\}$. Then $\langle p, f_i(\theta) \rangle = \langle p, f_j(\theta) \rangle$. Hence

$$f_j(\theta) = pg(\theta) + f_i(\theta)h(\theta)$$

for some $g(x), h(x) \in \mathbb{Z}[x]$. Applying v_i we obtain

$$g_j(\theta_i) = \bar{f}_j(\theta_i) = \bar{f}_i(\theta_i)\bar{h}(\theta_i) = g_i(\theta_i)\bar{h}(\theta_i) = 0,$$

so that $g_i(x) \mid g_j(x)$ in $\mathbb{Z}_p[x]$. Hence

$$g_j(x) = g_i(x)l(x)$$

for some $l(x) \in \mathbb{Z}_p[x]$. As $g_i(x)$ and $g_j(x)$ are both monic polynomials, which are irreducible in $\mathbb{Z}_p[x]$, we have $l(x) = 1$ so that $g_i(x) = g_j(x)$ and thus $i = j$.

We show next that

$$\langle p \rangle = P_1^{e_1} \cdots P_r^{e_r}.$$

For any ideals A, B_1, B_2 we have

$$(A + B_1)(A + B_2) \subseteq A + B_1 B_2,$$

so that

$$
\begin{aligned}
P_1^{e_1} \cdots P_r^{e_r} &= \langle p, f_1(\theta) \rangle^{e_1} \cdots \langle p, f_r(\theta) \rangle^{e_r} \\
&= (\langle p \rangle + \langle f_1(\theta) \rangle)^{e_1} \cdots (\langle p \rangle + \langle f_r(\theta) \rangle)^{e_r} \\
&\subseteq \langle p \rangle + \langle f_1(\theta) \rangle^{e_1} \cdots \langle f_r(\theta) \rangle^{e_r} \\
&= \langle p \rangle + \langle f_1(\theta)^{e_1} \cdots f_r(\theta)^{e_r} \rangle \\
&= \langle p \rangle + \langle f(\theta) \rangle \\
&= \langle p \rangle
\end{aligned}
$$

and so

$$\langle p \rangle \mid P_1^{e_1} \cdots P_r^{e_r}.$$

Now $P_i = \langle p, f_i(\theta) \rangle \supseteq \langle p \rangle$, so

$$P_i \mid \langle p \rangle, \quad i = 1, 2, \ldots, r.$$

Hence

$$\langle p \rangle = P_1^{k_1} \cdots P_r^{k_r},$$

where

$$k_i \in \{1, 2, \ldots, e_i\}, \quad i = 1, 2, \ldots, r. \tag{10.3.1}$$

Now

$$O_K / P_i = \mathbb{Z}[\theta]/P_i = \mathbb{Z}[\theta]/\ker \nu_i \simeq \nu_i(\mathbb{Z}[\theta]) = \mathbb{Z}_p[\theta_i],$$

so that

$$N(P_i) = \mathrm{card}(O_K / P_i) = \mathrm{card}(\mathbb{Z}_p[\theta_i]) = p^{d_i},$$

where

$$d_i = \deg g_i = \deg \bar{f}_i.$$

Hence we have

$$
\begin{aligned}
p^n = N(\langle p \rangle) &= N(P_1^{k_1} \cdots P_r^{k_r}) \\
&= N(P_1)^{k_1} \cdots N(P_r)^{k_r} \\
&= (p^{d_1})^{k_1} \cdots (p^{d_r})^{k_r} \\
&= p^{d_1 k_1 + \cdots + d_r k_r},
\end{aligned}
$$

so that

$$d_1 k_1 + \cdots + d_r k_r = n. \tag{10.3.2}$$

Comparing degrees in

$$\bar{f}(x) = \bar{f}_1(x)^{e_1} \cdots \bar{f}_r(x)^{e_r}$$

we obtain

$$d_1 e_1 + \cdots + d_r e_r = n. \tag{10.3.3}$$

From (10.3.1)–(10.3.3) we deduce that

$$k_i = e_i, \quad i = 1, 2, \ldots, r,$$

so that

$$\langle p \rangle = P_1^{e_1} \cdots P_r^{e_r},$$

as asserted.

Finally, we observe that

$$N(P_i) = p^{d_i} = p^{\deg \bar{f}_i} = p^{\deg f_i}, \quad i = 1, 2, \ldots, r. \qquad \blacksquare$$

Theorem 10.3.1 relates the factorization of a monic irreducible polynomial $f(x) \in \mathbb{Z}[x]$ modulo a prime p to the factorization of p into prime ideals in the algebraic number field K defined by a root of $f(x)$ when K is monogenic. Primes p for which the congruence $f(x) \equiv 0 \pmod{p}$ is solvable, so that $f(x)$ has at least one linear factor modulo p, are called prime divisors of f and the set of prime divisors of f is denoted by $P(f)$. Thus

$$P(x^2 + 1) = \{2, p \text{ (prime)} \equiv 1 \pmod 4\}.$$

The set $P(f)$ is discussed in the beautiful article by Gerst and Brillhart [4].

If $f(x)$ factors modulo p into a product of distinct linear factors, we say that $f(x)$ splits completely modulo p. The set of all primes p such that $f(x)$ splits completely modulo p is denoted by $\mathrm{Spl}(f)$. This set is discussed by Wyman in his classic article [7]. Thus for example

$$\mathrm{Spl}(x^3 - 31x + 62) = \{p \text{ (prime} > 2) \equiv 1, 2, 4, 8, 15, 16, 23, 27, 29, 30$$
$$\pmod{31}\}$$

(see [5]).

The next section will be devoted to numerical examples illustrating Theorem 10.3.1.

10.4 Some Factorizations in Cubic Fields

Example 10.4.1 *We factor $\langle 5 \rangle$ as a product of prime ideals in O_K, where $K = \mathbb{Q}(\sqrt[3]{2})$. Set $\theta = \sqrt[3]{2}$. We have seen in Example 7.1.6 that $\{1, \theta, \theta^2\}$ is an integral basis for $K = \mathbb{Q}(\theta)$ so that K is monogenic. The minimal polynomial of θ over $\mathbb{Q}$ is $x^3 - 2$. We have*

$$x^3 - 2 = (x + 2)(x^2 + 3x + 4) \pmod 5,$$

where $x + 2$ and $x^2 + 3x + 4$ are irreducible (mod 5). *Hence, by Theorem 10.3.1, we have*

$$\langle 5 \rangle = PQ,$$

where

$$P = \langle 5, \theta + 2 \rangle, \quad Q = \langle 5, \theta^2 + 3\theta + 4 \rangle$$

are distinct prime ideals with

$$N(P) = 5, \quad N(Q) = 5^2 = 25.$$

As a check on the calculation in Example 10.4.1 we compute PQ directly.

We have

$$
\begin{aligned}
PQ &= \langle 5, \theta + 2 \rangle \langle 5, \theta^2 + 3\theta + 4 \rangle \\
&= \langle 25, 5(\theta + 2), 5(\theta^2 + 3\theta + 4), \theta^3 + 5\theta^2 + 10\theta + 8 \rangle \\
&= \langle 25, 5(\theta + 2), 5(\theta^2 + 3\theta + 4), 5\theta^2 + 10\theta + 10 \rangle \\
&= \langle 5 \rangle \langle 5, \theta + 2, \theta^2 + 3\theta + 4, \theta^2 + 2\theta + 2 \rangle \\
&= \langle 5 \rangle
\end{aligned}
$$

as

$$
1 = 1 \cdot 5 + (2\theta + 2)(\theta + 2) - 2(\theta^2 + 3\theta + 4).
$$

Example 10.4.2 *Let $K = \mathbb{Q}(\theta)$, where $\theta^3 - 9\theta - 6 = 0$. It is known that $[K : \mathbb{Q}] = 3$, $\{1, \theta, \theta^2\}$ is an integral basis for K, and $d(K) = 2^3 \cdot 3^5$ (see Exercise 5 of Chapter 7). As $2 \mid d(K)$ and $3 \mid d(K)$, both 2 and 3 ramify in K by Dedekind's theorem (Theorem 10.1.5). We determine their prime ideal decompositions.*

We have

$$
x^3 - 9x - 6 \equiv x(x + 1)^2 \pmod{2}
$$

so that by Theorem 10.3.1 we have

$$
\langle 2 \rangle = PQ^2,
$$

where

$$
P = \langle 2, \theta \rangle, \quad Q = \langle 2, \theta + 1 \rangle
$$

are distinct prime ideals with $N(P) = N(Q) = 2$. In fact P and Q are both principal ideals as we now show. From

$$
\theta^3 - 9\theta - 6 = 0,
$$

we deduce that

$$
(\theta + 1)^3 - 3(\theta + 1)^2 - 6(\theta + 1) + 2 = 0,
$$

so that

$$
\theta + 1 \mid 2.
$$

Hence

$$
Q = \langle 2, \theta + 1 \rangle = \langle \theta + 1 \rangle
$$

and

$$P = \langle 2 \rangle Q^{-2}$$
$$= \langle \frac{2}{(\theta + 1)^2} \rangle$$
$$= \langle \frac{6(\theta + 1) + 3(\theta + 1)^2 - (\theta + 1)^3}{(\theta + 1)^2} \rangle$$
$$= \langle 3\frac{2}{\theta + 1} + 3 - (\theta + 1) \rangle$$
$$= \langle 3(6 + 3(\theta + 1) - (\theta + 1)^2) + 2 - \theta \rangle$$
$$= \langle 26 + 2\theta - 3\theta^2 \rangle.$$

As a check on this calculation we note that

$$PQ^2 = \langle 26 + 2\theta - 3\theta^2 \rangle \langle 1 + \theta \rangle^2$$
$$= \langle (26 + 2\theta - 3\theta^2)(1 + 2\theta + \theta^2) \rangle$$
$$= \langle 26 + 54\theta + 27\theta^2 - 4\theta^3 - 3\theta^4 \rangle$$
$$= \langle 2 \rangle,$$

as $\theta^3 = 9\theta + 6$ *and* $\theta^4 = 9\theta^2 + 6\theta$.

Turning to the prime 3 *we have*

$$x^3 - 9x - 6 \equiv x^3 \text{ (mod 3)},$$

so that by Theorem 10.3.1 we obtain

$$\langle 3 \rangle = R^3,$$

where $R = \langle 3, \theta \rangle$ *is a prime ideal with* $N(R) = 3$. *We show that* R *is a principal ideal. We have*

$$PR = \langle 2, \theta \rangle \langle 3, \theta \rangle = \langle 6, 2\theta, 3\theta, \theta^2 \rangle = \langle 6, \theta \rangle = \langle \theta^3 - 9\theta, \theta \rangle = \langle \theta \rangle,$$

so that

$$R = \langle \theta \rangle P^{-1}$$
$$= \langle \theta \rangle Q^2 (PQ^2)^{-1}$$
$$= \langle \frac{\theta(\theta + 1)^2}{2} \rangle$$
$$= \langle \frac{\theta^3 + 2\theta^2 + \theta}{2} \rangle = \langle \frac{2\theta^2 + 10\theta + 6}{2} \rangle$$
$$= \langle 3 + 5\theta + \theta^2 \rangle.$$

We now verify directly that $R^3 = \langle 3 \rangle$. We have

$$(3 + 5\theta + \theta^2)^2 = 9 + 30\theta + 31\theta^2 + 10\theta^3 + \theta^4$$
$$= 69 + 126\theta + 40\theta^2,$$

$$(3 + 5\theta + \theta^2)^3 = (3 + 5\theta + \theta^2)(69 + 126\theta + 40\theta^2)$$
$$= 207 + 723\theta + 819\theta^2 + 326\theta^3 + 40\theta^4$$
$$= 2163 + 3897\theta + 1179\theta^2$$
$$= 3(721 + 1299\theta + 393\theta^2).$$

To complete the verification of $R^3 = \langle 3 \rangle$ we must show that $721 + 1299\theta + 393\theta^2$ is a unit of $O_K = \mathbb{Z} + \mathbb{Z}\theta + \mathbb{Z}\theta^2$ so that

$$R^3 = \langle 3 + 5\theta + \theta^2 \rangle^3 = \langle 3 \rangle \langle 721 + 1299\theta + 393\theta^2 \rangle = \langle 3 \rangle.$$

We do this by seeking $a, b, c \in \mathbb{Z}$ such that

$$(721 + 1299\theta + 393\theta^2)(a + b\theta + c\theta^2) = 1.$$

Multiplying out the left-hand side, replacing θ^3 and θ^4 by $9\theta + 6$ and $9\theta^2 + 6\theta$ respectively, and equating the coefficients of $1, \theta, \theta^2$, we are led to the three linear equations in a, b, c :

$$721a + 2358b + 7794c = 1,$$
$$1299a + 4258b + 14049c = 0,$$
$$393a + 1299b + 4258c = 0.$$

Using the program MAPLE we find that

$$a = -119087, \ b = -9885, \ c = 14007,$$

which is easily checked directly. Hence

$$(721 + 1299\theta + 393\theta^2)^{-1} = -119087 - 9885\theta + 14007\theta^2,$$

so that $721 + 1299\theta + 393\theta^2$ is a unit. Alternatively, we could have used Theorem 9.2.4.

This example was considered in [2: p. 230]. However, the value of $N(\theta + 1)$ is given there incorrectly as -4 (see Eq. (7.16)). Its correct value is

$$N(\theta + 1) = (\theta + 1)(\theta' + 1)(\theta'' + 1)$$
$$= \theta\theta'\theta'' + (\theta\theta' + \theta'\theta'' + \theta''\theta) + (\theta + \theta' + \theta'') + 1$$
$$= 6 - 9 + 0 + 1$$
$$= -2,$$

where θ, θ', θ'' are the roots of $x^3 - 9x - 6$. The factorization into prime ideals of the principal ideals considered in [2] are

$$\langle 2 \rangle = PQ^2, \ \langle 3 \rangle = R^3, \ \langle \theta \rangle = PR, \ \langle \theta + 1 \rangle = Q, \ \langle \theta - 1 \rangle = QS,$$

where

$$P = \langle 2, \theta \rangle = \langle 26 + 2\theta - 3\theta^2 \rangle, \ Q = \langle 2, \theta + 1 \rangle = \langle \theta + 1 \rangle,$$
$$R = \langle 3, \theta \rangle = \langle 3 + 5\theta + \theta^2 \rangle, \ S = \langle 7 + \theta - \theta^2 \rangle.$$

10.5 Factoring Primes in an Arbitrary Number Field

Theorem 10.3.1 was actually proved by Dedekind in the following slightly stronger form. For all but at most a finite number of primes, Theorem 10.5.1 gives the factorization of a prime into prime ideals in an arbitrary algebraic number field.

Theorem 10.5.1 *Let $K = \mathbb{Q}(\theta)$ be an algebraic number field with $\theta \in O_K$. Let p be a rational prime. Let*

$$f(x) = \text{irr}_{\mathbb{Q}}(\theta) \in \mathbb{Z}[x].$$

Let $^-$ denote the natural map $: \mathbb{Z}[x] \longrightarrow \mathbb{Z}_p[x]$, where $\mathbb{Z}_p = \mathbb{Z}/p\mathbb{Z}$. Let

$$\bar{f}(x) = g_1(x)^{e_1} \cdots g_r(x)^{e_r},$$

where $g_1(x), \ldots, g_r(x)$ are distinct monic irreducible polynomials in $\mathbb{Z}_p[x]$ and $e_1, \ldots, e_r$ are positive integers. For $i = 1, 2, \ldots, r$ let $f_i(x)$ be any monic polynomial of $\mathbb{Z}[x]$ such that $\bar{f}_i = g_i$. Set

$$P_i = \langle p, f_i(\theta) \rangle, \ i = 1, 2, \ldots, r.$$

If $\text{ind}(\theta) \not\equiv 0 \,(\text{mod } p)$ then $P_1, \ldots, P_r$ are distinct prime ideals of O_K with

$$\langle p \rangle = P_1^{e_1} \cdots P_r^{e_r}$$

and

$$N(P_i) = p^{\deg f_i}, \ i = 1, 2, \ldots, r.$$

We leave the proof of Theorem 10.5.1 as an exercise (Exercise 2) since it can be modeled on the proof of Theorem 10.3.1. We note that Theorem 10.3.1 is the special case ind $\theta = 1$ of Theorem 10.5.1.

Example 10.5.1 *Let K be the cubic field $\mathbb{Q}(\theta)$, where $\theta^3 - \theta + 4 = 0$. An integral basis for K is $\{1, \theta, (\theta + \theta^2)/2\}$ (Example 7.3.7). This basis is not a power basis and at this stage we do not know whether K is monogenic or not. As ind $\theta = 2$ we can apply Theorem 10.5.1 to obtain the prime ideal factorization of any prime*

$p \neq 2$ *in* O_K. *The prime* 107 *ramifies in* O_K *as* $d(K) = -107$ (*Example 7.3.7*) *and we make use of Theorem 10.5.1 to find its precise prime decomposition. We have*

$$x^3 - x + 4 \equiv (x - 6)^2(x + 12) \,(\text{mod } 107),$$

so that

$$\langle 107 \rangle = P^2 Q,$$

where P *and* Q *are the distinct prime ideals given by*

$$P = \langle 107, \theta - 6 \rangle, \quad N(P) = 107,$$

and

$$Q = \langle 107, \theta + 12 \rangle, \quad N(Q) = 107.$$

We next show that K *is monogenic. Let* $\alpha = (\theta + \theta^2)/2$. *Then*

$$\alpha^2 = \left(\frac{\theta + \theta^2}{2}\right)^2 = \frac{\theta^2 + 2\theta^3 + \theta^4}{4}$$

$$= \frac{\theta^2 + 2(\theta - 4) + (\theta^2 - 4\theta)}{4} = -2 - \theta + \frac{\theta + \theta^2}{2},$$

so that

$$\theta = -2 + \alpha - \alpha^2.$$

Hence

$$O_K = \mathbb{Z} + \mathbb{Z}\theta + \mathbb{Z}\left(\frac{\theta + \theta^2}{2}\right) = \mathbb{Z} + \mathbb{Z}(-2 + \alpha - \alpha^2) + \mathbb{Z}\alpha = \mathbb{Z} + \mathbb{Z}\alpha + \mathbb{Z}\alpha^2,$$

proving that $\{1, \alpha, \alpha^2\}$ *is an integral basis for* K. *This basis is clearly a power basis, so* K *is monogenic.*

As K *is monogenic we can apply Theorem 10.3.1 to factor the prime* 2 *in* O_K. *By Example 7.3.2 we know that* $\alpha = (\theta + \theta^2)/2$ *is a root of* $x^3 - x^2 + 3x - 2 = 0$. *Thus* $K = \mathbb{Q}(\alpha)$, *where* $\alpha^3 - \alpha^2 + 3\alpha - 2 = 0$. *Now*

$$x^3 - x^2 + 3x - 2 \equiv x(x^2 + x + 1) \,(\text{mod } 2),$$

where $x^2 + x + 1$ *is irreducible* (mod 2), *so by Theorem 10.3.1 we have*

$$\langle 2 \rangle = P_1 Q_1,$$

where P_1 *and* Q_1 *are the distinct prime ideals given by*

$$P_1 = \langle 2, \alpha \rangle = \langle 2, \frac{\theta + \theta^2}{2} \rangle, \quad N(P) = 2,$$

$$Q_1 = \langle 2, 1 + \alpha + \alpha^2 \rangle = \langle 2, -1 + \theta^2 \rangle, \quad N(Q) = 4.$$

If we had in error applied Theorem 10.3.1 directly to the prime 2, we would have obtained the incorrect factorization

$$\langle 2 \rangle = \langle 2, \theta \rangle \langle 2, 1 + \theta \rangle^2,$$

as

$$x^3 - x + 4 \equiv x(x + 1)^2 \ (\mathrm{mod}\ 2),$$

showing that the condition $p \nmid \mathrm{ind}(\theta)$ is essential in Theorem 10.5.1.

If K is an algebraic number field of degree n and p is a rational prime such that $\langle p \rangle = P^n$ for some prime ideal P of O_K, then we say that p is completely ramified in K. If $K = \mathbb{Q}(\theta)$ with $\theta \in O_K$ and the polynomial $\mathrm{irr}_\mathbb{Q}\ \theta$ is p-Eisenstein then p completely ramifies.

Theorem 10.5.2 *Let $K = \mathbb{Q}(\theta)$ be an algebraic number field of degree n with $\theta \in O_K$. Let $x^n + a_{n-1}x^{n-1} + \cdots + a_1 x + a_0 \in \mathbb{Z}[x]$ be the minimal polynomial of θ over $\mathbb{Q}$. If p is a prime such that*

$$p \parallel a_0, \ p \mid a_1, \ldots, p \mid a_{n-1}$$

then

$$\langle p \rangle = P^n$$

in O_K for some prime ideal P.

Proof: Let P be a prime ideal of O_K that divides $\langle p \rangle$. As

$$\theta^n = -a_{n-1}\theta^{n-1} - \cdots - a_1\theta - a_0,$$

and each of $a_0, a_1, \ldots, a_{n-1}$ is divisible by p, we see that

$$P \mid \langle \theta \rangle^n.$$

As P is a prime ideal, we deduce that

$$P \mid \langle \theta \rangle.$$

Thus we can define positive integers r and s by

$$P^r \parallel \langle p \rangle, \ P^s \parallel \langle \theta \rangle.$$

Then, from

$$a_0 + \theta^n = -a_{n-1}\theta^{n-1} - \cdots - a_1\theta,$$

as each of $a_1, \ldots, a_{n-1}$ is divisible by p, we obtain

$$P^{r+1} \mid \langle a_0 + \theta^n \rangle = \langle a_0 \rangle + \langle \theta \rangle^n.$$

But $p \parallel a_0$, so $P^r \parallel \langle a_0 \rangle$ and thus $P^r \parallel \langle \theta \rangle^n$. Hence $r = sn$ and so $P^{sn} \parallel \langle p \rangle$. Thus $P^n \mid \langle p \rangle$ and by (10.1.3) we have $\langle p \rangle = P^n$ as asserted. ∎

Example 10.5.2 *Let K be the cubic field $\mathbb{Q}(\theta)$, where $\theta^3 - 2\theta + 2 = 0$. As $\mathrm{irr}_{\mathbb{Q}}\, \theta = x^3 - 2x + 2$ is 2-Eisenstein, by Theorem 10.5.2, we have $\langle 2 \rangle = P^3$ for some prime ideal P. Indeed $P = \langle 2, \theta \rangle$ as*

$$\langle 2, \theta \rangle^3 = \langle 8, 4\theta, 2\theta^2, \theta^3 \rangle = \langle 8, 4\theta, 2\theta^2, 2\theta - 2 \rangle$$
$$= \langle 2 \rangle \langle 4, 2\theta, \theta^2, \theta - 1 \rangle = \langle 2 \rangle = P^3,$$

as $1 = \theta^2 - (\theta + 1)(\theta - 1) \in \langle 4, 2\theta, \theta^2, \theta - 1 \rangle$.

10.6 Factoring Primes in a Cyclotomic Field

Let m be a positive integer and let ζ_m be a primitive mth root of unity. The cyclotomic field $\mathbb{Q}(\zeta_m)$ is denoted by K_m. We give (without proof) the decomposition of a rational prime p into prime ideals in O_{K_m}.

Theorem 10.6.1 *Let $m = p^r m_1$, where $r \in \mathbb{N} \cup \{0\}$, $m_1 \in \mathbb{N}$, and $p \nmid m_1$. Let h be the least positive integer such that $p^h \equiv 1 \pmod{m_1}$. Then $h \mid \phi(m_1)$ and*

$$\langle p \rangle = (P_1 P_2 \cdots P_{\phi(m_1)/h})^{\phi(p^r)},$$

where $P_1, P_2, \ldots, P_{\phi(m_1)/h}$ are distinct prime ideals with

$$N(P_i) = p^h, \quad i = 1, 2, \ldots, \phi(m_1)/h.$$

We refer the reader to Mann's book [6] for a proof of this theorem.

Example 10.6.1 *We determine the prime ideal decomposition of $\langle 3 \rangle$ in O_{K_9}. Here $p = 3$, $m = 9$, $\phi(m) = 6$, $r = 2$, $m_1 = 1$, and $h = 1$ so that by Theorem 10.6.1*

$$\langle 3 \rangle = P^6,$$

where P is a prime ideal with $N(P) = 3$.

Example 10.6.2 *We determine the prime ideal decomposition of $\langle 2 \rangle$ in O_{K_5}. Here $p = 2$, $m = 5$, $r = 0$, $m_1 = 5$, $\phi(m_1) = 4$, $\phi(p^r) = 1$, and $h = 4$ so that by Theorem 10.6.1*

$$\langle 2 \rangle = P,$$

where P is a prime ideal with $N(P) = 2^4$.

Example 10.6.3 *We determine the prime ideal decomposition of $\langle 2 \rangle$ in O_{K_7}. Here $p = 2$, $m = 7$, $r = 0$, $m_1 = 7$, $\phi(m_1) = 6$, $\phi(p^r) = 1$, and $h = 3$ so that by*

Theorem 10.6.1

$$\langle 2 \rangle = P_1 P_2,$$

where P_1 and P_2 are distinct prime ideals with

$$N(P_1) = N(P_2) = 2^3.$$

By Theorem 7.5.2 the cyclotomic field K_7 is monogenic so we can apply Theorem 10.3.1 to obtain P_1 and P_2 explicitly. We have

$$\mathrm{irr}_{\mathbb{Q}}(\zeta_7) = \frac{x^7 - 1}{x - 1} = x^6 + x^5 + x^4 + x^3 + x^2 + x + 1$$

and

$$x^6 + x^5 + x^4 + x^3 + x^2 + x + 1 \equiv (x^3 + x + 1)(x^3 + x^2 + 1) \,(\mathrm{mod}\ 2),$$

so that

$$P_1 = \langle 2, 1 + \zeta_7 + \zeta_7^3 \rangle, \quad P_2 = \langle 2, 1 + \zeta_7^2 + \zeta_7^3 \rangle.$$

Exercises

1. In Example 10.4.2 show that $721 + 1299\theta + 393\theta^2$ is a unit by finding its norm.
2. Factor $\langle 2 \rangle$ into prime ideals in $O_{\mathbb{Q}(\sqrt{47})}$.
3. Factor $\langle 6 \rangle$ into prime ideals in $O_{\mathbb{Q}(\sqrt{366})}$.
4. Factor $\langle 2 \rangle$ into prime ideals in $O_{\mathbb{Q}(\sqrt[3]{2})}$.
5. Factor $\langle 2 \rangle$ into prime ideals in $O_{\mathbb{Q}(\sqrt[3]{3})}$.
6. Factor $\langle 2 \rangle$ into prime ideals in $O_{\mathbb{Q}(\sqrt{2}+\sqrt{-1})}$.
7. Prove (10.1.2) and (10.1.3).
8. Is $\mathbb{Q}(\sqrt[3]{10})$ monogenic?
9. Modify the proof of Theorem 10.3.1 to prove Theorem 10.5.1.
10. Let $K = \mathbb{Q}(\sqrt[3]{3})$. In O_K we have $\langle 3 \rangle = P^3$, where $P = \langle \sqrt[3]{3} \rangle$ is a prime ideal of norm 3. Are there any rational primes $p \neq 3$ such that $\langle p \rangle = Q^3$ in O_K for some prime ideal Q?
11. Determine all rational primes p that ramify in $\mathbb{Q}(\sqrt[3]{6})$ together with their prime ideal factorizations.
12. Determine the prime ideal decomposition of the prime 47 in $\mathbb{Q}(\sqrt{2}, \sqrt{3})$.
13. Let $K = \mathbb{Q}(\theta)$, where $\theta^3 - \theta + 4 = 0$. The ideal $I = \langle 2, \theta \rangle$ is principal in O_K. Find a generator of I.
14. Factor $\langle 5 \rangle$ into prime ideals in O_{K_5}.
15. Factor $\langle 3 \rangle$ into prime ideals in O_{K_7}.
16. As ζ_m is a unit of $O_{\mathbb{Q}(\zeta_m)}$, we know that $N(\zeta_m) = \pm 1$. Show that the $+$ sign holds.
17. Prove that $1 + \zeta_m + \zeta_m^2 + \cdots + \zeta_m^{k-1}$ is a unit of $O_{\mathbb{Q}(\zeta_m)}$ if k is a positive integer coprime with m.
18. Let K_1 and K_2 be algebraic number fields. Suppose that the prime p is totally ramified in O_{K_1} and unramified in O_{K_2}. Prove that $K_1 \cap K_2 = \mathbb{Q}$.

19. Let $K = \mathbb{Q}(\theta)$, where $\theta^3 - \theta - 1 = 0$. Prove that $\sqrt{\theta} \notin \mathbb{Q}(\theta)$.

20. Let $\mathbb{Q}(\theta_1)$ and $\mathbb{Q}(\theta_2)$ be algebraic number fields. Prove that

$$[\mathbb{Q}(\theta_1, \theta_2) : \mathbb{Q}] \le [\mathbb{Q}(\theta_1) : \mathbb{Q}][\mathbb{Q}(\theta_2) : \mathbb{Q}].$$

21. If $[\mathbb{Q}(\theta_1) : \mathbb{Q}]$ and $[\mathbb{Q}(\theta_2) : \mathbb{Q}]$ are coprime, prove that equality holds in the inequality in Exercise 20.

22. Let p be an odd prime. Prove that

$$\sqrt{\frac{(1 - \zeta_p^j)(1 - \zeta_p^{-j})}{(1 - \zeta_p)(1 - \zeta_p^{-1})}}, \quad j = 1, 2, \ldots, p - 1,$$

are real units of $O_{\mathbb{Q}(\zeta_p)}$.

23. Let

$$\alpha = 1 + \zeta_{23}^2 + \zeta_{23}^4 + \zeta_{23}^5 + \zeta_{23}^6 + \zeta_{23}^{10} + \zeta_{23}^{11}$$

and

$$\beta = 1 + \zeta_{23} + \zeta_{23}^5 + \zeta_{23}^6 + \zeta_{23}^7 + \zeta_{23}^9 + \zeta_{23}^{11}.$$

Prove that 2 is not a prime in $O_{\mathbb{Q}(\zeta_{23})}$ by considering the divisibility of $\alpha\beta$ by 2.

24. Prove that 2 is an irreducible in $O_{\mathbb{Q}(\zeta_{23})}$.

25. What can you deduce from Exercises 23 and 24 about $O_{\mathbb{Q}(\zeta_{23})}$?

Suggested Reading

1. G. Bachman, *The decomposition of a rational prime ideal in cyclotomic fields*, American Mathematical Monthly 73 (1966), 494–497.

   An alternate proof of the way a rational prime ideal decomposes in the ring of integers of a cyclotomic field is given

2. Z. I. Borevich and I. R. Shafarevich, *Number Theory*, Academic Press, New York and London, 1966.

   Example 10.4.2 is based upon Example 2, p. 230.

3. R. Dedekind, *Über den Zusammenhang zwischen der Theorie der Ideale und der Theorie der höheren Kongruenzen*, Abh. Kgl. Ges. Wiss. Göttingen 23 (1878), 1–23. (*Gesammelte Mathematische Werke I*, pp. 202–232, Vieweg, Wiesbaden, 1930.)

   Theorem 10.3.1 is Theorem 1 on pages 212 and 213 of Dedekind's Collected Papers.

4. I. Gerst and J. Brillhart, *On the prime divisors of polynomials*, American Mathematical Monthly 78 (1971), 250–260.

   The set of all primes for which an irreducible polynomial has at least one linear factor (mod p) is considered.

5. J. G. Huard, B. K. Spearman, and K. S. Williams, *The primes for which an abelian cubic polynomial splits*, Tokyo Journal of Mathematics 17 (1994), 467–478.

   Let $x^3 + ax + b \in \mathbb{Z}[x]$ be an irreducible abelian cubic polynomial. Explicit integers $a_1, \ldots, a_n, m$ are determined such that $x^3 + ax + b \equiv 0 \pmod{p}$ has three solutions $\iff p \equiv a_1, \ldots, a_n$ (mod m) except for finitely many primes p.

6. H. B. Mann, *Introduction to Algebraic Number Theory*, Ohio State University Press, Columbus, Ohio, 1955.

   For a proof of Theorem 10.6.1, see Theorems 8.7 and 8.8 in Chapter 8.

7. B. F. Wyman, *What is a reciprocity law?*, American Mathematical Monthly 79 (1972), 571–586.

   The set of primes p for which an irreducible polynomial factors (mod p) into a product of linear factors is discussed.

11

Units in Real Quadratic Fields

11.1 The Units of $\mathbb{Z} + \mathbb{Z}\sqrt{2}$

In Theorem 5.4.3 we determined the unit group $U(O_K)$ for an imaginary quadratic field K. The objective of this chapter is to determine the structure of the unit group $U(O_K)$ for an arbitrary real quadratic field K. We show that

$$U(O_K) \simeq \mathbb{Z}_2 \times \mathbb{Z}$$

(see Theorems 11.5.1 and 11.5.2). This is accomplished by showing that there exists a unit ϵ in O_K such that every unit is of the form $\pm\epsilon^n$ ($n \in \mathbb{Z}$). We show further that there exists a unique unit $\epsilon > 1$ of O_K with this property. This unit is called the fundamental unit of O_K (or of K). In Section 11.6 we show how continued fractions can be used to determine the fundamental unit. In Chapter 13 we prove Dirichlet's unit theorem, which gives the structure of $U(O_K)$ for an arbitrary algebraic number field K.

To illustrate some of the ideas that will be involved, we begin by determining

$$U(O_{\mathbb{Q}(\sqrt{2})}) = U(\mathbb{Z} + \mathbb{Z}\sqrt{2}).$$

Theorem 11.1.1 *All the units of $\mathbb{Z} + \mathbb{Z}\sqrt{2}$ are given by $\pm(1 + \sqrt{2})^n$ ($n \in \mathbb{Z}$), so that*

$$U(\mathbb{Z} + \mathbb{Z}\sqrt{2}) \simeq \mathbb{Z}_2 \times \mathbb{Z}.$$

Proof: We begin by showing that there does not exist a unit λ of $\mathbb{Z} + \mathbb{Z}\sqrt{2}$ satisfying

$$1 < \lambda < 1 + \sqrt{2}. \tag{11.1.1}$$

Suppose on the contrary that such a unit λ exists having property (11.1.1). By Theorem 6.2.1 there are exactly two monomorphisms σ_1 and $\sigma_2 : \mathbb{Q}(\sqrt{2}) \to \mathbb{C}$. These monomorphisms are given by

$$\sigma_1(x + y\sqrt{2}) = x + y\sqrt{2}, \quad \sigma_2(x + y\sqrt{2}) = x - y\sqrt{2},$$

for all $x, y \in \mathbb{Q}$. As λ is a unit we have $\lambda \mid 1$ so that

$$1 = \lambda\mu \text{ for some } \mu \in \mathbb{Z} + \mathbb{Z}\sqrt{2}. \tag{11.1.2}$$

Set $\lambda' = \sigma_2(\lambda)$ and $\mu' = \sigma_2(\mu)$. Applying σ_2 to (11.1.2) we obtain

$$1 = \sigma_2(1) = \sigma_2(\lambda\mu) = \sigma_2(\lambda)\sigma_2(\mu) = \lambda'\mu'.$$

Hence

$$1 = (\lambda\lambda')(\mu\mu').$$

But $\lambda\lambda' \in \mathbb{Z}$ and $\mu\mu' \in \mathbb{Z}$ so that

$$\lambda\lambda' = \pm 1.$$

We consider two cases: (i) $\lambda\lambda' = 1$ and (ii) $\lambda\lambda' = -1$.

Case (i): $\lambda\lambda' = 1$. In this case by (11.1.1) we have

$$\sqrt{2} - 1 = \frac{1}{1 + \sqrt{2}} < \lambda' < 1$$

so that

$$\sqrt{2} < \lambda + \lambda' < 2 + \sqrt{2}$$

and thus

$$0.7 < \frac{1}{\sqrt{2}} < \frac{\lambda + \lambda'}{2} < 1 + \frac{1}{\sqrt{2}} < 1.8.$$

As $(\lambda + \lambda')/2 \in \mathbb{Z}$ we must have $(\lambda + \lambda')/2 = 1$. From $\lambda\lambda' = 1$ and $\lambda + \lambda' = 2$ we deduce that $\lambda = \lambda' = 1$, contradicting $\lambda > 1$.

Case (ii): $\lambda\lambda' = -1$. In this case by (11.1.1) we have

$$-1 < \lambda' < 1 - \sqrt{2}$$

so that

$$0 < \lambda + \lambda' < 2;$$

that is,

$$0 < \frac{\lambda + \lambda'}{2} < 1.$$

This is a contradiction as $(\lambda + \lambda')/2 \in \mathbb{Z}$.

This completes the proof that there are no units of $\mathbb{Z} + \mathbb{Z}\sqrt{2}$ between 1 and $1 + \sqrt{2}$.

Now let η be any unit > 1. Since there is no unit between 1 and $1 + \sqrt{2}$ we must have $\eta \geq 1 + \sqrt{2}$. Then there exists a unique positive integer n such that

$$(1 + \sqrt{2})^n \leq \eta < (1 + \sqrt{2})^{n+1}.$$

Thus

$$1 \leq \eta(1 + \sqrt{2})^{-n} < 1 + \sqrt{2}.$$

As $\eta(1 + \sqrt{-2})^{-n}$ is a unit of $\mathbb{Z} + \mathbb{Z}\sqrt{2}$, we have

$$\eta = (1 + \sqrt{2})^n, \quad n \in \mathbb{N}. \tag{11.1.3}$$

If η is a unit with $0 < \eta < 1$ then $1/\eta$ is a unit with $1/\eta > 1$. Hence, from $(11.1.3)$ we have

$$\frac{1}{\eta} = (1 + \sqrt{2})^n$$

for some $n \in \mathbb{N}$, so that

$$\eta = (1 + \sqrt{2})^{-n}, \quad n \in \mathbb{N}.$$

If η is a unit with $-1 < \eta < 0$ then $-1/\eta$ is a unit with $-1/\eta > 1$. Hence, by $(11.1.3)$, there exists $n \in \mathbb{N}$ such that

$$\frac{-1}{\eta} = (1 + \sqrt{2})^n,$$

so that

$$\eta = -(1 + \sqrt{2})^{-n}, \quad n \in \mathbb{N}.$$

If η is a unit with $\eta < -1$ then $-\eta$ is a unit with $-\eta > 1$. Hence, by $(11.1.3)$, there exists $n \in \mathbb{N}$ such that

$$-\eta = (1 + \sqrt{2})^n,$$

so that

$$\eta = -(1 + \sqrt{2})^n, \quad n \in \mathbb{N}.$$

Clearly

$$\pm 1 = \pm(1 + \sqrt{2})^0.$$

Hence every unit η is given by

$$\eta = \pm(1 + \sqrt{2})^k, \quad k \in \mathbb{Z}.$$

This completes the proof that

$$U(\mathbb{Z} + \mathbb{Z}\sqrt{2}) \simeq \mathbb{Z}_2 \times \mathbb{Z}.$$

∎

11.2 The Equation $x^2 - my^2 = 1$

In this section we show that there exist integers x and y with $(x, y) \neq (\pm 1, 0)$ such that $x^2 - my^2 = 1$, where m is a positive integer that is not a perfect square. This result tells us that

$$x + y\sqrt{m} \ (\neq \pm 1) \in U(O_K),$$

where $K = \mathbb{Q}(\sqrt{m})$.

   Euler (1707–1783) attributed to the English mathematician John Pell (1611–1685) a method of solving the equation $x^2 - my^2 = 1$ in integers x and y. Thus the equation has become known as the Pell equation. However, this method had been found by another English mathematician, William Brouncker (1620–1684), in a series of letters (1657–1658) to Pierre Fermat (1601–1665). Lagrange (1736–1813) was the first mathematician to prove that the equation $x^2 - my^2 = 1$ has infinitely many solutions in integers x and y.

Theorem 11.2.1 *Let m be a positive integer that is not a perfect square. Then there exist integers x and y with $(x, y) \neq (\pm 1, 0)$ such that*

$$x^2 - my^2 = 1.$$

Proof: Let N be a positive integer. We show first that there exist integers x and y such that

$$0 < |x - y\sqrt{m}| < \frac{1}{N}, \ 0 < y \leq N. \qquad (11.2.1)$$

We divide the interval $0 < x \leq 1$ into N subintervals $r/N < x \leq (r+1)/N$, $r = 0, 1, \ldots, N - 1$, each of the same length $1/N$. For $i = 0, 1, \ldots, N$ we define the integers x_i and y_i by

$$x_i = [i\sqrt{m}] + 1, \ y_i = i.$$

Now

$$[i\sqrt{m}] \leq i\sqrt{m} < [i\sqrt{m}] + 1$$

so that

$$x_i - 1 \leq y_i\sqrt{m} < x_i;$$

that is,

$$0 < x_i - y_i\sqrt{m} \leq 1, \ i = 0, 1, \ldots, N.$$

Thus we have $N + 1$ numbers $x_i - y_i\sqrt{m}$ lying in the interval $0 < x \leq 1$. Hence at least two of these numbers lie in the same subinterval $(r/N, (r+1)/N]$; that is, there exist integers i and j with $i \neq j$, $0 \leq i \leq N$, $0 \leq j \leq N$ such that

$$r/N < x_i - y_i \sqrt{m} \leq (r+1)/N$$

and

$$r/N < x_j - y_j \sqrt{m} \leq (r+1)/N.$$

Interchanging i and j, if necessary, we may suppose that

$$x_i - y_i \sqrt{m} \geq x_j - y_j \sqrt{m},$$

so that

$$0 \leq (x_i - y_i \sqrt{m}) - (x_j - y_j \sqrt{m}) < \frac{1}{N}.$$

We note that $y_i - y_j = i - j \neq 0$. We define the integers x and y by

$$(x, y) = \begin{cases} (x_i - x_j, y_i - y_j), & \text{if } y_i - y_j > 0, \\ (x_j - x_i, y_j - y_i), & \text{if } y_i - y_j < 0. \end{cases}$$

Thus

$$\frac{-1}{N} < x - y\sqrt{m} < \frac{1}{N}, \quad y > 0,$$

and

$$y = |y| = |y_i - y_j| = |i - j| \leq N.$$

This completes the proof of (11.2.1).

Next we show that there exist infinitely many pairs of integers (x, y) with $y \neq 0$ such that

$$0 < |x^2 - my^2| < 1 + 2\sqrt{m}. \tag{11.2.2}$$

Let N_1 be any positive integer. By (11.2.1) there exist integers x_1 and y_1 such that

$$0 < |x_1 - y_1 \sqrt{m}| < \frac{1}{N_1}, \quad 0 < y_1 \leq N_1.$$

Now let N_2 be any positive integer $> 1/|x_1 - y_1 \sqrt{m}|$. By (11.2.1) there exist integers x_2 and y_2 such that

$$0 < |x_2 - y_2 \sqrt{m}| < \frac{1}{N_2}, \quad 0 < y_2 \leq N_2.$$

Continuing in this way, after obtaining N_r, x_r, y_r $(r = 1, 2, \ldots, k-1)$, we choose N_k to be any integer $> 1/|x_{k-1} - y_{k-1}\sqrt{m}|$ and integers x_k and y_k (> 0) such that

$$0 < |x_k - y_k \sqrt{m}| < \frac{1}{N_k}, \quad 0 < y_k \leq N_k.$$

Clearly

$$0 < |x_k - y_k\sqrt{m}| < \frac{1}{N_k} < |x_{k-1} - y_{k-1}\sqrt{m}| < \frac{1}{N_{k-1}} < |x_{k-2} - y_{k-2}\sqrt{m}|$$

$$< \cdots < |x_2 - y_2\sqrt{m}| < \frac{1}{N_2} < |x_1 - y_1\sqrt{m}|$$

so that (x_k, y_k) $(k = 1, 2, \ldots)$ is an infinite sequence of pairs of integers satisfying

$$0 < |x_k - y_k\sqrt{m}| < \frac{1}{N_k} \leq \frac{1}{y_k}.$$

Hence, as m is not a perfect square and $y_k > 0$, we have

$$0 < |x_k + y_k\sqrt{m}| \leq |x_k - y_k\sqrt{m}| + 2y_k\sqrt{m} < \frac{1}{y_k} + 2y_k\sqrt{m}.$$

Then, as

$$|x_k^2 - my_k^2| = |x_k - y_k\sqrt{m}||x_k + y_k\sqrt{m}|,$$

we deduce that

$$0 < |x_k^2 - my_k^2| < \frac{1}{y_k}(\frac{1}{y_k} + 2y_k\sqrt{m}) = \frac{1}{y_k^2} + 2\sqrt{m} \leq 1 + 2\sqrt{m},$$

for $k = 1, 2, \ldots$, proving that there are infinitely many pairs of integers (x, y) with $y > 0$ satisfying (11.2.2), namely, $(x, y) = (x_k, y_k)$ $(k = 1, 2, \ldots)$.

From (11.2.2) we see that there is an integer t with $0 < |t| < 1 + 2\sqrt{m}$ for which the equation

$$x^2 - my^2 = t \qquad (11.2.3)$$

has infinitely many distinct solutions in integers x and y. Replacing x by $-x$ and y by $-y$ if necessary we see that (11.2.3) has infinitely many distinct solutions in positive integers x and y. Let t be such that (11.2.3) has infinitely many solutions in positive integers x and y, with $|t|$ minimal. As m is not a perfect square we see that $|t| > 0$. We show next that $(t, y) = 1$ for infinitely many (indeed, for all but finitely many) of these solutions. Suppose on the contrary that $(t, y) > 1$. As t has only finitely many prime factors, there exists at least one prime divisor p of t for which Eq. (11.2.3) has infinitely many solutions in integers x and y with $p \mid y$. For each such solution we have $p \mid x$ so that $p^2 \mid t$ and we conclude that the equation

$$x^2 - my^2 = t/p^2$$

has infinitely many solutions in positive integers x and y, contradicting that $|t|$ is the least such integer with this property.

Let (x, y) be one of the solutions of (11.2.3) in positive integers so that $(t, y) = 1$. Let u be the unique integer such that

$$uy \equiv 1 \pmod{t}, \quad 0 < u < |t|.$$

Since there are $|t|$ residue classes modulo t, we can find two such solutions, say (x_1, y_1) and (x_2, y_2), such that

$$u_1 x_1 \equiv u_2 x_2 \ (\text{mod } t).$$

Then we have

$$
\begin{aligned}
\frac{x_1 + \sqrt{m}\,y_1}{x_2 + \sqrt{m}\,y_2} &= \frac{(x_1 + \sqrt{m}\,y_1)(x_2 - \sqrt{m}\,y_2)}{x_2^2 - m y_2^2} \\
&= \frac{(x_1 x_2 - m y_1 y_2) + \sqrt{m}(x_2 y_1 - x_1 y_2)}{t} \\
&= x + \sqrt{m}\,y,
\end{aligned}
$$

where

$$x = (x_1 x_2 - m y_1 y_2)/t, \quad y = (x_2 y_1 - x_1 y_2)/t.$$

Clearly $x \in \mathbb{Q}$ and $y \in \mathbb{Q}$. We show that $x \in \mathbb{Z}$ and $y \in \mathbb{Z}$. We have

$$
\begin{aligned}
u_1 u_2 (x_2 y_1 - y_2 x_1) &= (u_1 y_1)(u_2 x_2) - (u_2 y_2)(u_1 x_1) \\
&\equiv u_2 x_2 - u_1 x_1 \\
&\equiv 0 \ (\text{mod } t).
\end{aligned}
$$

Now $(u_1 u_2, t) = 1$ so that

$$x_2 y_1 - y_2 x_1 \equiv 0 \ (\text{mod } t),$$

proving that $y \in \mathbb{Z}$. Similarly, we have (as $u_1 x_1 \equiv u_2 x_2 \ (\text{mod } t)$ and $u_1 y_1 \equiv u_2 y_2 \ (\text{mod } t)$)

$$
\begin{aligned}
u_1 u_2 (x_1 x_2 - m y_1 y_2) &= (u_1 x_1)(u_2 x_2) - m(u_1 y_1)(u_2 y_2) \\
&\equiv (u_1 x_1)^2 - m(u_1 y_1)^2 \\
&= u_1^2 (x_1^2 - m y_1^2) \\
&= u_1^2 t \\
&\equiv 0 \ (\text{mod } t),
\end{aligned}
$$

so that as $(u_1 u_2, t) = 1$ we have

$$x_1 x_2 - m y_1 y_2 \equiv 0 \ (\text{mod } t),$$

proving that $x \in \mathbb{Z}$. Hence

$$x_1 + \sqrt{m}\,y_1 = (x_2 + \sqrt{m}\,y_2)(x + \sqrt{m}\,y)$$

and so

$$t = x_1^2 - my_1^2 = (x_2^2 - my_2^2)(x^2 - my^2) = t(x^2 - my^2)$$

so that

$$x^2 - my^2 = 1.$$

Now if $(x, y) = (\pm 1, 0)$ then

$$x_1 + \sqrt{m}\,y_1 = \pm(x_2 + \sqrt{m}\,y_2)$$

so that

$$(x_1, y_1) = \pm(x_2, y_2).$$

But $x_1 > 0$ and $x_2 > 0$, so $(x_1, y_1) = (x_2, y_2)$, contradicting that (x_1, y_1) and (x_2, y_2) are distinct solutions of (11.2.3).

Hence we have shown the existence of a pair of integers $(x, y) \neq (\pm 1, 0)$ such that $x^2 - my^2 = 1$. ∎

11.3 Units of Norm 1

Let m be a positive squarefree integer. Theorem 11.2.1 tells us that there exist positive integers x and y such that $x^2 - my^2 = 1$. Hence $\lambda = x + y\sqrt{m}$ is a unit of O_K, where $K = \mathbb{Q}(\sqrt{m})$, such that $\lambda > 1$ and $N(\lambda) = 1$. Since $\lambda^n \to \infty$ as $n \to \infty$, O_K has infinitely many units of norm 1, namely $\{\lambda^n \mid n \in \mathbb{Z}\}$. All of these units are of the form $u + v\sqrt{m}$, where u and v are integers such that $u^2 - mv^2 = 1$. However, when $m \equiv 1 \pmod 4$, there may be units in O_K of the form $(u + v\sqrt{m})/2$, where u and v are both odd integers. For example $(3 + \sqrt{5})/2$ is a unit of norm 1 in $O_{\mathbb{Q}(\sqrt{5})}$. In contrast, $O_{\mathbb{Q}(\sqrt{17})}$ does not contain any units of the form $(u + v\sqrt{17})/2$, where u and v are both odd integers, since $u^2 - 17v^2 = \pm 4$ cannot hold modulo 8 for odd integers u and v.

Let $\lambda = x + y\sqrt{m}$ be a unit of O_K ($K = \mathbb{Q}(\sqrt{m})$) of norm 1 with x and y both integers or possibly in the case $m \equiv 1 \pmod 4$ both halves of odd integers. We now show how the signs of x and y determine to which of the four intervals $(-\infty, -1)$, $(-1, 0)$, $(0, 1)$, or $(1, \infty)$ λ belongs.

Theorem 11.3.1 *Let m be a positive squarefree integer. Let x and y both be integers or both halves of odd integers such that $x^2 - my^2 = 1$. Then*

$$x + y\sqrt{m} > 1 \Longleftrightarrow x > 0,\ y > 0, \tag{11.3.1}$$
$$0 < x + y\sqrt{m} < 1 \Longleftrightarrow x > 0,\ y < 0, \tag{11.3.2}$$
$$-1 < x + y\sqrt{m} < 0 \Longleftrightarrow x < 0,\ y > 0, \tag{11.3.3}$$
$$x + y\sqrt{m} < -1 \Longleftrightarrow x < 0,\ y < 0. \tag{11.3.4}$$

Proof: First we prove (11.3.1). We have

$$x > 0, \ y > 0 \Longrightarrow x \geq \frac{1}{2}, \ y \geq \frac{1}{2} \Longrightarrow x + y\sqrt{m} \geq \frac{1 + \sqrt{m}}{2} \geq \frac{1 + \sqrt{2}}{2} > 1.$$

Conversely, as $(x + y\sqrt{m})(x - y\sqrt{m}) = x^2 - my^2 = 1$, we have

$$x + y\sqrt{m} > 1 \Longrightarrow 0 < x - y\sqrt{m} < 1$$

$$\Longrightarrow \begin{cases} x = \dfrac{1}{2}((x + y\sqrt{m}) + (x - y\sqrt{m})) > \dfrac{1}{2} > 0, \\ y = \dfrac{1}{2\sqrt{m}}((x + y\sqrt{m}) - (x - y\sqrt{m})) > \dfrac{1 - 1}{2\sqrt{m}} = 0. \end{cases}$$

This proves (11.3.1).

Next we prove (11.3.2). We have

$$x > 0, \ y < 0 \Longrightarrow x > 0, -y > 0 \Longrightarrow x \geq \frac{1}{2}, \ -y \geq \frac{1}{2}$$

$$\Longrightarrow x - y\sqrt{m} \geq \frac{1 + \sqrt{m}}{2} \geq \frac{1 + \sqrt{2}}{2} > 1$$

$$\Longrightarrow 0 < x + y\sqrt{m} < 1,$$

as $(x - y\sqrt{m})(x + y\sqrt{m}) = 1$. Conversely,

$$0 < x + y\sqrt{m} < 1 \Longrightarrow x - y\sqrt{m} > 1$$

$$\Longrightarrow \begin{cases} x = \dfrac{1}{2}((x + y\sqrt{m}) + (x - y\sqrt{m})) > \dfrac{1}{2} > 0, \\ y = \dfrac{1}{2\sqrt{m}}((x + y\sqrt{m}) - (x - y\sqrt{m})) < \dfrac{1 - 1}{2\sqrt{m}} = 0. \end{cases}$$

This proves (11.3.2).

Finally, (11.3.4) follows from (11.3.1) and (11.3.3) follows from (11.3.2) by changing x to $-x$ and y to $-y$. ∎

Definition 11.3.1 (Fundamental unit of norm 1) *Let m be a positive squarefree integer. Let*

$$S_m = \{(x, y) \mid x \in \mathbb{N}, \ y \in \mathbb{N}\}, \ if \ m \equiv 2, 3 \ (\text{mod } 4),$$

and

$$S_m = \{(\frac{x}{2}, \frac{y}{2}) \mid x \in \mathbb{N}, \ y \in \mathbb{N}, x \equiv y \ (\text{mod } 2)\}, \ if \ m \equiv 1 \ (\text{mod } 4).$$

Let $(a, b) \in S_m$ be the solution of $a^2 - mb^2 = 1$ for which a is least. (Theorem 11.2.1 guarantees that (a, b) exists.) Let $\epsilon = a + b\sqrt{m}$ so that ϵ is a unit of $O_{\mathbb{Q}(\sqrt{m})}$ of norm 1. The unit ϵ is called the fundamental unit of norm 1 of $O_{\mathbb{Q}(\sqrt{m})}$.

We note that

$$\epsilon \geq 1 + \sqrt{m} \geq 1 + \sqrt{2}, \text{ if } m \equiv 2, 3 \ (\mathrm{mod}\ 4),$$

$$\epsilon \geq \frac{1 + \sqrt{m}}{2} \geq \frac{1 + \sqrt{5}}{2}, \text{ if } m \equiv 1 \ (\mathrm{mod}\ 4),$$

so that

$$\epsilon > 1.$$

Our next theorem shows how the units of norm 1 in $O_{\mathbb{Q}(\sqrt{m})}$ are related to the fundamental unit of norm 1.

Theorem 11.3.2 *Let m be a positive squarefree integer. Let ϵ be the fundamental unit of norm* 1 *of $O_{\mathbb{Q}(\sqrt{m})}$. Then*

(a) *ϵ is the smallest unit in $O_{\mathbb{Q}(\sqrt{m})}$ of norm* 1 *that is greater than* 1,
(b) *every unit in $O_{\mathbb{Q}(\sqrt{m})}$ of norm* 1 *is of the form $\pm\epsilon^n$ for some integer n, and*
(c) *if τ is a unit of norm* 1 *in $O_{\mathbb{Q}(\sqrt{m})}$ such that $\tau > 1$ and every unit in $O_{\mathbb{Q}(\sqrt{m})}$ of norm* 1 *is of the form $\pm\tau^k$ for some integer k then $\tau = \epsilon$.*

Proof: **(a)** As ϵ is the fundamental unit of $O_{\mathbb{Q}(\sqrt{m})}$ of norm 1, we have by Definition 11.3.1

$$\epsilon = a + b\sqrt{m}, \ (a, b) \in S_m, \ a^2 - mb^2 = 1, \ a \text{ least}.$$

Suppose that ϵ_1 is a unit of $O_{\mathbb{Q}(\sqrt{m})}$ of norm 1 with $1 < \epsilon_1 < \epsilon$. Then, by Theorems 5.4.2 and 11.3.1, we have

$$\epsilon_1 = a_1 + b_1\sqrt{m}, \ (a_1, b_1) \in S_m, \ a_1^2 - mb_1^2 = 1.$$

By the minimality of a we have

$$a < a_1$$

so that

$$b^2 = \frac{a^2 - 1}{m} < \frac{a_1^2 - 1}{m} = b_1^2,$$

and thus

$$b < b_1.$$

Hence

$$\epsilon = a + b\sqrt{m} < a_1 + b_1\sqrt{m} = \epsilon_1,$$

contradicting $\epsilon_1 < \epsilon$. Thus no such unit ϵ_1 exists, proving that ϵ is the smallest unit of $O_{\mathbb{Q}(\sqrt{m})}$ of norm 1 that is greater than 1.

(b) Let η be a unit of $O_{\mathbb{Q}(\sqrt{m})}$ of norm 1. Let η^* be the unit of $O_{\mathbb{Q}(\sqrt{m})}$ of norm 1 defined by

$$\eta^* = \begin{cases} \eta, & \text{if } \eta \geq 1, \\ 1/\eta, & \text{if } 0 < \eta < 1, \\ -1/\eta, & \text{if } -1 < \eta < 0, \\ -\eta, & \text{if } \eta \leq -1, \end{cases} \tag{11.3.5}$$

so that

$$\eta^* \geq 1.$$

Let k be the unique nonnegative integer such that

$$\epsilon^k \leq \eta^* < \epsilon^{k+1}.$$

Then $\eta^* \epsilon^{-k}$ is a unit of $O_{\mathbb{Q}(\sqrt{m})}$ of norm 1 satisfying

$$1 \leq \eta^* \epsilon^{-k} < \epsilon.$$

By part (a) there is no unit in $O_{\mathbb{Q}(\sqrt{m})}$ of norm 1 strictly between 1 and ϵ. Hence

$$\eta^* \epsilon^{-k} = 1$$

and so

$$\eta^* = \epsilon^k.$$

Then, from (11.3.5), we obtain

$$\eta = \pm \epsilon^n$$

for some choice of sign and some integer n.

(c) By assumption we have

$$\epsilon = \pm \tau^l,$$

for some integer l, and by part (b) we have

$$\tau = \pm \epsilon^n,$$

for some integer n. Hence

$$\epsilon = \pm(\pm \epsilon^n)^l = \pm \epsilon^{ln}$$

so that

$$\epsilon^{ln-1} = \pm 1$$

and thus

$$\epsilon^{2(ln-1)} = 1.$$

If $ln − 1 \neq 0$ then ϵ is a root of unity in $O_{\mathbb{Q}(\sqrt{m})}$. But $\mathbb{Q}(\sqrt{m})$ is a real field so the only roots of unity in $O_{\mathbb{Q}(\sqrt{m})}$ are ± 1. Hence $\epsilon = \pm 1$, contradicting $\epsilon > 1$. Thus $ln − 1 = 0$ and so

$$l = n = \pm 1,$$

showing that

$$\tau = \pm \epsilon \text{ or } \pm \epsilon^{-1}.$$

Since $\tau > 1$ and $\epsilon > 1$ we deduce that

$$\tau = \epsilon. \qquad \blacksquare$$

11.4 Units of Norm −1

Let m be a positive squarefree integer. We have already observed that the ring $O_{\mathbb{Q}(\sqrt{m})}$ of integers of the real quadratic field $\mathbb{Q}(\sqrt{m})$ may or may not contain units of norm $−1$. Indeed $O_{\mathbb{Q}(\sqrt{2})}$ has units such as $1 + \sqrt{2}$ of norm $−1$ whereas $O_{\mathbb{Q}(\sqrt{3})}$ does not contain any units of norm $−1$. We suppose that $O_{\mathbb{Q}(\sqrt{m})}$ contains units of norm $−1$ and show that there exists a unique unit $\sigma > 1$ in $O_{\mathbb{Q}(\sqrt{m})}$ of norm $−1$ such that all units in $O_{\mathbb{Q}(\sqrt{m})}$ of norm $−1$ are given by $\pm \sigma^{2k+1}$ ($k = 0, \pm 1, \pm 2, \ldots$) and all units in $O_{\mathbb{Q}(\sqrt{m})}$ of norm 1 are given by $\pm \sigma^{2k}$ ($k = 0, \pm 1, \pm 2, \ldots$).

Theorem 11.4.1 *Let m be a positive squarefree integer. Suppose that $O_{\mathbb{Q}(\sqrt{m})}$ contains units of norm* −1. *Then there exists a unique unit $\sigma > 1$ of norm* −1 *in $O_{\mathbb{Q}(\sqrt{m})}$ such that every unit in $O_{\mathbb{Q}(\sqrt{m})}$ is of the form $\pm\sigma^n$ for some integer n.*

Proof: Let ρ be a unit in $O_{\mathbb{Q}(\sqrt{m})}$ of norm $−1$. Let ρ' denote its conjugate. Then

$$\rho\rho' = N(\rho) = −1$$

so that

$$\rho^2 \rho'^2 = 1.$$

Thus ρ^2 is a unit of $O_{\mathbb{Q}(\sqrt{m})}$ of norm 1. Hence, by Theorem 11.3.2(b), we have

$$\rho^2 = \pm\epsilon^n,$$

for some integer n, where ϵ is the fundamental unit of $O_{\mathbb{Q}(\sqrt{m})}$ of norm 1. Clearly $\rho^2 > 0$ and $\epsilon^n > 0$ so that

$$\rho^2 = \epsilon^n.$$

If n is even, say $n = 2k$, then

$$\rho^2 = \epsilon^{2k}$$

so that

$$\rho = \pm\epsilon^k.$$

Hence

$$N(\rho) = N(\pm\epsilon^k) = N(\epsilon)^k = 1,$$

contradicting $N(\rho) = -1$. Thus n must be odd, say $n = 2l + 1$, and so

$$\rho^2 = \epsilon^{2l+1}.$$

Hence

$$\epsilon = (\rho\epsilon^{-l})^2.$$

Set

$$\sigma = \rho\epsilon^{-l}$$

so that σ is a unit of norm -1 such that

$$\epsilon = \sigma^2.$$

If μ is a unit of $O_{\mathbb{Q}(\sqrt{m})}$ of norm -1 then $\mu\rho^{-1}$ is a unit of $O_{\mathbb{Q}(\sqrt{m})}$ of norm 1 and thus by Theorem 11.3.2(b)

$$\mu\rho^{-1} = \pm\epsilon^k$$

for some $k \in \mathbb{Z}$. Hence, as $\rho = \epsilon^l\sigma$ and $\epsilon = \sigma^2$, we deduce that

$$\mu = \pm\epsilon^k\rho = \pm\epsilon^{k+l}\sigma = \pm\sigma^{2(k+l)+1}.$$

However, if μ is a unit of $O_{\mathbb{Q}(\sqrt{m})}$ of norm 1 then by Theorem 11.3.2(b)

$$\mu = \pm\epsilon^k$$

for some $k \in \mathbb{Z}$. Hence, as $\epsilon = \sigma^2$, we deduce that

$$\mu = \pm\epsilon^k = \pm\sigma^{2k}.$$

Thus every unit of $O_{\mathbb{Q}(\sqrt{m})}$ is of the form

$$\pm\sigma^n \ (n \in \mathbb{Z}).$$

Note that n even gives the units of norm 1 and n odd the units of norm -1.

Replacing σ by $1/\sigma$ if $0 < \sigma < 1$, by $-1/\sigma$ if $-1 < \sigma < 0$, and by $-\sigma$ if $\sigma < -1$, we may suppose that $\sigma > 1$. We show that σ is uniquely determined: For suppose σ and τ are two units > 1 of norm -1 in $O_{\mathbb{Q}(\sqrt{m})}$ such that every unit is of the form $\pm\sigma^n \ (n \in \mathbb{Z})$ and of the form $\pm\tau^q \ (q \in \mathbb{Z})$. Then there exist integers k and l such that

$$\sigma = \pm\tau^k, \ \tau = \pm\sigma^l.$$

Hence

$$\sigma = \pm \sigma^{kl}$$

and so

$$\sigma^2 = \sigma^{2kl},$$

giving

$$\sigma^{2(kl-1)} = 1.$$

Suppose $kl - 1 \neq 0$. Then σ is a root of unity. But $O_{\mathbb{Q}(\sqrt{m})}$ being a real field contains no roots of unity except ± 1. Thus $\sigma = \pm 1$, a contradiction. Hence $kl - 1 = 0$ and so

$$k = l = \pm 1.$$

Thus

$$\sigma = \pm \tau \text{ or } \pm \tau^{-1}.$$

But $\sigma > 1$ and $\tau > 1$ so

$$\sigma = \tau,$$

proving that σ is unique.   ∎

Definition 11.4.1 (Fundamental unit of norm −1**)** *Let m be a positive squarefree integer such that $O_{\mathbb{Q}(\sqrt{m})}$ contains units of norm* −1. *The unique unit $\sigma > 1$ of norm* −1 *such that every unit in $O_{\mathbb{Q}(\sqrt{m})}$ is of the form $\pm\sigma^n$ ($n \in \mathbb{Z}$) is called the fundamental unit of $O_{\mathbb{Q}(\sqrt{m})}$ of norm* −1.

We next relate the fundamental unit ϵ of norm 1 and the fundamental unit σ of norm −1 when $O_{\mathbb{Q}(\sqrt{m})}$ contains units of norm −1.

Theorem 11.4.2 *Let m be a positive squarefree integer such that $O_{\mathbb{Q}(\sqrt{m})}$ contains units of norm* −1. *Then the fundamental unit ϵ of norm 1 and the fundamental unit σ of norm* −1 *are related by*

$$\epsilon = \sigma^2.$$

Proof: By Theorem 11.4.1 we have $\epsilon = \pm \sigma^k$ for some $k \in \mathbb{Z}$. As $\epsilon > 1$ and $\sigma > 1$ the plus sign must hold so that

$$\epsilon = \sigma^k$$

for some $k \in \mathbb{Z}$. Then

$$1 = N(\epsilon) = N(\sigma^k) = N(\sigma)^k = (-1)^k,$$

so that k is even, say $k = 2g$, $g \in \mathbb{Z}$. Hence

$$\epsilon = \sigma^{2g}. \tag{11.4.1}$$

Now

$$N(\sigma^2) = N(\sigma)^2 = (-1)^2 = 1$$

so that σ^2 is a unit of norm 1 and thus, by Theorem 11.3.2(b), we have $\sigma^2 = \pm\epsilon^l$ for some $l \in \mathbb{Z}$. As $\sigma > 1$ and $\epsilon > 1$ the plus sign must hold so that

$$\sigma^2 = \epsilon^l \tag{11.4.2}$$

for some $l \in \mathbb{Z}$. From (11.4.1) and (11.4.2) we deduce that

$$\epsilon = \epsilon^{gl}$$

so that

$$\epsilon^{gl-1} = 1.$$

As ϵ is not a root of unity, we deduce that $gl - 1 = 0$, so that

$$\epsilon = \sigma^2 \text{ or } \sigma^{-2}.$$

As $\epsilon > 1$ and $\sigma > 1$ we have

$$\epsilon = \sigma^2$$

as asserted.   ∎

11.5 The Fundamental Unit

Theorems 11.3.2 and 11.4.1 show that all the units of $O_{\mathbb{Q}(\sqrt{m})}$ are given by $\pm\epsilon^n$ ($n \in \mathbb{Z}$) or by $\pm\sigma^n$ ($n \in \mathbb{Z}$) depending on whether $O_{\mathbb{Q}(\sqrt{m})}$ has only units of norm 1 or not. This enables us to define the "fundamental unit" of $O_{\mathbb{Q}(\sqrt{m})}$.

Definition 11.5.1 (Fundamental unit) *Let m be a positive squarefree integer. The fundamental unit η of $O_{\mathbb{Q}(\sqrt{m})}$ is defined to be σ if $O_{\mathbb{Q}(\sqrt{m})}$ contains units of norm -1 and to be ϵ otherwise. We note that $\eta > 1$.*

By Theorems 11.3.2 and 11.4.1 we have

Theorem 11.5.1 *Let m be a positive squarefree integer. Then every unit of $O_{\mathbb{Q}(\sqrt{m})}$ is of the form $\pm\eta^n$ ($n \in \mathbb{Z}$), where η is the fundamental unit of $O_{\mathbb{Q}(\sqrt{m})}$. If $O_{\mathbb{Q}(\sqrt{m})}$ contains units of norm -1 these are given by $\pm\eta^n$ with n odd and the ones of norm 1 by $\pm\eta^n$ with n even.*

From Theorem 11.5.1 we have immediately

Theorem 11.5.2 *Let K be a real quadratic field. Then*

$$U(O_K) \simeq \mathbb{Z}_2 \times \mathbb{Z}.$$

The following analogue of Theorem 11.3.2(a) is a simple consequence of Theorem 11.5.1.

Theorem 11.5.3 *Let K be a real quadratic field. The fundamental unit of O_K is the smallest unit of O_K greater than 1.*

Proof: Let η be the fundamental unit of O_K and suppose that there exists a unit θ of O_K with

$$1 < \theta < \eta.$$

By Theorem 11.5.1 we have

$$\theta = \pm \eta^n$$

for some $n \in \mathbb{Z}$. As θ and η are both positive, the positive sign must hold and we have

$$\theta = \eta^n.$$

If $n \geq 1$ then

$$\theta = \eta^n \geq \eta,$$

contradicting $\theta < \eta$. If $n \leq 0$ then

$$\theta = \eta^n \leq 1,$$

contradicting $\theta > 1$. Hence no such θ can exist, proving that η is the smallest unit greater than 1. ∎

Before proceeding to find the norm of the fundamental unit η of $O_{\mathbb{Q}(\sqrt{m})}$ for certain special values of m, we present in Table 4 the values of ϵ, σ, and η for squarefree positive integers $m < 40$.

We next determine the norm of the fundamental unit of $O_{\mathbb{Q}(\sqrt{m})}$ when m is an odd prime p. First we consider the case $p \equiv 1 \pmod 4$.

Theorem 11.5.4 *Let p be a prime with $p \equiv 1 \pmod 4$. Then the fundamental unit of $O_{\mathbb{Q}(\sqrt{p})}$ has norm -1.*

We give two proofs of this theorem, the first due to Hilbert and the second due to Peter Gustav Lejeune Dirichlet (1805–1859).

Table 4.   *Fundamental units of $O_{\mathbb{Q}(\sqrt{m})}$, $2 \le m < 40$, m squarefree*

m	Fundamental unit of norm 1 (ϵ)	Fundamental unit of norm -1 (σ)	Fundamental unit (η)	Norm $N(\eta)$
2	$3 + 2\sqrt{2}$	$1 + \sqrt{2}$	$1 + \sqrt{2}$	-1
3	$2 + \sqrt{3}$		$2 + \sqrt{3}$	1
5	$(3 + \sqrt{5})/2$	$(1 + \sqrt{5})/2$	$(1 + \sqrt{5})/2$	-1
6	$5 + 2\sqrt{6}$		$5 + 2\sqrt{6}$	1
7	$8 + 3\sqrt{7}$		$8 + 3\sqrt{7}$	1
10	$19 + 6\sqrt{10}$	$3 + \sqrt{10}$	$3 + \sqrt{10}$	-1
11	$10 + 3\sqrt{11}$		$10 + 3\sqrt{11}$	1
13	$(11 + 3\sqrt{13})/2$	$(3 + \sqrt{13})/2$	$(3 + \sqrt{13})/2$	-1
14	$15 + 4\sqrt{14}$		$15 + 4\sqrt{14}$	1
15	$4 + \sqrt{15}$		$4 + \sqrt{15}$	1
17	$33 + 8\sqrt{17}$	$4 + \sqrt{17}$	$4 + \sqrt{17}$	-1
19	$170 + 39\sqrt{19}$		$170 + 39\sqrt{19}$	1
21	$(5 + \sqrt{21})/2$		$(5 + \sqrt{21})/2$	1
22	$197 + 42\sqrt{22}$		$197 + 42\sqrt{22}$	1
23	$24 + 5\sqrt{23}$		$24 + 5\sqrt{23}$	1
26	$51 + 10\sqrt{26}$	$5 + \sqrt{26}$	$5 + \sqrt{26}$	-1
29	$(27 + 5\sqrt{29})/2$	$(5 + \sqrt{29})/2$	$(5 + \sqrt{29})/2$	-1
30	$11 + 2\sqrt{30}$		$11 + 2\sqrt{30}$	1
31	$1520 + 273\sqrt{31}$		$1520 + 273\sqrt{31}$	1
33	$23 + 4\sqrt{33}$		$23 + 4\sqrt{33}$	1
34	$35 + 6\sqrt{34}$		$35 + 6\sqrt{34}$	1
35	$6 + \sqrt{35}$		$6 + \sqrt{35}$	1
37	$73 + 12\sqrt{37}$	$6 + \sqrt{37}$	$6 + \sqrt{37}$	-1
38	$37 + 6\sqrt{38}$		$37 + 6\sqrt{38}$	1
39	$25 + 4\sqrt{39}$		$25 + 4\sqrt{39}$	1

First proof: Suppose that the fundamental unit η of $O_{\mathbb{Q}(\sqrt{p})}$ has norm 1. Then

$$N(\eta) = \eta\eta' = 1.$$

As $\eta > 1$ we have $0 < \eta' < 1$ so that $1 + \eta' \ne 0$. Hence

$$\eta = \frac{1 + \eta}{1 + \eta'}.$$

Let m be the largest positive integer such that

$$m \mid 1 + \eta, \quad m \mid 1 + \eta'$$

in $O_{\mathbb{Q}(\sqrt{p})}$. Set

$$\gamma = \frac{1+\eta}{m} \in O_{\mathbb{Q}(\sqrt{p})}$$

and

$$\gamma' = \frac{1+\eta'}{m} \in O_{\mathbb{Q}(\sqrt{p})}$$

so that

$$\eta = \frac{\gamma}{\gamma'}$$

and

$$k \mid \gamma, \ k \mid \gamma', \ k \in \mathbb{N} \Longrightarrow k = 1. \tag{11.5.1}$$

Now

$$\gamma = \eta\gamma',$$

where η is a unit of $O_{\mathbb{Q}(\sqrt{p})}$, so that

$$\langle \gamma \rangle = \langle \gamma' \rangle. \tag{11.5.2}$$

Let Q be any prime ideal of $O_{\mathbb{Q}(\sqrt{p})}$ such that

$$Q \mid \langle \gamma \rangle. \tag{11.5.3}$$

Then, by (11.5.2), we have

$$Q \mid \langle \gamma' \rangle. \tag{11.5.4}$$

Taking conjugate ideals in (11.5.4), we obtain

$$Q' \mid \langle \gamma \rangle. \tag{11.5.5}$$

As Q is a prime ideal, and the discriminant of $O_{\mathbb{Q}(\sqrt{p})}$ is p, by Theorem 10.2.2 we have

$$Q = Q' = \langle q \rangle, \ \text{where } q \text{ is a rational prime with } \left(\frac{p}{q}\right) = -1,$$

or

$$Q \neq Q', \ QQ' = \langle q \rangle, \ \text{where } q \text{ is a rational prime with } \left(\frac{p}{q}\right) = 1,$$

or

$$Q = Q', \ Q^2 = \langle q \rangle, \ \text{where } q \text{ is a rational prime with } \left(\frac{p}{q}\right) = 0.$$

In the first case, from (11.5.3) and (11.5.4), we deduce that $q \mid \gamma$, $q \mid \gamma'$, contradicting (11.5.1).

In the second case we have by (11.5.3) and (11.5.5) as Q and Q' are distinct prime ideals

$$\langle q \rangle = QQ' \mid \langle \gamma \rangle = \langle \gamma' \rangle.$$

Hence $q \mid \gamma$ and $q \mid \gamma'$, contradicting (11.5.1).

In the third case we have $q = p$ and $Q = \langle \sqrt{p} \rangle$. Hence $\langle \sqrt{p} \rangle$ is the only prime ideal that can divide $\langle \gamma \rangle$. Thus

$$\langle \gamma \rangle = \langle \sqrt{p} \rangle^j$$

for some nonnegative integer j. If $j \geq 2$ then $p \mid \gamma$ and $p \mid \gamma'$, contradicting (11.5.1). Hence $j = 0$ or 1. If $j = 0$ then

$$\langle \gamma \rangle = \langle 1 \rangle$$

and so

$$\gamma = \lambda$$

for some unit λ of $O_{\mathbb{Q}(\sqrt{p})}$. As $\lambda \lambda' = \pm 1$ we have

$$\eta = \frac{\gamma}{\gamma'} = \frac{\lambda}{\lambda'} = \frac{\lambda^2}{\lambda \lambda'} = \pm \lambda^2,$$

contradicting that η is the fundamental unit of $O_{\mathbb{Q}(\sqrt{p})}$. If $j = 1$ then

$$\langle \gamma \rangle = \langle \sqrt{p} \rangle$$

so that

$$\gamma = \lambda \sqrt{p}$$

for some unit λ of $O_{\mathbb{Q}(\sqrt{p})}$. Hence

$$\eta = \frac{\gamma}{\gamma'} = \frac{\lambda \sqrt{p}}{\lambda'(-\sqrt{p})} = \frac{-\lambda}{\lambda'} = \frac{-\lambda^2}{\lambda \lambda'} = \mp \lambda^2,$$

again contradicting that η is the fundamental unit of $O_{\mathbb{Q}(\sqrt{p})}$.

This completes the proof that the fundamental unit of $O_{\mathbb{Q}(\sqrt{p})}$ (p (prime) $\equiv 1$ (mod 4)) must have norm -1.   ∎

Second proof: Suppose that the fundamental unit η of $O_{\mathbb{Q}(\sqrt{p})}$ has norm 1. Then

$$\eta = \frac{x + y\sqrt{p}}{2},$$

where x and y are positive integers such that

$$x > 2, \quad x \equiv y \,(\text{mod } 2), \quad x^2 - py^2 = 4.$$

We first treat the case when $x \equiv y \equiv 0 \,(\text{mod } 2)$. Set $x = 2X$, $y = 2Y$ so that X and Y are positive integers such that

$$\eta = X + Y\sqrt{p}, \quad X^2 - pY^2 = 1, \quad X > 1.$$

As $p \equiv 1 \,(\text{mod } 4)$, X is odd and Y is even, so that $\frac{X-1}{2}$, $\frac{X+1}{2}$, and $\frac{Y}{2}$ are positive integers such that

$$\frac{(X-1)}{2} \cdot \frac{(X+1)}{2} = p\left(\frac{Y}{2}\right)^2.$$

As $\frac{X+1}{2} - \frac{X-1}{2} = 1$, the integers $\frac{X-1}{2}$ and $\frac{X+1}{2}$ are coprime. Since $p \mid \frac{X-1}{2} \cdot \frac{X+1}{2}$ either $p \mid \frac{X-1}{2}$ or $p \mid \frac{X+1}{2}$.

If $p \mid \frac{X-1}{2}$ then

$$\frac{(X-1)}{2p} \cdot \frac{(X+1)}{2} = \left(\frac{Y}{2}\right)^2$$

and thus there are positive coprime integers A and B such that

$$\frac{X-1}{2p} = A^2, \quad \frac{X+1}{2} = B^2, \quad \frac{Y}{2} = AB,$$

so that

$$X = 2pA^2 + 1 = 2B^2 - 1, \quad Y = 2AB.$$

Hence

$$B^2 - pA^2 = 1$$

so that $B + A\sqrt{p}$ is a unit of norm 1 in $O_{\mathbb{Q}(\sqrt{p})}$. Now

$$1 \leq B \leq B^2 \leq 2B^2 - 1 = X$$

and

$$1 \leq A < 2AB = Y$$

so that

$$1 < B + A\sqrt{p} < X + Y\sqrt{p} = \eta,$$

contradicting that η is the fundamental unit of $O_{\mathbb{Q}(\sqrt{p})}$.

If $p \mid \frac{X+1}{2}$ then

$$\frac{(X-1)}{2} \cdot \frac{(X+1)}{2p} = \left(\frac{Y}{2}\right)^2$$

and thus there are integers A and B such that

$$\frac{X-1}{2} = A^2, \quad \frac{X+1}{2p} = B^2, \quad \frac{Y}{2} = AB,$$

so that

$$X = 2A^2 + 1 = 2pB^2 - 1, \quad Y = 2AB.$$

Hence

$$A^2 - pB^2 = -1$$

so that $A + B\sqrt{p}$ is a unit of norm -1 in $O_{\mathbb{Q}(\sqrt{p})}$, contradicting that all the units of $O_{\mathbb{Q}(\sqrt{p})}$ have norm 1.

Now we turn to the case when $x \equiv y \equiv 1 \pmod 2$. Reducing $x^2 - py^2 = 4$ modulo 8 we see that $p \equiv 5 \pmod 8$. Now $x - 2$, $x + 2$, and y are positive odd integers such that

$$(x-2)(x+2) = py^2.$$

As $(x+2) - (x-2) = 4$ the integers $x - 2$ and $x + 2$ are coprime. Since $p \mid (x-2)(x+2)$ either $p \mid x - 2$ or $p \mid x + 2$.

If $p \mid x - 2$ then

$$\frac{(x-2)}{p} \cdot (x+2) = y^2$$

and there exist positive coprime odd integers A and B such that

$$\frac{x-2}{p} = A^2, \quad x + 2 = B^2, \quad y = AB,$$

so that

$$x = pA^2 + 2 = B^2 - 2, \quad y = AB.$$

Hence

$$B^2 - pA^2 = 4$$

so that $(B + A\sqrt{p})/2$ is a unit of norm 1 in $O_{\mathbb{Q}(\sqrt{p})}$. From $B^2 - pA^2 = 4$ we see that $B \neq 1$. Thus, as B is odd and positive, we must have $B \geq 3$. Hence

$$1 < B < B^2 - 2 = x$$

and

$$1 \leq A < AB = y$$

so that

$$1 < \frac{B + A\sqrt{p}}{2} < \frac{x + y\sqrt{p}}{2},$$

contradicting that $(x + y\sqrt{p})/2$ is the fundamental unit of norm 1 in $O_{\mathbb{Q}(\sqrt{p})}$ by Theorem 11.3.2(a).

If $p \mid x + 2$ then

$$(x - 2)\frac{(x + 2)}{p} = y^2$$

and thus there are positive coprime odd integers A and B such that

$$x - 2 = A^2, \quad \frac{x + 2}{p} = B^2, \quad y = AB,$$

so that

$$x = A^2 + 2 = pB^2 - 2, \quad y = AB.$$

Hence

$$A^2 - pB^2 = -4$$

so that $(A + B\sqrt{p})/2$ is a unit of norm -1 in $O_{\mathbb{Q}(\sqrt{p})}$, contradicting that the fundamental unit of $O_{\mathbb{Q}(\sqrt{p})}$ has norm 1.

This completes the proof of Theorem 11.5.4 using Dirichlet's method. ∎

When m is a prime $p \equiv 3 \,(\text{mod } 4)$ the fundamental unit of $O_{\mathbb{Q}(\sqrt{p})}$ has norm 1. This is a special case of the following theorem.

Theorem 11.5.5 *Let m be a positive squarefree integer. If there exists a prime $q \equiv 3$ (mod 4) dividing m then the fundamental unit of $O_{\mathbb{Q}(\sqrt{m})}$ has norm 1.*

Proof: Suppose that the fundamental unit η of $O_{\mathbb{Q}(\sqrt{m})}$ has norm -1. Then

$$\eta = \frac{x + y\sqrt{m}}{2},$$

where x and y are integers such that

$$\begin{cases} x \equiv y \equiv 0 \,(\text{mod } 2), & \text{if } m \equiv 2, 3 \,(\text{mod } 4), \\ x \equiv y \,(\text{mod } 2), & \text{if } m \equiv 1 \,(\text{mod } 4), \end{cases}$$

and

$$\frac{x^2 - my^2}{4} = N(\eta) = -1.$$

Hence

$$x^2 - my^2 = -4.$$

As $q \mid m$ we deduce that

$$x^2 \equiv -4 \,(\text{mod } q).$$

Thus

$$\left(\frac{-1}{q}\right) = \left(\frac{-4}{q}\right) = 1$$

so that $q \equiv 1 \pmod 4$, contradicting that $q \equiv 3 \pmod 4$. This proves that η must have norm 1. ∎

The following two theorems can be proved using Dirichlet's method in a similar manner to the second proof of Theorem 11.5.4.

Theorem 11.5.6 *Let p be a prime with $p \equiv 5 \pmod 8$. Then the fundamental unit of $O_{\mathbb{Q}(\sqrt{2p})}$ has norm -1.*

Theorem 11.5.7 *Let p and q be distinct primes such that*

$$p \equiv q \equiv 1 \pmod 4, \quad \left(\frac{p}{q}\right) = -1.$$

Then the fundamental unit of $O_{\mathbb{Q}(\sqrt{pq})}$ has norm -1.

Up to this point we have said almost nothing about calculating the fundamental unit η of $O_{\mathbb{Q}(\sqrt{m})}$ for a particular value of m. We address this problem in the next section.

11.6 Calculating the Fundamental Unit

Let m be a positive squarefree integer. The standard method of calculating the fundamental unit η of $O_{\mathbb{Q}(\sqrt{m})}$ is by means of the continued fraction expansion of $\sqrt{m}$. We assume that the reader is familiar with the basic properties of continued fractions as found for example in Chapter 7 of the book on elementary number theory by Niven, Zuckerman, and Montgomery [2]. We just recall the basic facts that we shall need and refer the reader to [2] for proofs.

Given a positive squarefree integer m, we define a sequence $\alpha_0, \alpha_1, \alpha_2, \ldots$ of real numbers by

$$\alpha_0 = \sqrt{m} \tag{11.6.1}$$

and

$$\alpha_{n+1} = \frac{1}{\alpha_n - [\alpha_n]}, \quad n = 0, 1, 2, \ldots. \tag{11.6.2}$$

Example 11.6.1 *If $m = 31$ we find that*

$$\alpha_0 = \sqrt{31},$$

$$\alpha_1 = \frac{1}{\alpha_0 - [\alpha_0]} = \frac{1}{\sqrt{31} - 5} = \frac{5 + \sqrt{31}}{6},$$

$$\alpha_2 = \frac{1}{\alpha_1 - [\alpha_1]} = \frac{1}{\dfrac{5 + \sqrt{31}}{6} - 1} = \frac{1 + \sqrt{31}}{5},$$

$$\alpha_3 = \frac{1}{\alpha_2 - [\alpha_2]} = \frac{1}{\dfrac{1 + \sqrt{31}}{5} - 1} = \frac{4 + \sqrt{31}}{3},$$

$$\alpha_4 = \frac{1}{\alpha_3 - [\alpha_3]} = \frac{1}{\dfrac{4 + \sqrt{31}}{3} - 3} = \frac{5 + \sqrt{31}}{2},$$

$$\alpha_5 = \frac{1}{\alpha_4 - [\alpha_4]} = \frac{1}{\dfrac{5 + \sqrt{31}}{2} - 5} = \frac{5 + \sqrt{31}}{3},$$

$$\alpha_6 = \frac{1}{\alpha_5 - [\alpha_5]} = \frac{1}{\dfrac{5 + \sqrt{31}}{3} - 3} = \frac{4 + \sqrt{31}}{5},$$

$$\alpha_7 = \frac{1}{\alpha_6 - [\alpha_6]} = \frac{1}{\dfrac{4 + \sqrt{31}}{5} - 1} = \frac{1 + \sqrt{31}}{6},$$

$$\alpha_8 = \frac{1}{\alpha_7 - [\alpha_7]} = \frac{1}{\dfrac{1 + \sqrt{31}}{6} - 1} = \frac{5 + \sqrt{31}}{1},$$

$$\alpha_9 = \frac{1}{\alpha_8 - [\alpha_8]} = \frac{1}{5 + \sqrt{31} - 10} = \frac{5 + \sqrt{31}}{6} = \alpha_1,$$

$$\alpha_{10} = \alpha_2, \ \alpha_{11} = \alpha_3, \ldots.$$

Clearly each $\alpha_n > 1$ and

$$\alpha_n = \frac{P_n + \sqrt{m}}{Q_n}, \quad n = 0, 1, 2, \ldots, \tag{11.6.3}$$

where $P_0 = 0$, $Q_0 = 1$, and P_n, Q_n are positive integers for $n \geq 1$. Moreover, it is known that there exists a positive integer l such that $\alpha_{l+1} = \alpha_1$. It follows from (11.6.2) that $\alpha_{l+2} = \alpha_2$, $\alpha_{l+3} = \alpha_3$, ... so that the sequence $\{\alpha_n\}_{n \geq 1}$ is purely periodic. We set

$$a_n = [\alpha_n], \quad n = 0, 1, 2, \ldots, \tag{11.6.4}$$

so that $\{a_n\}_{n\geq 1}$ is a sequence of positive integers. Since $\{\alpha_n\}_{n\geq 1}$ is a purely periodic sequence so is the sequence $\{a_n\}_{n\geq 1}$.

Example 11.6.1 (continued) *For* $m = 31$

$$\{a_n\}_{n\geq 0} = \{5, 1, 1, 3, 5, 3, 1, 1, 10, 1, 1, 3, 5, 3, 1, 1, 10, \ldots\}.$$

Next we define two further sequences of integers $\{h_n\}_{n\geq -1}$ and $\{k_n\}_{n\geq -1}$ by

$$h_{-1} = 1, \ h_0 = a_0, \ h_n = a_n h_{n-1} + h_{n-2}, \ n = 1, 2, \ldots \qquad (11.6.5)$$

and

$$k_{-1} = 0, \ k_0 = 1, \ k_n = a_n k_{n-1} + k_{n-2}, \ n = 1, 2, \ldots. \qquad (11.6.6)$$

All of the h_n and k_n are positive except for $k_{-1} = 0$.

Example 11.6.1 (continued) *For* $m = 31$

$$\{h_n\}_{n\geq -1} = \{1, 5, 6, 11, 39, 206, 657, 863, 1520, 16063, 17583, \ldots\},$$
$$\{k_n\}_{n\geq -1} = \{0, 1, 1, 2, 7, 37, 118, 155, 273, 2885, 3158, \ldots\}.$$

It is easily shown that $(h_n, k_n) = 1$ and

$$\frac{h_n}{k_n} = a_0 + \cfrac{1}{a_1 + \cfrac{1}{a_2 + \cdots + \cfrac{1}{a_{n-1} + \cfrac{1}{a_n}}}}, \ n = 0, 1, 2, \ldots. \qquad (11.6.7)$$

To save space we abbreviate the fraction on the right-hand side of (11.6.7) by the space-saving flat notation $[a_0, a_1, a_2, \ldots, a_n]$ so that (11.6.7) becomes

$$\frac{h_n}{k_n} = [a_0, a_1, a_2, \ldots, a_n], \ n = 0, 1, 2, \ldots. \qquad (11.6.8)$$

It is known that $\lim_{n\to\infty} h_n/k_n$ exists and is equal to $\sqrt{m}$.

Example 11.6.1 (continued) *For m = 31 we have to seven decimal places*

$$\frac{h_0}{k_0} = 5, \quad \frac{h_1}{k_1} = \frac{6}{1} = 6, \quad \frac{h_2}{k_2} = \frac{11}{2} = 5.5, \quad \frac{h_3}{k_3} = \frac{39}{7} = 5.5714285,$$

$$\frac{h_4}{k_4} = \frac{206}{37} = 5.5675675, \quad \frac{h_5}{k_5} = \frac{657}{118} = 5.5677966, \quad \frac{h_6}{k_6} = \frac{863}{155} = 5.5677419,$$

$$\frac{h_7}{k_7} = \frac{1520}{273} = 5.5677655, \quad \frac{h_8}{k_8} = \frac{16063}{2885} = 5.5677642,$$

$$\frac{h_9}{k_9} = \frac{17583}{3158} = 5.5677644,$$

and $\sqrt{31} = 5.5677643 \ldots$.

We write $[a_0, a_1, a_2, \ldots]$ for $\lim_{n \to \infty} [a_0, a_1, a_2, \ldots, a_n]$, and we say that $\sqrt{m}$ has the infinite continued fraction expansion

$$\sqrt{m} = [a_0, a_1, a_2, \ldots]. \tag{11.6.9}$$

The convergents of the infinite continued fraction $[a_0, a_1, a_2, \ldots]$ are the rational numbers h_n/k_n ($n = 0, 1, 2, \ldots$). As we have already mentioned, the sequence $\{a_n\}_{n \geq 1}$ is purely periodic. Let l be the least positive integer such that $\alpha_{l+n} = \alpha_n$ (equivalently, $P_{l+n} = P_n$, $Q_{l+n} = Q_n$) for all positive integers n. The integer l is called the period of the continued fraction expansion of $\sqrt{d}$. Then

$$\sqrt{m} = [a_0, a_1, \ldots, a_l, a_1, \ldots, a_l, a_1, \ldots, a_l, \ldots]$$

and we abbreviate this by

$$\sqrt{m} = [a_0, \overline{a_1, \ldots, a_l}]. \tag{11.6.10}$$

Example 11.6.1 (continued) *For m = 31 we have l = 8 and*

$$\sqrt{31} = [5, 1, 1, 3, 5, 3, 1, 1, 10, 1, 1, 3, 5, 3, 1, 1, 10, 1, \ldots]$$
$$= [5, \overline{1, 1, 3, 5, 3, 1, 1, 10}].$$

Clearly $a_0 = [\sqrt{m}]$ and it is further known that $a_l = 2a_0$. Putting

$$\alpha_n = \frac{P_n + \sqrt{m}}{Q_n}, \quad \alpha_{n-1} = \frac{P_{n-1} + \sqrt{m}}{Q_{n-1}},$$

and $[\alpha_{n-1}] = a_{n-1}$ in

$$\alpha_n = \frac{1}{\alpha_{n-1} - [\alpha_{n-1}]} \quad (n \geq 1),$$

we obtain

$$\frac{P_n + \sqrt{m}}{Q_n} = \frac{Q_{n-1}}{(P_{n-1} - a_{n-1}Q_{n-1}) + \sqrt{m}}.$$

Cross-multiplying and equating coefficients we see that

$$P_n(P_{n-1} - a_{n-1}Q_{n-1}) + m = Q_nQ_{n-1},$$
$$P_n + P_{n-1} - a_{n-1}Q_{n-1} = 0.$$

Thus $\alpha_n = (P_n + \sqrt{m})/Q_n$ $(n \geq 1)$ is determined recursively by

$$P_n = -P_{n-1} + a_{n-1}Q_{n-1},$$
$$Q_n = \frac{m - P_n^2}{Q_{n-1}},$$
$$a_n = \left[\frac{P_n + \sqrt{m}}{Q_n}\right],$$

with $P_0 = 0$, $Q_0 = 1$, $a_0 = [\sqrt{m}]$.

The central result that allows us to determine the fundamental unit of a real quadratic field is the following theorem.

Theorem 11.6.1 *Let m be a positive squarefree integer. Let h_n/k_n $(n = 0, 1, 2, \ldots)$ be the convergents of the infinite continued fraction expansion of $\sqrt{m}$. Let l be the period of the expansion.*

If l is even then $x^2 - my^2 = -1$ has no solutions in integers x and y and the solution of $x^2 - my^2 = 1$ in positive integers x and y with x least is $(x, y) = (h_{l-1}, k_{l-1})$.

If l is odd then $x^2 - my^2 = -1$ has solutions in integers x and y and the solution of $x^2 - my^2 = -1$ in positive integers x and y with x least is given by $(x, y) = (h_{l-1}, k_{l-1})$.

If $m \equiv 2$, $m \equiv 3 \pmod 4$, or $m \equiv 1 \pmod 8$ all the units of $O_{\mathbb{Q}(\sqrt{m})}$ are of the form $x + y\sqrt{m}$ with x and y integers, so that by Theorems 11.5.3 and 11.6.1 we see that the fundamental unit η of $O_{\mathbb{Q}(\sqrt{m})}$ is given by

$$\eta = h_{l-1} + k_{l-1}\sqrt{m}, \quad N(\eta) = (-1)^l.$$

If $m \equiv 5 \pmod 8$ there may or may not be units of $O_{\mathbb{Q}(\sqrt{m})}$ of the form $\frac{1}{2}(x + y\sqrt{m})$ with x and y odd integers. If there are no such units then $\eta \in \mathbb{Z} + \mathbb{Z}\sqrt{m}$ and, as in the previous case, we have

$$\eta = h_{l-1} + k_{l-1}\sqrt{m}, \quad N(\eta) = (-1)^l.$$

If there are such units then $\eta \notin \mathbb{Z} + \mathbb{Z}\sqrt{m}$ and it can be shown that $\eta^3 \in \mathbb{Z} + \mathbb{Z}\sqrt{m}$. In this case $\eta^3 = x + y\sqrt{m}$, where x and y are positive integers satisfying $x^2 - my^2 = \pm1$ with x least so that by Theorems 11.5.3 and 11.6.1

$$\eta^3 = h_{l-1} + k_{l-1}\sqrt{m}, \quad N(\eta) = (-1)^l.$$

If $\eta = (A + B\sqrt{m})/2$, where A and B are odd positive integers, then

$$\left(\frac{A + B\sqrt{m}}{2}\right)^3 = h_{l-1} + k_{l-1}\sqrt{m},$$

and so

$$A^3 + 3AB^2m = 8h_{l-1},$$
$$3A^2B + B^3m = 8k_{l-1}.$$

Hence

$$A \mid h_{l-1}, \quad 1 \le A < 2h_{l-1}^{1/3}$$

and

$$B \mid k_{l-1}, \quad 1 \le B < 2\left(\frac{k_{l-1}}{m}\right)^{1/3}.$$

This gives the following algorithm for determining the fundamental unit η of $O_{\mathbb{Q}(\sqrt{m})}$ for a positive squarefree integer m.

Step 1: $h_{-1} = 1, \ k_{-1} = 0.$

$$P_0 = 0, \ Q_0 = 1, \ a_0 = [\sqrt{m}], \ h_0 = [\sqrt{m}], \ k_0 = 1.$$

Step 2: Determine $P_n, \ Q_n, \ a_n, \ h_n, \ k_n \ (n = 1, 2, \ldots)$ recursively by means of

$$P_n = -P_{n-1} + a_{n-1}Q_{n-1}, \ n = 1, 2, \ldots,$$
$$Q_n = \frac{m - P_n^2}{Q_{n-1}}, \ n = 1, 2, \ldots,$$
$$a_n = \left[\frac{P_n + \sqrt{m}}{Q_n}\right], \ n = 1, 2, \ldots,$$
$$h_n = a_n h_{n-1} + h_{n-2}, \ n = 1, 2, \ldots,$$
$$k_n = a_n k_{n-1} + k_{n-2}, \ n = 1, 2, \ldots.$$

Stop at the first integer $N > 1$ such that

$$P_N = P_1, \ Q_N = Q_1.$$

Step 3: $l = N - 1.$

Step 4: If $m \equiv 2 \,(\text{mod } 4)$, $m \equiv 3 \,(\text{mod } 4)$, or $m \equiv 1 \,(\text{mod } 8)$ then

$$\eta = h_{l-1} + k_{l-1}\sqrt{m}, \ N(\eta) = (-1)^l.$$

Step 5: If $m \equiv 5 \,(\text{mod } 8)$ determine all positive odd divisors A of h_{l-1} less than $2h_{l-1}^{1/3}$ and all positive odd divisors B of k_{l-1} less than $2\,(k_{l-1}/m)^{1/3}$. If for some pair (A, B) we have

$$A^3 + 3AB^2m = 8h_{l-1}, \ 3A^2B + B^3m = 8k_{l-1},$$

then

$$\eta = \frac{A + B\sqrt{m}}{2}, \ N(\eta) = (-1)^l;$$

otherwise

$$\eta = h_{l-1} + k_{l-1}\sqrt{m}, \quad N(\eta) = (-1)^l.$$

We present several examples.

Example 11.6.1 (continued) $m = 31 \equiv 3 \,(\text{mod } 4)$. *Starting with*

$$h_{-1} = 1, k_{-1} = 0, P_0 = 0, Q_0 = 1, a_0 = 5, h_0 = 5, k_0 = 1,$$

we obtain successively the values of

$$P_n, Q_n, a_n, h_n, k_n, n = 1, 2, \ldots,$$

as in Step 2.

n	P_n	Q_n	a_n	h_n	k_n
-1				1	0
0	0	1	5	5	1
1	5	6	1	6	1
2	1	5	1	11	2
3	4	3	3	39	7
4	5	2	5	206	37
5	5	3	3	657	118
6	4	5	1	863	155
7	1	6	1	1520	273
8	5	1	10	16063	2885
9	5	6	1	17583	3158

As

$$P_9 = P_1 = 5, Q_9 = Q_1 = 6,$$

we see that

$$N = 9, l = N - 1 = 8, h_{l-1} = h_7 = 1520, k_{l-1} = k_7 = 273,$$
$$\eta = h_{l-1} + k_{l-1}\sqrt{m} = 1520 + 273\sqrt{31}, \quad N(\eta) = (-1)^l = 1.$$

The fundamental unit of $O_{Q(\sqrt{31})}$ *is* $1520 + 273\sqrt{31}$ *of norm 1.*

Example 11.6.2 $m = 41 \equiv 1 \,(\text{mod } 8)$. *Here*

n	P_n	Q_n	a_n	h_n	k_n
-1				1	0
0	0	1	6	6	1
1	6	5	2	13	2
2	4	5	2	32	5
3	6	1	12	397	62
4	6	5	2	826	129

As

$$P_4 = P_1 = 6, \ Q_4 = Q_1 = 5,$$

we have

$$N = 4, \ l = N - 1 = 3, \ h_{l-1} = h_2 = 32, \ k_{l-1} = k_2 = 5,$$
$$\eta = h_{l-1} + k_{l-1}\sqrt{m} = 32 + 5\sqrt{41}, \ N(\eta) = (-1)^l = -1.$$

The fundamental unit of $O_{Q(\sqrt{41})}$ is $32 + 5\sqrt{41}$ of norm -1.

Example 11.6.3 $m = 82 \equiv 2 \, (\text{mod } 4)$. *Here*

n	P_n	Q_n	a_n	h_n	k_n
-1				1	0
0	0	1	9	9	1
1	9	1	18	163	18
2	9	1	18	2943	325

As

$$P_2 = P_1 = 9, \ Q_2 = Q_1 = 1,$$

we have

$$N = 2, \ l = N - 1 = 1, \ h_{l-1} = h_0 = 9, \ k_{l-1} = k_0 = 1,$$
$$\eta = h_{l-1} + k_{l-1}\sqrt{m} = 9 + \sqrt{82}, \ N(\eta) = (-1)^l = -1.$$

The fundamental unit of $O_{Q(\sqrt{82})}$ is $9 + \sqrt{82}$ of norm -1.

Example 11.6.4 $m = 13 \equiv 5 \, (\text{mod } 8)$. *Here*

n	P_n	Q_n	a_n	h_n	k_n
-1				1	0
0	0	1	3	3	1
1	3	4	1	4	1
2	1	3	1	7	2
3	2	3	1	11	3
4	1	4	1	18	5
5	3	1	6	119	33
6	3	4	1	137	38

As

$$P_6 = P_1 = 3, \ Q_6 = Q_1 = 4,$$

we deduce that

$$N = 6, \ l = N - 1 = 5, \ h_{l-1} = h_4 = 18, \ k_{l-1} = k_4 = 5.$$

Next,

$$A \text{ odd}, \ A \mid h_{l-1}, \ 1 \le A < 2h_{l-1}^{1/3} \Longrightarrow A \mid 9, \ 1 \le A < 5.3 \Longrightarrow A = 1 \text{ or } 3,$$

$$B \text{ odd}, \ B \mid k_{l-1}, \ 1 \le B < 2 \left(\frac{k_{l-1}}{m} \right)^{1/3} \Longrightarrow B \mid 5, \ 1 \le B < 1.5 \Longrightarrow B = 1.$$

Of the pairs $(A, B) = (1, 1), \ (3, 1)$ *only the latter satisfies the pair of equations*

$$A^3 + 39AB^2 = 144, \ 3A^2B + 13B^3 = 40.$$

Hence the fundamental unit $\eta(>1)$ *of* $O_{\mathbb{Q}(\sqrt{13})}$ *is*

$$\eta = \frac{3 + \sqrt{13}}{2}, \quad N(\eta) = -1.$$

Example 11.6.5 $m = 37 \equiv 5 \,(\text{mod } 8)$. *Here*

n	P_n	Q_n	a_n	h_n	k_n
-1				1	0
0	0	1	6	6	1
1	6	1	12	73	12
2	6	1	12	882	145

As

$$P_2 = P_1 = 6, \ Q_2 = Q_1 = 1,$$

we have

$$N = 2, \ l = N - 1 = 1, \ h_{l-1} = h_0 = 6, \ k_{l-1} = k_0 = 1.$$

Clearly the pair of equations

$$A^3 + 111AB^2 = 48, \ 3A^2B + 37B^3 = 8$$

has no solutions in positive integers, and so the fundamental unit $\eta(>1)$ *of* $O_{\mathbb{Q}(\sqrt{37})}$ *is*

$$\eta = 6 + \sqrt{37}, \ N(\eta) = -1.$$

11.7 The Equation $x^2 - my^2 = N$

Let m be a positive squarefree integer. The following theorem from the theory of continued fractions assists us in finding the solutions (if any) of the equation $x^2 - my^2 = N$ when N is a nonzero integer satisfying $|N| < \sqrt{m}$.

Theorem 11.7.1 *Let m be a positive nonsquare integer. Let $\{h_n\}_{n \ge -1}$ and $\{k_n\}_{n \ge -1}$ be defined as in (11.6.5) and (11.6.6). Let $g_n = h_n^2 - mk_n^2, \ n = -1, 0, 1, \ldots$. Let* *l be the period of the continued fraction expansion of $\sqrt{m}$. Let N be an integer satisfying $0 < |N| < \sqrt{m}$. Then the equation $x^2 - my^2 = N$ is solvable in coprime integers x and y if and only if $N = g_r$ for some $r \in \{0, 1, 2, \ldots, sl - 1\}$, where $s = 1$ if l is even and $s = 2$ if l is odd in which case a solution is $(x, y) = (h_r, k_r)$.*

Example 11.7.1 *We choose $m = 31$. Here $\sqrt{31} = 5.567\ldots$. Thus the integers N satisfying $0 < |N| < \sqrt{m}$ are*

$$N = -5, -4, -3, -2, -1, 1, 2, 3, 4, 5.$$

From Example 11.6.1 we have $l = 8$, so that $s = 1$ and $sl - 1 = 7$, and the values of g_n $(n = 0, 1, \ldots, 7)$ are given in the following table:

n	h_n	k_n	$g_n = h_n^2 - 31k_n^2$
0	5	1	-6
1	6	1	5
2	11	2	-3
3	39	7	2
4	206	37	-3
5	657	118	5
6	863	155	-6
7	1520	273	1

By Theorem 11.7.1 the equation $x^2 - 31y^2 = N$ $(0 < |N| < \sqrt{31})$ is solvable in coprime integers x and y for

$$N = -3, 1, 2, 5$$

and is not solvable in coprime integers for

$$N = -5, -4, -2, -1, 3, 4.$$

Example 11.7.2 *We saw in Example 11.7.1 that the equation $x^2 - 31y^2 = -3$ has the solutions $(x, y) = (11, 2)$ and $(206, 37)$. We now determine all solutions of $x^2 - 31y^2 = -3$ in integers x and y. (Notice that if x and y are integers satisfying $x^2 - 31y^2 = -3$ then x and y are necessarily coprime as -3 is squarefree.) Let $(x, y) \in \mathbb{Z}^2$ be a solution of $x^2 - 31y^2 = -3$. Thus, as $(x, y) = (11, 2)$ is a solution of this equation, we have in $O_{\mathbb{Q}(\sqrt{31})}$*

$$\langle x + y\sqrt{31}\rangle\langle x - y\sqrt{31}\rangle = \langle 11 + 2\sqrt{31}\rangle\langle 11 - 2\sqrt{31}\rangle.$$

As $\langle 11 + 2\sqrt{31}\rangle$ and $\langle 11 - 2\sqrt{31}\rangle$ are prime ideals of $O_{\mathbb{Q}(\sqrt{31})}$ by Theorem 10.1.6(a) since

$$N(\langle 11 + 2\sqrt{31}\rangle) = |N(11 + 2\sqrt{31})| = |-3| = 3 \text{ (a prime)},$$

appealing to Theorem 8.3.2 we see that

$$\langle x + y\sqrt{31}\rangle = \langle 11 \pm 2\sqrt{31}\rangle.$$

Hence by Theorem 1.3.1 we obtain

$$x + y\sqrt{31} = u(11 \pm 2\sqrt{31}),$$

where $u \in U(O_{\mathbb{Q}(\sqrt{31})})$. The fundamental unit of $O_{\mathbb{Q}(\sqrt{31})}$ is $1520 + 273\sqrt{31}$ (Example 11.6.1) so that by Theorem 11.5.1

$$u = \pm(1520 + 273\sqrt{31})^n$$

for some $n \in \mathbb{Z}$. Hence all the solutions of $x^2 - 31y^2 = -3$ are given by

$$x + y\sqrt{31} = \pm(1520 + 273\sqrt{31})^n(11 \pm 2\sqrt{31}), \quad n = 0, \pm 1, \pm 2, \ldots.$$

It is easily checked that these are solutions of $x^2 - 31y^2 = -3$. In particular the solution $(x, y) = (206, 37)$ is given by

$$-(1520 + 273\sqrt{31})(11 - 2\sqrt{31}).$$

Example 11.7.3   *We answer the question "Is the equation $x^2 - 41y^2 = 2$ solvable in integers x and y?"*

As $2 < \sqrt{41}$ we can apply Theorem 11.7.1. From Example 11.6.2 we have $l = 3$, so that $s = 2$ and $sl - 1 = 5$, and the values of g_n $(n = 0, 1, 2, 3, 4, 5)$ are as follows:

n	h_n	k_n	g_n
0	6	1	-5
1	13	2	5
2	32	5	-1
3	397	62	5
4	826	129	-5
5	2049	320	1

As $2 \neq g_n$ $(n = 0, 1, 2, 3, 4, 5)$ the equation $x^2 - 41y^2 = 2$ has no solution in coprime integers x and y and thus (as 2 is squarefree) no solution in integers x and y.

Example 11.7.4   *We determine all the solutions in integers x and y of the equation $x^2 - 10y^2 = 10$. In this case we cannot apply Theorem 11.7.1 directly as $m = N = 10$ and $|N| \not< \sqrt{m}$. Thus we proceed differently.*

Let $(x, y) \in \mathbb{Z}^2$ be a solution of $x^2 - 10y^2 = 10$. Then $10 \mid x^2$ and as 10 is squarefree we deduce that $10 \mid x$. Setting $x = 10z$ in the equation we obtain

$y^2 - 10z^2 = -1$. *Thus $y + z\sqrt{10}$ is a unit of $O_{\mathbb{Q}(\sqrt{10})}$ of norm -1. As the fundamental unit of $O_{\mathbb{Q}(\sqrt{10})}$ is $3 + \sqrt{10}$ (of norm -1), we have by Theorem 11.4.1*

$$y + z\sqrt{10} = \pm(3 + \sqrt{10})^{2n+1}$$

for some $n \in \mathbb{Z}$. Hence all the solutions of $x^2 - 10y^2 = 10$ are given by

$$x + y\sqrt{10} = \pm(3 + \sqrt{10})^{2n+1}\sqrt{10}, \ n \in \mathbb{Z}.$$

It is easily verified that these are solutions of $x^2 - 10y^2 = 10$.

Exercises

1. Prove Theorem 11.5.6.
2. Prove Theorem 11.5.7.
3. Determine the fundamental unit of $O_{\mathbb{Q}(\sqrt{11})}$.
4. Determine the fundamental unit of $O_{\mathbb{Q}(\sqrt{17})}$.
5. Prove that the fundamental unit of $O_{\mathbb{Q}(\sqrt{94})}$ is $\eta = 2143295 + 221064\sqrt{94}$.
6. Prove that the fundamental unit of $O_{\mathbb{Q}(\sqrt{163})}$ is $\eta = 64080026 + 5019135\sqrt{163}$.
7. Prove that the fundamental unit of $O_{\mathbb{Q}(\sqrt{165})}$ is $\eta = (13 + \sqrt{165})/2$.
8. Show that $\sqrt{1790} = [42, \overline{3, 4, 8, 4, 3, 84}]$.
9. Show that $\sqrt{925} = [30, \overline{2, 2, 2, 2, 60}]$.
10. Determine the length of the continued fraction expansion of $\sqrt{850}$.
11. Determine the norm of the fundamental unit of $O_{\mathbb{Q}(\sqrt{1378})}$.
12. Let m be a positive squarefree integer such that $O_{\mathbb{Q}(\sqrt{m})}$ contains units of norm -1. Let σ be the fundamental unit of $O_{\mathbb{Q}(\sqrt{m})}$ of norm -1. Prove that σ is the smallest unit > 1 of norm -1 in $O_{\mathbb{Q}(\sqrt{m})}$.
13. Let η be the fundamental unit of $O_{\mathbb{Q}(\sqrt{134})}$. Determine $\alpha \in O_{\mathbb{Q}(\sqrt{134})}$ such that $2\eta = \alpha^2$.
14. Let p be a prime $\equiv 3 \pmod{8}$. Let $t + u\sqrt{p}$ be the fundamental unit of $O_{\mathbb{Q}(\sqrt{p})}$, which necessarily is of norm 1. Starting from $t^2 - pu^2 = 1$, and using Dirichlet's method of proving Theorem 11.5.4, prove that the equation $x^2 - py^2 = -2$ is solvable in integers x and y.
15. Let p be a prime $\equiv 7 \pmod{8}$. Prove that the equation $x^2 - py^2 = 2$ is solvable in integers x and y.
16. Let p be a prime $\equiv 9 \pmod{16}$ for which the congruence $x^4 \equiv 2 \pmod{p}$ is insolvable. Prove that the norm of the fundamental unit of $O_{\mathbb{Q}(\sqrt{2p})}$ is -1.
17. Let p and q be distinct primes with $p \equiv q \equiv 1 \pmod 4$ and $\left(\frac{p}{q}\right) = 1$ (so that $\left(\frac{q}{p}\right) = 1$ by the law of quadratic reciprocity). Suppose that the congruences $x^4 \equiv p \pmod{q}$ and $y^4 \equiv q \pmod{p}$ are insolvable. Prove that the norm of the fundamental unit of $O_{\mathbb{Q}(\sqrt{pq})}$ is -1.
18. Is the equation $x^2 - 82y^2 = 2$ solvable in integers x and y?
19. Determine all solutions of $x^2 - 96y^2 = 161$.

20. Prove that all solutions of $x^2 - 10y^2 = 10$ (see Example 11.7.1) are given recursively by $\pm(x_k, y_k)$, where

$$x_{k+1} = 19x_k + 60y_k, \quad y_{k+1} = 6x_k + 19y_k, \quad k = 0, \pm1, \pm2, \ldots,$$

and $x_0 = 10$, $y_0 = 3$.

21. Let m be a positive integer such that $m - 1$ and m are not perfect squares but $4m + 1$ is a perfect square. Prove that the equation $x^2 - my^2 = -1$ is insolvable in integers x and y.

Suggested Reading

1. H. W. Lenstra Jr., *Solving the Pell equation*, Notices of the American Mathematical Society 49 (2002), 182–192.

   This up-to-date article describes the use of smooth numbers to solve Pell's equation $x^2 - my^2 = 1$.

2. I. Niven, H. S. Zuckerman, and H. L. Montgomery, *An Introduction to the Theory of Numbers*, fifth edition, Wiley, New York, 1991.

   Chapter 7 contains a comprehensive treatment of continued fractions.

3. W. Patz, *Tafel der Regelmässigen Kettenbrüche und ihrer Vollständigen Quotienten für die Quadratwurzeln aus der Natürlichen Zahlen von 1–10000*, Akademie-Verlag, Berlin, 1955.

   This book gives the continued fraction expansions of $\sqrt{m}$ for all nonsquares m up to 10,000.

Biographies

1. E. T. Bell, *Men of Mathematics*, Simon and Schuster, New York, 1937.

   Chapters 9 and 10 are devoted to Leonhard Euler (1707–1783) and Joseph-Louis Lagrange (1736–1813) respectively.

2. J. J. Burckhardt, *Leonhard Euler, 1707–1783*, Mathematics Magazine 56 (1983), 262–273.

   A brief overview of the work of Euler is given.

3. H. Davenport, *Dirichlet*, Mathematical Gazette 43 (1959), 268–269.

   A short biography of Dirichlet is given.

4. H. Koch, *Gustav Peter Lejeune Dirichlet*, in H. G. W. Begehr, H. Koch, J. Kramer, N. Schappacher, and E. J. Thiele (Eds.), *Mathematics in Berlin*, 33–40, Birkhäuser Verlag, Berlin, 1998.

   The book in which this biography of Dirichlet is included describes the many facets of Berlin's role in mathematics. It includes biographies of many famous mathematicians connected with Berlin.

5. The website

   http://www-groups.dcs.st-and.ac.uk/~history/

   has biographies of William Brouncker, Lejeune Dirichlet, Leonhard Euler, Joseph-Louis Lagrange, and John Pell.

12

The Ideal Class Group

12.1 Ideal Class Group

We have already seen that the nonzero integral and fractional ideals of the ring O_K of integers of an algebraic number field K form a group $I(K)$ under multiplication (Theorem 8.3.4). The principal ideals in $I(K)$ are of the form $\langle \alpha \rangle = \{ r\alpha \mid r \in O_K \}$ for some $\alpha \in K^*$ and they form a subgroup $P(K)$ of $I(K)$ as

$$\langle \alpha \rangle \langle \beta \rangle^{-1} = \langle \alpha \beta^{-1} \rangle \in P(K).$$

The group $I(K)$ is an Abelian group so $P(K)$ is a normal subgroup of $I(K)$ and the factor group $I(K)/P(K)$ is well defined and Abelian.

Definition 12.1.1 (Ideal class group) *Let K be an algebraic number field. Let $I(K)$ be the group of nonzero fractional and integral ideals of O_K. Let $P(K)$ be the subgroup of principal ideals of $I(K)$. Then the factor group $I(K)/P(K)$ is called the ideal class group of K and is denoted by $H(K)$.*

It is an important result that $H(K)$ is always a finite group. This is proved in Section 12.5 as a consequence of some theorems of Hermann Minkowski (1864–1909) in the geometry of numbers.

Definition 12.1.2 (Class number) *Let K be an algebraic number field. The order of the ideal class group $H(K)$ is called the class number of K and is denoted by $h(K)$.*

If two nonzero ideals A and B of O_K are in the same class of $H(K) = I(K)/P(K)$, we say that they are equivalent and write $A \sim B$. Clearly

$$
\begin{aligned}
A \sim B &\Longleftrightarrow AP(K) = BP(K) \\
&\Longleftrightarrow A^{-1}B \in P(K) \\
&\Longleftrightarrow A^{-1}B = \langle \alpha \rangle \text{ for some } \alpha \in K^* \\
&\Longleftrightarrow B = A\langle \alpha \rangle \text{ for some } \alpha \in K^* \\
&\Longleftrightarrow \langle a \rangle A = \langle b \rangle B \text{ for some } a, b \in O_K \setminus \{0\}.
\end{aligned}
$$

Example 12.1.1 *In* $\mathbb{Q}(\sqrt{-5})$ *we have*

$$\langle 2, 1 - \sqrt{-5} \rangle \sim \langle 3, 1 + \sqrt{-5} \rangle$$

as

$$
\begin{aligned}
\langle 3 \rangle \langle 2, 1 - \sqrt{-5} \rangle &= \langle 6, 3(1 - \sqrt{-5}) \rangle \\
&= \langle (1 + \sqrt{-5})(1 - \sqrt{-5}), 3(1 - \sqrt{-5}) \rangle \\
&= \langle 1 - \sqrt{-5} \rangle \langle 1 + \sqrt{-5}, 3 \rangle.
\end{aligned}
$$

Theorem 12.1.1 *Let K be an algebraic number field. Then*

$$h(K) = 1 \iff O_K \text{ is a principal ideal domain}$$
$$\iff O_K \text{ is a unique factorization domain.}$$

Proof: If $h(K) = 1$ then $[I(K) : P(K)] = \text{card}(I(K)/P(K)) = \text{card } H(K) = h(K) = 1$ so that $P(K) = I(K)$. Hence every ideal of O_K is principal and so O_K is a principal ideal domain, and thus a unique factorization domain, by Theorem 3.3.1.

Conversely, if O_K is a unique factorization domain, since it is a Dedekind domain, it is a principal ideal domain (Exercise 13 of Chapter 8). Hence every ideal of O_K is principal and so $I(K) = P(K)$ and thus

$$h(K) = \text{card } H(K) = \text{card}(I(K)/P(K)) = [I(K) : P(K)] = 1. \qquad \blacksquare$$

Leonard Carlitz (1907–1999) has shown that $h(K) = 1$ or 2 if and only if whenever a nonzero nonunit $\alpha \in O_K$ can be written $\alpha = u\pi_1 \cdots \pi_s = u'\pi_1' \cdots \pi_t'$ with u, u' units and $\pi_1, \ldots, \pi_s, \pi_1', \ldots, \pi_t'$ prime elements of O_K then $s = t$ [2].

In the next three sections we prove three theorems of Minkowski from which we can deduce that the class number is always finite.

12.2 Minkowski's Translate Theorem

Let $\mathbb{R}^n$ denote the vector space of all n-tuples $(x_1, x_2, \ldots, x_n)$ with $x_1, x_2, \ldots, x_n \in \mathbb{R}$. We let $\mathbb{Z}^n$ be the subset of $\mathbb{R}^n$ given by

$$\mathbb{Z}^n = \{(x_1, \ldots, x_n) \in \mathbb{R}^n \mid x_1, \ldots, x_n \in \mathbb{Z}\}.$$

The elements of $\mathbb{Z}^n$ are called lattice points and $\mathbb{Z}^n$ is called a lattice. Clearly $\mathbb{Z}^n$ is a group under addition. For $\alpha = (a_1, \ldots, a_n) \in \mathbb{R}^n$ we set

$$\|\alpha\| = \max_{1 \le i \le n} |a_i| \ (\in \mathbb{R}).$$

Definition 12.2.1 (Translate) *If S is a subset of $\mathbb{R}^n$ and $\alpha \in \mathbb{R}^n$ we let*

$$S_\alpha = \{\alpha + \beta \mid \beta \in S\}.$$

The set $S_\alpha (\subseteq \mathbb{R}^n)$ is called a translate of S in $\mathbb{R}^n$.

Clearly $S_{\underline{0}} = S$, where $\underline{0} = (0, \ldots, 0)$.

Definition 12.2.2 (Magnification) *If S is a subset of $\mathbb{R}^n$ and $a \in \mathbb{R}^+$ we let*

$$aS = \{a\beta \mid \beta \in S\}.$$

The set $aS (\subseteq \mathbb{R}^n)$ is called a magnification of S in $\mathbb{R}^n$.

Definition 12.2.3 (Bounded set) *A subset S of $\mathbb{R}^n$ is said to be bounded if there exists $B \in \mathbb{R}^+$ such that*

$$\|\alpha\| \leq B \text{ for all } \alpha \in S.$$

Definition 12.2.4 (Closed set) *Let $\alpha \in \mathbb{R}^n$. Let $r \in \mathbb{R}^+$. The set*

$$\{\beta \in \mathbb{R}^n \mid \|\alpha - \beta\| < r\}$$

is called a neighborhood of α. The point α is called a limit point of the subset S of $\mathbb{R}^n$ if every neighborhood of α contains a point $\beta \neq \alpha$ such that $\beta \in S$. The set S is said to be closed if every limit point of S is a point of S.

Definition 12.2.5 (Convex set) *A subset S of $\mathbb{R}^n$ is said to be convex if*

$$t\beta + (1 - t)\gamma \in S \text{ for all } \beta, \gamma \in S \text{ and all } t \in \mathbb{R} \text{ with } 0 \leq t \leq 1.$$

Clearly S is convex if it contains the line segment joining β and γ for all points β and γ in S.

Definition 12.2.6 (Convex body) *A closed, bounded, convex subset of $\mathbb{R}^n$ is called a convex body.*

If S is a convex body so are S_α ($\alpha \in \mathbb{R}^n$) and aS ($a \in \mathbb{R}^+$). Moreover, $aS \subseteq bS$ if $0 < a < b$.

A theorem of Minkowski asserts that every convex body S has a volume, which we denote by $V(S)$ and which has the following properties:

(i) $0 \leq V(S) < \infty$,
(ii) if S_i ($i = 1, 2, \ldots, k$) are disjoint convex bodies with $S_i \subseteq S$ ($i = 1, 2, \ldots, k$) then $\sum_{i=1}^{k} V(S_i) \leq V(S)$,

(iii) $V(S) = V(S_\alpha)$ for all $\alpha \in \mathbb{R}^n$,
(iv) $V(aS) = a^n V(S)$ for all $a \in \mathbb{R}^+$.

If $S_1, \ldots, S_k$ are disjoint convex bodies we define

$$V(S_1 \cup S_2 \cup \cdots \cup S_k) = \sum_{i=1}^{k} V(S_i).$$

The volume $V(S)$ of a convex body S is defined by means of the multiple integral

$$V(S) = \int \cdots \int_S dx_1 \cdots dx_n \, .$$

Definition 12.2.7 (Hypercube H_t) *The hypercube H_t ($t \in \mathbb{R}^+$) in $\mathbb{R}^n$ is defined by*

$$H_t = \{\beta \in \mathbb{R}^n \mid ||\beta|| \leq t\}.$$

It is easy to check that the hypercube H_t is a convex body. Its volume is given by

$$V(H_t) = \int_{\beta_1 = -t}^{t} \cdots \int_{\beta_n = -t}^{t} d\beta_1 \cdots d\beta_n = \left(\int_{\beta = -t}^{t} d\beta \right)^n = (2t)^n.$$

Theorem 12.2.1 (Minkowski's translate theorem) *Let S be a convex body in $\mathbb{R}^n$ that contains the origin $\underline{0} = (0, 0, \ldots, 0)$. If $V(S) \geq 1$ then for at least one $\alpha \in \mathbb{Z}^n \setminus \{\underline{0}\}$*

$$S \cap S_\alpha \neq \phi.$$

Proof: We first treat the case $V(S) > 1$. Suppose on the contrary that

$$S \cap S_\alpha = \phi \text{ for all } \alpha \in \mathbb{Z}^n \setminus \{\underline{0}\}.$$

First we prove that

$$S_\beta \cap S_\gamma = \phi \text{ for all } \beta, \gamma \in \mathbb{Z}^n, \ \beta \neq \gamma. \tag{12.2.1}$$

Let $x \in S_\beta \cap S_\gamma$. Then $x \in S_\beta$ and $x \in S_\gamma$. Hence $x - \gamma \in S_{\beta - \gamma}$ and $x - \gamma \in S$. Thus $x - \gamma \in S_{\beta - \gamma} \cap S$, contradicting our assumption. This proves (12.2.1).

Now let N be an arbitrary positive integer, and let

$$T = \{\alpha \in \mathbb{Z}^n \mid ||\alpha|| \leq N\}.$$

Clearly

$$\text{card } T = (2N + 1)^n.$$

Our second step is to show that

$$V \left(\bigcup_{\alpha \in T} S_\alpha \right) = (2N + 1)^n V(S). \tag{12.2.2}$$

We have

$$V\left(\bigcup_{\alpha \in T} S_\alpha\right) = \sum_{\alpha \in T} V(S_\alpha) \text{ (by (12.2.1))}$$

$$= \sum_{\alpha \in T} V(S)$$

$$= V(S)\text{card } T$$

$$= (2N + 1)^n V(S),$$

as asserted.

As S is a bounded set, we can define the diameter $d \in \mathbb{R}^+$ of S by

$$d = \max_{s_1, s_2 \in S} ||s_1 - s_2||.$$

We let H be the hypercube H_{N+d}, that is,

$$H = H_{N+d} = \{\beta \in \mathbb{R}^n \mid ||\beta|| \le N + d\}.$$

Clearly

$$V(H) = (2N + 2d)^n.$$

Our third step is to show that

$$S_\alpha \subseteq H \text{ for all } \alpha \in T. \tag{12.2.3}$$

Let $\alpha \in T$ and $\beta \in S_\alpha$. Then $\beta = \alpha + s$, where $s \in S$. As $\alpha \in T$ we have $||\alpha|| \le N$. As $\underline{0} \in S$ we have $\alpha = \alpha + \underline{0} \in S_\alpha$ so that $s = \beta - \alpha$, where $\alpha, \beta \in S_\alpha$. Thus

$$||s|| = ||\beta - \alpha|| \le \max_{t_1, t_2 \in S_\alpha} ||t_1 - t_2|| = \max_{s_1, s_2 \in S} ||(\alpha + s_1) - (\alpha + s_2)||$$

$$= \max_{s_1, s_2 \in S} ||s_1 - s_2|| = d.$$

Hence

$$||\beta|| = ||\alpha + s|| \le ||\alpha|| + ||s|| \le N + d$$

so that $\beta \in H$. Hence $S_\alpha \subseteq H$ for $\alpha \in T$.

We are now in a position to complete the proof. From (12.2.3) we have

$$\bigcup_{\alpha \in T} S_\alpha \subseteq H$$

so that

$$V\left(\bigcup_{\alpha \in T} S_\alpha\right) \le V(H)$$

and thus

$$(2N + 1)^n V(S) \le (2N + 2d)^n.$$

Therefore

$$V(S) \le \left(\frac{2N + 2d}{2N + 1}\right)^n = \left(\frac{1 + \dfrac{d}{N}}{1 + \dfrac{1}{2N}}\right)^n.$$

As d and n are fixed, letting $N \to \infty$ we obtain $V(S) \le 1$, contradicting $V(S) > 1$. Hence there is at least one $\alpha \in \mathbb{Z}^n \setminus \{\underline{0}\}$ with $S \cap S_\alpha \ne \phi$.

We now turn to the case $V(S) = 1$. Let k be a positive integer. We consider the convex body

$$S(k) = \left(1 + \frac{1}{k}\right) S$$

obtained by magnifying the convex body S by a factor $1 + 1/k$. As S contains $\underline{0}$ so does $S(k)$. The volume of $S(k)$ satisfies

$$V(S(k)) = V\left(\left(1 + \frac{1}{k}\right) S\right) = \left(1 + \frac{1}{k}\right)^n V(S) = \left(1 + \frac{1}{k}\right)^n > 1.$$

Thus the first part of the theorem applies to the convex body $S(k)$. Hence there exists a translate $(S(k))_{\alpha_k}$ $(\alpha_k \in \mathbb{Z}^n \setminus \{\underline{0}\})$ of $S(k)$ such that

$$S(k) \cap (S(k))_{\alpha_k} \ne \phi.$$

Let $x_k \in S(k) \cap (S(k))_{\alpha_k}$.

Next we consider the set

$$A_k = \{\beta_k \in \mathbb{Z}^n \setminus \{\underline{0}\} \mid x_k \in S(k) \cap (S(k))_{\beta_k}\}.$$

Clearly $A_k \ne \phi$ as $\alpha_k \in A_k$. For $\beta_k \in A_k$ we have

$$x_k \in S(k) \subseteq S(1)$$

and

$$x_k \in (S(k))_{\beta_k} \text{ so } x_k - \beta_k \in S(k) \subseteq S(1).$$

Let d_1 denote the diameter of $S(1)$. Then

$$\begin{aligned}
\|\beta_k\| &= \|x_k - (x_k - \beta_k)\| \le \|x_k\| + \|x_k - \beta_k\| \\
&= \|x_k - \underline{0}\| + \|(x_k - \beta_k) - \underline{0}\| \\
&\le d_1 + d_1 \\
&= 2d_1,
\end{aligned}$$

as $\underline{0} \in S(1)$. Thus β_k lies in a bounded set, namely,

$$\beta_k \in H_{2d_1},$$

so that

$$A_k \subseteq H_{2d_1}.$$

But

$$A_k \subseteq \mathbb{Z}^n$$

so

$$A_k \subseteq H_{2d_1} \cap \mathbb{Z}^n.$$

Thus every A_k is contained in the finite set $H_{2d_1} \cap \mathbb{Z}^n$. Hence we can find a subsequence of the sequence $\{x_k\}$ for which the corresponding $\beta_k \in A_k$ is constant, say, equal to α. After relabeling we may therefore assume that

$$x_k \in S(k) \cap (S(k))_\alpha.$$

As each $x_k \in S(1)$, the infinite sequence $\{x_k\}$ is bounded. Hence, by the Bolzano–Weierstrass theorem, the sequence $\{x_k\}$ has at least one limit point, say x. Let l be an arbitrary positive integer. We have

$$x_l \in S(l),$$
$$x_{l+1} \in S(l+1) \subset S(l)$$
$$x_{l+2} \in S(l+2) \subset S(l+1) \subset S(l),$$
$$\cdots$$

so that the infinite sequence $\{x_k\}_{k \geq l}$ lies in $S(l)$. As x is a limit point of $\{x_k\}_{k \geq l}$ and $S(l)$ is closed, we deduce that $x \in S(l)$ for every positive integer l. Thus

$$x \in \bigcap_{l=1}^{\infty} S(l) = \bigcap_{l=1}^{\infty} \left(1 + \frac{1}{l}\right) S = S.$$

Similarly,

$$x \in \bigcap_{l=1}^{\infty} (S(l))_\alpha = S_\alpha.$$

Hence

$$x \in S \cap S_\alpha$$

so that $S \cap S_\alpha \neq \phi$ for $\alpha \in \mathbb{Z}^n \setminus \{0\}$. ∎

12.3 Minkowski's Convex Body Theorem

Minkowski's famous convex body theorem asserts that if the volume of a convex body, which is symmetrical about the origin, is large enough then the convex body must contain at least one lattice point different from the origin. We begin by making the notion "symmetrical about the origin" precise.

Definition 12.3.1 (Centrally symmetric set) *A subset S of $\mathbb{R}^n$ is said to be centrally symmetric if*

$$-\alpha \in S \text{ for all } \alpha \in S.$$

We note that a subset S of $\mathbb{R}^n$, which is both centrally symmetric and convex, must contain the origin $\underline{0} = (0, 0, \ldots, 0)$. To see this take any $\alpha \in S$. As S is centrally symmetric, we have $-\alpha \in S$. Then, as S is convex, we have $\frac{1}{2}(\alpha) + \frac{1}{2}(-\alpha) \in S$, that is, $\underline{0} \in S$.

We now use Minkowski's translate theorem to prove his convex body theorem.

Theorem 12.3.1 (Minkowski's convex body theorem) *Let $S \,(\subseteq \mathbb{R}^n)$ be a centrally symmetric convex body of volume $V(S) \geq 2^n$. Then S contains a lattice point $\neq \underline{0}$.*

Proof: Let T be the magnification of S given by

$$T = \frac{1}{2}S = \{\frac{1}{2}\alpha \mid \alpha \in S\}.$$

As S is a centrally symmetric convex body so is T. As T is both centrally symmetric and convex it contains $\underline{0}$. Moreover,

$$V(T) = V(\tfrac{1}{2}S) = \tfrac{1}{2^n}V(S) \geq 1,$$

so that by Minkowski's translate theorem (Theorem 12.2.1) there exists $\alpha \in \mathbb{Z}^n \setminus \{\underline{0}\}$ such that

$$T \cap T_\alpha \neq \phi.$$

Let $x \in T \cap T_\alpha$. Then $x \in T_\alpha$ and thus $x - \alpha \in T$. Since T is centrally symmetric, we have $\alpha - x = -(x - \alpha) \in T$. Hence, as $x \in T$ and T is convex, we have

$$\frac{1}{2}(\alpha - x) + \frac{1}{2}x \in T,$$

so that $\alpha/2 \in T$ and thus $\alpha \in S$. This proves that S contains the lattice point $\alpha \neq \underline{0}$.   ∎

12.4 Minkowski's Linear Forms Theorem

Let n be a positive integer and let r and s be nonnegative integers such that $r + 2s = n$. Let a_{jk} $(j, k = 1, 2, \ldots, n)$ be n^2 complex numbers with $\det a_{jk} \neq 0$ and

$$a_{jk} \in \mathbb{R} \text{ for } j = 1, 2, \ldots, r; \ k = 1, 2, \ldots, n, \tag{12.4.1}$$

for each $j = r + 1, \ldots, n$ there exists $k_j \in \{1, 2, \ldots, n\}$
such that $a_{jk_j} \in \mathbb{C} \setminus \mathbb{R}$, (12.4.2)

and

$$a_{j+s\ k} = \overline{a_{jk}} \text{ for } j = r + 1, \ldots, r + s; \ k = 1, 2, \ldots, n. \quad (12.4.3)$$

Here $\bar{z}$ denotes the complex conjugate of the complex number z. Set

$$L_j = L_j(\underline{x}) = \sum_{k=1}^{n} a_{jk} x_k, \ j = 1, 2, \ldots, n, \quad (12.4.4)$$

so that $L_1, \ldots, L_n$ are linear forms such that $L_1, \ldots, L_r$ are real, $L_{r+1}, \ldots, L_{n=r+2s}$ are nonreal, and $L_{r+s+1} = \overline{L_{r+1}}, \ldots, L_{r+2s} = \overline{L_{r+s}}$.

Theorem 12.4.1 (Minkowski's linear forms theorem) *Let $\delta_1, \ldots, \delta_n$ be n positive real numbers such that*

$$\delta_1 \cdots \delta_n \geq \left(\frac{2}{\pi}\right)^s |\det(a_{jk})|, \quad (12.4.5)$$

where the a_{jk} are defined in (12.4.1)–(12.4.3), and

$$\delta_j = \delta_{j+s}, \ j = r + 1, \ldots, r + s. \quad (12.4.6)$$

Then there exist integers $y_1, \ldots, y_n$, not all zero, such that

$$\left| \sum_{k=1}^{n} a_{jk} y_k \right| \leq \delta_j, \ j = 1, 2, \ldots, n.$$

Proof: We define n real linear forms $M_j = M_j(\underline{x})$ $(j = 1, 2, \ldots, n)$ in terms of the $L_j(\underline{x})$ $(j = 1, 2, \ldots, n)$ by

$$M_j = \begin{cases} L_j, & j = 1, 2, \ldots, r, \\ \frac{1}{2}(L_j + L_{j+s}), & j = r + 1, \ldots, r + s, \\ \frac{1}{2i}(L_{j-s} - L_j), & j = r + s + 1, \ldots, n. \end{cases} \quad (12.4.7)$$

From (12.4.7) we see that the $n \times n$ matrix $\left(\frac{\partial M_j}{\partial L_k}\right)$ is given by

$$\left(\frac{\partial M_j}{\partial L_k}\right) = \begin{bmatrix} I_r & O_{r,s} & O_{r,s} \\ O_{s,r} & \frac{1}{2}I_s & \frac{1}{2}I_s \\ O_{s,r} & \frac{1}{2i}I_s & \frac{-1}{2i}I_s \end{bmatrix},$$

where I_l denotes the $l \times l$ identity matrix and $O_{l,m}$ denotes the $l \times m$ zero matrix. Adding the $(r + s + k)$th column to the $(r + k)$th column in the matrix $\left(\frac{\partial M_j}{\partial L_k}\right)$ for

$k = 1, 2, \ldots, s$, we obtain the upper triangular matrix

$$A = \begin{bmatrix} I_r & O_{r,s} & O_{r,s} \\ O_{s,r} & I_s & \frac{1}{2}I_s \\ O_{s,r} & O_s & \frac{-1}{2i}I_s \end{bmatrix},$$

where $O_s = O_{s,s}$. The determinant of a triangular matrix is the product of its diagonal entries so that

$$\det A = \left(\frac{-1}{2i} \right)^s = \left(\frac{i}{2} \right)^s.$$

As the elementary column operations used to obtain A do not change the value of the determinant of the matrix, we have

$$\det \left(\frac{\partial M_j}{\partial L_k} \right) = \det A = \left(\frac{i}{2} \right)^s.$$

The quantity $\det \left(\frac{\partial M_j}{\partial L_k} \right)$ is the Jacobian of the M_j with respect to the L_k so that

$$\frac{\partial(M_1, \ldots, M_n)}{\partial(L_1, \ldots, L_n)} = \left(\frac{i}{2} \right)^s.$$

From (12.4.4) we deduce that

$$\frac{\partial L_j}{\partial x_k} = a_{jk}, \quad j, k = 1, 2, \ldots, n,$$

so that

$$\frac{\partial(L_1, \ldots, L_n)}{\partial(x_1, \ldots, x_n)} = \det \left(\frac{\partial L_j}{\partial x_k} \right) = \det(a_{jk}).$$

Hence

$$\frac{\partial(M_1, \ldots, M_n)}{\partial(x_1, \ldots, x_n)} = \frac{\partial(M_1, \ldots, M_n)}{\partial(L_1, \ldots, L_n)} \frac{\partial(L_1, \ldots, L_n)}{\partial(x_1, \ldots, x_n)} = \left(\frac{i}{2} \right)^s \det(a_{jk}),$$

so that

$$\left| \frac{\partial(x_1, \ldots, x_n)}{\partial(M_1, \ldots, M_n)} \right| = \frac{2^s}{|\det(a_{jk})|}. \tag{12.4.8}$$

Now let S be the subset of $\mathbb{R}^n$ given by

$$S = \{\underline{x} \in \mathbb{R}^n \mid |L_j(\underline{x})| \le \delta_j, \; j = 1, 2, \ldots, r + s\}.$$

It is easily checked that S is a centrally symmetric convex body. The volume of S is given by

$$V(S) = \int \cdots \int\limits_{\substack{|L_1(\underline{x})| \leq \delta_1 \\ \vdots \\ |L_{r+s}(\underline{x})| \leq \delta_{r+s}.}} dx_1 \cdots dx_{r+s}.$$

Now, as

$$|L_{r+j}(\underline{x})| = |\overline{L_{r+j}(\underline{x})}| = |L_{r+s+j}(\underline{x})|, \ \ j = 1, 2, \ldots, s,$$

and

$$\delta_{r+j} = \delta_{r+s+j}, \ \ j = 1, 2, \ldots, s,$$

we deduce that

$$\begin{aligned}
|L_{r+j}(\underline{x})| \leq \delta_{r+j} &\iff |L_{r+j}(\underline{x})|^2 \leq \delta_{r+j}^2 \\
&\iff |L_{r+j}(\underline{x}) L_{r+s+j}(\underline{x})|^2 \leq \delta_{r+j}^2 \\
&\iff |(M_{r+j}(\underline{x}) + i\, M_{r+s+j}(\underline{x}))(M_{r+j}(\underline{x}) - i\, M_{r+s+j}(\underline{x}))|^2 \leq \delta_{r+j}^2 \\
&\iff M_{r+j}(\underline{x})^2 + M_{r+s+j}(\underline{x})^2 \leq \delta_{r+j}^2
\end{aligned}$$

for $j = 1, 2, \ldots, s$. Hence

$$V(S) = \int \cdots \int\limits_{\substack{|M_1(\underline{x})| \leq \delta_1 \\ \vdots \\ |M_r(\underline{x})| \leq \delta_r \\ M_{r+1}(\underline{x})^2 + M_{r+s+1}(\underline{x})^2 \leq \delta_{r+1}^2 \\ \vdots \\ M_{r+s}(\underline{x})^2 + M_n(\underline{x})^2 \leq \delta_{r+s}^2}} dx_1 \cdots dx_n.$$

Making the change of variable

$$M_j = M_j(\underline{x}), \ \ j = 1, 2, \ldots, n,$$

in the integral, we obtain

$$V(S) = \int \cdots \int\limits_{\substack{|M_1| \leq \delta_1 \\ \vdots \\ |M_r| \leq \delta_r \\ M_{r+1}^2 + M_{r+s+1}^2 \leq \delta_{r+1}^2 \\ \vdots \\ M_{r+s}^2 + M_n^2 \leq \delta_{r+s}^2}} \left| \frac{\partial(x_1, \ldots, x_n)}{\partial(M_1, \ldots, M_n)} \right| dM_1 \cdots dM_n.$$

Appealing to (12.4.8), we have

$$V(S) = \frac{2^s}{|\det(a_{jk})|} \int_{|M_1| \le \delta_1} \cdots \int dM_1 \cdots dM_n.$$

$$\vdots$$
$$|M_r| \le \delta_r$$
$$M_{r+1}^2 + M_{r+s+1}^2 \le \delta_{r+1}^2$$
$$\vdots$$
$$M_{r+s}^2 + M_n^2 \le \delta_{r+s}^2$$

Expressing the integral in terms of repeated integrals, we deduce that

$$V(S) = \frac{2^s}{|\det(a_{jk})|} \prod_{j=1}^{r} \left(\int_{-\delta_j}^{\delta_j} dM_j \right) \prod_{k=1}^{s} \left(\int \cdots \int_{M_{r+k}^2 + M_{r+s+k}^2 \le \delta_{r+k}^2} dM_{r+k} dM_{r+s+k} \right).$$

Now

$$\int_{-\delta}^{\delta} dM = 2\delta$$

and

$$\int \cdots \int_{M^2 + N^2 \le \delta^2} dM \, dN = \text{area of a circle of radius } \delta = \pi\delta^2,$$

so that

$$V(S) = \frac{2^s}{|\det(a_{jk})|} \prod_{j=1}^{r} 2\delta_j \prod_{k=1}^{s} \pi\delta_{r+k}^2$$

$$= \frac{2^{r+s}\pi^s \delta_1 \cdots \delta_r (\delta_{r+1} \cdots \delta_{r+s})^2}{|\det(a_{jk})|}$$

$$= \frac{2^{r+s}\pi^s \delta_1 \cdots \delta_n}{|\det(a_{jk})|} \quad \text{(by (12.4.6))}$$

$$\ge \frac{2^{r+s}\pi^s}{|\det(a_{jk})|} \left(\frac{2}{\pi}\right)^s |\det(a_{jk})| \quad \text{(by (12.4.5))}$$

$$= 2^{r+2s} = 2^n.$$

Hence, by Minkowski's convex body theorem (Theorem 12.3.1), S contains a lattice point $\underline{y} = (y_1, \ldots, y_n) \ne (0, \ldots, 0)$. Thus

$$|L_j(\underline{y})| \le \delta_j, \quad j = 1, 2, \ldots, r+s,$$

from which the asserted result follows by (12.4.3), (12.4.4), and (12.4.6). ∎

12.5 Finiteness of the Ideal Class Group

In this section we use Minkowski's linear forms theorem to show that every class in the ideal class group $H(K)$ of an algebraic number field K contains an integral ideal of O_K with norm less than a certain bound, called the Minkowski bound, that depends only on the degree of the field K and the discriminant of K. For a particular algebraic number field K these ideas give a method of determining the ideal class group $H(K)$.

Theorem 12.5.1 *Let $K = \mathbb{Q}(\theta)$ be an algebraic number field of degree $n = r + 2s$, where θ has r real conjugates and s pairs of nonreal complex conjugates. Let A be an integral or fractional ideal of O_K. Then there exists an element α ($\neq 0$) $\in A$ such that*

$$|N(\alpha)| \leq \left(\frac{2}{\pi}\right)^s N(A)\sqrt{|d(K)|}.$$

Proof: Let $\theta_1, \theta_2, \ldots, \theta_n$ be the conjugates of θ. We reorder $\theta_1, \theta_2, \ldots, \theta_n$ in such a way that $\theta_1, \theta_2, \ldots, \theta_r \in \mathbb{R}$ and $\theta_{r+1}, \theta_{r+2}, \ldots, \theta_n \in \mathbb{C} \setminus \mathbb{R}$. As the complex conjugate of any conjugate of θ is also a conjugate of θ (Exercise 20 of Chapter 5), we can further order $\theta_{r+1}, \theta_{r+2}, \ldots, \theta_n$ so that $\theta_{r+s+1} = \overline{\theta_{r+1}}, \ldots, \theta_n = \theta_{r+2s} = \overline{\theta_{r+s}}$, where $r + 2s = n$. Let $\sigma_1, \ldots, \sigma_n$ be the n monomorphisms : $K \to \mathbb{C}$ chosen so that $\sigma_i(\theta) = \theta_i$. Hence $\sigma_{r+s+t} = \overline{\sigma_{r+t}}$ ($t = 1, \ldots, s$).

Let $\{\alpha_1, \ldots, \alpha_n\}$ be a basis for A. We define n linear forms $L_j(\underline{x})$ ($j = 1, 2, \ldots, n$) by

$$L_j(\underline{x}) = \sum_{k=1}^{n} \sigma_j(\alpha_k)x_k.$$

These forms satisfy (12.4.1)–(12.4.4) with $a_{jk} = \sigma_j(\alpha_k)$ ($j, k = 1, 2, \ldots, n$). Moreover,

$$|\det(a_{jk})| = |\det(\sigma_j(\alpha_k))| = \sqrt{|D(A)|} = N(A)\sqrt{|d(K)|} \neq 0.$$

Let

$$\delta_j = \left(\frac{2}{\pi}\right)^{s/n} N(A)^{1/n}|d(K)|^{1/2n}, \quad j = 1, 2, \ldots, n.$$

Then

$$\delta_1 \cdots \delta_n = \left(\frac{2}{\pi}\right)^s N(A)|d(K)|^{1/2} = \left(\frac{2}{\pi}\right)^s |\det(a_{jk})|$$

so, by Minkowski's linear forms theorem (Theorem 12.4.1), there exist integers $y_1, \ldots, y_n$, not all zero, such that

$$|L_j(\underline{y})| \leq \left(\frac{2}{\pi}\right)^{s/n} N(A)^{1/n} |d(K)|^{1/2n}, \quad j = 1, 2, \ldots, n.$$

Choose $m \in \{1, 2, \ldots, n\}$ such that $\sigma_m = 1$, where 1 denotes the identity monomorphism from K to K. Set

$$\alpha = L_m(\underline{y}) = \sum_{k=1}^{n} \sigma_m(\alpha_k) y_k = \sum_{k=1}^{n} \alpha_k y_k,$$

so that $\alpha \in A$ and $\alpha \neq 0$. The conjugates of α are

$$\sigma_j(\alpha) = \sum_{k=1}^{n} \sigma_j(\alpha_k) y_k = L_j(\underline{y}), \quad j = 1, 2, \ldots, n.$$

Hence

$$|\sigma_j(\alpha)| \leq \left(\frac{2}{\pi}\right)^{s/n} N(A)^{1/n} |d(K)|^{1/2n}, \quad j = 1, 2, \ldots, n,$$

and so

$$|N(\alpha)| = |\sigma_1(\alpha) \cdots \sigma_n(\alpha)| \leq \left(\frac{2}{\pi}\right)^{s} N(A)|d(K)|^{1/2}$$

as asserted.   ∎

Theorem 12.5.2 *Let $K = \mathbb{Q}(\theta)$ be an algebraic number field of degree $n = r + 2s$, where θ has r real conjugates and s pairs of nonreal complex conjugates. Let $C \in H(K)$. Then C contains an integral ideal $B \neq \langle 0 \rangle$ with*

$$N(B) \leq \left(\frac{2}{\pi}\right)^{s} \sqrt{|d(K)|}.$$

Proof: Let A be an ideal in the class C^{-1} of $H(K)$. Then $A^{-1} \in C$. By Theorem 12.5.1 there exists $\alpha \, (\neq 0) \in A$ such that

$$|N(\alpha)| \leq \left(\frac{2}{\pi}\right)^{s} N(A)\sqrt{|d(K)|}.$$

As $\alpha \in A$ we have $\langle \alpha \rangle \subseteq A$ so that, by Theorem 8.4.1, $A \mid \langle \alpha \rangle$. Hence $B = \langle \alpha \rangle A^{-1}$ is an integral ideal of O_K. As $\alpha \neq 0$ we have $B \neq \langle 0 \rangle$. Also, $B \in C$ as $A^{-1} \in C$. Finally,

$$N(B) = N(\langle \alpha \rangle A^{-1}) = N(\langle \alpha \rangle)N(A^{-1})$$
$$= |N(\alpha)|N(A)^{-1} \leq \left(\frac{2}{\pi}\right)^{s} \sqrt{|d(K)|}. \qquad \blacksquare$$

We next establish that there are only finitely many integral ideals in the ring of integers of an algebraic number field having a given norm.

Theorem 12.5.3 *Let K be an algebraic number field. Let k be a positive integer. There are only finitely many integral ideals A of O_K with $N(A) = k$.*

Proof: Let A be an integral ideal of O_K with $N(A) = k$. By Theorem 9.1.3

$$\text{card}(O_K/A) = N(A) = k.$$

Hence

$$k + A = k(1 + A) = 0 + A$$

so that $k \in A$. Thus $\langle k \rangle \subseteq A$ and so $A \mid \langle k \rangle$. By Theorem 8.3.1 there exist distinct prime ideals $P_1, \ldots, P_r$ and nonnegative integers $a_1, \ldots, a_r$ such that

$$\langle k \rangle = P_1^{a_1} \cdots P_r^{a_r}.$$

Hence, as $A \mid \langle k \rangle$, we have

$$A = P_1^{c_1} \cdots P_r^{c_r},$$

where

$$c_i \in \{0, 1, \ldots, a_i\} \text{ for } i = 1, 2, \ldots, r.$$

Thus there are at most

$$(a_1 + 1)(a_2 + 1) \cdots (a_r + 1)$$

possibilities for A.   ∎

We now use Theorems 12.5.2 and 12.5.3 to show that the ideal class group of an algebraic number field is finite.

Theorem 12.5.4 *Let K be an algebraic number field. Then the ideal class group $H(K)$ of K is a finite group (so that the class number $h(K) = \text{card } H(K)$ is finite).*

Proof: By Theorem 12.5.3 there are only finitely many integral ideals B of O_K with $N(B) \leq \left(\frac{2}{\pi}\right)^s \sqrt{|d(K)|}$. It follows from Theorem 12.5.2 that each ideal class is represented by an integral ideal of O_K from a finite set. Thus there are only finitely many ideal classes. Hence $H(K)$ is a finite group and $h(K)$ is finite. ∎

The quantity on the right-hand side of the inequality in Theorem 12.5.2 is called the Minkowski bound for the number field K.

Definition 12.5.1 (Minkowski bound) *Let* $K = \mathbb{Q}(\theta)$ *be an algebraic number field of degree n. Let r denote the number of real conjugates of* θ *and s the number of complex conjugate pairs of nonreal conjugates of* θ *so that* $r + 2s = n$. *The Minkowski bound for K is denoted by* M_K *and is given by*

$$M_K = \left(\frac{2}{\pi}\right)^s \sqrt{|d(K)|}.$$

The significance of the Minkowski bound M_K is that Theorem 12.5.2 guarantees that every ideal class of K contains a nonzero integral ideal with norm less than or equal to M_K. In fact by more detailed reasoning it can be shown that every ideal class contains a nonzero integral ideal with norm less than or equal to

$$\left(\frac{4}{\pi}\right)^s \frac{n!}{n^n} \sqrt{|d(K)|}, \tag{12.5.1}$$

but we will not prove this here. As the norm of a nonzero integral ideal is at least 1, we have the inequality

$$\sqrt{|d(K)|} \geq \left(\frac{\pi}{4}\right)^s \frac{n^n}{n!}. \tag{12.5.2}$$

Now $s \leq \frac{n}{2}$ and $\frac{n^n}{n!} \geq 2^{n-1}$ so for $n \geq 2$ we have

$$\sqrt{|d(K)|} \geq \left(\frac{\pi}{4}\right)^{n/2} 2^{n-1} = \frac{1}{2}\pi^{n/2} \geq \frac{\pi}{2} > 1$$

so that

$$|d(K)| > 1 \quad \text{for} \quad K \neq \mathbb{Q}.$$

12.6 Algorithm to Determine the Ideal Class Group

The results of the previous section give us a method of determining all the ideal classes of a given algebraic number field K. To determine representatives of the ideal classes, we need only look at the integral ideals of O_K with norm less than or equal to the Minkowski bound M_K. If A is such an ideal then $N(P) \leq M_K$ for every prime ideal P dividing A. Now $N(P) = p^f$ for some rational prime p and some positive integer f so the prime ideals occurring in the prime factorizations of the various integral ideals A are all factors of rational primes $p \leq M_K$. Thus if we take each rational prime $p \leq M_K$, determine the prime ideal factorization of $\langle p \rangle$ in O_K, and form all possible products of the prime ideal factors of these various rational primes that yield ideals with norm $\leq M_K$ then we are sure to have at least one representative of every ideal class.

In particular, if every rational prime $\leq M_K$ factors into a product of prime ideals of O_K, each of which is a principal ideal, then K has class number $h(K) = 1$. For in this case every ideal of the type described here will also be principal.

Algorithm to find the ideal class group $H(K)$ **of an algebraic number field** K:

Input. Algebraic number field $K = \mathbb{Q}(\theta)$.

Step 1. Determine $n = [K : \mathbb{Q}]$.

Step 2. Determine r the number of real conjugates of θ. Then $s = \frac{1}{2}(n - r)$.

Step 3. Determine $d(K)$.

Step 4. Compute the Minkowski bound $M_K = (2/\pi)^s \sqrt{|d(K)|}$.

Step 5. Determine all rational primes $p \leq M_K$.

Step 6. Determine the prime ideal factorization of each principal ideal $\langle p \rangle$ in O_K with p as in Step 5.

Step 7. Determine all products of these prime ideals having norm $\leq M_K$.

Step 8. Determine the generators of $H(K)$ from the classes of these products.

Output. $H(K)$.

We illustrate this algorithm by finding the ideal class group of several algebraic number fields.

We denote the class containing the ideal A by $[A]$ and the class of principal ideals by 1.

Example 12.6.1 *We show that* $K = \mathbb{Q}(\sqrt{-19})$ *has class number* $h(K) = 1$. *Here*

$$n = 2, \ r = 0, \ s = 1, \ d(K) = -19.$$

The Minkowski bound is

$$M_K = \left(\frac{2}{\pi}\right)^s \sqrt{|d(K)|} = \frac{2}{\pi}\sqrt{19} < \frac{2}{3} \cdot 5 < 4,$$

so that the primes $p \leq M_K$ *are* $p = 2$ *and* 3. *As*

$$\left(\frac{-19}{2}\right) = \left(\frac{-19}{3}\right) = -1,$$

the principal ideals $\langle 2 \rangle$ *and* $\langle 3 \rangle$ *are both prime ideals in* O_K. *This is the situation described just before the algorithm and so* $h(K) = h(\mathbb{Q}(\sqrt{-19})) = 1$. *Hence the ring of integers* $\mathbb{Z} + \mathbb{Z}\left(\frac{1+\sqrt{-19}}{2}\right)$ *of* $\mathbb{Q}(\sqrt{-19})$ *is a principal ideal domain and thus a unique factorization domain.*

Example 12.6.2 *We show that* $K = \mathbb{Q}(\sqrt{-163})$ *has class number* $h(K) = 1$. *Here*

$$n = 2, \ r = 0, \ s = 1, \ d(K) = -163.$$

The Minkowski bound is

$$M_K = \left(\frac{2}{\pi}\right)^s \sqrt{|d(K)|} = \left(\frac{2}{\pi}\right)\sqrt{163} < \frac{2}{3} \cdot 13 < 9,$$

so that the primes $p \leq M_K$ are $p = 2, 3, 5,$ and 7. As

$$\left(\frac{-163}{2}\right) = \left(\frac{-163}{3}\right) = \left(\frac{-163}{5}\right) = \left(\frac{-163}{7}\right) = -1,$$

the principal ideals $\langle 2 \rangle, \langle 3 \rangle, \langle 5 \rangle, \langle 7 \rangle$ are all prime ideals in O_K. Hence $h(K) = h(\mathbb{Q}(\sqrt{-163})) = 1$. Thus the ring of integers $\mathbb{Z} + \mathbb{Z}\left(\frac{1+\sqrt{-163}}{2}\right)$ of $\mathbb{Q}(\sqrt{-163})$ is a principal ideal domain and so a unique factorization domain.

Example 12.6.3 *We show that* $K = \mathbb{Q}(\sqrt{23})$ *has class number* $h(K) = 1$. *Here*

$$n = 2, \ r = 2, \ s = 0, \ d(K) = 92.$$

The Minkowski bound is

$$M_K = \left(\frac{2}{\pi}\right)^s \sqrt{|d(K)|} = \sqrt{92} < 10,$$

so that the primes $p \leq M_K$ *are* $p = 2, 3, 5,$ *and* 7. *As*

$$\left(\frac{92}{3}\right) = \left(\frac{-1}{3}\right) = -1,$$

$\langle 3 \rangle$ *is a prime ideal in* O_K. *As*

$$\left(\frac{92}{5}\right) = \left(\frac{2}{5}\right) = -1,$$

$\langle 5 \rangle$ *is also a prime ideal in* O_K. *As*

$$\left(\frac{92}{2}\right) = 0,$$

$\langle 2 \rangle$ *ramifies in* O_K. *Indeed*

$$\langle 2 \rangle = \langle 2, \ 1 + \sqrt{23} \rangle^2.$$

The prime ideal $\langle 2, \ 1 + \sqrt{23} \rangle$ *is principal as*

$$\langle 2, \ 1 + \sqrt{23} \rangle = \langle 2, \ 5 + \sqrt{23} \rangle \ (as \ 5 + \sqrt{23} = 2 \cdot 2 + (1 + \sqrt{23}))$$
$$= \langle 5 + \sqrt{23} \rangle \ (as \ 5 + \sqrt{23} \mid 2).$$

Finally, as

$$\left(\frac{92}{7}\right) = \left(\frac{1}{7}\right) = 1,$$

$\langle 7 \rangle$ *splits in* O_K. *We have*

$$\langle 7 \rangle = \langle 7, \ 3 + \sqrt{23} \rangle \langle 7, \ 3 - \sqrt{23} \rangle.$$

The prime ideal $\langle 7, \ 3 + \sqrt{23} \rangle$ *is principal as*

$$\langle 7, \ 3 + \sqrt{23} \rangle = \langle 7, \ 4 - \sqrt{23} \rangle \quad (\text{as } 4 - \sqrt{23} = 7 - (3 + \sqrt{23}))$$
$$= \langle 4 - \sqrt{23} \rangle \quad (\text{as } 4 - \sqrt{23} \mid 7).$$

Similarly, $\langle 7, \ 3 - \sqrt{23} \rangle = \langle 4 + \sqrt{23} \rangle$.

Hence every prime ideal in O_K *dividing a rational prime* $\leq M_K$ *is principal and so* O_K *is a principal ideal domain, that is,*

$$h(\mathbb{Q}(\sqrt{23})) = 1.$$

Before continuing we expand on that part of the calculation in Example 11.6.3 that shows that the prime ideal $P = \langle 2, 1 + \sqrt{23} \rangle$ is principal. We must find $\alpha \in P$ such that $P = \langle \alpha \rangle$. By Exercise 8 of Chapter 9 it suffices to find an $\alpha \in P$ such that $|N(\alpha)| = N(P) = 2$, equivalently, $N(\alpha) = \pm 2$. As α is an integer of $\mathbb{Q}(\sqrt{23})$ we have $\alpha = x + y\sqrt{23}$ $(x, y \in \mathbb{Z})$ so we wish to solve $x^2 - 23y^2 = \pm 2$. This can be done by means of Theorem 11.7.1. However, in this case $x = 5$, $y = 1$ is an obvious solution and $\alpha = 5 + \sqrt{23} = (2)2 + (1)(1 + \sqrt{23}) \in P$. Thus $P = \langle \alpha \rangle = \langle 5 + \sqrt{23} \rangle$.

Example 12.6.4 *We show that*

$$H(\mathbb{Q}(\sqrt{-14})) \simeq \mathbb{Z}_4.$$

Here

$$K = \mathbb{Q}(\sqrt{-14}), \ n = 2, \ r = 0, \ s = 1, \ d(K) = -56.$$

The Minkowski bound is

$$M_K = \left(\frac{2}{\pi}\right)^s \sqrt{|d(K)|} = \left(\frac{2}{\pi}\right)\sqrt{56} < \frac{1}{3}\sqrt{224} < 5.$$

The rational primes $p \leq M_K$ *are* $p = 2$ *and* 3. *As*

$$\left(\frac{-56}{2}\right) = 0 \ and \ \left(\frac{-56}{3}\right) = 1,$$

we have

$$\langle 2 \rangle = P^2, \ \langle 3 \rangle = P_1 P_2,$$

where the prime ideals P, P_1, P_2 *are given by*

$$P = \langle 2, \sqrt{-14} \rangle, \ P_1 = \langle 3, 1 + \sqrt{-14} \rangle, \ P_2 = \langle 3, 1 - \sqrt{-14} \rangle.$$

The norms of these ideals are

$$N(P) = 2, \ N(P_1) = N(P_2) = 3.$$

Clearly $2 - \sqrt{-14} \in P$ and $2 - \sqrt{-14} \in P_1$. Hence $\langle 2 - \sqrt{-14} \rangle \subseteq P$ and $\langle 2 - \sqrt{-14} \rangle \subseteq P_1$. Thus $P \mid \langle 2 - \sqrt{-14} \rangle$ and $P_1 \mid \langle 2 - \sqrt{-14} \rangle$. As P and P_1 are distinct prime ideals, we have $P P_1 \mid \langle 2 - \sqrt{-14} \rangle$. Hence there exists an integral ideal B of O_K such that

$$\langle 2 - \sqrt{-14} \rangle = P P_1 B.$$

Taking norms we obtain

$$18 = N(\langle 2 - \sqrt{-14} \rangle) = N(P)N(P_1)N(B) = 6N(B),$$

so that $N(B) = 3$. Hence $B = P_1$ or P_2. If $B = P_2$ then

$$\langle 2 - \sqrt{-14} \rangle = P P_1 P_2 = P \langle 3 \rangle,$$

so that $\langle 3 \rangle \mid \langle 2 - \sqrt{-14} \rangle$, which is impossible. Hence $B = P_1$ and

$$\langle 2 - \sqrt{-14} \rangle = P P_1^2.$$

Thus

$$[P][P_1]^2 = [P P_1^2] = [\langle 2 - \sqrt{-14} \rangle] = 1.$$

As $[P]^2 = [P^2] = 1$ we deduce that $[P] = [P_1]^2$ and $[P_1]^4 = 1$. Also, $[P_1][P_2] = [P_1 P_2] = 1 = [P_1]^4$ so that $[P_2] = [P_1]^3$. Thus 1, $[P_1]$, $[P_1]^2$, and $[P_1]^3$ comprise all the ideal classes. We show that these four ideal classes are in fact distinct. We do this by proving that $[P_1]^2 \neq 1$. Suppose that $[P_1]^2 = 1$. Then $[P] = 1$ so that the ideal $P = \langle 2, \sqrt{-14} \rangle$ is principal, say $P = \langle x + y\sqrt{-14} \rangle$, where $x, y \in \mathbb{Z}$. Then

$$2 = N(\langle 2, \sqrt{-14} \rangle) = N(\langle x + y\sqrt{-14} \rangle) = x^2 + 14y^2,$$

which is impossible. This proves that $H(\mathbb{Q}(\sqrt{-14}))$ is a cyclic group of order 4 generated by the class of P_1.

It is often useful when determining the ideal class group of an algebraic number field $K = \mathbb{Q}(\theta)$ to calculate $N(k + \theta)$ for $k = 0, 1, 2, \ldots$ and use those values that only involve the primes $p \leq M_K$ to find relations among the ideal classes.

This is illustrated in the next example.

Example 12.6.5 *We show that*

$$H(\mathbb{Q}(\sqrt{-65})) \simeq \mathbb{Z}_2 \times \mathbb{Z}_4.$$

In this example we have

$$K = \mathbb{Q}(\sqrt{-65}), \ n = 2, \ r = 0, \ s = 1, \ d(K) = -260.$$

The Minkowski bound is

$$M_K = \left(\frac{2}{\pi}\right)^s \sqrt{|d(K)|} = \left(\frac{2}{\pi}\right) \sqrt{260} < 11.$$

The primes $p \le M_K$ are $p = 2, 3, 5,$ and 7. As

$$\left(\frac{-260}{2}\right) = 0, \quad \left(\frac{-260}{3}\right) = \left(\frac{1}{3}\right) = 1,$$

$$\left(\frac{-260}{5}\right) = 0, \quad \left(\frac{-260}{7}\right) = \left(\frac{-1}{7}\right) = -1,$$

we have

$$\langle 2 \rangle = P_1^2, \quad \langle 3 \rangle = Q_1 Q_2, \quad \langle 5 \rangle = P_2^2, \quad \langle 7 \rangle = \text{prime ideal,}$$

where

$$P_1 = \langle 2, 1 + \sqrt{-65} \rangle, \quad Q_1 = \langle 3, 1 + \sqrt{-65} \rangle,$$
$$Q_2 = \langle 3, 1 - \sqrt{-65} \rangle, \quad P_2 = \langle 5, \sqrt{-65} \rangle$$

are distinct prime ideals. Thus

$$[P_1]^2 = 1, \quad [Q_1][Q_2] = 1, \quad [P_2]^2 = 1.$$

Next we calculate the values of $N(k + \sqrt{-65})$ for $k = 0, 1, 2, \ldots$, retaining those that only involve the primes 2, 3, and 5, until we have enough values to find all the relations between $[P_1]$, $[Q_1]$, $[Q_2]$, and $[P_2]$. The first relevant value is

$$N(4 + \sqrt{-65}) = 81 = 3^4.$$

We have

$$4 + \sqrt{-65} = 3 + (1 + \sqrt{-65}) \in Q_1,$$

so that

$$\langle 4 + \sqrt{-65} \rangle \subseteq Q_1$$

and thus

$$Q_1 \mid \langle 4 + \sqrt{-65} \rangle.$$

Suppose that $Q_2 \mid \langle 4 + \sqrt{-65} \rangle$; then

$$\langle 3 \rangle = Q_1 Q_2 \mid \langle 4 + \sqrt{-65} \rangle,$$

which is impossible. Hence $Q_2 \nmid \langle 4 + \sqrt{-65} \rangle$. Let r be the unique positive integer such that

$$Q_1^r \| \langle 4 + \sqrt{-65} \rangle.$$

Then

$$\langle 4 + \sqrt{-65} \rangle = Q_1^r B$$

for some integral ideal B of O_K with $Q_1 \nmid B$ and $Q_2 \nmid B$. As $Q_1 \nmid B$ and $Q_2 \nmid B$ we see that $N(B)$ is not a power of 3. Taking norms we obtain

$$3^4 = 81 = N(\langle 4 + \sqrt{-65} \rangle) = N(Q_1^r B) = N(Q_1)^r N(B) = 3^r N(B),$$

so that $r = 4$ and $N(B) = 1$. Hence $B = \langle 1 \rangle$ and

$$\langle 4 + \sqrt{-65} \rangle = Q_1^4,$$

showing that

$$[Q_1]^4 = 1.$$

Then, from $[Q_1][Q_2] = 1 = [Q_1]^4$, we deduce that

$$[Q_2] = [Q_1]^3.$$

Next we find that

$$N(5 + \sqrt{-65}) = 90 = 2 \cdot 3^2 \cdot 5.$$

We have

$$5 + \sqrt{-65} = 2(2) + 1(1 + \sqrt{-65}) \in P_1,$$

$$5 + \sqrt{-65} = 2(3) - 1(1 - \sqrt{-65}) \in Q_2,$$

$$5 + \sqrt{-65} = 1(5) + 1(\sqrt{-65}) \in P_2,$$

so that

$$P_1 \mid \langle 5 + \sqrt{-65} \rangle, \quad Q_2 \mid \langle 5 + \sqrt{-65} \rangle, \quad P_2 \mid \langle 5 + \sqrt{-65} \rangle.$$

Hence

$$\langle 5 + \sqrt{-65} \rangle = P_1^r Q_2^s P_2^t B,$$

for positive integers r, s, t and an integral ideal B of O_K with $P_1 \nmid B$, $Q_2 \nmid B$, $P_2 \nmid B$. We show next that $Q_1 \nmid B$. Suppose that $Q_1 \mid B$. Then

$$\langle 3 \rangle = Q_1 Q_2 \mid \langle 5 + \sqrt{-65} \rangle,$$

which is impossible. Hence $Q_1 \nmid B$. Since $Q_1 \nmid B$ and $Q_2 \nmid B$ the norm of B cannot be a power of 3. Taking norms we obtain

$$2 \cdot 3^2 \cdot 5 = 90 = N(\langle 5 + \sqrt{-65} \rangle) = 2^r \cdot 3^s \cdot 5^t N(B),$$

so that $r = 1$, $t = 1$, *and* $N(B) = 3^{2-s}$. *As* $N(B)$ *is not a power of 3, we have* $s = 2$, $N(B) = 1$, $B = \langle 1 \rangle$. *Hence*

$$\langle 5 + \sqrt{-65} \rangle = P_1 Q_2^2 P_2.$$

Thus

$$[P_1][Q_2]^2[P_2] = 1.$$

Hence

$$[P_2] = [P_1][Q_2]^2[P_2]^2 = [P_1][Q_1]^6 = [P_1][Q_1]^2.$$

Thus all the ideal classes of K lie among

$$1, [Q_1], [Q_1]^2, [Q_1]^3, [P_1], [P_1][Q_1], [P_1][Q_1]^2, [P_1][Q_1]^3.$$

We claim that all these classes are distinct. We show first that

$$1, [Q_1], [Q_1]^2, [Q_1]^3$$

are distinct. As $[Q_1]^4 = 1$ *it suffices to show that* $[Q_1]^2 \neq 1$. *If* $[Q_1]^2 = 1$ *then* Q_1^2 *is a principal ideal, say,*

$$Q_1^2 = \langle x + y\sqrt{-65} \rangle, \ x, y \in \mathbb{Z}.$$

Taking norms we deduce that

$$9 = x^2 + 65y^2,$$

so that $x = \pm 3$, $y = 0$. *Then*

$$Q_1^2 = \langle 3 \rangle = Q_1 Q_2$$

and so $Q_1 = Q_2$, *contradicting that* Q_1 *and* Q_2 *are distinct ideals.*

  Thus $H = \{1, [Q_1], [Q_1]^2, [Q_1]^3\}$ *is a subgroup of order 4 of* $G = H(\mathbb{Q}(\sqrt{-65}))$, *so by Lagrange's theorem,* $4 \mid |G|$. *If* $|G| < 8$ *then* $|G| = 4$ *and so* $G = H$. *Hence* $[P_1] = 1$, $[Q_1], [Q_1]^2$, *or* $[Q_1]^3$. *Now* $[P_1] \neq 1$ *as* $2 \neq x^2 + 65y^2$ *for integers x and y. Thus* $\mathrm{ord}[P_1] = 2$. *As* $\mathrm{ord}[Q_1] = \mathrm{ord}[Q_1]^3 = 4$ *we must have* $[P_1] = [Q_1]^2$. *Hence* $[P_1 Q_1^2] = 1$ *so that* $PQ_1^2 = \langle x + y\sqrt{-65} \rangle$ *for integers x and y. Then*

$$18 = N(P_1 Q_1^2) = x^2 + 65y^2,$$

which is impossible. This proves that

$$H(\mathbb{Q}(\sqrt{-65})) = \{1, [Q_1], [Q_1]^2, [Q_1]^3, [P_1], [P_1][Q_1], [P_1][Q_1]^2, [P_1][Q_1]^3\},$$

where

$$[Q_1]^4 = [P_1]^2 = 1,$$

Table 5. *Nontrivial ideal class groups*
$H(\mathbb{Q}(\sqrt{k}))$, $-30 < k < 0$, k squarefree

k	$H(\mathbb{Q}(\sqrt{k}))$
-5	$\{1, A\} \simeq \mathbb{Z}_2$, $A = [\langle 2, 1 + \sqrt{-5}\rangle]$, $A^2 = 1$
-6	$\{1, A\} \simeq \mathbb{Z}_2$, $A = [\langle 2, \sqrt{-6}\rangle]$, $A^2 = 1$
-10	$\{1, A\} \simeq \mathbb{Z}_2$, $A = [\langle 2, \sqrt{-10}\rangle]$, $A^2 = 1$
-13	$\{1, A\} \simeq \mathbb{Z}_2$, $A = [\langle 2, 1 + \sqrt{-13}\rangle]$, $A^2 = 1$
-14	$\{1, A, A^2, A^3\} \simeq \mathbb{Z}_4$, $A = [\langle 3, 1 + \sqrt{-14}\rangle]$, $A^2 = [\langle 2, \sqrt{-14}\rangle]$, $A^3 = [\langle 3, 1 - \sqrt{-14}\rangle]$, $A^4 = 1$
-15	$\{1, A\} \simeq \mathbb{Z}_2$, $A = [\langle 2, \frac{1}{2}(3 + \sqrt{-15})\rangle]$, $A^2 = 1$
-17	$\{1, A, A^2, A^3\} \simeq \mathbb{Z}_4$, $A = [\langle 3, 1 + \sqrt{-17}\rangle]$, $A^2 = [\langle 2, 1 + \sqrt{-17}\rangle]$, $A^3 = [\langle 3, 1 - \sqrt{-17}\rangle]$, $A^4 = 1$
-21	$\{1, A, B, AB\} \simeq \mathbb{Z}_2 \times \mathbb{Z}_2$, $A = [\langle 2, 1 + \sqrt{-21}\rangle]$, $B = [\langle 3, \sqrt{-21}\rangle]$, $AB = [\langle 5, 3 + \sqrt{-21}\rangle]$, $A^2 = B^2 = (AB)^2 = 1$
-22	$\{1, A\} \simeq \mathbb{Z}_2$, $A = [\langle 2, \sqrt{-22}\rangle]$, $A^2 = 1$
-23	$\{1, A, A^2\} \simeq \mathbb{Z}_3$, $A = [\langle 2, \frac{1}{2}(1 + \sqrt{-23})\rangle]$, $A^2 = [\langle 2, \frac{1}{2}(1 - \sqrt{-23})\rangle]$, $A^3 = 1$
-26	$\{1, A, A^2, A^3, A^4, A^5\} \simeq \mathbb{Z}_6$, $A = [\langle 5, 2 + \sqrt{-26}\rangle]$, $A^2 = [\langle 3, 1 - \sqrt{-26}\rangle]$, $A^3 = [\langle 2, \sqrt{-26}\rangle]$, $A^4 = [\langle 3, 1 + \sqrt{-26}\rangle]$, $A^5 = [\langle 5, 2 - \sqrt{-26}\rangle]$, $A^6 = 1$
-29	$\{1, A, A^2, A^3, A^4, A^5\} \simeq \mathbb{Z}_6$, $A = [\langle 3, 1 + \sqrt{-29}\rangle]$, $A^2 = [\langle 5, 1 + \sqrt{-29}\rangle]$, $A^3 = [\langle 2, 1 + \sqrt{-29}\rangle]$, $A^4 = [\langle 5, 1 - \sqrt{-29}\rangle]$, $A^5 = [\langle 3, 1 - \sqrt{-29}\rangle]$, $A^6 = 1$
-30	$\{1, A, B, AB\} \simeq \mathbb{Z}_2 \times \mathbb{Z}_2$, $A = [\langle 2, \sqrt{-30}\rangle]$, $B = [\langle 3, \sqrt{-30}\rangle]$, $AB = [\langle 5, \sqrt{-30}\rangle]$, $A^2 = B^2 = (AB)^2 = 1$

Note: Excluded fields have class number 1.

so that

$$H(\mathbb{Q}(\sqrt{-65})) \simeq \mathbb{Z}_4 \times \mathbb{Z}_2.$$

Using the method illustrated in Examples 12.6.1–12.6.5 we can construct Tables 5 and 6 of class groups.

For a quadratic field K, Dirichlet has given an explicit formula for $h(K)$. We refer the reader to [1, p. 342] for a proof.

Theorem 12.6.1 *Let K be a quadratic field of discriminant d. Then*

$$h(K) = \frac{-w(d)}{2|d|} \sum_{r=1}^{|d|-1} r\left(\frac{d}{r}\right), \quad if \, d < 0,$$

Table 6.   *Nontrivial ideal class groups*
$H(\mathbb{Q}(\sqrt{k}))$, $2 \le k < 100$, k *squarefree*

k	$H(\mathbb{Q}(\sqrt{k}))$
10	$\{1, A\} \simeq \mathbb{Z}_2$, $A = [\langle 2, \sqrt{10}\rangle]$, $A^2 = 1$
15	$\{1, A\} \simeq \mathbb{Z}_2$, $A = [\langle 2, 1 + \sqrt{15}\rangle]$, $A^2 = 1$
26	$\{1, A\} \simeq \mathbb{Z}_2$, $A = [\langle 2, \sqrt{26}\rangle]$, $A^2 = 1$
30	$\{1, A\} \simeq \mathbb{Z}_2$, $A = [\langle 2, \sqrt{30}\rangle]$, $A^2 = 1$
34	$\{1, A\} \simeq \mathbb{Z}_2$, $A = [\langle 3, 1 + \sqrt{34}\rangle]$, $A^2 = 1$
35	$\{1, A\} \simeq \mathbb{Z}_2$, $A = [\langle 2, 1 + \sqrt{35}\rangle]$, $A^2 = 1$
39	$\{1, A\} \simeq \mathbb{Z}_2$, $A = [\langle 2, 1 + \sqrt{39}\rangle]$, $A^2 = 1$
42	$\{1, A\} \simeq \mathbb{Z}_2$, $A = [\langle 2, \sqrt{42}\rangle]$, $A^2 = 1$
51	$\{1, A\} \simeq \mathbb{Z}_2$, $A = [\langle 3, \sqrt{51}\rangle]$, $A^2 = 1$
55	$\{1, A\} \simeq \mathbb{Z}_2$, $A = [\langle 2, 1 + \sqrt{55}\rangle]$, $A^2 = 1$
58	$\{1, A\} \simeq \mathbb{Z}_2$, $A = [\langle 2, \sqrt{58}\rangle]$, $A^2 = 1$
65	$\{1, A\} \simeq \mathbb{Z}_2$, $A = [\langle 5, \sqrt{65}\rangle]$, $A^2 = 1$
66	$\{1, A\} \simeq \mathbb{Z}_2$, $A = [\langle 3, \sqrt{66}\rangle]$, $A^2 = 1$
70	$\{1, A\} \simeq \mathbb{Z}_2$, $A = [\langle 2, \sqrt{70}\rangle]$, $A^2 = 1$
74	$\{1, A\} \simeq \mathbb{Z}_2$, $A = [\langle 2, \sqrt{74}\rangle]$, $A^2 = 1$
78	$\{1, A\} \simeq \mathbb{Z}_2$, $A = [\langle 2, \sqrt{78}\rangle]$, $A^2 = 1$
79	$\{1, A, A^2\} \simeq \mathbb{Z}_3$, $A = [\langle 3, 1 + \sqrt{79}\rangle]$, $A^2 = [\langle 3, 1 - \sqrt{79}\rangle]$, $A^3 = 1$
82	$\{1, A, A^2, A^3\} \simeq \mathbb{Z}_4$, $A = [\langle 3, 2 + \sqrt{82}\rangle]$, $A^2 = [\langle 2, \sqrt{82}\rangle]$, $A^3 = [\langle 3, 2 - \sqrt{82}\rangle]$, $A^4 = 1$
85	$\{1, A\} \simeq \mathbb{Z}_2$, $A = [\langle 5, \sqrt{85}\rangle]$, $A^2 = 1$
87	$\{1, A\} \simeq \mathbb{Z}_2$, $A = [\langle 2, 1 + \sqrt{87}\rangle]$, $A^2 = 1$
91	$\{1, A\} \simeq \mathbb{Z}_2$, $A = [\langle 2, 1 + \sqrt{91}\rangle]$, $A^2 = 1$
95	$\{1, A\} \simeq \mathbb{Z}_2$, $A = [\langle 2, 1 + \sqrt{95}\rangle]$, $A^2 = 1$

Note: Excluded fields have class number 1.

and

$$h(K) = \frac{-1}{\log \eta} \sum_{1 \le r < d/2} \left(\frac{d}{r}\right) \log \sin \frac{\pi r}{d}, \quad \textit{if } d > 0.$$

Here $w(d)$ denotes the number of roots of unity in $O_{\mathbb{Q}(\sqrt{d})}$ ($d < 0$) so that

$$w(d) = \begin{cases} 6, & \text{if } d = -3, \\ 4, & \text{if } d = -4, \\ 2, & \text{if } d < -4, \end{cases}$$

where $\left(\frac{d}{n}\right)$ ($n \in \mathbb{N}$) is the Kronecker symbol and η is the fundamental unit of $O_{\mathbb{Q}(\sqrt{d})}$ ($d > 0$).

Example 12.6.6 *We use Dirichlet's formula to show that* $h(\mathbb{Q}(\sqrt{-15})) = 2$. *Here* $d = -15$, $w(d) = 2$, *and Theorem 12.6.1 gives*

$$h(\mathbb{Q}(\sqrt{-15})) = -\frac{1}{15}\sum_{r=1}^{14} r\left(\frac{-15}{r}\right)$$

$$= \frac{-1}{15}(1(1) + 2(1) + 3(0) + 4(1) + 5(0) + 6(0) + 7(-1) + 8(1)$$
$$+ 9(0) + 10(0) + 11(-1) + 12(0) + 13(-1) + 14(-1))$$

$$= \frac{-1}{15}(1 + 2 + 4 - 7 + 8 - 11 - 13 - 14)$$

$$= \frac{-1}{15}(-30) = 2.$$

Example 12.6.7 *We use Dirichlet's formula to show that* $h(\mathbb{Q}(\sqrt{5})) = 1$. *In this case* $d = 5$, $\eta = (1 + \sqrt{5})/2$, *and Theorem 12.6.1 gives*

$$h(\mathbb{Q}(\sqrt{5})) = \frac{-1}{\log\left(\frac{1 + \sqrt{5}}{2}\right)}\sum_{r=1}^{2}\left(\frac{5}{r}\right)\log\sin\frac{\pi r}{5}$$

$$= \frac{-1}{\log\left(\frac{1 + \sqrt{5}}{2}\right)}\left(\log\sin\frac{\pi}{5} - \log\sin\frac{2\pi}{5}\right)$$

$$= \frac{1}{\log\left(\frac{1 + \sqrt{5}}{2}\right)}\left(\log\sin\frac{2\pi}{5} - \log\sin\frac{\pi}{5}\right)$$

$$= \frac{1}{\log\left(\frac{1 + \sqrt{5}}{2}\right)}\left(\log\left(\frac{\sqrt{10 + 2\sqrt{5}}}{4}\right) - \log\left(\frac{\sqrt{10 - 2\sqrt{5}}}{4}\right)\right)$$

$$= \frac{1}{2\log\left(\frac{1 + \sqrt{5}}{2}\right)}\log\left(\frac{10 + 2\sqrt{5}}{10 - 2\sqrt{5}}\right) = \frac{\log\left(\frac{3 + \sqrt{5}}{2}\right)}{2\log\left(\frac{1 + \sqrt{5}}{2}\right)} = 1.$$

Tables 7 and 8, which give the class numbers of quadratic fields $\mathbb{Q}(\sqrt{k})$ with k squarefree between -195 and 197, can be constructed using Dirichlet's formula.

We conclude this section by determining the ideal class group for two cubic fields and a quartic field.

Table 7.   *Class numbers of imaginary quadratic fields*
$$K = \mathbb{Q}(\sqrt{k}), \quad -195 \le k < 0, \ k \ squarefree$$

k	$h(K)$	k	$h(K)$	k	$h(K)$	k	$h(K)$	k	$h(K)$
-1	1	-38	6	-78	4	-115	2	-158	8
-2	1	-39	4	-79	5	-118	6	-159	10
-3	1	-41	8	-82	4	-119	10	-161	16
-5	2	-42	4	-83	3	-122	10	-163	1
-6	2	-43	1	-85	4	-123	2	-165	8
-7	1	-46	4	-86	10	-127	5	-166	10
-10	2	-47	5	-87	6	-129	12	-167	11
-11	1	-51	2	-89	12	-130	4	-170	12
-13	2	-53	6	-91	2	-131	5	-173	14
-14	4	-55	4	-93	4	-133	4	-174	12
-15	2	-57	4	-94	8	-134	14	-177	4
-17	4	-58	2	-95	8	-137	8	-178	8
-19	1	-59	3	-97	4	-138	8	-179	5
-21	4	-61	6	-101	14	-139	3	-181	10
-22	2	-62	8	-102	4	-141	8	-182	12
-23	3	-65	8	-103	5	-142	4	-183	8
-26	6	-66	8	-105	8	-143	10	-185	16
-29	6	-67	1	-106	6	-145	8	-186	12
-30	4	-69	8	-107	3	-146	16	-187	2
-31	3	-70	4	-109	6	-149	14	-190	4
-33	4	-71	7	-110	12	-151	7	-191	13
-34	4	-73	4	-111	8	-154	8	-193	4
-35	2	-74	10	-113	8	-155	4	-194	20
-37	2	-77	8	-114	8	-157	6	-195	4

Example 12.6.8 *We show that $H(K)$ is trivial for the cubic field $K = \mathbb{Q}(\theta)$, where $\theta^3 + \theta + 1 = 0$ (see Example 7.1.3). Thus O_K is a principal ideal domain and so is a unique factorization domain. Here*

$$D(\theta) = -4 \cdot 1^3 - 27 \cdot 1^2 = -31$$

is negative so that $x^3 + x + 1 = 0$ has one real root and two nonreal roots. Thus

$$r = 1, \ s = 1.$$

As $D(\theta)$ is squarefree, $K = \mathbb{Z} + \mathbb{Z}\theta + \mathbb{Z}\theta^2$ and

$$d(K) = D(\theta) = -31.$$

The Minkowski bound is

$$M_K = \left(\frac{2}{\pi}\right)^s \sqrt{|d(K)|} = \left(\frac{2}{\pi}\right) \sqrt{31} < \frac{2}{3} \cdot 6 = 4,$$

so that the primes $p \le M_K$ are $p = 2$ and 3. The polynomial $x^3 + x + 1$ is

Table 8. *Class numbers of real quadratic fields*
$K = \mathbb{Q}(\sqrt{k}), \;\; 0 < k \le 197, \; k \; squarefree$

k	$h(K)$	k	$h(K)$	k	$h(K)$	k	$h(K)$	k	$h(K)$
2	1	39	2	79	3	118	1	159	2
3	1	41	1	82	4	119	2	161	1
5	1	42	2	83	1	122	2	163	1
6	1	43	1	85	2	123	2	165	2
7	1	46	1	86	1	127	1	166	1
10	2	47	1	87	2	129	1	167	1
11	1	51	2	89	1	130	4	170	4
13	1	53	1	91	2	131	1	173	1
14	1	55	2	93	1	133	1	174	2
15	2	57	1	94	1	134	1	177	1
17	1	58	2	95	2	137	1	178	2
19	1	59	1	97	1	138	2	179	1
21	1	61	1	101	1	139	1	181	1
22	1	62	1	102	2	141	1	182	2
23	1	65	2	103	1	142	3	183	2
26	2	66	2	105	2	143	2	185	2
29	1	67	1	106	2	145	4	186	2
30	2	69	1	107	1	146	2	187	2
31	1	70	2	109	1	149	1	190	2
33	1	71	1	110	2	151	1	191	1
34	2	73	1	111	2	154	2	193	1
35	2	74	2	113	1	155	2	194	2
37	1	77	1	114	2	157	1	195	4
38	1	78	2	115	2	158	1	197	1

Gauss conjectured that $\mathbb{Q}(\sqrt{k})$ has class number 1 for infinitely many squarefree $k \in \mathbb{N}$. It is still not known whether this conjecture is true or false.

irreducible (mod 2), *so that the principal ideal* $\langle 2 \rangle$ *is prime in* O_K. *The factorization of* $x^3 + x + 1$ *into irreducibles* (mod 3) *is*

$$x^3 + x + 1 \equiv (x - 1)(x^2 + x - 1) \,(\text{mod } 3),$$

so that by Theorem 10.3.1 the factorization of $\langle 3 \rangle$ *into prime ideals in* O_K *is*

$$\langle 3 \rangle = PQ,$$

where

$$P = \langle 3, \theta - 1 \rangle, \;\; N(P) = 3,$$

$$Q = \langle 3, \theta^2 + \theta - 1 \rangle, \;\; N(Q) = 3^2.$$

Now

$$(\theta - 1)^3 + 3(\theta - 1)^2 + 4(\theta - 1) + 3 = 0,$$

so that

$$\theta - 1 \mid 3,$$

and thus

$$P = \langle 3, \theta - 1 \rangle = \langle \theta - 1 \rangle.$$

Further,

$$\frac{3}{\theta - 1} = -4 - 3(\theta - 1) - (\theta - 1)^2 = -2 - \theta - \theta^2$$

so that

$$Q = \langle 3 \rangle P^{-1} = \langle 3 \rangle \langle \theta - 1 \rangle^{-1} = \langle \frac{3}{\theta - 1} \rangle$$

$$= \langle -2 - \theta - \theta^2 \rangle = \langle 2 + \theta + \theta^2 \rangle.$$

Hence all the prime ideals dividing the principal ideals $\langle p \rangle$ (p (prime) $\leq M_K$) are principal so that the ideal class group $H(K)$ is trivial.

Example 12.6.9 *We show that $H(\mathbb{Q}(\sqrt[3]{2}))$ is trivial. Let $\theta = \sqrt[3]{2}$ and $K = \mathbb{Q}(\theta) = \mathbb{Q}(\sqrt[3]{2})$. Clearly, $\mathrm{irr}_\mathbb{Q}\ \theta = x^3 - 2$, which has one real root (namely θ) and two nonreal roots (namely $\omega\theta$ and $\omega^2\theta$, where ω is a complex cube root of unity). Thus*

$$r = 1, \ s = 1.$$

It was shown in Example 7.1.6 that $\{1, \theta, \theta^2\}$ is an integral basis for K and $d(K) = -108$. The Minkowski bound is

$$M_K = \left(\frac{2}{\pi}\right)^s \sqrt{|d(K)|} = \frac{2}{\pi} \sqrt{108} < \frac{2}{3} \cdot \frac{21}{2} = 7.$$

Thus the primes $p \leq M_K$ are $p = 2, 3,$ and 5. Clearly,

$$\langle 2 \rangle = P^3,$$

where $P = \langle \theta \rangle$ is a principal prime ideal of norm 2. Also,

$$\langle 3 \rangle = Q^3,$$

where $Q = \langle \theta + 1 \rangle$ is a principal ideal of norm 3. This is clear as

$$Q^3 = \langle \theta + 1 \rangle^3 = \langle (\theta + 1)^3 \rangle = \langle \theta^3 + 3\theta^2 + 3\theta + 1 \rangle = \langle 3 + 3\theta + 3\theta^2 \rangle$$
$$= \langle 3(1 + \theta + \theta^2) \rangle = \langle 3 \rangle,$$

since $1 + \theta + \theta^2$ is a unit of O_K as

$$(1 + \theta + \theta^2)(-1 + \theta) = -1 + \theta^3 = 1.$$

Finally, as

$$x^3 - 2 = (x + 2)(x^2 - 2x - 1) \,(\text{mod } 5),$$

by Theorem 10.3.1 we have

$$\langle 5 \rangle = PQ,$$

where P and Q are distinct prime ideals with

$$P = \langle 5, 2 + \theta \rangle, \quad N(P) = 5,$$
$$Q = \langle 5, -1 - 2\theta + \theta^2 \rangle, \quad N(Q) = 5^2.$$

Now

$$5 = 4 + 1 = \theta^6 + 1 = (\theta^2 + 1)(\theta^4 - \theta^2 + 1) = (\theta^2 + 1)(1 + 2\theta - \theta^2),$$

so that $1 + 2\theta - \theta^2 \mid 5$ and thus

$$Q = \langle 1 + 2\theta - \theta^2 \rangle$$

and

$$P = \langle 5 \rangle Q^{-1} = \langle 5 \rangle \langle 1 + 2\theta - \theta^2 \rangle^{-1}$$
$$= \langle 5 \rangle \langle (1 + 2\theta - \theta^2)^{-1} \rangle = \langle 5(1 + 2\theta - \theta^2)^{-1} \rangle$$
$$\langle \frac{5}{1 + 2\theta - \theta^2} \rangle = \langle 1 + \theta^2 \rangle.$$

Since all the prime factors of 2, 3, and 5 are principal, $H(\mathbb{Q}(\sqrt[3]{2}))$ is trivial. Hence $h(\mathbb{Q}(\sqrt[3]{2})) = 1$.

The class numbers of $\mathbb{Q}(\sqrt[3]{k})$ for cubefree positive integers k up to 101 are given in Table 9. Note that $\mathbb{Q}(\sqrt[3]{-k}) = \mathbb{Q}(\sqrt[3]{k})$ and $\mathbb{Q}(\sqrt[3]{k^2}) = \mathbb{Q}(\sqrt[3]{k})$.

Example 12.6.10 *We show that the ideal class group $H(K)$ of the quartic field $K = \mathbb{Q}(\sqrt{2} + i)$ is trivial. We have already observed that K is a cyclotomic field, namely, $K = \mathbb{Q}(\zeta_8) = K_8$. Thus, by Theorem 7.5.2, K is a monogenic field. Indeed $O_K = \mathbb{Z} + \mathbb{Z}\zeta_8 + \mathbb{Z}\zeta_8^2 + \mathbb{Z}\zeta_8^3$, by Theorem 7.5.1. Set $\theta = \zeta_8 = (\sqrt{2} + \sqrt{-2})/2$.*

The minimal polynomial of θ is irr$_\mathbb{Q}\,\theta = x^4 + 1$, which has four nonreal roots, namely, $\frac{1}{2}(\pm\sqrt{2} \pm i\sqrt{2})$. Thus $r = 0$, $s = 2$. It was shown in Example 7.1.7 that $d(K) = 256$. Thus the Minkowski bound M_K satisfies

$$M_K = \left(\frac{2}{\pi}\right)^s \sqrt{|d(K)|} = \left(\frac{2}{\pi}\right)^2 \sqrt{256} = \frac{64}{\pi^2} < 7.$$

Hence the primes $p \le M_K$ are $p = 2, 3,$ and 5.

The factorization of $x^4 + 1$ into irreducible polynomials modulo 2 is given by

$$x^4 + 1 \equiv (x + 1)^4 \,(\text{mod } 2)$$

Table 9.   *Class numbers of*
$\mathbb{Q}(\sqrt[3]{k})$, $2 \leq k \leq 101$, k *cubefree*

k	$h(\mathbb{Q}(\sqrt[3]{k}))$	k	$h(\mathbb{Q}(\sqrt[3]{k}))$	k	$h(\mathbb{Q}(\sqrt[3]{k}))$
2	1	37	3	69	1
3	1	38	3	70	9
5	1	39	6	71	1
6	1	41	1	73	3
7	3	42	3	74	3
10	1	43	12	76	6
11	2	44	1	77	3
12	1	45	1	78	3
13	3	46	1	79	6
14	3	47	2	82	1
15	2	51	3	83	2
17	1	52	3	84	3
19	3	53	1	85	3
20	3	55	1	86	9
21	3	57	6	87	1
22	3	58	6	89	2
23	1	59	1	90	3
26	3	60	3	91	9
28	3	61	6	92	3
29	1	62	3	93	3
30	3	63	6	94	3
31	3	65	18	95	3
33	1	66	6	97	3
34	3	67	6	99	1
35	3	68	3	101	2

so that by Theorem 10.3.1 the factorization of the principal ideal $\langle 2 \rangle$ into prime ideals in O_K is

$$\langle 2 \rangle = P_1^4,$$

where

$$P_1 = \langle 2, 1 + \theta \rangle, \quad N(P_1) = 2.$$

Now

$$(1 + \theta)^4 - 4(1 + \theta)^3 + 6(1 + \theta)^2 - 4(1 + \theta) + 2 = 0,$$

so that $1 + \theta \mid 2$. Hence

$$P_1 = \langle 2, 1 + \theta \rangle = \langle 1 + \theta \rangle$$

is principal.

The factorization of $x^4 + 1$ into irreducible polynomials modulo 3 is given by

$$x^4 + 1 \equiv (x^2 + x - 1)(x^2 - x - 1) \,(\mathrm{mod}\ 3),$$

so that by Theorem 10.3.1 the factorization of the principal ideal $\langle 3 \rangle$ into prime ideals in O_K is

$$\langle 3 \rangle = P_2 P_3,$$

where

$$P_2 = \langle 3, 1 - \theta - \theta^2 \rangle, \ \ P_3 = \langle 3, 1 + \theta - \theta^2 \rangle, \ \ N(P_2) = N(P_3) = 3^2, \ \ P_2 \neq P_3.$$

As $\theta^4 + 1 = 0$ we see that $\theta \mid 1$, so that θ is a unit of O_K. Further, as

$$(1 - \theta - \theta^2)(1 + \theta - \theta^2) = -3\theta^2,$$

we deduce that

$$1 - \theta - \theta^2 \mid 3 \text{ and } 1 + \theta - \theta^2 \mid 3,$$

so that the ideals

$$P_2 = \langle 1 - \theta - \theta^2 \rangle, \ \ P_3 = \langle 1 + \theta - \theta^2 \rangle$$

are principal.

The factorization of $x^4 + 1$ into irreducible polynomials modulo 5 is given by

$$x^4 + 1 \equiv (x^2 + 2)(x^2 - 2) \,(\mathrm{mod}\ 5),$$

so that by Theorem 10.3.1 the factorization of the principal ideal $\langle 5 \rangle$ into prime ideals in O_K is

$$\langle 5 \rangle = P_4 P_5,$$

where

$$P_4 = \langle 5, 2 + \theta^2 \rangle, \ \ P_5 = \langle 5, -2 + \theta^2 \rangle, \ \ N(P_4) = N(P_5) = 5^2, \ \ P_4 \neq P_5.$$

Now

$$(2 + \theta^2)(-2 + \theta^2) = -4 + \theta^4 = -5,$$

so that

$$2 + \theta^2 \mid 5, \ \ -2 + \theta^2 \mid 5.$$

Hence the ideals

$$P_4 = \langle 2 + \theta^2 \rangle, \ \ P_5 = \langle -2 + \theta^2 \rangle$$

are principal.

Table 10.   *Class numbers of*
cyclotomic fields K_m, $3 \leq m \leq 45$,
$m \not\equiv 2 \,(\mathrm{mod}\ 4)$

m	$h(K_m)$	m	$h(K_m)$	m	$h(K_m)$
3	1	17	1	32	1
4	1	19	1	33	1
5	1	20	1	35	1
7	1	21	1	36	1
8	1	23	3	37	37
9	1	24	1	39	2
11	1	25	1	40	1
12	1	27	1	41	121
13	1	28	1	43	211
15	1	29	8	44	1
16	1	31	9	45	1

Note: $K_{2n} = K_n$ for n odd.

We have shown that all the prime ideals dividing the principal ideals $\langle p \rangle$, where p is a prime $\leq M_K$, are principal so that the ideal class group $H(K)$ is trivial. Hence $h(K_8) = 1$.

We conclude this section with a short table of class numbers of cyclotomic fields (Table 10).

12.7 Applications to Binary Quadratic Forms

Let m be a squarefree integer. Let K be the quadratic field $\mathbb{Q}(\sqrt{m})$. The discriminant $d(K)$ is given by

$$d(K) = 2^{2\delta} m, \tag{12.7.1}$$

where

$$\delta = \begin{cases} 0, & \text{if } m \equiv 1 \ (\mathrm{mod}\ 4), \\ 1, & \text{if } m \equiv 2 \text{ or } 3 \ (\mathrm{mod}\ 4). \end{cases} \tag{12.7.2}$$

Let p be an odd prime such that

$$\left(\frac{d(K)}{p} \right) = 1. \tag{12.7.3}$$

We observe that (12.7.3) is equivalent to $\left(\frac{m}{p} \right) = 1$ as $p \neq 2$. Then

$$\langle p \rangle = P_1 P_2,$$

where P_1 and P_2 are distinct conjugate prime ideals of O_K. Let h denote the class number $h(K)$. Then

$$\langle p^h \rangle = \langle p \rangle^h = P_1^h P_2^h.$$

As $[P_1] \in H(K)$ and $\operatorname{card}(H(K)) = h$, we have

$$[P_1^h] = [P_1]^h = 1.$$

Thus P_1^h is a principal ideal, say,

$$P_1^h = \langle \frac{x + y\sqrt{m}}{2^{1-\delta}} \rangle,$$

where x and y are rational integers with $x \equiv y \pmod{2}$ if $m \equiv 1 \pmod{4}$. As P_2^h is the conjugate ideal of P_1^h, we have

$$P_2^h = \langle \frac{x - y\sqrt{m}}{2^{1-\delta}} \rangle,$$

so that

$$\langle p^h \rangle = \langle \frac{x^2 - my^2}{4^{1-\delta}} \rangle.$$

Thus

$$p^h = \theta \left(\frac{x^2 - my^2}{4^{1-\delta}} \right)$$

for some $\theta \in U(O_K)$. But

$$\theta = \frac{4^{1-\delta} p^h}{x^2 - my^2} \in \mathbb{Q}$$

so that $\theta \in U(\mathbb{Z}) = \{\pm 1\}$. Hence

$$4^{1-\delta} p^h = \pm(x^2 - my^2), \tag{12.7.4}$$

showing that $4^{1-\delta} p^h$ is represented by one or both of the binary quadratic forms $x^2 - my^2$ and $-x^2 + my^2$.

We prove that

$$(x, y) = \begin{cases} 1, & \text{if } m \equiv 2 \text{ or } 3 \pmod{4}, \\ 1 \text{ or } 2, & \text{if } m \equiv 1 \pmod{4}. \end{cases} \tag{12.7.5}$$

Suppose that q is an odd prime with $q \mid (x, y)$. Then $x = qx_1$ and $y = qy_1$ for integers x_1 and y_1 with $x_1 \equiv y_1 \pmod{2}$ if $m \equiv 1 \pmod{4}$. From (12.7.4) we deduce that $q^2 \mid 4^{1-\delta} p^h$. As $q \neq 2$ we must have $q = p$. Thus

$$P_1^h = \langle p \rangle \langle \frac{x_1 + y_1\sqrt{m}}{2^{1-\delta}} \rangle = P_1 P_2 \langle \frac{x_1 + y_1\sqrt{m}}{2^{1-\delta}} \rangle,$$

so that $P_2 \mid P_1^h$, contradicting that P_1 and P_2 are distinct prime ideals. Hence there are no odd primes dividing (x, y) and so

$$(x, y) = 2^w$$

for some nonnegative integer w. From (12.7.4) we deduce that $2^{2w} \mid 2^{2(1-\delta)}$ so that $0 \le w \le 1 - \delta$. If $m \equiv 2$ or $3 \pmod 4$ then $\delta = 1$ and $w = 0$. If $m \equiv 1 \pmod 4$ then $\delta = 0$ and $w = 0$ or 1.

If m is negative then $x^2 - my^2 > 0$, so the plus sign holds in (12.7.4). If m is positive and there exist integers T and U such that $T^2 - mU^2 = -1$ then

$$-(x^2 - my^2) = (T^2 - mU^2)(x^2 - my^2) = (Tx + mUy)^2 - m(Ty + Ux)^2$$

and the plus sign holds in (12.7.4).

If $m \equiv 1 \pmod 8$ then $\delta = 0$ (by (12.7.2)) and (12.7.4) gives

$$x^2 - y^2 \equiv x^2 - my^2 = \pm 4p^h \equiv 4 \pmod 8,$$

so that $x \equiv y \equiv 0 \pmod 2$. Setting $x = 2u$, $y = 2v$ $(u, v \in \mathbb{Z})$, we obtain from (12.7.4)

$$p^h = \pm(u^2 - mv^2).$$

From (12.7.5) we deduce that $(u, v) = 1$.

If $m \equiv 5 \pmod 8$ then $\delta = 0$ (by (12.7.2)) and $x \equiv y \pmod 2$. Set

$$x = v + 2u, \quad y = v,$$

where $u, v \in \mathbb{Z}$. Then (12.7.4) becomes

$$\pm 4p^h = x^2 - my^2 = (v + 2u)^2 - mv^2 = 4u^2 + 4uv + (1 - m)v^2$$

so that

$$p^h = \pm(u^2 + uv + (\tfrac{1-m}{4})v^2).$$

From (12.7.5) we have

$$(2u, v) = (v + 2u, v) = (x, y) = 1 \text{ or } 2,$$

so that $(u, v) = 1$ or 2. But $(u, v)^2 \mid p^h$, so that $(u, v) = p^t$ for some nonnegative integer t. Hence $t = 0$ and $(u, v) = 1$. We have proved the following result.

Theorem 12.7.1 *Let m be a squarefree integer. Let p be an odd prime with $\left(\frac{m}{p}\right) = 1$. Let h denote the class number of the quadratic field $\mathbb{Q}(\sqrt{m})$.*

If m is negative or m is positive and there are integers T and U such that
$T^2 - mU^2 = -1$ *then there exist coprime integers u and v such that*

$$p^h = \begin{cases} u^2 - mv^2, & \text{if } m \equiv 1 \pmod 8 \text{ or } m \equiv 2, 3 \pmod 4, \\ u^2 + uv + \dfrac{1}{4}(1 - m)v^2, & \text{if } m \equiv 5 \pmod 8. \end{cases}$$

Otherwise there exist coprime integers u and v such that

$$p^h = \begin{cases} u^2 - mv^2 \text{ or } -u^2 + mv^2, & \text{if } m \equiv 1 \pmod 8 \text{ or } m \equiv 2, 3 \pmod 4, \\ u^2 + uv + \dfrac{1}{4}(1 - m)v^2 \text{ or } -u^2 - uv - \dfrac{1}{4}(1 - m)v^2, & \text{if } m \equiv 5 \pmod 8. \end{cases}$$

The reader should compare this theorem with Theorems 1.4.4 and 1.4.5.

In the opposite direction to Theorem 12.7.1 we have the following simple result.

Theorem 12.7.2 *Let $ax^2 + bxy + cy^2$ be an integral binary quadratic form of discriminant d. Let p be an odd prime with $p \nmid a$. Let k be a positive integer. If $\left(\dfrac{d}{p}\right) = -1$ then there do not exist coprime integers u and v such that*

$$p^k = au^2 + buv + cv^2. \tag{12.7.6}$$

Proof: Suppose on the contrary that $\left(\dfrac{d}{p}\right) = -1$ and there are coprime integers u and v satisfying (12.7.6). Then, as $d = b^2 - 4ac$, we have

$$4ap^k = (2au + bv)^2 - dv^2. \tag{12.7.7}$$

From (12.7.7) we see that

$$p \mid v \implies p \mid 2au + bv \implies p \mid u, \text{ as } p \nmid 2a,$$

contradicting that $(u, v) = 1$. Hence $p \nmid v$. Then there exists an integer w such that $vw \equiv 1 \pmod p$. Thus

$$((2au + bv)w)^2 \equiv dv^2w^2 \equiv d \pmod p,$$

so that $\left(\dfrac{d}{p}\right) = 0$ or 1, contradicting that $\left(\dfrac{d}{p}\right) = -1$. This proves that no such integers u and v exist. ∎

In the next three examples we apply Theorems 12.7.1 and 12.7.2 in the cases $m = -1, -2$, and -3. We recover Theorems 2.5.1, 2.5.2, and 2.5.3 respectively (see Exercises 12, 14, and 16 of Chapter 2).

Example 12.7.1 $m = -1$. *Here $h = h(\mathbb{Q}(\sqrt{-1})) = 1$. If p is an odd prime with $\left(\dfrac{-1}{p}\right) = 1$, by Theorem 12.7.1 there exist (coprime) integers u and v such that $p = u^2 + v^2$. Conversely, by Theorem 12.7.2, if there exist integers u and v such* *that* $p = u^2 + v^2$ *(so that* $(u, v) = 1$*) then* $\left(\frac{-4}{p}\right) = 0$ *or* 1*, that is,* $\left(\frac{-1}{p}\right) = 1$ *as* $p \neq 2$. *Since* $\left(\frac{-1}{p}\right) = 1 \Longleftrightarrow p \equiv 1 \pmod 4$*, and* $2 = 1^2 + 1^2$*, we deduce that for a prime* p

$$p = u^2 + v^2 \Longleftrightarrow p = 2 \text{ or } p \equiv 1 \pmod 4$$

(see Theorem 2.5.1 and Exercise 12 of Chapter 2).

Example 12.7.2 $m = -2$. *Here* $h = h(\mathbb{Q}(\sqrt{-2})) = 1$. *Let* p *be an odd prime. If* $\left(\frac{-2}{p}\right) = 1$*, then, by Theorem 12.7.1, there exist (coprime) integers* u *and* v *such that* $p = u^2 + 2v^2$. *Conversely, by Theorem 12.7.2, if there exist integers* u *and* v *such that* $p = u^2 + 2v^2$ *then* $\left(\frac{-8}{p}\right) = 0$ *or* 1*, so that* $\left(\frac{-2}{p}\right) = 1$ *as* p *is odd. Since* $\left(\frac{-2}{p}\right) = 1 \Longleftrightarrow p \equiv 1, 3 \pmod 8$*, and* $2 = 0^2 + 2 \cdot 1^2$*, we deduce that for a prime* p

$$p = u^2 + 2v^2 \Longleftrightarrow p = 2 \text{ or } p \equiv 1, 3 \pmod 8$$

(see Theorem 2.5.2 and Exercise 14 of Chapter 2).

Example 12.7.3 $m = -3$. *Here* $h = h(\mathbb{Q}(\sqrt{-3})) = 1$. *Let* p *be an odd prime. If* $\left(\frac{-3}{p}\right) = 1$*, then, by Theorem 12.7.1, there exist (coprime) integers* u *and* v *such that* $p = u^2 + uv + v^2$. *Conversely, by Theorem 12.7.2, if there exist integers* u *and* v *such that* $p = u^2 + uv + v^2$ *then* $\left(\frac{-3}{p}\right) = 0$ *or* 1*, so that* $p = 3$ *or* $\left(\frac{-3}{p}\right) = 1$. *Since* $\left(\frac{-3}{p}\right) = 1 \Longleftrightarrow p \equiv 1 \pmod 3$*,* $2 \neq u^2 + uv + v^2$*, and* $3 = 1^2 + 1 \cdot 1 + 1^2$*, we have for a prime* p

$$p = u^2 + uv + v^2 \Longleftrightarrow p = 3 \text{ or } p \equiv 1 \pmod 3$$

(see Theorem 2.5.3 and Exercise 16 of Chapter 2).

We now give an example with $h(\mathbb{Q}(\sqrt{m})) > 1$.

Example 12.7.4 $m = -5$. *Here* $h = h(\mathbb{Q}(\sqrt{-5})) = 2$. *If* p *is an odd prime with* $\left(\frac{-5}{p}\right) = 1$*, by Theorem 12.7.1 there exist coprime integers* u *and* v *such that* $p^2 = u^2 + 5v^2$. *Conversely, by Theorem 12.7.2, if there exist coprime integers* u *and* v *such that* $p^2 = u^2 + 5v^2$*, where* p *is an odd prime, then* $\left(\frac{-5}{p}\right) = 0$ *or* 1. *Hence* $p = 5$ *or* $p \equiv 1, 3, 7,$ *or* $9 \pmod{20}$. *Clearly* $2^2, 5^2 \neq u^2 + 5v^2$ *with* $(u, v) = 1$*, so that for a prime* p

$$p^2 = u^2 + 5v^2, \ (u, v) = 1 \Longleftrightarrow p \equiv 1, 3, 7, 9 \pmod{20}.$$

We note that

$$29^2 = 11^2 + 5 \cdot 12^2 \text{ and } 29 = 3^2 + 5 \cdot 2^2,$$
$$3^2 = 2^2 + 5 \cdot 1^2 \text{ but } 3 \neq u^2 + 5v^2.$$

We return to the representability of p by the form $u^2 + 5v^2$ in Example 12.7.5.

Theorem 12.7.1 is concerned with the representability of p^h by a binary quadratic form of discriminant m or $4m$. But what about the representability of p itself by such a form? To tackle this problem we must use our knowledge of the generators of the ideal class group of $\mathbb{Q}(\sqrt{m})$. We illustrate the ideas involved with two examples.

Example 12.7.5 *The ideal class group of $K = \mathbb{Q}(\sqrt{-5})$ is*

$$H(K) = H(\mathbb{Q}(\sqrt{-5})) = \{1, [Q]\} \simeq \mathbb{Z}_2,$$

where $[Q]$ is the class of the ideal Q given by

$$Q = \langle 2, 1 + \sqrt{-5} \rangle, \quad Q^2 = \langle 2 \rangle, \quad N(Q) = 2$$

(see Table 5). Let $p \neq 2, 5$ be a prime such that $\left(\frac{-5}{p}\right) = 1$, so that

$$\langle p \rangle = P_1 P_2,$$

where P_1 and P_2 are distinct conjugate prime ideals of O_K. Hence $[P_1] = 1$ or $[P_1] = [Q]$. In the first case P_1 is a principal ideal, say, $P_1 = \langle x + y\sqrt{-5} \rangle$, where $x, y \in \mathbb{Z}$, so that

$$p = N(P_1) = N(\langle x + y\sqrt{-5} \rangle) = |N(x + y\sqrt{-5})| = x^2 + 5y^2.$$

In the second case $[P_1 Q] = [P_1][Q] = [Q]^2 = [Q^2] = [\langle 2 \rangle] = 1$, so that $P_1 Q$ is a principal ideal, say, $P_1 Q = \langle x + y\sqrt{-5} \rangle$, where $x, y \in \mathbb{Z}$, and

$$2p = N(P_1)N(Q) = N(P_1 Q) = N(\langle x + y\sqrt{-5} \rangle) = x^2 + 5y^2.$$

Hence we have shown that for a prime $p \neq 2, 5$

$$\left(\frac{-5}{p}\right) = 1 \implies p \text{ or } 2p = x^2 + 5y^2 \text{ for integers } x \text{ and } y.$$

Conversely, if $p \neq 2, 5$ is a prime with p or $2p = x^2 + 5y^2$ for some integers x and y then $p \nmid y$ so that $yz \equiv 1 \pmod{p}$ for some integer z and thus, as $x^2 \equiv -5y^2 \pmod{p}$, we have

$$(xz)^2 \equiv -5(yz)^2 \equiv -5 \pmod{p},$$

so that $\left(\frac{-5}{p}\right) = 0$ or 1. As $p \neq 5$ we have $\left(\frac{-5}{p}\right) = 1$. Hence, for a prime $p \neq 2, 5$, we have shown that

$$p \text{ or } 2p = x^2 + 5y^2 \iff \left(\frac{-5}{p} \right) = 1.$$

Now suppose that $p = x^2 + 5y^2$. Working modulo 4, we obtain

$$p = x^2 + 5y^2 \implies p \equiv x^2 + y^2 \pmod{4} \implies p \equiv 1 \pmod{4} \implies \left(\frac{-1}{p} \right) = 1$$

and modulo 5, we obtain

$$p = x^2 + 5y^2 \implies p \equiv x^2 \pmod{5} \implies p \equiv 1, 4 \pmod{5} \implies \left(\frac{p}{5} \right) = 1.$$

However, if $2p = x^2 + 5y^2$ then modulo 8 we deduce

$$2p = x^2 + 5y^2 \implies 2p \equiv 6 \pmod{8} \implies p \equiv 3 \pmod{4} \implies \left(\frac{-1}{p} \right) = -1$$

and modulo 5 we get

$$2p = x^2 + 5y^2 \implies 2p \equiv x^2 \pmod{5} \implies 2p \equiv 1 \text{ or } 4 \pmod{5}$$
$$\implies p \equiv 2, 3 \pmod{5} \implies \left(\frac{p}{5} \right) = -1.$$

By the law of quadratic reciprocity, we have

$$\left(\frac{-1}{p} \right) \left(\frac{p}{5} \right) = \left(\frac{-1}{p} \right) \left(\frac{5}{p} \right) = \left(\frac{-5}{p} \right) = 1.$$

Hence, for a prime $p \neq 2, 5$, we have proved that

$$p = x^2 + 5y^2 \iff \left(\frac{-1}{p} \right) = \left(\frac{p}{5} \right) = 1,$$
$$2p = x^2 + 5y^2 \iff \left(\frac{-1}{p} \right) = \left(\frac{p}{5} \right) = -1.$$

When p is a prime such that

$$\left(\frac{-1}{p} \right) = \left(\frac{p}{5} \right) = -1$$

(for example a prime $p \equiv 3 \pmod{20}$) $2p$ is represented by a binary quadratic form of discriminant -20, namely $x^2 + 5y^2$. But what about the representation of p itself by a binary quadratic form of discriminant -20? It cannot be the form $x^2 + 5y^2$ but maybe there is some other form of discriminant -20 that represents p. We show that this is indeed the case and that the form can be taken to be $2x^2 + 2xy + 3y^2$.

Suppose that $p \neq 2, 5$ is a prime such that

$$\left(\frac{-1}{p}\right) = \left(\frac{p}{5}\right) = -1.$$

Then there exist integers y and z such that $2p = z^2 + 5y^2$. Clearly $z \equiv y \pmod 2$. Thus we can define an integer x by $z = y + 2x$. Then

$$2p = (y + 2x)^2 + 5y^2 = 4x^2 + 4xy + 6y^2$$

so that

$$p = 2x^2 + 2xy + 3y^2.$$

The form $2x^2 + 2xy + 3y^2$ has discriminant $= 2^2 - 4 \cdot 2 \cdot 3 = -20$. Conversely, if $p = 2x^2 + 2xy + 3y^2$ for integers x and y, then $2p = X^2 + 5Y^2$ with $X = 2x + y$ and $Y = y$. Hence p satisfies

$$\left(\frac{-1}{p}\right) = \left(\frac{p}{5}\right) = -1.$$

We have shown that for a prime $p \neq 2, 5$

$$p = x^2 + 5y^2 \iff \left(\frac{-1}{p}\right) = \left(\frac{p}{5}\right) = 1 \iff p \equiv 1, 9 \pmod{20},$$

$$p = 2x^2 + 2xy + 3y^2 \iff \left(\frac{-1}{p}\right) = \left(\frac{p}{5}\right) = -1 \iff p \equiv 3, 7 \pmod{20}.$$

This result provides a refinement of Example 12.7.4. Note that

$$p = x^2 + 5y^2 \implies p^2 = u^2 + 5v^2, \ (u, v) = 1,$$
$$\text{with } u = x^2 - 5y^2, \ v = 2xy,$$

and

$$p = 2x^2 + 2xy + 3y^2 \implies p^2 = u^2 + 5v^2, \ (u, v) = 1,$$
$$\text{with } u = 2x^2 + 2xy - 2y^2, \ v = 2xy + y^2.$$

We conclude by noting that the prime 2 is represented by $2x^2 + 2xy + 3y^2$ but not by $x^2 + 5y^2$ and that 5 is represented by $x^2 + 5y^2$ but not by $2x^2 + 2xy + 3y^2$.

Example 12.7.6 *The ideal class group of $K = \mathbb{Q}(\sqrt{-21})$ is*

$$H(K) = H(\mathbb{Q}(\sqrt{-21})) = \{1, [A], [B], [A][B]\} \simeq \mathbb{Z}_2 \times \mathbb{Z}_2,$$

where $[A]$ is the class of the ideal A given by

$$A = \langle 2, 1 + \sqrt{-21} \rangle, \ A^2 = \langle 2 \rangle, \ N(A) = 2,$$

[B] is the class of the ideal B given by

$$B = \langle 3, \sqrt{-21} \rangle, \quad B^2 = \langle 3 \rangle, \quad N(B) = 3,$$

and

$$[A][B] = [AB] = [\langle 5, 3 + \sqrt{-21} \rangle], [A]^2 = [B]^2 = [AB]^2 = 1$$

(see Table 5).

Let $p \neq 2, 3, 7$ *be a prime such that* $\left(\frac{-21}{p} \right) = 1$. *By the law of quadratic reciprocity we have*

$$\left(\frac{-21}{p} \right) = \left(\frac{p}{3} \right) \left(\frac{p}{7} \right) \left(\frac{-1}{p} \right)$$

so that

$$\left(\left(\frac{p}{3} \right), \left(\frac{p}{7} \right), \left(\frac{-1}{p} \right) \right) = (1, 1, 1), \ (-1, 1, -1), \ (1, -1, -1), \ or \ (-1, -1, 1).$$

As $\left(\frac{-21}{p} \right) = 1$ *we have*

$$\langle p \rangle = P_1 P_2,$$

where P_1 and P_2 are distinct conjugate prime ideals in O_K. Hence

$$[P_1] = 1, \ [A], \ [B], \ or \ [AB].$$

If $[P_1] = 1$ then P_1 is a principal ideal, say, $P_1 = \langle x + y\sqrt{-21} \rangle$, where $x, y \in \mathbb{Z}$, so that

$$p = N(P_1) = N(\langle x + y\sqrt{-21} \rangle) = x^2 + 21y^2.$$

If $[P_1] = [A]$ then $[AP_1] = [A][P_1] = [A]^2 = 1$, so that AP_1 is a principal ideal, say, $AP_1 = \langle x + y\sqrt{-21} \rangle$, where $x, y \in \mathbb{Z}$, so that

$$2p = N(AP_1) = N(\langle x + y\sqrt{-21} \rangle) = x^2 + 21y^2.$$

Similarly, if $[P_1] = [B]$ we find that $3p = x^2 + 21y^2$ ($x, y \in \mathbb{Z}$) and if $[P_1] = [AB]$ then $5p = x^2 + 21y^2$ ($x, y \in \mathbb{Z}$).

Next if $p = x^2 + 21y^2$ then we have

$$\left(\frac{p}{3} \right) = \left(\frac{x^2 + 21y^2}{3} \right) = \left(\frac{x^2}{3} \right) = 1,$$

$$\left(\frac{p}{7} \right) = \left(\frac{x^2 + 21y^2}{7} \right) = \left(\frac{x^2}{7} \right) = 1,$$

$$\left(\frac{-1}{p} \right) = 1, \ as \ p \equiv x^2 + y^2 \equiv 1 \ (\mathrm{mod} \ 4).$$

Similarly,

$$2p = x^2 + 21y^2 \implies \left(\frac{p}{3}\right) = -1, \ \left(\frac{p}{7}\right) = 1, \ \left(\frac{-1}{p}\right) = -1,$$

$$3p = x^2 + 21y^2 \implies \left(\frac{p}{3}\right) = 1, \ \left(\frac{p}{7}\right) = -1, \ \left(\frac{-1}{p}\right) = -1,$$

$$5p = x^2 + 21y^2 \implies \left(\frac{p}{3}\right) = -1, \ \left(\frac{p}{7}\right) = -1, \ \left(\frac{-1}{p}\right) = 1.$$

However, if $p \neq 2, 3, 7$ is a prime such that $\left(\frac{-21}{p}\right) = -1$ then $p, 2p, 3p, 5p \neq x^2 + 21y^2$, so we have proved the following: If $p \neq 2, 3, 7$ is a prime then

$$p = x^2 + 21y^2 \iff \left(\frac{p}{3}\right) = 1, \ \left(\frac{p}{7}\right) = 1, \ \left(\frac{-1}{p}\right) = 1,$$

$$2p = x^2 + 21y^2 \iff \left(\frac{p}{3}\right) = -1, \ \left(\frac{p}{7}\right) = 1, \ \left(\frac{-1}{p}\right) = -1,$$

$$3p = x^2 + 21y^2 \iff \left(\frac{p}{3}\right) = 1, \ \left(\frac{p}{7}\right) = -1, \ \left(\frac{-1}{p}\right) = -1,$$

$$5p = x^2 + 21y^2 \implies \left(\frac{p}{3}\right) = -1, \ \left(\frac{p}{7}\right) = -1, \ \left(\frac{-1}{p}\right) = 1.$$

Easy calculations show that

$$2p = x^2 + 21y^2 \iff p = 2u^2 + 2uv + 11v^2,$$
$$3p = x^2 + 21y^2 \iff p = 3u^2 + 7v^2,$$
$$5p = x^2 + 21y^2 \iff p = 5u^2 + 4uv + 5v^2.$$

Thus, for a prime $p \neq 2, 3, 7$, we have

$$p = x^2 + 21y^2 \iff p = 1, 25, 37 \ (\mathrm{mod}\ 84),$$
$$p = 2x^2 + 2xy + 11y^2 \iff p = 11, 23, 71 \ (\mathrm{mod}\ 84),$$
$$p = 3x^2 + 7y^2 \iff p = 19, 31, 55 \ (\mathrm{mod}\ 84),$$
$$p = 5x^2 + 4xy + 5y^2 \iff p = 5, 17, 41 \ (\mathrm{mod}\ 84).$$

We leave it to the reader to determine which forms represent the primes 2, 3, and 7.

These examples suggest a theorem of the following type: If $ax^2 + bxy + cy^2$ is a form of discriminant D then there exist positive integers $s, a_1, \ldots, a_s, m$ (depending on a, b, c) such that for a prime $p \neq 2$ not dividing D

$$p = ax^2 + bxy + cy^2 \iff p \equiv a_1, \ldots, a_s \ (\mathrm{mod}\ m).$$

However, such a result does not hold for every form $ax^2 + bxy + cy^2$. This is proved in [6], where it is shown that every arithmetic progression either contains no primes of the form $x^2 + 14y^2$ or contains primes of both forms $x^2 + 14y^2$ and $2x^2 + 7y^2$, proving that congruences cannot be used to distinguish the representability of a prime by $x^2 + 14y^2$ from that by $2x^2 + 7y^2$. By the methods used in Examples 12.7.5 and 12.7.6 we can prove that for a prime $p \neq 2, 7$

$$p = x^2 + 14y^2 \text{ or } 2x^2 + 7y^2 \iff \left(\frac{p}{7}\right) = \left(\frac{2}{p}\right) = 1$$
$$\iff p \equiv 1, 9, 15, 23, 25, 39 \ (\mathrm{mod}\ 56)$$

and

$$p = 3x^2 + 2xy + 5y^2 \iff \left(\frac{p}{7}\right) = \left(\frac{2}{p}\right) = -1$$
$$\iff p \equiv 3, 5, 13, 19, 27, 45 \ (\mathrm{mod}\ 56).$$

Muskat [5] has shown how to distinguish the representations $p = x^2 + 14y^2$ and $p = 2x^2 + 7y^2$ as follows. Let p be a prime with $p \equiv 1, 9, 15, 23, 25,$ $39 \ (\mathrm{mod}\ 56)$. Then, as in Example 12.7.1, we can show that $p = u^2 + 7v^2$ for some integers u and v. If $p \equiv 1 \ (\mathrm{mod}\ 8)$ then u is odd and $v \equiv 0 \ (\mathrm{mod}\ 4)$, and replacing u by $-u$ if necessary we may suppose that $u \equiv 1 \ (\mathrm{mod}\ 4)$; if $p \equiv 7 \ (\mathrm{mod}\ 8)$ then $u \equiv 0 \ (\mathrm{mod}\ 4)$ and v is odd, and replacing v by $-v$ if necessary we may suppose that $v \equiv 1 \ (\mathrm{mod}\ 4)$. Thus in both cases we have

$$2p + u + v \equiv 3 \ (\mathrm{mod}\ 4)$$

and Muskat has proved that

$$p = x^2 + 14y^2 \iff 2p + u + v \equiv 3 \ (\mathrm{mod}\ 8),$$
$$p = 2x^2 + 7y^2 \iff 2p + u + v \equiv 7 \ (\mathrm{mod}\ 8).$$

Exercises

1. Prove that $H(\mathbb{Q}(\sqrt{-6})) = \{1, [\langle 2, \sqrt{-6}\rangle]\} \simeq \mathbb{Z}_2$.
2. Prove that $h(\mathbb{Q}(\sqrt{-7})) = 1$.
3. Prove that $h(\mathbb{Q}(\sqrt{-11})) = 1$.
4. Prove that $H(\mathbb{Q}(\sqrt{-13})) \simeq \mathbb{Z}_2$.
5. Prove that $H(\mathbb{Q}(\sqrt{-15})) \simeq \mathbb{Z}_2$.
6. Prove that $H(\mathbb{Q}(\sqrt{-17})) \simeq \mathbb{Z}_4$.
7. Prove that $H(\mathbb{Q}(\sqrt{-23})) \simeq \mathbb{Z}_3$.
8. Prove that $H(\mathbb{Q}(\sqrt{-26})) \simeq \mathbb{Z}_6$.
9. Prove that $H(\mathbb{Q}(\sqrt{-30})) \simeq \mathbb{Z}_2 \times \mathbb{Z}_2$.
10. Prove that $H(\mathbb{Q}(\sqrt{-47})) \simeq \mathbb{Z}_5$.
11. Prove that $h(\mathbb{Q}(\sqrt{6})) = 1$.
12. Prove that $h(\mathbb{Q}(\sqrt{10})) = 2$.

13. Prove that $H(\mathbb{Q}(\sqrt{15})) = \{1, [\langle 2, 1 + \sqrt{15}\rangle]\} \simeq \mathbb{Z}_2$.
14. Let $K = \mathbb{Q}(\theta)$, where $\theta^3 - 4\theta + 2 = 0$. Prove that $h(K) = 1$.
15. Prove that $h(\mathbb{Q}(\sqrt[3]{3})) = 1$.
16. Prove that $h(\mathbb{Q}(\sqrt[3]{5})) = 1$.
17. Determine $h(\mathbb{Q}(\sqrt[4]{2}))$.
18. Let p be a prime $\neq 2, 5$. Prove that

$$p = x^2 + 10y^2 \Longleftrightarrow p \equiv 1, 9, 11, 19 \,(\mathrm{mod}\, 40),$$
$$p = 2x^2 + 5y^2 \Longleftrightarrow p \equiv 7, 13, 23, 37 \,(\mathrm{mod}\, 40).$$

19. Let p be a prime $\neq 3, 13$. Prove that

$$p = 2x^2 + xy + 5y^2 \Longleftrightarrow \left(\frac{p}{3}\right) = \left(\frac{p}{13}\right) = -1.$$

20. Let p be a prime $\neq 3, 5$. Prove that

$$p = x^2 + xy + 4y^2 \Longleftrightarrow \left(\frac{p}{3}\right) = \left(\frac{p}{5}\right) = 1,$$

$$p = 2x^2 + xy + 2y^2 \Longleftrightarrow \left(\frac{p}{3}\right) = \left(\frac{p}{5}\right) = -1.$$

21. Determine exactly which primes p are represented by $x^2 + xy + 5y^2$.
22. Determine exactly which primes p are represented by $x^2 + xy + 11y^2$.
23. Let p be a prime $\neq 2, 31$. Prove that

$$p = x^2 + 62y^2, \ 2x^2 + 31y^2, \ \text{or}\ 7x^2 + 2xy + 9y^2 \Longleftrightarrow \left(\frac{2}{p}\right) = \left(\frac{p}{31}\right) = 1$$

and

$$p = 3x^2 + 2xy + 21y^2 \ \text{or}\ 11x^2 + 4xy + 6y^2 \Longleftrightarrow \left(\frac{2}{p}\right) = \left(\frac{p}{31}\right) = -1.$$

24. Let K be a quadratic field. Let I be an ideal of O_K. If I^2 is a principal ideal prove that I is a equivalent to its conjugate ideal I'.
25. Let K be an imaginary quadratic field with discriminant $d < -4$. Use Dirichlet's class number formula to prove that

$$h(K) = \frac{1}{2 - \left(\dfrac{d}{2}\right)} \sum_{1 \leq r \langle \frac{|d|}{2}} \left(\frac{d}{r}\right).$$

26. Let p be a prime $\equiv 3 \,(\mathrm{mod}\, 4)$. Use Dirichlet's class number formula to prove that $h(\mathbb{Q}(\sqrt{-p})) \equiv 1 \,(\mathrm{mod}\, 2)$.
27. Let $K = \mathbb{Q}(\sqrt{n})$, where n is a squarefree integer > 1 with $n \equiv 1 \,(\mathrm{mod}\, 4)$. Let m be a positive integer dividing n. Prove that

$$\langle m, \sqrt{n}\rangle = \langle m, \frac{m + \sqrt{n}}{2}\rangle$$

in O_K.

28. Let p be a prime with $p \equiv 3 \,(\text{mod } 4)$. It is known that $h(\mathbb{Q}(\sqrt{p}))$ is odd. Use this fact to prove that there exist integers a and b such that

$$a^2 - pb^2 = (-1)^{(p+1)/4} 2.$$

[Hint: Consider the ideal $\langle 2, 1 + \sqrt{p} \rangle$.]

Suggested Reading

1. Z. I. Borevich and I. R. Shafarevich, *Number Theory*, Academic Press, New York and London, 1966.

   Dirichlet's formula for the class number of a quadratic field is proved in Chapter 5.

2. L. Carlitz, *A characterization of algebraic number fields with class number two*, Proceedings of the American Mathematical Society 11 (1960), 391–392.

   It is proved that an algebraic number field K has class number $h(K) \leq 2$ if and only if whenever a nonzero, nonunit $\alpha \in O_K$ can be written as $\alpha = u\pi_1 \cdots \pi_s = u'\pi'_1 \cdots \pi'_t$ with u, u' units and $\pi_1, \ldots, \pi_s, \pi'_1, \ldots, \pi'_t$ are primes in O_K then $s = t$.

3. D. A. Marcus, *Number Fields*, Springer-Verlag, New York, Heidelberg, Berlin, 1977.

   Chapter 5 contains a proof of (12.5.1).

4. J. M. Masley and H. L. Montgomery, *Cyclotomic fields with unique factorization*, Journal für die reine und angewandte Mathematik 286/287 (1976), 248–256.

   The authors prove that there are precisely 29 distinct cyclotomic fields K_m ($m \not\equiv 2 \,(\text{mod } 4)$) with $h(K_m) = 1$, namely those given by

   $$m = 3, 4, 5, 7, 8, 9, 11, 12, 13, 15, 16, 17, 19, 20, 21, 24, 25, 27, 28, 32, 33, 35, 36,$$
   $$40, 44, 45, 48, 60, 84.$$

5. J. B. Muskat, *On simultaneous representations of primes by binary quadratic forms*, Journal of Number Theory 19 (1984), 263–282.

   It is shown how the representability of primes by the forms $x^2 + 14y^2$ and $2x^2 + 7y^2$ can be distinguished.

6. B. K. Spearman and K. S. Williams, *Representing primes by binary quadratic forms*, American Mathematical Monthly 99 (1992), 423–426.

   It is shown that the representability of primes by the forms $x^2 + 14y^2$ and $2x^2 + 7y^2$ cannot be decided by congruence considerations alone.

7. H. M. Stark, *A complete determination of the complex quadratic fields of class number one*, Michigan Mathematical Journal 14 (1967), 1–27.

   The author shows that $h(\mathbb{Q}(\sqrt{k})) = 1$, where k is a negative squarefree integer, if and only if $k = -1, -2, -3, -7, -11, -19, -43, -67, -163$.

Biographies

1. J. V. Brawley, *In memoriam: Leonard Carlitz (1907–1999)*, Finite Fields and Applications 6 (2000), 203–206.

   A brief biography of Carlitz is given.

2. F. T. Howard, *In memoriam—Leonard Carlitz*, Fibonacci Quarterly 38 (2000), 316.

   Another brief biography of Carlitz is given.

13

Dirichlet's Unit Theorem

13.1 Valuations of an Element of a Number Field

Let K be an algebraic number field of degree $n \geq 2$ over $\mathbb{Q}$. Let $\{\sigma_1, \ldots, \sigma_n\}$ be the set of all monomorphisms : $K \to \mathbb{C}$. If $\sigma_i(K) \subseteq \mathbb{R}$ we say that σ_i is a real embedding; otherwise σ_i is said to be a complex embedding. As usual $\bar{\alpha}$ denotes the complex conjugate of $\alpha \in \mathbb{C}$. We define for all $\alpha \in K$

$$\bar{\sigma}_i(\alpha) = \overline{\sigma_i(\alpha)}.$$

Since complex conjugation is an automorphism of $\mathbb{C}$, $\bar{\sigma}_i$ is a monomorphism : $K \to \mathbb{C}$. Hence $\bar{\sigma}_i = \sigma_j$ for some j. Now $\sigma_i = \bar{\sigma}_i$ if and only if σ_i is real, and $\bar{\bar{\sigma}}_i = \sigma_i$ so that complex monomorphisms occur as conjugate pairs. We enumerate the monomorphisms in such a way that $\sigma_1, \ldots, \sigma_r$ are real, $\sigma_{r+1}, \ldots, \sigma_{r+s}$ are complex, and $\sigma_{r+s+1} = \bar{\sigma}_{r+1}, \ldots, \sigma_n = \sigma_{r+2s} = \bar{\sigma}_{r+s}$. The conjugate fields of K are $K^{(i)} = \sigma_i(K)$, $i = 1, 2, \ldots, n$. The r conjugate fields $K^{(1)}, \ldots, K^{(r)}$ are real and the $n - r$ fields $K^{(r+1)}, \ldots, K^{(n)}$ are nonreal with $K^{(r+s+1)} = \overline{K^{(r+1)}}, \ldots, K^{(n)} = K^{(r+2s)} = \overline{K^{(r+s)}}$. We note that

$$n = r + 2s \tag{13.1.1}$$

and

$$r + s \geq \frac{1}{2}(r + 2s) = \frac{n}{2} \geq 1.$$

If $s = 0$ then all the conjugate fields of K are real and K is said to be a totally real field. If $r = 0$ then K and all its conjugate fields are nonreal and K is said to be a totally complex or totally imaginary field. If K is a normal field then K is either totally real or totally complex, since all the conjugate fields of K coincide.

Example 13.1.1 *The cubic polynomial $x^3 - 6x + 2 \in \mathbb{Z}[x]$ is 2-Eisenstein and has discriminant $-4(-6)^3 - 27(2)^2 = 756 > 0$. Thus it is irreducible and has three real* *roots. Thus the cubic field*

$$K = \mathbb{Q}(\theta), \ where \ \theta^3 - 6\theta + 2 = 0,$$

is a totally real field.

Example 13.1.2 *The field* $\mathbb{Q}(\sqrt{2} + i)$ *is totally complex as the conjugates of* $\sqrt{2} + i$ *are* $\sqrt{2} + i, \ \sqrt{2} - i, \ -\sqrt{2} + i, \ -\sqrt{2} - i.$

We next define the valuations of an element of an algebraic number field.

Definition 13.1.1 (Valuations of a field element) *For* $a \in K$ *we define*

$$\beta_i(a) = |\sigma_i(a)|, \ i = 1, 2, \ldots, r + s.$$

The $r + s$ *quantities* $\beta_i(a)$ $(i = 1, 2, \ldots, r + s)$ *are called the valuations of* a.

Clearly $\sigma_i(a) \in K^{(i)}$ and $[K^{(i)} : \mathbb{Q}] = [K : \mathbb{Q}] = n$, so that $\sigma_i(a)$ is an algebraic number of degree at most n. Thus $\overline{\sigma_i(a)}$ is also an algebraic number of degree at most n. Hence $\sigma_i(a)\overline{\sigma_i(a)}$ is an algebraic number of degree at most n^2. This proves that each valuation

$$\beta_i(a) = |\sigma_i(a)| = \sqrt{\sigma_i(a)\overline{\sigma_i(a)}}$$

is a nonnegative real algebraic number of degree at most $2n^2$.

Example 13.1.3 *Let* $K = \mathbb{Q}(\sqrt[3]{2})$. *Here* $n = 3$, $r = 1$, $s = 1$. *The three monomorphisms* : $K \longrightarrow \mathbb{C}$ *are given by*

$$\sigma_1(a + b\sqrt[3]{2} + c(\sqrt[3]{2})^2) = a + b\sqrt[3]{2} + c(\sqrt[3]{2})^2,$$
$$\sigma_2(a + b\sqrt[3]{2} + c(\sqrt[3]{2})^2) = a + b\omega\sqrt[3]{2} + c\omega^2(\sqrt[3]{2})^2,$$
$$\sigma_3(a + b\sqrt[3]{2} + c(\sqrt[3]{2})^2) = a + b\omega^2\sqrt[3]{2} + c\omega(\sqrt[3]{2})^2,$$

where $\omega = e^{2\pi i/3} = (-1 + i\sqrt{3})/2$ *is a complex cube root of unity, so that* $\sigma_3 = \overline{\sigma_2}$. *Then*

$$\beta_1(\sqrt[3]{2} + (\sqrt[3]{2})^2) = |\sigma_1(\sqrt[3]{2} + (\sqrt[3]{2})^2)| = |\sqrt[3]{2} + (\sqrt[3]{2})^2| = \sqrt[3]{2} + (\sqrt[3]{2})^2$$

and

$$\beta_2(\sqrt[3]{2} + (\sqrt[3]{2})^2) = |\sigma_2(\sqrt[3]{2} + (\sqrt[3]{2})^2)| = |\omega\sqrt[3]{2} + \omega^2(\sqrt[3]{2})^2|$$

$$= \left|\left(\frac{-1 + i\sqrt{3}}{2}\right)\sqrt[3]{2} + \left(\frac{-1 - i\sqrt{3}}{2}\right)(\sqrt[3]{2})^2\right|$$

$$= \left|\frac{-1}{2}(\sqrt[3]{2} + (\sqrt[3]{2})^2) + i\frac{\sqrt{3}}{2}(\sqrt[3]{2} - (\sqrt[3]{2})^2)\right|$$

$$= \sqrt{\frac{(\sqrt[3]{2} + (\sqrt[3]{2})^2)^2 + 3(\sqrt[3]{2} - (\sqrt[3]{2})^2)^2}{4}}$$

$$= \sqrt{-2 + 2\sqrt[3]{2} + (\sqrt[3]{2})^2}.$$

In the next section we develop the properties of the valuations of an element of an algebraic number field K. Using these properties we prove later in the chapter the famous theorem of Dirichlet concerning the units of O_K.

Theorem 13.1.1 (Dirichlet's unit theorem) *Let K be an algebraic number field of degree n. Let r be the number of real conjugate fields of K and 2s the number of complex conjugate fields of K so that r and s satisfy (13.1.1). Then O_K contains $r + s - 1$ units $\epsilon_1, \ldots, \epsilon_{r+s-1}$ such that each unit of O_K can be expressed uniquely in the form $\rho\epsilon_1^{n_1} \cdots \epsilon_{r+s-1}^{n_{r+s-1}}$, where ρ is a root of unity in O_K and $n_1, \ldots, n_{r+s-1}$ are integers.*

13.2 Properties of Valuations

In this section we develop the properties of valuations that we shall need to prove Dirichlet's unit theorem. We fix once and for all an integral basis $\{\omega_1, \ldots, \omega_n\}$ for K. If $a \in O_K$ the coordinates of a are the uniquely determined rational integers $c_1, \ldots, c_n$ given by

$$a = c_1\omega_1 + \cdots + c_n\omega_n.$$

We set

$$M = \max_{1 \le i, j \le n} |\sigma_i(\omega_j)| \tag{13.2.1}$$

and

$$D = \det(\sigma_i(\omega_j)). \tag{13.2.2}$$

As $\{\omega_1, \ldots, \omega_n\}$ is an integral basis for K, we have

$$D^2 = d(K)$$

so that

$$|D| = |d(K)|^{1/2} \tag{13.2.3}$$

and

$$D \neq 0. \tag{13.2.4}$$

Lemma 13.2.1 *If $m \in \mathbb{Z}$ then*

$$\beta_i(m) = |m|, \; i = 1, 2, \ldots, r + s.$$

Proof: For $i = 1, 2, \ldots, r + s$, each $\sigma_i \; : \; K \to \mathbb{C}$ is a monomorphism so that $\sigma_i(a) = a$ for all $a \in \mathbb{Q}$. Hence for $m \in \mathbb{Z}$ we have

$$\beta_i(m) = |\sigma_i(m)| = |m|, \; i = 1, 2, \ldots, r + s. \qquad \blacksquare$$

Lemma 13.2.2 *If $a, b \in K$ then*

$$\beta_i(a)\beta_i(b) = \beta_i(ab), \; i = 1, 2, \ldots, r + s.$$

Proof: For $i = 1, 2, \ldots, r + s$ and $a, b \in K$ we have

$$\beta_i(ab) = |\sigma_i(ab)| = |\sigma_i(a)\sigma_i(b)| = |\sigma_i(a)||\sigma_i(b)| = \beta_i(a)\beta_i(b). \qquad \blacksquare$$

Lemma 13.2.3 *If $a \in O_K$ is such that its coordinates c_i $(i = 1, 2, \ldots, n)$ satisfy $|c_i| \leq C$ then*

$$\beta_i(a) \leq nCM, \; i = 1, 2, \ldots, r + s.$$

Proof: We have

$$a = c_1\omega_1 + \cdots + c_n\omega_n$$

so that for $i = 1, 2, \ldots, n$

$$\sigma_i(a) = c_1\sigma_i(\omega_1) + \cdots + c_n\sigma_i(\omega_n)$$

and thus for $i = 1, 2, \ldots, r + s$ we have

$$\begin{aligned}
\beta_i(a) = |\sigma_i(a)| &= |c_1\sigma_i(\omega_1) + \cdots + c_n\sigma_i(\omega_n)| \\
&\leq |c_1||\sigma_i(\omega_1)| + \cdots + |c_n||\sigma_i(\omega_n)| \\
&\leq CM + \cdots + CM \\
&= nCM. \qquad \blacksquare
\end{aligned}$$

Lemma 13.2.3 tells us that the integers of K with bounded coordinates have bounded valuations.

Lemma 13.2.4 *If $a \in O_K$ is such that*

$$\beta_i(a) \leq L, \ i = 1, 2, \ldots, r + s,$$

then the coordinates c_i $(i = 1, 2, \ldots, n)$ of a satisfy

$$|c_i| \leq \frac{n! L M^{n-1}}{|d(K)|^{1/2}}, \ i = 1, 2, \ldots, n.$$

Proof: We have

$$a = c_1 \omega_1 + \cdots + c_n \omega_n$$

so that

$$\sigma_i(a) = c_1 \sigma_i(\omega_1) + \cdots + c_n \sigma_i(\omega_n), \ i = 1, 2, \ldots, n.$$

Hence, by Cramer's rule, we have

$$c_i = \frac{N_i}{D}, \ i = 1, 2, \ldots, n, \tag{13.2.5}$$

where the determinant D is defined in (13.2.2) and the determinant N_i is formed from D by replacing the ith column by the column consisting of $\sigma_1(a), \ \sigma_2(a), \ldots, \sigma_n(a)$. Expanding N_i by its ith column we obtain

$$N_i = \sum_{k=1}^{n} \sigma_k(a)(-1)^{k+i} \triangle_k,$$

where each $\triangle_k$ is an $(n-1) \times (n-1)$ determinant whose entries $\in \{\sigma_p(\omega_q) \mid p, q = 1, 2, \ldots, n\}$. As $|\sigma_p(\omega_q)| \leq M$ for all $p, q \in \{1, 2, \ldots, n\}$, we see that

$$|\triangle_k| \leq (n-1)! M^{n-1},$$

so that for $i = 1, 2, \ldots, n$

$$|N_i| \leq \sum_{k=1}^{n} \beta_k(a)|\triangle_k| \leq L n! M^{n-1}. \tag{13.2.6}$$

Finally, from (13.2.5), (13.2.6), and (13.2.3), we deduce that

$$|c_i| = \frac{|N_i|}{|D|} \leq \frac{n! L M^{n-1}}{|d(K)|^{1/2}}, \ i = 1, 2, \ldots, n. \qquad \blacksquare$$

Lemma 13.2.4 tells us that the integers of K with bounded valuations have bounded coordinates. The next lemma is an immediate consequence of this fact.

Lemma 13.2.5 *There are only finitely many $a \in O_K$, all of whose valuations $\beta_i(a)$ $(i = 1, 2, \ldots, r + s)$ lie below a given limit.*

Proof: Let $a \in O_K$ be such that $\beta_i(a) \leq L$, $i = 1, 2, \ldots, r + s$. Then, by Lemma 13.2.4, we have

$$a = c_1 \omega_1 + \cdots + c_n \omega_n,$$

where each $c_i \in \mathbb{Z}$ and $|c_i| \leq n! L M^{n-1}/|d(K)|^{1/2}$. The number of possible choices for each c_i is

$$2 \left[\frac{n! L M^{n-1}}{|d(K)|^{1/2}} \right] + 1,$$

so the number of $a \in O_K$ with $\beta_i(a) \leq L$ ($i = 1, 2, \ldots, r+s$) is at most

$$\left(2 \left[\frac{n! L M^{n-1}}{|d(K)|^{1/2}} \right] + 1 \right)^n.$$
∎

Lemma 13.2.6 *Let $a \in K$. Then*

$$N(\langle a \rangle) = \prod_{i=1}^{r+s} \beta_i(a)^{d_i},$$

where

$$d_i = \begin{cases} 1, & i = 1, \ldots, r \\ 2, & i = r+1, \ldots, r+s. \end{cases} \tag{13.2.7}$$

Proof: We have

$$N(\langle a \rangle) = |N(a)|$$

$$= \left| \prod_{i=1}^{n} \sigma_i(a) \right|$$

$$= \left| \prod_{i=1}^{r+s} \sigma_i(a) \prod_{i=r+s+1}^{r+2s} \sigma_i(a) \right|$$

$$= \left| \prod_{i=1}^{r+s} \sigma_i(a) \prod_{i=r+1}^{r+s} \sigma_{i+s}(a) \right|$$

$$= \left| \prod_{i=1}^{r+s} \sigma_i(a) \prod_{i=r+1}^{r+s} \bar{\sigma}_i(a) \right|$$

$$= \left| \prod_{i=1}^{r} \sigma_i(a) \prod_{i=r+1}^{r+s} \sigma_i(a)\overline{\sigma_i(a)} \right|$$

$$= \left| \prod_{i=1}^{r} \sigma_i(a) \prod_{i=r+1}^{r+s} |\sigma_i(a)|^2 \right|$$

$$= \prod_{i=1}^{r} |\sigma_i(a)| \prod_{i=r+1}^{r+s} |\sigma_i(a)|^2$$

$$= \prod_{i=1}^{r+s} \beta_i(a)^{d_i},$$

where d_i is given by (13.2.7). ∎

Lemma 13.2.7 *If ϵ is a unit of O_K then*

$$\prod_{i=1}^{r+s} \beta_i(\epsilon)^{d_i} = 1,$$

where d_i is defined in (13.2.7).

Proof: This result follows immediately from Lemma 13.2.6 as $N(\langle \epsilon \rangle) = N(\langle 1 \rangle) = 1$. ∎

Lemma 13.2.8 *Let p_i, q_i ($i = 1, 2, \ldots, r + s$) be rational numbers such that*

$$0 \le p_i < q_i, \quad i = 1, 2, \ldots, r + s.$$

Then there exists $a \in K$ such that

$$p_i < \beta_i(a) < q_i, \quad i = 1, 2, \ldots, r + s.$$

Proof: For $i = 1, 2, \ldots, r + s$ choose $h_i = \frac{1}{2}(p_i + q_i) \in \mathbb{Q}$ so that $p_i < h_i < q_i$. For $i = r + s + 1, \ldots, r + 2s$ set

$$h_i = h_{i-s}.$$

Thus h_i is defined for $i = 1, 2, \ldots, n$ and $h_{r+s+j} = h_{r+j}$ for $j = 1, 2, \ldots, s$. Consider the system of n linear equations in the n unknowns $b_1, \ldots, b_n$ given by

$$b_1 \sigma_i(\omega_1) + \cdots + b_n \sigma_i(\omega_n) = h_i, \quad i = 1, 2, \ldots, n. \qquad (13.2.8)$$

All the constant terms in this system are real. The determinant of the coefficient matrix of this system is $D \ne 0$ (see (13.2.2)), so that the system has a unique solution $(b_1, \ldots, b_n) \in \mathbb{C}^n$. The first r equations in the system (13.2.8) have real coefficients and the last $n - r = 2s$ equations occur in complex conjugate pairs. Hence $(b_1, \ldots, b_n) \in \mathbb{R}^n$.

Now let

$$\delta = \min_{1 \le i \le r+s} \left(\frac{q_i - p_i}{2Mn} \right),$$

so that

$$0 < \delta \le \frac{q_i - p_i}{2Mn}, \quad i = 1, 2, \ldots, r + s. \qquad (13.2.9)$$

Next choose $c_i \in \mathbb{Q}$ such that

$$|b_i - c_i| < \delta, \quad i = 1, 2, \ldots, n.$$

Set

$$a = c_1\omega_1 + \cdots + c_n\omega_n \in K.$$

Then

$$\sigma_i(a) = c_1\sigma_i(\omega_1) + \cdots + c_n\sigma_i(\omega_n), \ i = 1, 2, \ldots, n,$$

so that

$$\sigma_i(a) - h_i = (c_1 - b_1)\sigma_i(\omega_1) + \cdots + (c_n - b_n)\sigma_i(\omega_n), \ i = 1, 2, \ldots, n.$$

Hence for $i = 1, 2, \ldots, r + s$ we have

$$\begin{aligned} |\sigma_i(a) - h_i| &\leq |c_1 - b_1||\sigma_i(\omega_1)| + \cdots + |c_n - b_n||\sigma_i(\omega_n)| \\ &\leq M(|c_1 - b_1| + \cdots + |c_n - b_n|) \\ &< Mn\delta \\ &\leq \frac{q_i - p_i}{2}, \end{aligned}$$

so that

$$h_i - \frac{(q_i - p_i)}{2} < |\sigma_i(a)| < h_i + \frac{(q_i - p_i)}{2}, \ i = 1, 2, \ldots, r + s,$$

that is,

$$p_i < \beta_i(a) < q_i, \ i = 1, 2, \ldots, r + s. \qquad \blacksquare$$

Lemma 13.2.9 *Let k be a positive integer. Let A be a nonzero (integral or fractional) ideal of O_K with $N(A) \leq k^n$. Then A contains an element $a \neq 0$ with $\beta_i(a) \leq nMk$ ($i = 1, 2, \ldots, r + s$).*

Proof: First we consider the case when A is an integral ideal. Let

$$S = \{b \in O_K \mid b = b_1\omega_1 + \cdots + b_n\omega_n, \ b_1, \ldots, b_n \in \{0, 1, 2, \ldots, k\}\}.$$

Clearly card $S = (k + 1)^n > k^n \geq N(A)$, so that there exist $b' \in S$, $b'' \in S$, $b' \neq b''$ such that

$$b' \equiv b'' \pmod{A}.$$

Set $a = b' - b''$ so that $a \neq 0$, $a \in O_K$, and $a \equiv 0 \pmod{A}$. The latter condition is equivalent to $a \in A$. The coordinates $a_1, \ldots, a_n$ of a satisfy

$$|a_i| = |b_i' - b_i''| \leq k, \ i = 1, 2, \ldots, n,$$

where $b_1', \ldots, b_n'$ are the coordinates of b' and $b_1'', \ldots, b_n''$ the coordinates of b'', so that by Lemma 13.2.3

$$\beta_i(a) \leq nkM, \ i = 1, 2, \ldots, r + s.$$

Now we treat the case when A is a fractional ideal of O_K satisfying $N(A) \leq k^n$. Let γ be a common denominator for A. Let $\gamma_1 = \gamma, \gamma_2, \ldots, \gamma_n$ be the conjugates of γ so that $N(\gamma) = \gamma_1 \cdots \gamma_n \in \mathbb{Z} \setminus \{0\}$ and $N(\gamma)A = \gamma_2 \cdots \gamma_n(\gamma A)$ is an integral ideal of O_K. Set $m = |N(\gamma)|$. Then m is a positive integer such that $B = mA$ is an integral ideal of O_K with

$$N(B) = N(mA) = m^n N(A) \leq (mk)^n.$$

By the previous case there exists $b(\neq 0) \in B$ such that

$$\beta_i(b) \leq mkn M, \quad i = 1, 2, \ldots, r + s.$$

As $b \in B$ there exists $a \in A$ such that $b = ma$. Clearly $a \neq 0$ and

$$\beta_i(a) = \beta_i \left(\frac{b}{m} \right) = \frac{\beta_i(b)}{\beta_i(m)} \leq \frac{mkn M}{m} = kn M, \quad i = 1, 2, \ldots, r + s,$$

by Lemmas 13.2.1 and 13.2.2.   ∎

Lemma 13.2.10 *There exists a fixed bound $B > 0$ such that for each number $a \in K$ with $\frac{1}{2} < N(\langle a \rangle) \leq 1$ there exists a unit $\epsilon \in O_K$ such that*

$$\beta_j(\epsilon a) \leq B, \quad j = 1, 2, \ldots, r + s.$$

Proof: Let $a \in K$ satisfy $\frac{1}{2} < N(\langle a \rangle) \leq 1$. Set $I = \langle a \rangle$ so that $\frac{1}{2} < N(I) \leq 1$. Let S be the set of all such distinct principal ideals I. By Lemma 13.2.9, for each $I = \langle a \rangle$ in S there exists $b(\neq 0) \in I$ such that

$$\beta_i(b) \leq n M, \quad i = 1, 2, \ldots, r + s.$$

As $b \in \langle a \rangle$ we have $b = qa$ for some $q(\neq 0) \in O_K$. Then

$$\frac{N(\langle q \rangle)}{2} < N(\langle q \rangle)N(I) = N(\langle q \rangle)N(\langle a \rangle) = N(\langle q \rangle \langle a \rangle)$$

$$= N(\langle qa \rangle) = N(\langle b \rangle) = \prod_{i=1}^{r+s} \beta_i(b)^{d_i}$$

$$\leq \prod_{i=1}^{r+s} (nM)^{d_i} = (nM)^n$$

so that

$$N(\langle q \rangle) < 2(nM)^n.$$

Hence among the principal ideals $\langle q \rangle$ there are only finitely many different ones, say,

$$\langle q_1 \rangle, \ldots, \langle q_t \rangle.$$

Thus each $q = \epsilon q_j$ for some unit ϵ of O_K and some $j \in \{1, 2, \ldots, t\}$. Set

$$l = \max_{\substack{i = 1, \ldots, r+s \\ j = 1, \ldots, t}} \beta_i(q_j^{-1}).$$

Then, for $i = 1, \ldots, r+s$ and $j = 1, \ldots, t$, we have

$$1 = \beta_i(1) = \beta_i(q_j q_j^{-1}) = \beta_i(q_j)\beta_i(q_j^{-1}) \leq l\beta_i(q_j).$$

Thus

$$\beta_i(\epsilon a) \leq l\beta_i(q_j)\beta_i(\epsilon a) = l\beta_i(q_j\epsilon a) = l\beta_i(qa) = l\beta_i(b) \leq lnM,$$

that is $\beta_i(\epsilon a) \leq B$ ($i = 1, 2, \ldots, r+s$) with $B = lnM$. ∎

Lemma 13.2.11 *For each $j \in \{1, 2, \ldots, r+s-1\}$ there exists $a \in K$ with $\frac{1}{2} < N(\langle a \rangle) \leq 1$ such that*

$$\beta_i(a) > B, \ i = 1, 2, \ldots, r+s, \ i \neq j.$$

Proof: If $r+s = 1$ there is nothing to do so we may suppose that $r+s \geq 2$. By Lemma 13.2.8 there exists $a \in K$ such that

$$B < \beta_i(a) < 2^{1/n}B, \ i = 1, 2, \ldots, r+s, \ i \neq j,$$

$$\frac{1}{2B^{n-1}} < \beta_j(a) < \frac{1}{2^{1-\frac{1}{n}}B^{n-1}}, \ j \in \{1, 2, \ldots, r\},$$

$$\frac{1}{2^{\frac{1}{2}}B^{\frac{n-2}{2}}} < \beta_j(a) < \frac{1}{2^{\frac{1}{2}-\frac{1}{n}}B^{\frac{n-2}{2}}}, \ j \in \{r+1, \ldots, r+s-1\}.$$

Clearly,

$$\beta_i(a) > B, \ i = 1, 2, \ldots, r+s, \ i \neq j.$$

If $j \in \{1, 2, \ldots, r\}$ then

$$\beta_1(a)\cdots\beta_r(a)\beta_{r+1}(a)^2 \cdots \beta_{r+s}(a)^2 > B^{r-1}\frac{1}{2B^{n-1}}\left(B^2\right)^s = \frac{B^{r+2s-1}}{2B^{n-1}} = \frac{1}{2}$$

and

$$\beta_1(a)\cdots\beta_r(a)\beta_{r+1}(a)^2 \cdots \beta_{r+s}(a)^2 < \left(2^{1/n}B\right)^{r-1}\frac{1}{2^{1-\frac{1}{n}}B^{n-1}}\left(2^{1/n}B\right)^{2s}$$

$$= \frac{2^{\frac{r+2s-1}{n}}B^{r+2s-1}}{2^{1-\frac{1}{n}}B^{n-1}} = 1,$$

so that by Lemma 13.2.6

$$\frac{1}{2} < N(\langle a \rangle) < 1.$$

If $j \in \{r+1, \ldots, r+s-1\}$ then

$$\beta_1(a) \cdots \beta_r(a)\beta_{r+1}(a)^2 \cdots \beta_{r+s}(a)^2 > B^r \left(B^2\right)^{s-1} \left(\frac{1}{2^{1/2} B^{\frac{n-2}{2}}}\right)^2$$

$$= \frac{B^{r+2s-2}}{2 B^{n-2}} = \frac{1}{2}$$

and

$$\beta_1(a) \cdots \beta_r(a)\beta_{r+1}(a)^2 \cdots \beta_{r+s}(a)^2 < \left(2^{1/n} B\right)^r \left(2^{2/n} B^2\right)^{s-1} \left(\frac{1}{2^{\frac{1}{2}-\frac{1}{n}} B^{\frac{n-2}{2}}}\right)^2$$

$$= \frac{2^{\frac{r+2s-2}{n}} B^{r+2s-2}}{2^{1-\frac{2}{n}} B^{n-2}} = 1,$$

so that by Lemma 13.2.6

$$\frac{1}{2} < N(\langle a \rangle) < 1.$$

∎

Lemma 13.2.12 *For each* $j \in \{1, 2, \ldots, r+s-1\}$ *there exists a unit* $\epsilon_j \in O_K$ *such that*

$$\beta_i(\epsilon_j) < 1, \ i = 1, 2, \ldots, r+s, \ i \neq j,$$
$$\beta_j(\epsilon_j) > 1.$$

Proof: If $r+s = 1$ there is nothing to do so we may suppose that $r+s \geq 2$. By Lemma 13.2.11, for each $j \in \{1, 2, \ldots, r+s-1\}$, there exists $a \in K$ with $\frac{1}{2} N(\langle a \rangle) \leq 1$ such that

$$\beta_i(a) > B, \ i = 1, 2, \ldots, r+s, \ i \neq j.$$

By Lemma 13.2.10 there exists a unit $\epsilon_j \in O_K$ such that

$$\beta_i(\epsilon_j a) \leq B, \ i = 1, 2, \ldots, r+s.$$

Hence by Lemma 13.2.2

$$\beta_i(\epsilon_j) = \frac{\beta_i(\epsilon_j a)}{\beta_i(a)} < \frac{B}{B} = 1, \ i = 1, 2, \ldots, r+s, \ i \neq j.$$

Then, by Lemma 13.2.7, we obtain

$$1 = \prod_{i=1}^{r+s} \beta_i(\epsilon_j)^{d_i} < \beta_j(\epsilon_j)^{d_j}$$

so that $\beta_j(\epsilon_j) > 1$.

∎

Definition 13.2.1 (Independent units) *Let* K *be an algebraic number field. Let* $\epsilon_1, \ldots, \epsilon_k$ $(k \geq 1)$ *be units of* O_K. *The units* $\epsilon_1, \ldots, \epsilon_k$ *are said to be independent* *if and only if*

$$\epsilon_1^{r_1} \cdots \epsilon_k^{r_k} = 1 \ (r_1, \ldots, r_k \in \mathbb{Z}) \Longrightarrow r_1 = \cdots = r_k = 0.$$

Our next lemma shows that the units $\epsilon_1, \epsilon_2, \ldots, \epsilon_{r+s-1}$ constructed in Lemma 13.2.12 are independent.

Lemma 13.2.13 *The units* $\epsilon_1, \ldots, \epsilon_{r+s-1}$ *of Lemma 13.2.12 are independent.*

Proof: If $r + s = 1$ there is nothing to do so we may suppose that $r + s \geq 2$. We suppose that there exist integers $\rho_1, \ldots, \rho_{r+s-1}$, not all zero, such that

$$\prod_{j=1}^{r+s-1} \epsilon_j^{\rho_j} = 1. \tag{13.2.10}$$

As

$$\prod_{j=1}^{r+s-1} \epsilon_j^{-\rho_j} = 1$$

we can replace $(\rho_1, \ldots, \rho_{r+s-1})$ by $(-\rho_1, \ldots, -\rho_{r+s-1})$, if necessary, to ensure that at least one of $\rho_1, \ldots, \rho_{r+s-1}$ is positive. Relabeling $\epsilon_1, \ldots, \epsilon_{r+s-1}$, if necessary, we may suppose that $\rho_1, \ldots, \rho_k$ ($k \geq 1$) are positive and $\rho_{k+1}, \ldots, \rho_{r+s-1}$ are nonpositive. From the valuations $\beta_1, \ldots, \beta_k$ we form the product

$$\beta = \beta_1^{d_1} \cdots \beta_k^{d_k}$$

and from the remaining valuations we form

$$\beta' = \beta_{k+1}^{d_{k+1}} \cdots \beta_{r+s}^{d_{r+s}}.$$

By Lemmas 13.2.1 and 13.2.2 we have

$$\beta(1) = \beta'(1) = 1, \ \beta(a)\beta(b) = \beta(ab), \ \beta'(a)\beta'(b) = \beta'(ab), \tag{13.2.11}$$

for all a and b in K. By Lemma 13.2.7 we obtain

$$\beta(\epsilon)\beta'(\epsilon) = 1,$$

and thus

$$\beta'(\epsilon) = \beta^{-1}(\epsilon) \tag{13.2.12}$$

for every unit ϵ of O_K. For $j = 1, 2, \ldots, k$ we have

$$\beta'(\epsilon_j) = \beta_{k+1}^{d_{k+1}} \cdots \beta_{r+s}^{d_{r+s}}(\epsilon_j) = \beta_{k+1}(\epsilon_j)^{d_{k+1}} \cdots \beta_{r+s}(\epsilon_j)^{d_{r+s}},$$

so that

$$\beta'(\epsilon_j) < 1, \ j = 1, 2, \ldots, k,$$

by Lemma 13.2.12. For $j = k + 1, \ldots, r + s - 1$ we have

$$\beta(\epsilon_j) = \beta_1^{d_1} \cdots \beta_k^{d_k}(\epsilon_j) = \beta_1(\epsilon_j)^{d_1} \cdots \beta_k(\epsilon_j)^{d_k},$$

so that

$$\beta(\epsilon_j) < 1, \quad j = k + 1, \ldots, r + s - 1,$$

by Lemma 13.12.12. Next, by (13.2.10) and (13.2.11), we obtain

$$1 = \beta'(1) = \beta' \left(\prod_{j=1}^{r+s-1} \epsilon_j^{\rho_j} \right) = \prod_{j=1}^{r+s-1} \beta'(\epsilon_j)^{\rho_j},$$

so that by (13.2.12)

$$\prod_{j=1}^{k} \beta'(\epsilon_j)^{\rho_j} \prod_{j=k+1}^{r+s-1} \beta(\epsilon_j)^{-\rho_j} = 1.$$

However, all of the factors on the left-hand side are ≤ 1, and the first k of them are < 1. This is the required contradiction. ∎

For $j = 1, 2, \ldots, r + s - 1$ we set

$$a_j = \beta_j(\epsilon_j), \tag{13.2.13}$$

so that by Lemma 13.2.12 we have

$$a_j > 1, \quad j = 1, 2, \ldots, r + s - 1. \tag{13.2.14}$$

Lemma 13.2.14 *For each unit $\epsilon \in O_K$ with*

$$\beta_v(\epsilon) \leq 1, \quad v = 1, 2, \ldots, r + s - 1,$$

there exist integers $\rho_1, \ldots, \rho_{r+s-1}$ such that the unit

$$\eta = \epsilon \epsilon_1^{\rho_1} \cdots \epsilon_{r+s-1}^{\rho_{r+s-1}}$$

satisfies

$$1 < \beta_v(\eta) \leq a_v, \quad v = 1, 2, \ldots, r + s - 1, \quad \beta_{r+s}(\eta) \leq 1.$$

Proof: If $r + s = 1$ then by Lemma 13.2.7 we have

$$\beta_{r+s}(\eta)^{d_{r+s}} = 1$$

for any unit η of O_K, so that

$$\beta_{r+s}(\eta) = 1,$$

and this is all we require in this case. Thus we may suppose that $r + s \geq 2$. Let ϵ be a fixed unit of O_K satisfying

$$\beta_v(\epsilon) \leq 1, \quad v = 1, 2, \ldots, r + s - 1.$$

We consider all units η of O_K of the form

$$\eta = \epsilon \epsilon_1^{k_1} \cdots \epsilon_{r+s-1}^{k_{r+s-1}}$$

with

$$k_v \geq 0, \quad v = 1, 2, \ldots, r + s - 1,$$

and

$$\beta_v(\eta) \leq a_v, \quad v = 1, 2, \ldots, r + s - 1.$$

We note that ϵ is such a unit in view of (13.2.14). For these units we have by Lemma 13.2.12

$$\beta_{r+s}(\eta) = \beta_{r+s}(\epsilon \epsilon_1^{k_1} \cdots \epsilon_{r+s-1}^{k_{r+s-1}})$$

$$= \beta_{r+s}(\epsilon) \prod_{i=1}^{r+s-1} \beta_{r+s}(\epsilon_i)^{k_i}$$

$$\leq \beta_{r+s}(\epsilon),$$

so that all the valuations $\beta_v(\eta)$ $(v = 1, 2, \ldots, r + s)$ are bounded. Thus, by Lemma 13.2.5, there are only finitely many η of the type considered. Among these finitely many η, we choose η to be such that $\beta_{r+s}(\eta)$ is least. For this η we must have

$$1 < \beta_v(\eta), \quad v = 1, 2, \ldots, r + s - 1.$$

Otherwise, for some $v_0 \in \{1, 2, \ldots, r + s - 1\}$ we have

$$1 \geq \beta_{v_0}(\eta).$$

Then, for $v = 1, 2, \ldots, r + s - 1, \ v \neq v_0$, we have

$$\beta_v(\epsilon_{v_0} \eta) = \beta_v(\epsilon_{v_0}) \beta_v(\eta) < \beta_v(\eta) \leq a_v;$$

for $v = v_0$ we have

$$\beta_{v_0}(\epsilon_{v_0} \eta) = \beta_{v_0}(\epsilon_{v_0}) \beta_{v_0}(\eta) \leq \beta_{v_0}(\epsilon_{v_0}) = a_{v_0};$$

and for $v = r + s$ we have

$$\beta_{r+s}(\epsilon_{v_0} \eta) = \beta_{r+s}(\epsilon_{v_0}) \beta_{r+s}(\eta) < \beta_{r+s}(\eta),$$

contradicting the minimality of η. ∎

Lemma 13.2.15 *There exists a unit*

$$\epsilon_0 = \epsilon_1^{\sigma_1} \cdots \epsilon_{r+s-1}^{\sigma_{r+s-1}} \in O_K$$

with

$$1 < \beta_\nu(\epsilon_0), \quad \nu = 1, 2, \ldots, r+s-1.$$

Proof: This is the special case $\epsilon = 1$ of Lemma 13.2.14. ∎

Lemma 13.2.16 *For each unit* $\epsilon \in O_K$ *there exist integers* $\tau_1, \ldots, \tau_{r+s-1}$ *such that the unit*

$$\eta = \epsilon \epsilon_1^{\tau_1} \cdots \epsilon_{r+s-1}^{\tau_{r+s-1}}$$

satisfies

$$1 < \beta_\nu(\eta) \le a_\nu, \quad \nu = 1, 2, \ldots, r+s-1,$$

and

$$\beta_{r+s}(\eta) \le 1.$$

Proof: The case $r + s = 1$ follows as in the proof of Lemma 13.2.14. Thus we may suppose that $r + s \ge 2$. Let ϵ be a unit of O_K. Set

$$X = \max_{1 \le \nu \le r+s-1} \beta_\nu(\epsilon).$$

By Lemma 13.2.15 there exists a unit $\epsilon_0 = \epsilon_1^{\sigma_1} \cdots \epsilon_{r+s-1}^{\sigma_{r+s-1}}$ of O_K satisfying

$$1 < \beta_\nu(\epsilon_0), \quad \nu = 1, 2, \ldots, r+s-1.$$

Set

$$Y = \min_{1 \le \nu \le r+s-1} \beta_\nu(\epsilon_0),$$

so that

$$Y > 1.$$

We may choose $k \in \mathbb{N}$ so that

$$Y^k \ge X.$$

Then

$$\beta_\nu(\epsilon_0)^k \ge \beta_\nu(\epsilon), \quad \nu = 1, 2, \ldots, r+s-1.$$

Hence the unit $\lambda = \epsilon \epsilon_0^{-k}$ of O_K satisfies

$$\beta_\nu(\lambda) = \beta_\nu(\epsilon \epsilon_0^{-k}) = \frac{\beta_\nu(\epsilon)}{\beta_\nu(\epsilon_0)^k} \le 1, \quad \nu = 1, 2, \ldots, r+s-1.$$

Thus, by Lemma 13.2.14, there exist integers $\rho_1, \ldots, \rho_{r+s-1}$ such that the unit

$$\eta = \lambda \epsilon_1^{\rho_1} \cdots \epsilon_{r+s-1}^{\rho_{r+s-1}}$$

satisfies

$$1 < \beta_v(\eta) \le a_v, \quad v = 1, 2, \ldots, r+s-1.$$

We observe that

$$\eta = \epsilon \epsilon_0^{-k} \epsilon_1^{\rho_1} \cdots \epsilon_{r+s-1}^{\rho_{r+s-1}} = \epsilon \epsilon_1^{\tau_1} \cdots \epsilon_{r+s-1}^{\tau_{r+s-1}}$$

with

$$\tau_j = \rho_j - k\sigma_j, \quad j = 1, 2, \ldots, r+s-1.$$

Finally, as $\beta_v(\eta) > 1$ for $v = 1, 2, \ldots, r+s-1$, we deduce from Lemma 13.2.7 that

$$\beta_{r+s}(\eta) \le 1. \qquad \blacksquare$$

Lemma 13.2.17 *There exist finitely many units $\eta_1, \ldots, \eta_h$ of O_K such that every unit ϵ of O_K is of the form*

$$\epsilon = \eta_j \epsilon_1^{\rho_1} \cdots \epsilon_{r+s-1}^{\rho_{r+s-1}}$$

for some $j \in \{1, 2, \ldots, h\}$ and some $\rho_1, \ldots, \rho_{r+s-1} \in \mathbb{Z}$.

Proof: By Lemma 13.2.16 each unit ϵ of O_K can be expressed in the form

$$\epsilon = \eta \epsilon_1^{-\tau_1} \cdots \epsilon_{r+s-1}^{-\tau_{r+s-1}}$$

for some integers $\tau_1, \ldots, \tau_{r+s-1}$ and some unit $\eta \in O_K$ satisfying

$$\beta_v(\eta) \le a_v, \quad v = 1, 2, \ldots, r+s-1,$$
$$\beta_{r+s}(\eta) \le 1.$$

Hence, by Lemma 13.2.5, there are only finitely many such η, say $\eta_1, \ldots, \eta_h$. Thus

$$\epsilon = \eta_j \epsilon_1^{\rho_1} \cdots \epsilon_{r+s-1}^{\rho_{r+s-1}}$$

for some $j \in \{1, 2, \ldots, h\}$ and some integers $\rho_1, \ldots, \rho_{r+s-1}$. $\qquad \blacksquare$

We are now in a position to complete the proof of Dirichlet's unit theorem in the next section.

13.3 Proof of Dirichlet's Unit Theorem

By Lemma 13.2.17 the unit group $U(O_K)$ is generated by the units

$$\epsilon_1, \ldots, \epsilon_{r+s-1}, \eta_1, \ldots, \eta_h,$$

that is,

$$U(O_K) = \langle \epsilon_1, \ldots, \epsilon_{r+s-1}, \eta_1, \ldots, \eta_h \rangle.$$

Let H be the subgroup of $U(O_K)$ given by

$$H = \langle \epsilon_1, \ldots, \epsilon_{r+s-1} \rangle.$$

By Lemma 13.2.17 there are h distinct cosets of H in $U(O_K)$, so that the factor group $U(O_K)/H$ has order h. Hence for $\epsilon \in U(O_K)$ we have

$$(\epsilon H)^h = H,$$

so that

$$\epsilon^h \in H.$$

Thus for each $\epsilon \in U(O_K)$ there exist $a_1, \ldots, a_{r+s-1} \in \mathbb{Z}$ such that

$$\epsilon^h = \epsilon_1^{a_1} \cdots \epsilon_{r+s-1}^{a_{r+s-1}}.$$

Let $\lambda_1, \ldots, \lambda_m$ be m units of O_K, where $m \geq r + s$. By the previous observation there exist integers $a_{11}, \ldots, a_{1\,r+s-1}, \ldots, a_{m1}, \ldots, a_{m\,r+s-1}$ such that

$$\lambda_1^h = \epsilon_1^{a_{11}} \cdots \epsilon_{r+s-1}^{a_{1\,r+s-1}},$$

$$\cdots$$

$$\lambda_m^h = \epsilon_1^{a_{m1}} \cdots \epsilon_{r+s-1}^{a_{m\,r+s-1}}.$$

Consider the homogeneous system of $r + s - 1$ integral linear equations in the m unknowns $x_1, \ldots, x_m$:

$$a_{11}x_1 + \cdots + a_{m1}x_m = 0,$$

$$\cdots$$

$$a_{1\,r+s-1}x_1 + \cdots + a_{m\,r+s-1}x_m = 0.$$

As $m > r + s - 1$ this system has a solution $(x_1, \ldots, x_m) \in \mathbb{Q}^m$ with $(x_1, \ldots, x_m) \neq (0, \ldots, 0)$. Multiplying each x_i by the least common multiple of the denominators of $x_1, \ldots, x_m$, we may suppose that $(x_1, \ldots, x_m) \in \mathbb{Z}^m$. Then

$$\begin{aligned}
\lambda_1^{hx_1} \cdots \lambda_m^{hx_m} &= \left(\epsilon_1^{a_{11}} \cdots \epsilon_{r+s-1}^{a_{1\,r+s-1}} \right)^{hx_1} \cdots \left(\epsilon_1^{a_{m1}} \cdots \epsilon_{r+s-1}^{a_{m\,r+s-1}} \right)^{hx_m} \\
&= \epsilon_1^{h(a_{11}x_1 + \cdots + a_{m1}x_m)} \cdots \epsilon_{r+s-1}^{h(a_{1\,r+s-1}x_1 + \cdots + a_{m\,r+s-1}x_m)} \\
&= \epsilon_1^0 \cdots \epsilon_{r+s-1}^0 \\
&= 1.
\end{aligned}$$

This proves that any m units of O_K with $m \geq r + s$ are not independent. Therefore there are no more than $r + s - 1$ independent units in O_K. But by Lemma 13.2.13 the $r + s - 1$ units $\epsilon_1, \ldots, \epsilon_{r+s-1}$ are independent. Hence, by the main theorem of finitely generated Abelian groups, $U(O_K)$ is the direct product of cyclic groups, $r + s - 1$ of which have infinite order and the remaining ones finite order. The elements of a cyclic group of finite order however are roots of unity. This proves that every unit of O_K can be expressed in the form $\eta \epsilon_1^{x_1} \cdots \epsilon_{r+s-1}^{x_{r+s-1}}$, where η is a root of unity and $x_1, \ldots, x_{r+s-1} \in \mathbb{Z}$. Finally, we show that this representation is unique. Suppose

$$\eta \epsilon_1^{x_1} \cdots \epsilon_{r+s-1}^{x_{r+s-1}} = \theta \epsilon_1^{y_1} \cdots \epsilon_{r+s-1}^{y_{r+s-1}},$$

where η, θ are roots of unity and $x_1, \ldots, x_{r+s-1}, y_1, \ldots, y_{r+s-1} \in \mathbb{Z}$. Then

$$\eta \theta^{-1} = \epsilon_1^{y_1 - x_1} \cdots \epsilon_{r+s-1}^{y_{r+s-1} - x_{r+s-1}}.$$

As η, θ are roots of unity so is $\eta \theta^{-1}$ and thus there is a positive integer k such that $\left(\eta \theta^{-1}\right)^k = 1$. Hence

$$\epsilon_1^{k(y_1 - x_1)} \cdots \epsilon_{r+s-1}^{k(y_{r+s-1} - x_{r+s-1})} = 1.$$

As $\epsilon_1, \ldots, \epsilon_{r+s-1}$ are independent we deduce that

$$k(y_1 - x_1) = \cdots = k(y_{r+s-1} - x_{r+s-1}) = 0$$

so that

$$x_1 = y_1, \ldots, x_{r+s-1} = y_{r+s-1},$$

and thus $\eta = \theta$. ∎

13.4 Fundamental System of Units

Definition 13.4.1 (Fundamental system of units) *Let K be an algebraic number field of degree n with r real embeddings and $2s$ complex embeddings (so that $r + 2s = n$). If $\epsilon_1, \ldots, \epsilon_{r+s-1}$ are $r + s - 1$ units of O_K such that*

$$\epsilon_1, \ldots, \epsilon_{r+s-1} \quad \text{are independent} \tag{13.4.1}$$

and

$$\text{every unit } \epsilon \text{ of } O_K \text{ can be expressed in the form}$$
$$\epsilon = \eta \epsilon_1^{a_1} \cdots \epsilon_{r+s-1}^{a_{r+s-1}}, \text{ where } \eta \text{ is a root of unity in } K, \tag{13.4.2}$$

then $\{\epsilon_1, \ldots, \epsilon_{r+s-1}\}$ is called a fundamental system of units of O_K.

Theorem 13.1.1 guarantees that O_K always possesses a fundamental system of units for any algebraic number field K. By the final argument in the proof of Dirichlet's unit theorem, we see that the representation (13.4.2) of a unit in terms of a fundamental system of units is unique.

If $r + s = 1$ the fundamental system of units is empty and every unit of O_K is a root of unity. Our next theorem tells us for which fields K this occurs.

Theorem 13.4.1 *Let K be an algebraic number field. Then every unit in O_K is a root of unity if and only if $K = \mathbb{Q}$ or K is an imaginary quadratic field.*

Proof: Let K be of degree n. Let r be the number of real embeddings of K and $2s$ the number of complex embeddings so that $r + 2s = n$. By Theorem 13.1.1 every unit in O_K is a root of unity

$$\Longleftrightarrow r + s = 1$$
$$\Longleftrightarrow r = 1, \ s = 0 \ \text{ or } \ r = 0, \ s = 1$$
$$\Longleftrightarrow K = \mathbb{Q} \text{ or } K = \text{ imaginary quadratic field.} \qquad \blacksquare$$

This theorem is consistent with what we already know, namely,

$$U(O_\mathbb{Q}) = U(\mathbb{Z}) = \{\pm 1\}$$

and for m squarefree and negative (Theorem 5.4.3)

$$U(O_{\mathbb{Q}(\sqrt{m})}) = \begin{cases} \{\pm 1, \ \pm i\}, & \text{if } m = -1, \\ \{\pm 1, \ \pm \omega, \ \pm \omega^2\}, & \text{if } m = -3, \\ \{\pm 1\}, & \text{otherwise,} \end{cases}$$

where ω is a complex cube root of unity.

If $r + s = 2$, a fundamental system of units consists of exactly one unit.

Definition 13.4.2 (Fundamental unit) *Let K be an algebraic number field with $r + s = 2$. Then any unit $\epsilon \in O_K$ such that $\{\epsilon\}$ is a fundamental system of units for O_K is called a fundamental unit.*

If ϵ is a fundamental unit of O_K then every unit of O_K is expressible uniquely in the form $\eta \epsilon^k$, where η is a root of unity in O_K and $k \in \mathbb{Z}$. Moreover, if ϵ and ϵ_1 are two fundamental units for O_K, then either $\epsilon_1 = \lambda \epsilon$ or $\epsilon_1 = \lambda \epsilon^{-1}$ for some root of unity λ in O_K.

Our next theorem tells us exactly which fields possess a fundamental unit.

Theorem 13.4.2 *Let K be an algebraic number field. Then K possesses a fundamental unit if and only if K is a real quadratic field, a cubic field with exactly one real embedding, or a totally imaginary quartic field.*

Proof: Let K be of degree n. Let r be the number of real embeddings of K and $2s$ the number of complex embeddings, so that $r + 2s = n$. By Theorem 13.1.1

$U(O_K)$ possesses a fundamental unit

$$\Longleftrightarrow r + s = 2$$
$$\Longleftrightarrow r = 2,\ s = 0,\ n = 2$$
$$\text{or}$$
$$r = 1,\ s = 1,\ n = 3$$
$$\text{or}$$
$$r = 0,\ s = 2,\ n = 4$$
$$\Longleftrightarrow K = \text{ real quadratic field}$$
$$\text{or}$$
$$K = \text{ cubic field with one real embedding}$$
$$\text{or}$$
$$K = \text{ totally imaginary quartic field.} \qquad\blacksquare$$

13.5 Roots of Unity

The theorems of this section give us some information about which roots of unity can belong to the ring of integers O_K of an algebraic number field K.

We recall that if ζ_k is a primitive kth root of unity then

$$[\mathbb{Q}(\zeta_k) : \mathbb{Q}] = \phi(k), \tag{13.5.1}$$

where Euler's phi function ϕ is defined by

$$\phi(k) = \text{ number of integers } m \text{ satisfying} \tag{13.5.2}$$
$$1 \leq m \leq k \text{ with } (m, k) = 1.$$

It is shown in texts on elementary number theory (see, for example, [3, Theorem 2.19, p. 69]) that ϕ is multiplicative; that is,

$$\phi(kl) = \phi(k)\phi(l) \tag{13.5.3}$$

whenever k and l are coprime positive integers. If p is a prime there are $p^a - 1$ positive integers less than p^a ($a \geq 1$) of which $p^{a-1} - 1$ are multiples of p and the remainder coprime with p. Hence

$$\phi(p^a) = (p^a - 1) - (p^{a-1} - 1) = p^a - p^{a-1} = p^{a-1}(p - 1). \tag{13.5.4}$$

Thus if $k = p_1^{a_1} \cdots p_r^{a_r}$ is the factorization of k into powers of distinct primes $p_1, \ldots, p_r$ then by (13.5.3) and (13.5.4) we deduce that

$$\phi(k) = p_1^{a_1-1} \cdots p_r^{a_r-1}(p_1 - 1) \cdots (p_r - 1). \tag{13.5.5}$$

Using the prime power decompositions of the positive integers up to 40 in conjunction with (13.5.5), we obtain the following table of values of $\phi(k)$, $k = 1, 2, \ldots, 40$.

k	$\phi(k)$	k	$\phi(k)$	k	$\phi(k)$	k	$\phi(k)$
1	1	11	10	21	12	31	30
2	1	12	4	22	10	32	16
3	2	13	12	23	22	33	20
4	2	14	6	24	8	34	16
5	4	15	8	25	20	35	24
6	2	16	8	26	12	36	12
7	6	17	16	27	18	37	36
8	4	18	6	28	12	38	18
9	6	19	18	29	28	39	24
10	4	20	8	30	8	40	16

In the next six lemmas we prove elementary results about $\phi(k)$ that will be used in the proofs of our theorems giving information about the roots of unity in the ring of integers of an algebraic number field (Theorems 13.5.1–13.5.4).

Lemma 13.5.1 *For all positive integers* n

$$\phi(n) \geq \sqrt{\frac{n}{2}}.$$

Proof: If p is an odd prime and k is a positive integer then

$$\phi(p^k) = \begin{cases} p - 1 \geq p^{1/2}, & \text{if } k = 1, \\ p^{k-1}(p - 1) > p^{k-1} \geq p^{k/2}, & \text{if } k \geq 2, \end{cases}$$

so that

$$\phi(p^k) \geq \sqrt{p^k}.$$

Hence, if N is an odd positive integer, as ϕ is multiplicative we have

$$\phi(N) \geq \sqrt{N}.$$

Let n be a positive integer. Set $n = 2^\alpha N$, where α is a nonnegative integer and N is an odd positive integer. If $\alpha = 0$ or 1 then

$$\phi(n) = \phi(N) \geq \sqrt{N} \geq \sqrt{\frac{n}{2}}$$

and if $\alpha \geq 2$

$$\phi(n) = 2^{\alpha-1}\phi(N) \geq 2^{\alpha/2}\phi(N) \geq 2^{\alpha/2}\sqrt{N} = \sqrt{n}$$

so that

$$\phi(n) \geq \sqrt{\frac{n}{2}}$$

for all positive integers n. ∎

Lemma 13.5.2 *Let n be a positive integer. If $\phi(k) \leq n$ then $k \leq 2n^2$.*

Proof: By Lemma 13.5.1 we have

$$\sqrt{\frac{k}{2}} \leq \phi(k) \leq n$$

so that

$$k \leq 2n^2.$$ ∎

Lemma 13.5.3 $\phi(k) = 1$ *if and only if $k = 1, 2$.*

Proof: By Lemma 13.5.2

$$\phi(k) = 1 \Longrightarrow k \leq 2$$

and since $\phi(1) = \phi(2) = 1$ the result follows. ∎

Lemma 13.5.4 $\phi(k) = 2$ *if and only if $k = 3, 4, 6$.*

Proof: By Lemma 13.5.2

$$\phi(k) = 2 \Longrightarrow k \leq 8$$

and since $\phi(1) = 1$, $\phi(2) = 1$, $\phi(3) = 2$, $\phi(4) = 2$, $\phi(5) = 4$, $\phi(6) = 2$, $\phi(7) = 6$, and $\phi(8) = 4$ the result follows. ∎

Lemma 13.5.5 $\phi(k) = 4$ *if and only if $k = 5, 8, 10, 12$.*

Proof: By Lemma 13.5.2

$$\phi(k) = 4 \Longrightarrow k \leq 32$$

and the result follows by appealing to the table of values of the Euler phi function preceding Lemma 13.5.1. ∎

Lemma 13.5.6 *If $n \geq 3$ then $\phi(n)$ is even.*

Proof: If $n \geq 3$ then either there exists an odd prime p dividing n or $n = 2^\alpha$ with $\alpha \geq 2$. In the former case $n = p^\beta n_1$, where $\beta \geq 1$ and n_1 is not divisible by p, so that

$$\phi(n) = p^{\beta-1}(p-1)\phi(n_1) \equiv 0 \,(\mathrm{mod}\ 2),$$

as $2 \mid p - 1$. In the latter case

$$\phi(n) = 2^{\alpha-1} \equiv 0 \,(\mathrm{mod}\ 2),$$

as $\alpha \geq 2$. ∎

We now use Lemma 13.5.2 to show that the ring of integers of an algebraic number field can only contain finitely many roots of unity.

Theorem 13.5.1 *Let K be an algebraic number field. Then O_K contains only finitely many roots of unity.*

Proof: Let $[K : \mathbb{Q}] = n$. Let ζ_k be a primitive kth root of unity in O_K. Then $\zeta_k \in K$ so that

$$\mathbb{Q}(\zeta_k) \subseteq K,$$

and thus

$$[\mathbb{Q}(\zeta_k) : \mathbb{Q}] \leq [K : \mathbb{Q}],$$

that is,

$$\phi(k) \leq n.$$

Hence, by Lemma 13.5.2, we have

$$k \in \{1, 2, \ldots, 2n^2\},$$

proving that there are only finitely many roots of unity in O_K. ∎

If K has odd degree then we can say exactly which roots of unity are in O_K.

Theorem 13.5.2 *Let K be an algebraic number field of odd degree n. Then the only roots of unity in O_K are ± 1.*

Proof: Let ζ_k be a primitive kth root of unity in O_K. Then $\mathbb{Q}(\zeta_k) \subseteq K$ and so $[\mathbb{Q}(\zeta_k) : \mathbb{Q}] \mid [K : \mathbb{Q}]$, that is, $\phi(k) \mid n$. But n is odd so that $\phi(k)$ is odd. By Lemma 13.5.6, we must have $k \leq 2$, that is, $k = 1, 2$. Clearly $\zeta_1 = 1$ and $\zeta_2 = -1$ belong in O_K. ∎

Taking $n = 3$ in Theorem 13.5.2 we obtain immediately the following result.

Theorem 13.5.3 *The only roots of unity in the ring of integers of a cubic field are* ± 1.

The situation is much more complicated when n is even. We just determine the roots of a unity in the ring of integers of a quartic field.

Theorem 13.5.4 *Let K be a quartic field. Then the only possible roots of unity $\neq \pm 1$ in O_K are*

$$\zeta_3, \ \zeta_4, \ \zeta_5, \ \zeta_6, \ \zeta_8, \ \zeta_{10}, \ \zeta_{12},$$

and their powers. Moreover,

$$\zeta_3 \in O_K \Longleftrightarrow K \supseteq \mathbb{Q}(\sqrt{-3}),$$
$$\zeta_4 \in O_K \Longleftrightarrow K \supseteq \mathbb{Q}(\sqrt{-1}),$$
$$\zeta_5 \in O_K \Longleftrightarrow K = \mathbb{Q}(\sqrt{-10 - 2\sqrt{5}}),$$
$$\zeta_6 \in O_K \Longleftrightarrow K \supseteq \mathbb{Q}(\sqrt{-3}),$$
$$\zeta_8 \in O_K \Longleftrightarrow K = \mathbb{Q}(\sqrt{2}, \sqrt{-1}),$$
$$\zeta_{10} \in O_K \Longleftrightarrow K = \mathbb{Q}(\sqrt{-10 - 2\sqrt{5}}),$$
$$\zeta_{12} \in O_K \Longleftrightarrow K = \mathbb{Q}(\sqrt{3}, \sqrt{-1}).$$

Proof: Let ζ_k be a primitive kth root of unity in O_K. Then $\mathbb{Q}(\zeta_k) \subseteq K$ and thus $[\mathbb{Q}(\zeta_k) : \mathbb{Q}] \mid [K : \mathbb{Q}]$, that is, $\phi(k) \mid 4$. Hence $\phi(k) = 1, 2$, or 4. Thus, by Lemmas 13.5.3–13.5.5, we have $k = 1, 2, 3, 4, 5, 6, 8, 10$, or 12. Hence the only possible roots of unity in O_K are $\zeta_1 = 1$, $\zeta_2 = -1$, $\zeta_3, \zeta_4, \zeta_5, \zeta_6, \zeta_8, \zeta_{10}, \zeta_{12}$, and their powers.

As $e^{2\pi i/3} = (-1 + i\sqrt{3})/2$, we have $\mathbb{Q}(\zeta_3) = \mathbb{Q}(e^{2\pi i/3}) = \mathbb{Q}(\sqrt{-3})$, so that

$$\zeta_3 \in O_K \Longleftrightarrow K \supseteq \mathbb{Q}(\zeta_3) \Longleftrightarrow K \supseteq \mathbb{Q}(\sqrt{-3}).$$

Similarly, we can show that $\zeta_4 \in O_K \Longleftrightarrow K \supseteq \mathbb{Q}(\sqrt{-1})$ and $\zeta_6 \in O_K \Longleftrightarrow K \supseteq \mathbb{Q}(\sqrt{-3})$.

Next, as

$$e^{2\pi i/5} = \frac{1}{4}(\sqrt{5} - 1 + i\sqrt{10 + 2\sqrt{5}}),$$

we have

$$\mathbb{Q}(\zeta_5) = \mathbb{Q}(e^{2\pi i/5}) = \mathbb{Q}(\sqrt{-10 - 2\sqrt{5}}),$$

so that

$$\zeta_5 \in O_K \Longleftrightarrow K \supseteq \mathbb{Q}(\zeta_5)$$
$$\Longleftrightarrow K = \mathbb{Q}(\zeta_5) \text{ (as } [K : \mathbb{Q}] = [\mathbb{Q}(\zeta_5) : \mathbb{Q}] = 4)$$
$$\Longleftrightarrow K = \mathbb{Q}(\sqrt{-10 - 2\sqrt{5}}).$$

Similarly, we can show that

$$\zeta_{10} \in O_K \Longleftrightarrow K = \mathbb{Q}(\sqrt{-10 - 2\sqrt{5}}).$$

Finally,

$$e^{2\pi i/8} = \frac{1 + i}{\sqrt{2}},$$

so that

$$\mathbb{Q}(\zeta_8) = \mathbb{Q}(e^{2\pi i/8}) = \mathbb{Q}\left(\frac{1 + i}{\sqrt{2}}\right) = \mathbb{Q}(\sqrt{2}, \sqrt{-1}).$$

Thus

$$\zeta_8 \in O_K \Longleftrightarrow K \supseteq \mathbb{Q}(\zeta_8)$$
$$\Longleftrightarrow K = \mathbb{Q}(\zeta_8) \text{ (as } [K : \mathbb{Q}] = [\mathbb{Q}(\zeta_8) : \mathbb{Q}] = 4)$$
$$\Longleftrightarrow K = \mathbb{Q}(\sqrt{2}, \sqrt{-1}).$$

The corresponding result for ζ_{12} can be shown similarly. ∎

Example 13.5.1 *There are infinitely many quartic fields K such that $\zeta_4 \in O_K$.*
  Let

$$\mathbb{P} = \{2, 3, 5, 7, 11, 13, 17, \ldots\}$$

*be the set of prime numbers. It is a theorem going back to Euclid that $\mathbb{P}$ is an infinite
set. For $p \in \mathbb{P}$ let*

$$K_p = \mathbb{Q}(\sqrt{-1}, \sqrt{p}).$$

An easy calculation shows that

$$[K_p : \mathbb{Q}] = 4 \text{ for all } p \in \mathbb{P}.$$

Moreover, the only quadratic subfields of K_p ($p \in \mathbb{P}$) are

$$\mathbb{Q}(\sqrt{-1}), \ \mathbb{Q}(\sqrt{p}), \ \mathbb{Q}(\sqrt{-p}).$$

*Let $p, q \in \mathbb{P}$ with $p \neq q$. Then $\mathbb{Q}(\sqrt{q})$ is a quadratic subfield of K_q but not of
K_p as $\mathbb{Q}(\sqrt{q}) \neq \mathbb{Q}(\sqrt{-1}), \mathbb{Q}(\sqrt{p}), \mathbb{Q}(\sqrt{-p})$. Hence $K_p \neq K_q$. This shows that
$\{K_p \mid p \in \mathbb{P}\}$ is an infinite set of distinct quartic fields. The ring of integers of each
K_p ($p \in \mathbb{P}$) contains ζ_4 by Theorem 13.5.4.*

Example 13.5.2 *There are infinitely many quartic fields K such that the only roots of unity in their rings of integers O_K are ± 1. Let*

$$\mathbb{P}_{3,4} = \{3, 7, 11, 19, 23, 31, \ldots\}$$

be the set of prime numbers $\equiv 3 \pmod 4$. For $p \in \mathbb{P}_{3,4}$ let

$$\theta_p = \sqrt{1 + \sqrt{p}}, \quad K_p = \mathbb{Q}(\theta_p).$$

Clearly θ_p is a root of the polynomial

$$f_p(x) = x^4 - 2x^2 + (1 - p) \in \mathbb{Z}[x].$$

As $2 \parallel 1 - p$ for $p \in \mathbb{P}_{3,4}$, $f_p(x)$ is 2-Eisenstein and thus irreducible in $\mathbb{Z}[x]$. Hence $f_p(x)$ is the minimal polynomial of θ_p over $\mathbb{Q}$ and so

$$[K_p : \mathbb{Q}] = 4.$$

If $q \in \mathbb{P}_{3,4}$ is such that $q \neq p$ then an easy calculation shows that $\sqrt{q} \notin K_p$. But $\sqrt{q} \in K_q$ so that $K_p \neq K_q$. Hence, as there are infinitely many primes $p \equiv 3 \pmod 4$, $\{K_p \mid p \in \mathbb{P}_{3,4}\}$ is an infinite set of distinct quartic fields. As $K_p \subseteq \mathbb{R}$, none of the roots of unity ζ_k, $k \in \{3, 4, 5, 6, 8, 10, 12\}$, belongs to O_{K_p}. Thus, by Theorem 13.5.4, the only roots of unity in O_{K_p} are ± 1.

13.6 Fundamental Units in Cubic Fields

Let K be a cubic field with exactly one real embedding. By Theorem 13.4.2 we know that K possesses a fundamental unit η. Suppose further that K is a real field. Then $\eta \in \mathbb{R}$. By Theorem 13.5.3 the only roots of unity in K are ± 1. Hence the only fundamental units are $\pm\eta$ and $\pm\eta^{-1}$. Exactly one of these four units is greater than 1. Thus K has a unique fundamental unit $\eta > 1$. We determine η for $K = \mathbb{Q}(\sqrt[3]{2})$ and $K = \mathbb{Q}(\sqrt[3]{3})$. The main tool is Theorem 13.6.3, which gives a lower bound for the fundamental unit in terms of the discriminant of the field K.

We first prove two elementary inequalities needed in the proof of Theorem 13.6.3.

Lemma 13.6.1 *For all $x \in \mathbb{R}$ and all $\theta \in \mathbb{R}$*

$$\sin^2 \theta (x - 2\cos\theta)^2 < x^2 + 4.$$

Proof: For all $\theta \in \mathbb{R}$ we have

$$1 - \sin^2\theta \cos^2\theta - \sin^4\theta = 1 - \sin^2\theta(\cos^2\theta + \sin^2\theta)$$
$$= 1 - \sin^2\theta = \cos^2\theta \geq 0$$

with equality if and only if $\theta = (2k+1)\pi/2$, $k \in \mathbb{Z}$. Thus, for all $x \in \mathbb{R}$ and all $\theta \in \mathbb{R}$, we have

$$(x\cos\theta + 2\sin^2\theta)^2 + 4(1 - \sin^2\theta\cos^2\theta - \sin^4\theta) > 0 \qquad (13.6.1)$$

as

$$x\cos\left(\frac{(2k+1)\pi}{2}\right) + 2\sin^2\left(\frac{(2k+1)\pi}{2}\right) = 2.$$

Expanding the square in (13.6.1), we obtain

$$x^2\cos^2\theta + 4x\sin^2\theta\cos\theta + 4 - 4\sin^2\theta\cos^2\theta > 0,$$

so that

$$-4x\sin^2\theta\cos\theta + 4\sin^2\theta\cos^2\theta < x^2\cos^2\theta + 4.$$

Thus

$$\sin^2\theta(x - 2\cos\theta)^2 = x^2\sin^2\theta - 4x\sin^2\theta\cos\theta + 4\sin^2\theta\cos^2\theta$$
$$< x^2\sin^2\theta + x^2\cos^2\theta + 4 = x^2 + 4. \qquad \blacksquare$$

Lemma 13.6.2 *For $x \geq 33$*

$$\sqrt{\left(\frac{x}{8} - 3\right)^2 - 1} > \frac{x}{8} - \frac{15}{4}.$$

Proof: We have

$$\left(\frac{x}{8} - 3\right)^2 - \left(\frac{x}{8} - \frac{15}{4}\right)^2 = \frac{3}{4}\left(\frac{x}{4} - \frac{27}{4}\right) \geq \frac{3}{4}\left(\frac{33}{4} - \frac{27}{4}\right) = \frac{9}{8} > 1,$$

so that

$$\left(\frac{x}{8} - 3\right)^2 - 1 > \left(\frac{x}{8} - \frac{15}{4}\right)^2$$

and thus

$$\sqrt{\left(\frac{x}{8} - 3\right)^2 - 1} > \frac{x}{8} - \frac{15}{4}. \qquad \blacksquare$$

Theorem 13.6.1 *Let K be a real cubic field with two complex embeddings. Let $\eta > 1$ be the fundamental unit of O_K. If $|d(K)| \geq 33$ then*

$$\eta^3 > \frac{|d(K)| - 27}{4}.$$

Proof: Let η, $\rho e^{i\theta}$, $\rho e^{-i\theta}$ be the conjugates of η, where $\rho \in \mathbb{R}^+$. Then, as η is a unit, we have $N(\eta) = \pm 1$ so that (recalling Definition 9.2.1)

$$\eta \rho e^{i\theta} \rho e^{-i\theta} = \pm 1,$$

that is,

$$\eta \rho^2 = \pm 1.$$

As $\eta > 0$ and $\rho^2 > 0$ we must have $\eta \rho^2 = 1$ so that

$$\eta = \rho^{-2}.$$

Next we determine $D(\eta)$. We have

$$D(\eta) = D(1, \eta, \eta^2) = \begin{vmatrix} 1 & \eta & \eta^2 \\ 1 & \rho e^{i\theta} & \rho^2 e^{2i\theta} \\ 1 & \rho e^{-i\theta} & \rho^2 e^{-2i\theta} \end{vmatrix}^2.$$

Next

$$\begin{vmatrix} 1 & \eta & \eta^2 \\ 1 & \rho e^{i\theta} & \rho^2 e^{2i\theta} \\ 1 & \rho e^{-i\theta} & \rho^2 e^{-2i\theta} \end{vmatrix} = \begin{vmatrix} 1 & \rho^{-2} & \rho^{-4} \\ 0 & \rho e^{i\theta} - \rho^{-2} & \rho^2 e^{2i\theta} - \rho^{-4} \\ 0 & \rho e^{-i\theta} - \rho^{-2} & \rho^2 e^{-2i\theta} - \rho^{-4} \end{vmatrix}$$

$$= (\rho e^{i\theta} - \rho^{-2})(\rho^2 e^{-2i\theta} - \rho^{-4}) - (\rho e^{-i\theta} - \rho^{-2})(\rho^2 e^{2i\theta} - \rho^{-4})$$

$$= \rho^3 e^{-i\theta} - e^{-2i\theta} - \rho^{-3} e^{i\theta} + \rho^{-6} - \rho^3 e^{i\theta} + e^{2i\theta} + \rho^{-3} e^{-i\theta} - \rho^{-6}$$

$$= -\rho^3 (e^{i\theta} - e^{-i\theta}) + (e^{2i\theta} - e^{-2i\theta}) - \rho^{-3}(e^{i\theta} - e^{-i\theta})$$

$$= -\rho^3 2i \sin\theta + 2i \sin 2\theta - \rho^{-3} 2i \sin\theta$$

$$= -2i \sin\theta (\rho^3 + \rho^{-3} - 2\cos\theta),$$

so that

$$D(\eta) = -4 \sin^2\theta (\rho^3 + \rho^{-3} - 2\cos\theta)^2.$$

Hence, by Lemma 13.6.1, we obtain

$$|D(\eta)| = 4 \sin^2\theta (\rho^3 + \rho^{-3} - 2\cos\theta)^2$$
$$< 4((\rho^3 + \rho^{-3})^2 + 4)$$
$$= 4(\rho^6 + \rho^{-6} + 6)$$
$$= 4(\eta^3 + \eta^{-3} + 6).$$

Now, as K is a cubic field, by Theorem 7.1.16 we have

$$|d(K)| \leq |D(\alpha, \beta, \gamma)|$$

for any $\alpha, \beta, \gamma \in O_K$ with $D(\alpha, \beta, \gamma) \neq 0$, so that in particular we have

$$|d(K)| \leq |D(1, \eta, \eta^2)| = |D(\eta)|.$$

Hence

$$|d(K)| < 4(\eta^3 + \eta^{-3} + 6).$$

Thus

$$\eta^3 + \eta^{-3} > \frac{|d(K)|}{4} - 6$$

and so

$$\eta^6 - \left(\frac{|d(K)|}{4} - 6\right)\eta^3 + 1 > 0.$$

Completing the square, we obtain, as $|d(K)| \geq 33$,

$$\left(\eta^3 - \left(\frac{|d(K)|}{8} - 3\right)\right)^2 > \left(\frac{|d(K)|}{8} - 3\right)^2 - 1 > \left(\frac{|d(K)|}{8} - \frac{15}{4}\right)^2,$$

so that

$$\left|\eta^3 - \left(\frac{|d(K)|}{8} - 3\right)\right| > \frac{|d(K)|}{8} - \frac{15}{4}.$$

If

$$\eta^3 - \left(\frac{|d(K)|}{8} - 3\right) < 0$$

then

$$\left(\frac{|d(K)|}{8} - 3\right) - \eta^3 > \frac{|d(K)|}{8} - \frac{15}{4}$$

so that

$$\eta^3 < \frac{15}{4} - 3 = \frac{3}{4},$$

contradicting $\eta > 1$. Hence

$$\eta^3 - \left(\frac{|d(K)|}{8} - 3\right) \geq 0$$

so that

$$\eta^3 - \left(\frac{|d(K)|}{8} - 3\right) > \frac{|d(K)|}{8} - \frac{15}{4},$$

which gives

$$\eta^3 > \frac{|d(K)| - 27}{4}. \qquad \blacksquare$$

We next use Theorem 13.6.1 to determine the fundamental unit > 1 of the ring of integers of $\mathbb{Q}(\sqrt[3]{2})$.

Theorem 13.6.2 *The fundamental unit* > 1 *of* $O_{\mathbb{Q}(\sqrt[3]{2})}$ *is* $1 + \sqrt[3]{2} + (\sqrt[3]{2})^2$.

Proof: We set $\alpha = \sqrt[3]{2}$ and $K = \mathbb{Q}(\alpha)$. The ring of integers of K is

$$O_K = \mathbb{Z} + \mathbb{Z}\alpha + \mathbb{Z}\alpha^2,$$

and the discriminant of K is

$$d(K) = -108$$

(see Table 1). For the field K, we have $n = 3$, $r = 1$, $s = 1$, so, by Theorems 13.4.2 and 13.5.3, K possesses a unique fundamental unit $\eta > 1$. We show that

$$\eta = 1 + \alpha + \alpha^2.$$

Set

$$u = 1 + \alpha + \alpha^2 \in O_K.$$

Clearly, as $0 < \alpha < 2$ and $7^3 = 343 < 400 = 20^2$, we have

$$1 < u < 1 + 2 + 4 = 7 < 20^{2/3}.$$

Moreover, u is a unit of O_K as

$$\frac{1}{u} = \frac{1}{1 + \alpha + \alpha^2} = \frac{-1 + \alpha}{(1 + \alpha + \alpha^2)(-1 + \alpha)} = \frac{-1 + \alpha}{-1 + \alpha^3} = \frac{-1 + \alpha}{-1 + 2}$$
$$= -1 + \alpha \in O_K.$$

Appealing to Theorem 13.6.1, we obtain, as $|d(K)| = 108 > 33$,

$$\eta^3 > \frac{108 - 27}{4} = \frac{81}{4} > 20$$

so that

$$\eta > 20^{1/3}.$$

Hence we have shown that

$$1 < u < \eta^2. \tag{13.6.2}$$

Since η is a fundamental unit of O_K, and the only roots of unity in O_K are ± 1, we have by Dirichlet's unit theorem

$$u = \pm \eta^k, \text{ for some } k \in \mathbb{Z}. \tag{13.6.3}$$

As $\eta > 1$ we deduce from (13.6.2) and (13.6.3) that the plus sign holds in (13.6.3) and $k = 1$; that is, $u = \eta$ as asserted. ■

In the next theorem we determine in a similar manner the fundamental unit > 1 of the ring of integers of $\mathbb{Q}(\sqrt[3]{3})$.

Theorem 13.6.3 *The fundamental unit of* $O_{\mathbb{Q}(\sqrt[3]{3})}$ *is* $4 + 3\sqrt[3]{3} + 2(\sqrt[3]{3})^2$.

Proof: We set $\alpha = \sqrt[3]{3}$ and $K = \mathbb{Q}(\alpha)$. The ring of integers of K is

$$O_K = \mathbb{Z} + \mathbb{Z}\alpha + \mathbb{Z}\alpha^2$$

and the discriminant of K is

$$d(K) = -243$$

(see Table 1). For the field K we have $n = 3$, $r = 1$, $s = 1$, so, by Theorems 13.4.2 and 13.5.3, O_K possesses a unique fundamental unit $\eta > 1$. We show that

$$\eta = 4 + 3\alpha + 2\alpha^2.$$

Set

$$u = 4 + 3\alpha + 2\alpha^2 \in O_K.$$

As $0 < \alpha < 3/2$ we have

$$1 < u < 4 + 3(\frac{3}{2}) + 2(\frac{3}{2})^2 = 13.$$

Next we show that u is a unit of O_K. We seek $r, s, t \in \mathbb{Z}$ such that

$$(4 + 3\alpha + 2\alpha^2)(r + s\alpha + t\alpha^2) = 1.$$

Multiplying out the left-hand side and making use of the relations $\alpha^3 = 3, \alpha^4 = 3\alpha$, we obtain

$$(4r + 6s + 9t) + (3r + 4s + 6t)\alpha + (2r + 3s + 4t)\alpha^2 = 1,$$

so that

$$4r + 6s + 9t = 1,$$
$$3r + 4s + 6t = 0,$$
$$2r + 3s + 4t = 0.$$

Solving this system of linear equations, we obtain

$$r = -2, \ s = 0, \ t = 1.$$

Hence

$$\frac{1}{4 + 3\alpha + 2\alpha^2} = -2 + \alpha^2 \in O_K,$$

so that $4 + 3\alpha + 2\alpha^2 \in U(O_K)$. By Theorem 13.6.1 we have

$$\eta^3 > \frac{243 - 27}{4} = 54,$$

Table 11. *Fundamental unit* (> 1) *of*
$\mathbb{Q}(\sqrt[3]{m})$ *for a few values of* $m \in \mathbb{N}$

m	Fundamental unit > 1 of $\mathbb{Q}(\sqrt[3]{m})$
2	$1 + \sqrt[3]{2} + (\sqrt[3]{2})^2$
3	$4 + 3\sqrt[3]{3} + 2(\sqrt[3]{3})^2$
5	$41 + 24\sqrt[3]{5} + 14(\sqrt[3]{5})^2$
6	$109 + 60\sqrt[3]{6} + 33(\sqrt[3]{6})^2$
7	$4 + 2\sqrt[3]{7} + (\sqrt[3]{7})^2$
11	$89 + 40\sqrt[3]{11} + 18(\sqrt[3]{11})^2$
12	$55 + 24\sqrt[3]{12} + \frac{21}{2}(\sqrt[3]{12})^2$

so that

$$\eta^6 > 54^2 = 2916 > 2744 = 14^3$$

and thus

$$\eta^2 > 14.$$

Hence we have shown that

$$1 < u < \eta^2.$$

Then, exactly as in the proof of Theorem 13.6.2, we deduce that $u = \eta$. ■

Table 11 gives the fundamental unit > 1 for a few pure cubic fields $\mathbb{Q}(\sqrt[3]{m})$, $m \in \mathbb{N}$.

Table 12 gives the fundamental unit of the first thirty cubic fields K with exactly one real embedding arranged in order of increasing $|d(K)|$.

Table 13 gives a fundamental system of units $\{\epsilon_1, \epsilon_2\}$ of O_K for the first thirty cubic fields $K = \mathbb{Q}(\theta)$ having three real embeddings arranged in order of increasing $d(K)$.

In Chapter 14 we make use of our knowledge of the units of O_K, where $K = \mathbb{Q}(\theta)$, $\theta^3 - 4\theta + 2 = 0$, to determine all the solutions in integers of the equation

$$y(y + 1) = x(x + 1)(x + 2).$$

In the next example we determine a fundamental system of units for O_K.

Example 13.6.1 *The polynomial* $x^3 - 4x + 2 \in \mathbb{Z}[x]$ *is 2-Eisenstein and so is irreducible. Hence* $K = \mathbb{Q}(\theta)$, *where* $\theta^3 - 4\theta + 2 = 0$, *is a cubic field. The*

Table 12.   *Fundamental unit of cubic fields K with
exactly one real embedding and* $-268 \leq d(K) < 0$

$d(K)$	$K = \mathbb{Q}(\theta)$	Fundamental unit
-23	$x^3 + x^2 - 1$	$\theta + \theta^2$
-31	$x^3 - x^2 - 1$	θ
-44	$x^3 - x^2 - x - 1$	θ
-59	$x^3 + 2x - 1$	$2 + \theta^2$
-76	$x^3 - 2x - 2$	$1 + \theta$
-83	$x^3 - x^2 + x - 2$	$1 + \theta^2$
-87	$x^3 + x^2 + 2x - 1$	$2 + \theta + \theta^2$
-104	$x^3 - x - 2$	$1 + \theta + \theta^2$
-107	$x^3 - x^2 + 3x - 2$	$3 + \theta^2$
-108	$x^3 - 2$	$1 + \theta + \theta^2$
-116	$x^3 - x^2 - 2$	$1 + \theta + \theta^2$
-135	$x^3 + 3x - 1$	$3 + \theta^2$
-139	$x^3 + x^2 + x - 2$	$3 + 2\theta + \theta^2$
-140	$x^3 + 2x - 2$	$3 + \theta + \theta^2$
-152	$x^3 - x^2 - 2x - 2$	$1 + \theta + \theta^2$
-172	$x^3 + x^2 - x - 3$	$2 + 2\theta + \theta^2$
-175	$x^3 - x^2 + 2x - 3$	$2 + \theta^2$
-199	$x^3 - x^2 + 4x - 1$	$4 - \theta + \theta^2$
-200	$x^3 + x^2 + 2x - 2$	$9 + 5\theta + 3\theta^2$
-204	$x^3 - x^2 + x - 3$	$4 + \theta + 2\theta^2$
-211	$x^3 - 2x - 3$	$2 + 2\theta + \theta^2$
-212	$x^3 - x^2 + 4x - 2$	$15 - 2\theta + 4\theta^2$
-216	$x^3 + 3x - 2$	$17 + 3\theta + 5\theta^2$
-231	$x^3 + x^2 - 3$	$2 + 2\theta + \theta^2$
-239	$x^3 - x - 3$	$2 + 2\theta + \theta^2$
-243	$x^3 - 3$	$4 + 3\theta + 2\theta^2$
-244	$x^3 + x^2 - 4x - 6$	$5 + 6\theta + 2\theta^2$
-247	$x^3 + x - 3$	$2 + \theta + \theta^2$
-255	$x^3 - x^2 - 3$	$2 + \theta + \theta^2$
-268	$x^3 + x^2 - 3x - 5$	$3 + 3\theta + \theta^2$

discriminant of $x^3 - 4x + 2$ *is*

$$-4(-4)^3 - 27(2)^2 = 148 = 2^2 \cdot 37,$$

which is positive, so that K has three real embeddings. Furthermore, we have

$$d(K) = 148 \ or \ 37.$$

But the smallest discriminant of a cubic field with three real embeddings is 49
(see Table 13). Hence $d(K) \neq 37$ *and so* $d(K) = 148$. *Thus K must be the third
field listed in Table 13. Hence* $K = \mathbb{Q}(\phi)$, *where* $\phi^3 + \phi^2 - 3\phi - 1 = 0$. *The*

Table 13.   *Units of totally real cubic fields K with* $0 < d(K) \leq 1101$

$d(K)$	$K = \mathbb{Q}(\theta)$	ϵ_1	ϵ_2
49	$x^3 + x^2 - 2x - 1$	$-1 + \theta + \theta^2$	$2 - \theta^2$
81	$x^3 - 3x - 1$	$2 + \theta - \theta^2$	$-\theta$
148	$x^3 + x^2 - 3x - 1$	θ	$2 - \theta^2$
169	$x^3 - x^2 - 4x - 1$	$2 + 2\theta - \theta^2$	$-\theta$
229	$x^3 - 4x - 1$	θ	$2 + \theta$
257	$x^3 - 5x - 3$	$4 + \theta - \theta^2$	$5 + \theta - \theta^2$
316	$x^3 + x^2 - 4x - 2$	$-3 + \theta + \theta^2$	$-5 + \theta + \theta^2$
321	$x^3 + x^2 - 4x - 1$	$-\theta$	$-1 + 2\theta + \theta^2$
361	$x^3 + x^2 - 6x - 7$	$4 + \theta - \theta^2$	$5 - \theta^2$
404	$x^3 - x^2 - 5x - 1$	$-\theta$	$1 - \theta - \theta^2$
469	$x^3 + x^2 - 5x - 4$	$-1 - \theta$	$-1 + 2\theta + \theta^2$
473	$x^3 - 5x - 1$	$-\theta$	$-2 - \theta$
564	$x^3 + x^2 - 5x - 3$	$-2 + \theta$	$-1 - \theta + \theta^2$
568	$x^3 - x^2 - 6x - 2$	$-5 - \theta + \theta^2$	$-7 - 4\theta + 2\theta^2$
621	$x^3 - 6x - 3$	$-2 - \theta$	$1 + 2\theta$
697	$x^3 - x^2 - 8x - 5$	$6 + 2\theta - \theta^2$	$7 + 2\theta - \theta^2$
733	$x^3 + x^2 - 7x - 8$	$1 + \theta$	$-5 - 2\theta$
756	$x^3 - 6x - 2$	$5 - \theta^2$	$11 + \theta - 2\theta^2$
761	$x^3 - x^2 - 6x - 1$	θ	$2 + \theta$
785	$x^3 + x^2 - 6x - 5$	$1 + \theta$	$-4 + \theta + \theta^2$
788	$x^3 - x^2 - 7x - 3$	$2 + \theta$	$-1 - 2\theta$
837	$x^3 - 6x - 1$	$-\theta$	$-3 - 6\theta - 2\theta^2$
892	$x^3 + x^2 - 8x - 10$	$3 + \theta - \theta^2$	$1 + 3\theta + \theta^2$
940	$x^3 - 7x - 4$	$-11 - 2\theta + 2\theta^2$	$-3 + \theta + \theta^2$
961	$x^3 + x^2 - 10x - 8$	$-1 + 2\theta + 2\theta^2$	$3 + 4\theta - 2\theta^2$
985	$x^3 + x^2 - 6x - 1$	θ	$-2 + \theta$
993	$x^3 + x^2 - 6x - 3$	$5 - \theta - \theta^2$	$5 - \theta^2$
1016	$x^3 + x^2 - 6x - 2$	$7 - \theta - \theta^2$	$-11 - \theta + \theta^2$
1076	$x^3 - 8x - 6$	$1 + \theta$	$-7 - 3\theta$
1101	$x^3 + x^2 - 9x - 12$	$5 + 2\theta - \theta^2$	$-7 - 4\theta + 2\theta^2$

relationship between θ and ϕ is given by

$$\phi = \frac{1}{\theta - 1}.$$

From Table 13 we see that a fundamental system of units for O_K is

$$\{\phi, 2 - \phi^2\}.$$

In terms of θ, we deduce that

$$\{\theta - 1, 2\theta^2 - 4\theta + 1\}$$

Table 14. *Fundamental unit of some pure quartic fields* $\mathbb{Q}(\sqrt[4]{-m})$

m	Fundamental unit of $\mathbb{Q}(\theta)$, $\theta = \sqrt[4]{-m}$
1	$\theta - \theta^2 + \theta^3$
2	$-1 + \theta^2 - \theta^3$
5	$-2 + 2\theta - \theta^2$
6	$1 + 4\theta - 4\theta^2 + 2\theta^3$
10	$-27 + 12\theta - \theta^2 - 3\theta^3$

is a fundamental system of units for O_K. Finally, as

$$(2\theta - 1)(2\theta^2 - 4\theta + 1) = 4\theta^3 - 10\theta^2 + 6\theta - 1$$
$$= \theta(\theta^3 - 4\theta + 2) - (\theta^4 - 4\theta^3 + 6\theta^2 - 4\theta + 1)$$
$$= -(\theta - 1)^4,$$

we see that

$$\{\theta - 1, 2\theta - 1\}$$

is a fundamental system of units for O_K.

Table 14 gives a fundamental unit for a few pure quartic fields $\mathbb{Q}(\sqrt[4]{-m})$, $m \in \mathbb{N}$. Such fields are totally imaginary quartic fields.

13.7 Regulator

Let $\{\epsilon_1, \ldots, \epsilon_{r+s-1}\}$ and $\{\epsilon_1', \ldots, \epsilon_{r+s-1}'\}$ be any two fundamental systems of units for the ring of integers O_K of an algebraic number field K. As $\{\epsilon_1, \ldots, \epsilon_{r+s-1}\}$ is a fundamental system of units, we have

$$\epsilon_j' = \zeta^{b_j}\epsilon_1^{a_{1j}} \cdots \epsilon_{r+s-1}^{a_{r+s-1\,j}}, \quad j = 1, 2, \ldots, r + s - 1, \tag{13.7.1}$$

where ζ is a root of unity in K and a_{ij}, $b_j \in \mathbb{Z}$. Similarly, as $\{\epsilon_1', \ldots, \epsilon_{r+s-1}'\}$ is also a fundamental system of units, we have

$$\epsilon_j = \rho^{b_j'}\epsilon_1'^{a_{1j}'} \cdots \epsilon_{r+s-1}'^{a_{r+s-1\,j}'}, \quad j = 1, 2, \ldots, r + s - 1, \tag{13.7.2}$$

where $a'_{ij}, b'_j \in \mathbb{Z}$. Hence, for $j = 1, 2, \ldots, r + s - 1$, we have by (13.7.1) and (13.7.2)

$$\epsilon'_j = \zeta^{b_j} \prod_{k=1}^{r+s-1} \epsilon_k^{a_{kj}} = \zeta^{b_j} \prod_{k=1}^{r+s-1} \left(\zeta^{b'_k} \prod_{l=1}^{r+s-1} \epsilon_l^{'a'_{lk}} \right)^{a_{kj}}$$

$$= \zeta^{b_j + \sum_{k=1}^{r+s-1} a_{kj} b'_k} \prod_{l=1}^{r+s-1} \epsilon_l^{'\sum_{k=1}^{r+s-1} a'_{lk} a_{kj}}.$$

By the uniqueness of the representation of units, we have

$$\sum_{k=1}^{r+s-1} a'_{lk} a_{kj} = \begin{cases} 1, & \text{if } l = j, \\ 0, & \text{if } l \neq j. \end{cases} \tag{13.7.3}$$

Next we define the $(r + s - 1) \times (r + s - 1)$ matrices A and A' by

$$A = [a_{ij}], \quad A' = [a'_{ij}].$$

From (13.7.3) we see that

$$A'A = I_{r+s-1}.$$

Thus

$$\det A' \cdot \det A = \det (A'A) = \det I_{r+s-1} = 1.$$

As the matrices A and A' have integral entries, both $\det A$ and $\det A'$ are integers so that $\det A' = \det A = \pm 1$ and hence

$$|\det A| = |\det A'| = 1. \tag{13.7.4}$$

Let σ_k $(k = 1, 2, \ldots, n)$ be the $n = [K : \mathbb{Q}]$ distinct monomorphisms: $K \to \mathbb{C}$ with $\sigma_1, \ldots, \sigma_r$ real, $\sigma_{r+1}, \ldots, \sigma_{r+s}$ complex, and $\sigma_{r+s+1} = \overline{\sigma_{r+1}}, \ldots, \sigma_n = \sigma_{r+2s} = \overline{\sigma_{r+s}}$. For $j, k = 1, 2, \ldots, r + s - 1$ we have

$$\sigma_k(\epsilon'_j) = \sigma_k \left(\zeta^{b_j} \prod_{l=1}^{r+s-1} \epsilon_l^{a_{lj}} \right) = \sigma_k(\zeta)^{b_j} \prod_{l=1}^{r+s-1} \sigma_k(\epsilon_l)^{a_{lj}}$$

so that

$$|\sigma_k(\epsilon'_j)| = \left| \sigma_k(\zeta)^{b_j} \prod_{l=1}^{r+s-1} \sigma_k(\epsilon_l)^{a_{lj}} \right| = |\sigma_k(\zeta)|^{b_j} \prod_{l=1}^{r+s-1} |\sigma_k(\epsilon_l)|^{a_{lj}} = \prod_{l=1}^{r+s-1} |\sigma_k(\epsilon_l)|^{a_{lj}}$$

and thus

$$\log |\sigma_k(\epsilon'_j)| = \sum_{l=1}^{r+s-1} a_{lj} \log |\sigma_k(\epsilon_l)|. \tag{13.7.5}$$

Let E and E' denote the $(r + s - 1) \times (r + s - 1)$ matrices $[\log |\sigma_i(\epsilon_j)|]$ and $[\log |\sigma_i(\epsilon'_j)|]$ respectively. Then, from (12.7.5), we deduce that

$$E' = AE,$$

and so

$$\det E' = \det (AE) = \det A \cdot \det E.$$

Finally,

$$|\det E'| = |\det A \cdot \det E| = |\det A||\det E| = |\det E|,$$

by (13.7.4).

We have shown that the nonnegative real number

$$|\det (\log |\sigma_i(\epsilon_j)|)|$$

is independent of the choice of fundamental system of units $\{\epsilon_1, \ldots, \epsilon_{r+s-1}\}$ of O_K. We can therefore introduce the following concept.

Definition 13.7.1 (Regulator) *Let K be an algebraic number field of degree n over $\mathbb{Q}$. Let r be the number of real embeddings of K and $2s$ the number of non-real embeddings of K so that $n = r + 2s$. Let σ_i $(i = 1, 2, \ldots, n)$ be the n distinct monomorphisms $: K \to \mathbb{C}$ with $\sigma_1, \ldots, \sigma_r$ real, $\sigma_{r+1}, \ldots, \sigma_{r+s}$ complex, and $\sigma_{r+s+1} = \overline{\sigma_{r+1}}, \ldots, \sigma_{r+2s} = \overline{\sigma_{r+s}}$. Let $\{\epsilon_1, \ldots, \epsilon_{r+s-1}\}$ be any fundamental system of units of O_K. Let E denote the $(r + s - 1) \times (r + s - 1)$ matrix whose entry in the (i, j) place is*

$$\log |\sigma_i(\epsilon_j)|, \ i, j = 1, 2, \ldots, r + s - 1.$$

Then the nonnegative real number

$$R(K) = |\det E|$$

is called the regulator of K.

If K is either $\mathbb{Q}$ or an imaginary quadratic field then $r + s - 1 = 0$ and the set comprising a fundamental system of units of O_K is empty. In this case we understand $R(K)$ to be zero. Otherwise $R(K) > 0$. We now determine the regulator of a real quadratic field.

Theorem 13.7.1 *Let K be a real quadratic field. Then*

$$R(K) = \log \eta,$$

where η is the fundamental unit (> 1) of K.

Proof: As K is a real quadratic field we have $n = r = 2$ and $s = 0$ so that $r + s - 1 = 1$. Thus a fundamental system of units of O_K is $\{\eta\}$ and

$$R(K) = |\det (\log |\eta|)| = |\log |\eta|| = |\log \eta| = \log \eta,$$

as $\eta > 1$ ensures $\log \eta > 0$. ∎

Example 13.7.1 *The fundamental unit of O_K, where $K = \mathbb{Q}(\sqrt{2})$, is $\eta = 1 + \sqrt{2}$ so that $R(K) = \log(1 + \sqrt{2})$.*

Example 13.7.2 *Let K be the cubic field given by*

$$K = \mathbb{Q}(\theta), \quad \theta^3 - 4\theta + 2 = 0.$$

The discriminant of the polynomial $x^3 - 4x + 2$ is positive as

$$-4(-4)^3 - 27(2)^2 = 148,$$

so that the three roots θ, θ', θ'' of $x^3 - 4x + 2 = 0$ are all real. Hence

$$n = r = 3, \quad s = 0, \quad r + s - 1 = 2.$$

Thus a fundamental system of units of O_K comprises two units, and it was shown in Example 13.6.1 that these can be taken to be $\theta - 1$ and $2\theta - 1$.
  We choose the roots θ, θ', θ'' of $x^3 - 4x + 2 = 0$ so that $\theta < \theta' < \theta''$. Thus

$$\theta \simeq -2.2143, \quad \theta' \simeq 0.5391, \quad \theta'' \simeq 1.6751.$$

Then

$$
\begin{aligned}
|\theta - 1| &\simeq 3.2143, &\quad \log |\theta - 1| &\simeq 1.1676, \\
|2\theta - 1| &\simeq 5.4286, &\quad \log |2\theta - 1| &\simeq 1.6916, \\
|\theta' - 1| &\simeq 0.4609, &\quad \log |\theta' - 1| &\simeq -0.7745, \\
|2\theta' - 1| &\simeq 0.0782, &\quad \log |2\theta' - 1| &\simeq -2.5484, \\
|\theta'' - 1| &\simeq 0.6751, &\quad \log |\theta'' - 1| &\simeq -0.3928, \\
|2\theta'' - 1| &\simeq 2.3502, &\quad \log |2\theta'' - 1| &\simeq 0.8545.
\end{aligned}
$$

Hence

$$
\begin{aligned}
R(K) &= \left| \det \begin{bmatrix} \log |\theta - 1| & \log |\theta' - 1| \\ \log |2\theta - 1| & \log |2\theta' - 1| \end{bmatrix} \right| \\
&= |\log |\theta - 1| \log |2\theta' - 1| - \log |\theta' - 1| \log |2\theta - 1|| \\
&\simeq |(1.1676)(-2.5484) + (0.7745)(1.6916)| \\
&\simeq |-2.9755 + 1.3101| \\
&= |-1.6654| = 1.6654.
\end{aligned}
$$

It should be noted that we also have

$$R(K) = \left|\det \begin{bmatrix} \log|\theta - 1| & \log|\theta'' - 1| \\ \log|2\theta - 1| & \log|2\theta'' - 1| \end{bmatrix}\right|$$

$$= \left|\log|\theta - 1|\log|2\theta'' - 1| - \log|\theta'' - 1|\log|2\theta - 1|\right|$$

$$\simeq |(1.1676)(0.8545) + (0.3928)(1.6916)|$$

$$\simeq |0.9977 + 0.6644|$$

$$= 1.6621$$

and

$$R(K) = \left|\det \begin{bmatrix} \log|\theta' - 1| & \log|\theta'' - 1| \\ \log|2\theta' - 1| & \log|2\theta'' - 1| \end{bmatrix}\right|$$

$$= \left|\log|\theta' - 1|\log|2\theta'' - 1| - \log|\theta'' - 1|\log|2\theta' - 1|\right|$$

$$\simeq |-(0.7745)(0.8545) - (0.3928)(2.5484)|$$

$$\simeq |-0.6618 - 1.0010|$$

$$= |1.6628| = 1.6628.$$

We close by remarking that some authors use a slightly different definition of the regulator.

Exercises

1. Prove that there do not exist $a, b, c \in \mathbb{Z}$ such that

$$\sqrt{-2 + 2\sqrt[3]{2} + (\sqrt[3]{2})^2} = a + b\sqrt[3]{2} + c(\sqrt[3]{2})^2$$

(see Example 13.1.3).

2. Show that if θ is a root of a monic polynomial $f(x) \in \mathbb{Z}[x]$ and $n \in \mathbb{Z}$ is such that $f(n) = \pm 1$ then $\theta - n \in U(\mathbb{Q}(\theta))$.

3. Prove that $x^3 - 2x - 2$ has only one real root θ and that θ satisfies

$$1.7 < \theta < 1.8.$$

Use Theorem 13.6.1 to show that the fundamental unit (> 1) of $O_{\mathbb{Q}(\theta)}$ is $1 + \theta$.

4. Prove that $x^3 - x - 2$ has only one real root θ and that θ satisfies

$$1.5 < \theta < 1.6.$$

Use Theorem 13.6.1 to prove that $1 + \theta + \theta^2$ is the fundamental unit (> 1) of $O_{\mathbb{Q}(\theta)}$.

5. Prove that $x^3 - x^2 + x - 2$ has only one real root θ and that θ satisfies

$$1.3 < \theta < 1.4.$$

Use Theorem 13.6.1 to show that the fundamental unit (> 1) of $O_{\mathbb{Q}(\theta)}$ is $1 + \theta^2$.

6. Prove that $41 + 24\sqrt[3]{5} + 14(\sqrt[3]{5})^2$ is the fundamental unit (> 1) of $O_{\mathbb{Q}(\sqrt[3]{5})}$.

7. Prove that $109 + 60\sqrt[3]{6} + 33(\sqrt[3]{6})^2$ is the fundamental unit (> 1) of $O_{\mathbb{Q}(\sqrt[3]{6})}$.

8. Prove that $4 + 2\sqrt[3]{7} + (\sqrt[3]{7})^2$ is the fundamental unit (> 1) of $O_{\mathbb{Q}(\sqrt[3]{7})}$.

9. Prove that

$$K = \mathbb{Q}(\theta), \ \theta^3 + \theta^2 - 2\theta - 1 = 0,$$

is a totally real field. Prove that $-1 + \theta + \theta^2$ and $2 - \theta^2$ are independent units of K.

10. Prove that

$$1 + 2\sqrt[4]{34} - (\sqrt[4]{34})^2 \text{ and } 35 - 6(\sqrt[4]{34})^2$$

are independent units in $\mathbb{Q}(\sqrt[4]{34})$.

11. Prove that

$$\left\{ 1 + \sqrt{2}, \ \sqrt{2} + \sqrt{3}, \ \frac{1}{2}\left(\sqrt{2} + \sqrt{6}\right) \right\}$$

is a fundamental system of units for O_K, where $K = \mathbb{Q}(\sqrt{2}, \sqrt{3})$.

12. Let K be an algebraic number field such that $U(O_K)$ contains a nonreal root of unity. Prove that $N(\alpha) > 0$ for every $\alpha \in K \setminus \{0\}$.

13. Let $K = \mathbb{Q}(\zeta_m)$, where ζ_m is a primitive mth root of unity, $m \geq 3$. Determine r and s for K.

14. Let K be a totally imaginary quartic field containing a real quadratic field k. By Theorem 13.4.2 K possesses a fundamental unit. Give conditions under which this fundamental unit can be taken to be the fundamental unit of k.

15. Prove that

$$\{1 + \sqrt[4]{2}, 1 + (\sqrt[4]{2})^2\}$$

is a fundamental system of units for O_K, where $K = \mathbb{Q}(\sqrt[4]{2})$.

Suggested Reading

1. H. Cohen, *A Course in Computational Number Theory*, Springer-Verlag, Berlin, Heidelberg, New York, 1996.

   This book describes 148 algorithms that are fundamental for number theoretic computations. Algorithms 4.9.9 and 4.9.10 calculate the roots of unity in the ring of integers of an arbitrary algebraic number field. Algorithm 6.5.8 computes a fundamental system of units.

2. C. Levesque, *Systemes fondamentaux d'unites de certains composes de deux corps quadratiques*, I, Canadian Journal of Mathematics 33 (1981), 937–945.

   The author determines a fundamental system of units for certain quartic fields $\mathbb{Q}(\sqrt{m}, \sqrt{n})$, where m and n are positive integers.

3. I. Niven, H. S. Zuckerman, and H. L. Montgomery, *An Introduction to the Theory of Numbers*, fifth edition, Wiley, New York, 1991.

   A proof that Euler's phi function $\phi(n)$ is multiplicative can be found on page 69.

4. B. L. van der Waerden, *Ein Logarithmenfreier Beweis des Dirichletschen Einheitensatzes*, Abhandlungen aus dem Mathematischen Seminar der Universität Hamburg 6 (1928), 259–262.

   The proof of Dirichlet's unit theorem given in this chapter is based upon the approach in this paper.

Biographies

1. G. Frei, *Bartel Leendert van der Waerden*, Historia Mathematica 20 (1993), 5–11.

   A brief biography of van der Waerden (1903–1996) is given.

2. G. Frei, J. Top, and L. Walling, *A short biography of B. L. van der Waerden*, Nieuw Archief voor Wiskunde 12 (1994), 137–144.

   A well-written biography of van der Waerden is given.

14

Applications to Diophantine Equations

An equation that is to be solved in integers is called a Diophantine equation in honor of Diophantus (ca. 200–ca. 284), who proposed in his chief work *Arithmetic* many problems to be solved in rational numbers or integers. Not much is known about Diophantus. He lived in Alexandria, probably was not a Greek, and likely did most of his work during the latter half of the third century.

In this chapter we apply algebraic number theory to solve some Diophantine equations. We will be principally interested in the Diophantine equation

$$y^2 = x^3 + k,$$

where k is a given integer. This equation is often called Bachet's equation, after the French mathematician Claude Gaspard Bachet de Méziriac (1581–1638), who showed how to find solutions of $y^2 = x^3 - 2$ in rationals x and y from the solution $(x, y) = (3, 5)$. In 1917 Axel Thue (1863–1922) showed that for any given nonzero integer k, Bachet's equation has at most finitely many solutions in integers x and y. Deep estimates from transcendental number theory give bounds for the sizes of the solutions x and y. Hence the problem of finding all solutions in integers of Bachet's equation for a given nonzero integer k is reduced to a finite search. In Section 14.1 we use elementary congruence considerations to give classes of k for which Bachet's equation has no solutions in integers. In Section 14.2 we use the arithmetic of quadratic fields to determine all the solutions in integers (if any) of Bachet's equation for certain classes of k. In particular when $k = -2$ we show that $(x, y) = (3, \pm 5)$ are the only solutions in integers of $y^2 = x^3 - 2$, a result first stated by Fermat. In Section 14.3 we find all the solutions in integers x and y of the equation $y(y + 1) = x(x + 1)(x + 2)$.

14.1 Insolvability of $y^2 = x^3 + k$ Using Congruence Considerations

In this section, using only simple congruence arguments, we give four classes of integers k for which Bachet's equation $y^2 = x^3 + k$ has no solutions in integers x and y.

385

Theorem 14.1.1 *Let M and N be integers such that*

$$M \equiv 3 \ (\mathrm{mod}\ 4), \quad N \equiv 2 \ (\mathrm{mod}\ 4),$$
$$p \ (prime) \mid N/2 \Longrightarrow p \equiv 1 \ (\mathrm{mod}\ 4).$$

Set

$$k = M^3 - N^2.$$

Then the equation $y^2 = x^3 + k$ has no solutions in integers x and y.

Proof: Suppose that $(x, y) \in \mathbb{Z}^2$ is a solution of $y^2 = x^3 + k$. As $k \equiv -1 \ (\mathrm{mod}\ 4)$ we have

$$y^2 \equiv x^3 - 1 \ (\mathrm{mod}\ 4). \tag{14.1.1}$$

Now $y^2 \equiv 0$ or $1 \ (\mathrm{mod}\ 4)$ for every integer y, so (14.1.1) cannot be satisfied if x is even or $x \equiv 3 \ (\mathrm{mod}\ 4)$. Hence we must have $x \equiv 1 \ (\mathrm{mod}\ 4)$. As $k = M^3 - N^2$ we see that

$$y^2 + N^2 = x^3 + M^3 = (x + M)(x^2 - Mx + M^2). \tag{14.1.2}$$

Since $x \equiv 1 \ (\mathrm{mod}\ 4)$ and $M \equiv 3 \ (\mathrm{mod}\ 4)$ we deduce that

$$x^2 - Mx + M^2 \equiv 3 \ (\mathrm{mod}\ 4). \tag{14.1.3}$$

Hence $x^2 - Mx + M^2$ is odd and (14.1.3) shows that it has at least one prime factor $p \equiv 3 \ (\mathrm{mod}\ 4)$. Thus $y^2 \equiv -N^2 \ (\mathrm{mod}\ p)$. By assumption $p \nmid N$. Hence

$$\left(\frac{-1}{p}\right) = \left(\frac{-N^2}{p}\right) = \left(\frac{y^2}{p}\right) = 1,$$

contradicting $p \equiv 3 \ (\mathrm{mod}\ 4)$. This proves that the Diophantine equation $y^2 = x^3 + k$ has no solutions ∎

The following table gives some values of k (with $|k| < 100$) covered by Theorem 14.1.1.

M	-1	15	3	3
N	2	58	2	10
k	-5	11	23	-73

Theorem 14.1.2 *Let M and N be integers such that*

$$M \equiv 2 \ (\mathrm{mod}\ 4), \quad N \equiv 1 \ (\mathrm{mod}\ 2),$$
$$p \ (prime) \mid N \Longrightarrow p \equiv 1 \ (\mathrm{mod}\ 4).$$

Set

$$k = M^3 - N^2.$$

Then the equation $y^2 = x^3 + k$ *has no solutions in integers x and y.*

Proof: Suppose that $(x, y) \in \mathbb{Z}^2$ is a solution of $y^2 = x^3 + k$. Considering the equation modulo 4, we obtain

$$y^2 \equiv x^3 - 1 \pmod{4}.$$

Hence, as in the proof of Theorem 14.1.1, we must have $x \equiv 1 \pmod{4}$. As $k = M^3 - N^2$ we have

$$y^2 + N^2 = x^3 + M^3 = (x + M)(x^2 - Mx + M^2).$$

Since $x \equiv 1 \pmod{4}$ and $M \equiv 2 \pmod{4}$ we obtain

$$x^2 - Mx + M^2 \equiv 3 \pmod{4}.$$

The rest of the proof is the same as that of Theorem 14.1.1. ∎

The following table gives some values of k ($|k| < 100$) for which Theorem 14.1.2 applies.

M	2	-2	2	-2	6	14	6
N	1	1	5	5	13	53	17
k	7	-9	-17	-33	47	-65	-73

Theorem 14.1.3 *Let M and N be integers such that*

$$M \equiv 4, 6 \pmod{8}, \quad N \equiv 1 \pmod{2},$$
$$p \ (prime) \mid N \implies p \equiv \pm 1 \pmod{8}.$$

Set

$$k = M^3 + 2N^2.$$

Then the equation $y^2 = x^3 + k$ *has no solutions in integers x and y.*

Proof: Suppose that $(x, y) \in \mathbb{Z}^2$ is a solution of $y^2 = x^3 + k$. As $k = M^3 + 2N^2 \equiv 2 \pmod{4}$ we have

$$y^2 \equiv x^3 + 2 \pmod{4}.$$

Hence $x \not\equiv 0 \pmod{2}$, $x \not\equiv 1 \pmod{4}$, and so $x \equiv 3 \pmod{4}$. Next

$$y^2 - 2N^2 = x^3 + M^3 = (x + M)(x^2 - Mx + M^2).$$

If $x \equiv 3 \pmod 8$ then

$$x^2 - Mx + M^2 \equiv 1 - 3M + M^2 \equiv \pm 3 \pmod 8.$$

Hence $x^2 - Mx + M^2$ is odd and at least one of its prime factors p is $\equiv \pm 3 \pmod 8$. Thus $p \nmid N$, $y^2 \equiv 2N^2 \pmod p$, and so

$$\left(\frac{2}{p}\right) = \left(\frac{2N^2}{p}\right) = \left(\frac{y^2}{p}\right) = 1,$$

contradicting $p \equiv \pm 3 \pmod 8$.

If $x \equiv 7 \pmod 8$ then

$$x + M \equiv 7 + M \equiv \pm 3 \pmod 8.$$

Hence $x + M$ is odd and at least one of its prime factors p is $\equiv \pm 3 \pmod 8$. Thus $p \nmid N$, $y^2 \equiv 2N^2 \pmod p$, and so

$$\left(\frac{2}{p}\right) = \left(\frac{2N^2}{p}\right) = \left(\frac{y^2}{p}\right) = 1,$$

contradicting $p \equiv \pm 3 \pmod 8$.

This proves the insolvability of $y^2 = x^3 + k$ in integers x and y. ∎

The following table gives some values of k ($|k| < 100$) for which Theorem 14.1.3 applies.

M	-2	-4	-10	-4	4	-2
N	1	7	23	1	1	7
k	-6	34	58	-62	66	90

Theorem 14.1.4 *Let M and N be integers such that*

$$M \equiv 4 \pmod 8, \quad N \equiv 1 \pmod 2,$$
$$p \ (prime) \mid N \implies p \equiv 1, 3 \pmod 8.$$

Set

$$k = M^3 - 2N^2.$$

Then the equation $y^2 = x^3 + k$ has no solutions in integers x and y.

Proof: Suppose that $(x, y) \in \mathbb{Z}^2$ is a solution of $y^2 = x^3 + k$. As $k = M^3 - 2N^2 \equiv 2 \pmod 4$ we have

$$y^2 \equiv x^3 + 2 \pmod 4.$$

Hence $x \not\equiv 0 \,(\text{mod } 2)$, $x \not\equiv 1 \,(\text{mod } 4)$, and so $x \equiv 3 \,(\text{mod } 4)$. Further, as $k \equiv -2$ (mod 8), we have

$$y^2 \equiv x^3 - 2 \,(\text{mod } 8)$$

so that $x \not\equiv 7 \,(\text{mod } 8)$. Hence $x \equiv 3 \,(\text{mod } 8)$. Next

$$y^2 + 2N^2 = x^3 + M^3 = (x + M)(x^2 - Mx + M^2).$$

As $x \equiv 3 \,(\text{mod } 8)$ and $M \equiv 4 \,(\text{mod } 8)$, we see that $x + M \equiv 7 \,(\text{mod } 8)$. Hence $x + M$ is odd and has at least one prime factor $p \equiv 5$ or $7 \,(\text{mod } 8)$. Thus $p \nmid N$, $y^2 \equiv -2N^2 \,(\text{mod } p)$, and so

$$\left(\frac{-2}{p}\right) = \left(\frac{-2N^2}{p}\right) = \left(\frac{y^2}{p}\right) = 1,$$

contradicting $p \equiv 5$ or $7 \,(\text{mod } 8)$. ∎

The following table gives some values of k ($|k| < 100$) for which Theorem 14.1.4 applies.

M	4	4	-4	-4	4
N	3	1	1	3	9
k	46	62	-66	-82	-98

14.2 Solving $y^2 = x^3 + k$ Using Algebraic Numbers

In this section we make use of results from algebraic number theory to determine all the solutions in integers x and y of $y^2 = x^3 + k$ for certain classes of integers k. The principal results that we use are the following two theorems.

Theorem 14.2.1 *Let D be a Dedekind domain. Let A, B, C be nonzero integral ideals of D such that A and B are coprime and*

$$AB = C^n,$$

where n is a positive integer. Then there exist ideals A_1 and B_1 of D such that

$$A = A_1^n, \ B = B_1^n, \ C = A_1 B_1.$$

Proof: As D is a Dedekind domain, every nonzero integral ideal of D can be expressed uniquely as a product of prime ideals (Theorem 8.3.1). Thus

$$C = P_1^{a_1} \cdots P_r^{a_r},$$

where $P_1, \ldots, P_r$ are $r(\geq 0)$ distinct prime ideals and $a_1, \ldots, a_r$ are positive integers. Hence

$$AB = P_1^{na_1} \cdots P_r^{na_r}.$$

As A and B are coprime ideals, each prime power $P_i^{na_i}$ $(i = 1, \ldots, r)$ divides either A or B but not both. Hence by relabeling if necesssary we have

$$A = P_1^{na_1} \cdots P_s^{na_s}, \ B = P_{s+1}^{na_{s+1}} \cdots P_r^{na_r},$$

for some integer s with $0 \leq s \leq r$. Set $A_1 = P_1^{a_1} \cdots P_s^{a_s}$ and $B_1 = P_{s+1}^{a_{s+1}} \cdots P_r^{a_r}$; then $A = A_1^n$, $B = B_1^n$, and $C = A_1 B_1$ as asserted. ∎

As the ring of integers of an algebraic number field is a Dedekind domain (Theorem 8.1.1), Theorem 14.2.1 applies in this case.

Theorem 14.2.2 *Let K be an algebraic number field. Let h denote the class number of K. Let A be an integral ideal of O_K such that A^k is a principal ideal for some positive integer k coprime with h. Then A is a principal ideal.*

Proof: Let $[A]$ denote the class of A. As the order of $H(K)$ is h, we have $[A]^h = I$ so that A^h is a principal ideal. Since $(h, k) = 1$ there exist integers r and s such that $rh + sk = 1$. Then, as A^k is a principal ideal, so is

$$A = A^{rh+sk} = \left(A^h\right)^r \left(A^k\right)^s$$

as asserted. ∎

We now sketch the ideas involved in using Theorems 14.2.1 and 14.2.2 to obtain classes of rational integers k for which we can find the solutions (if any) of the Diophantine equation $y^2 = x^3 + k$.

We begin by supposing that the equation $y^2 = x^3 + k$ has a solution in integers x and y, so that

$$x^3 = (y + \sqrt{k})(y - \sqrt{k}),$$

where $y + \sqrt{k}$ and $y - \sqrt{k}$ are integers of the quadratic field $K = \mathbb{Q}(\sqrt{k})$. We assume that k is squarefree and that $k \equiv 2$ or $3 \pmod 4$ so that $O_K = \mathbb{Z} + \mathbb{Z}\sqrt{k}$. (The latter condition avoids 2's in the denominators of the integers of $K = \mathbb{Q}(\sqrt{k})$.) Passing to ideals, we obtain

$$\langle x \rangle^3 = \langle y + \sqrt{k} \rangle \langle y - \sqrt{k} \rangle.$$

If the values of k are chosen so that the principal ideals $\langle y + \sqrt{k} \rangle$ and $\langle y - \sqrt{k} \rangle$ are coprime, then we can deduce from Theorem 14.2.1 that

$$\langle y + \sqrt{k} \rangle = A^3$$

for some ideal A of O_K. Further, if the class number of K is not divisible by 3, we know by Theorem 14.2.2 that A is a principal ideal, say

$$A = \langle a + b\sqrt{k} \rangle$$

for some integers a and b. Thus

$$\langle y + \sqrt{k} \rangle = \langle a + b\sqrt{k} \rangle^3 = \langle (a + b\sqrt{k})^3 \rangle$$

and so by Theorem 1.3.1

$$y + \sqrt{k} = \epsilon(a + b\sqrt{k})^3,$$

where ϵ is a unit of O_K. Two cases arise depending on whether k is negative or positive. If k is negative then there are only finitely many possibilities for ϵ. Indeed if $k \neq -1$ then $\epsilon = \pm 1$ and if $k = -1$ then $\epsilon = \pm 1, \pm i$ (Theorem 5.4.3). Since cubes can be absorbed into $(a + b\sqrt{k})^3$, and $-1 = (-1)^3$, $i = (-i)^3$, and $-i = i^3$, the equation becomes

$$y + \sqrt{k} = (a + b\sqrt{k})^3.$$

Equating coefficients of $\sqrt{k}$, we obtain

$$1 = 3a^2b + kb^3 = b(3a^2 + kb^2),$$

so that $b = \pm 1$. It is now an easy matter to determine the possibilities for a, and then the solutions x, y in integers of $y^2 = x^3 + k$ (see Theorem 14.2.3). If k is positive then there are infinitely many possibilities for ϵ. Indeed $\epsilon = \pm \eta^l$, where $\eta = T + U\sqrt{k}$ (> 1) is the fundamental unit of O_K and $l \in \mathbb{Z}$ (Theorem 11.5.1). Absorbing the cubes $-1 = (-1)^3$ and $\zeta^{3m} = (\zeta^m)^3$ into $(a + b\sqrt{k})^3$ we see that we have only to examine the three equations

$$y + \sqrt{k} = (a + b\sqrt{k})^3,$$
$$y + \sqrt{k} = \eta(a + b\sqrt{k})^3,$$

and

$$y + \sqrt{k} = \eta^2(a + b\sqrt{k})^3.$$

The first of these equations can be treated as in the case $k < 0$. For the other two equations it is convenient to impose congruence conditions on k, T, and U to ensure that they do not have any solutions. This is illustrated in Theorem 14.2.4. It should be noted that absorbed cubes must be taken into account when seeking all solutions of $y^2 = x^3 + k$.

Theorem 14.2.3 *Let k be an integer such that*

$$k < -1,$$

k is squarefree,

$$k \equiv 2, 3 \pmod 4,$$

$$h(\mathbb{Q}(\sqrt{k})) \not\equiv 0 \pmod 3.$$

(a) *If there exists an integer a such that*

$$k = 1 - 3a^2$$

then the only solutions in integers of $y^2 = x^3 + k$ are

$$x = 4a^2 - 1, \ y = \pm(3a - 8a^3).$$

(b) *If there exists an integer a such that*

$$k = -1 - 3a^2$$

then the only solutions in integers of $y^2 = x^3 + k$ are

$$x = 4a^2 + 1, \ y = \pm(3a + 8a^3).$$

(c) *If $k \neq \pm 1 - 3a^2$ for any integer a then $y^2 = x^3 + k$ has no solutions in integers x and y.*

Proof: We suppose that $y^2 = x^3 + k$ has a solution in integers x and y and show that either case (a) or case (b) holds. (We note that in case (a) $k \equiv 1 \pmod 3$ and in case (b) $k \equiv 2 \pmod 3$, so that cases (a) and (b) are exclusive.)

First we show that $x \equiv 1 \pmod 2$. As $y^2 \equiv 0, 1 \pmod 4$ and $k \equiv 2, 3 \pmod 4$, we see that $x^3 = y^2 - k \equiv 1, 2, 3 \pmod 4$. But $x^3 \not\equiv 2 \pmod 4$ so $x \equiv 1 \pmod 2$.

Next we prove that $(x, k) = 1$. Suppose not. Then there exists a prime p such that $p \mid x$ and $p \mid k$. As k is squarefree we have $p \parallel k$. Hence $p \parallel x^3 + k$ and so $p \parallel y^2$, a contradiction.

From $x \equiv 1 \pmod 2$ and $(x, k) = 1$ we deduce that $(x, 2k) = 1$ so that there are integers l and m such that

$$lx + m(2k) = 1. \tag{14.2.1}$$

Now let $K = \mathbb{Q}(\sqrt{k})$ so that K is an imaginary quadratic field. As $k \equiv 2, 3 \pmod 4$ the ring O_K of integers of K is $\{u + v\sqrt{k} \mid u, v \in \mathbb{Z}\}$. We now show that the principal ideals $\langle y + \sqrt{k} \rangle$ and $\langle y - \sqrt{k} \rangle$ of O_K are coprime. Suppose not. Then there exists a prime ideal P such that

$$P \mid \langle y + \sqrt{k} \rangle, \ P \mid \langle y - \sqrt{k} \rangle.$$

Hence

$$y + \sqrt{k} \in P, \ y - \sqrt{k} \in P.$$

Thus

$$2\sqrt{k} = (y + \sqrt{k}) - (y - \sqrt{k}) \in P$$

and so

$$2k = \sqrt{k}(2\sqrt{k}) \in P. \tag{14.2.2}$$

Now

$$\langle y + \sqrt{k} \rangle \langle y - \sqrt{k} \rangle = \langle y^2 - k \rangle = \langle x^3 \rangle = \langle x \rangle^3$$

so that

$$P \mid \langle x \rangle^3.$$

As P is a prime ideal, we deduce that

$$P \mid \langle x \rangle.$$

Thus

$$x \in P. \tag{14.2.3}$$

From (14.2.1)–(14.2.3), we see that $1 \in P$, contradicting that P is a prime ideal.

We have shown that $\langle y + \sqrt{k} \rangle$ and $\langle y - \sqrt{k} \rangle$ are coprime ideals of O_K with $\langle y + \sqrt{k} \rangle \langle y - \sqrt{k} \rangle = \langle x \rangle^3$. As K is an algebraic number field, O_K is a Dedekind domain by Theorem 8.1.1, and thus by Theorem 14.2.1 there exists an ideal A of O_K such that

$$\langle y + \sqrt{k} \rangle = A^3.$$

Thus A^3 is a principal ideal and, as $h(\mathbb{Q}(\sqrt{k})) \neq 0 \,(\mathrm{mod}\ 3)$, by Theorem 14.2.2, A is a principal ideal, say,

$$A = \langle a + b\sqrt{k} \rangle,$$

where $a, b \in \mathbb{Z}$. Hence

$$\langle y + \sqrt{k} \rangle = \langle a + b\sqrt{k} \rangle^3 = \langle (a + b\sqrt{k})^3 \rangle.$$

By Theorem 1.3.1 there exists a unit $\epsilon \in O_K$ such that

$$y + \sqrt{k} = \epsilon (a + b\sqrt{k})^3. \tag{14.2.4}$$

As $k < -1$ and $k \equiv 2, 3 \,(\mathrm{mod}\ 4)$ by Theorem 5.4.3 we have $\epsilon = \pm 1$. Taking conjugates in (14.2.4), we obtain

$$y - \sqrt{k} = \epsilon (a - b\sqrt{k})^3. \tag{14.2.5}$$

Thus

$$x^3 = y^2 - k = (y + \sqrt{k})(y - \sqrt{k}) = (a + b\sqrt{k})^3(a - b\sqrt{k})^3$$
$$= ((a + b\sqrt{k})(a - b\sqrt{k}))^3 = (a^2 - kb^2)^3$$

so that

$$x = a^2 - kb^2. \tag{14.2.6}$$

Adding and subtracting (14.2.4) and (14.2.5), we obtain

$$2y = \epsilon((a + b\sqrt{k})^3 + (a - b\sqrt{k})^3)$$

and

$$2\sqrt{k} = \epsilon((a + b\sqrt{k})^3 - (a - b\sqrt{k})^3),$$

so that

$$y = \epsilon(a^3 + 3kab^2), \quad 1 = \epsilon(3a^2b + kb^3).$$

From $1 = \epsilon b(3a^2 + kb^2)$ we see that $b = \pm 1$, so that $b = \pm \epsilon$. If $b = \epsilon$ then

$$x = a^2 - k, \quad y = \epsilon(a^3 + 3ka), \quad 1 = 3a^2 + k,$$

so that

$$k = 1 - 3a^2$$

and

$$x = 4a^2 - 1, \quad y = \pm(3a - 8a^3).$$

Clearly

$$x^3 + k = (4a^2 - 1)^3 + (1 - 3a^2) = 64a^6 - 48a^4 + 9a^2 = (8a^3 - 3a)^2 = y^2.$$

If $b = -\epsilon$ then

$$x = a^2 - k, \quad y = \epsilon(a^3 + 3ka), \quad 1 = -3a^2 - k,$$

so that

$$k = -1 - 3a^2$$

and

$$x = 4a^2 + 1, \quad y = \pm(3a + 8a^3).$$

Clearly

$$x^3 + k = (4a^2 + 1)^3 - 1 - 3a^2 = 64a^6 + 48a^4 + 9a^2 = (8a^3 + 3a)^2 = y^2.$$

This completes the proof of the theorem   ∎

Example 14.2.1 *The integer $k = -2 = 1 - 3 \cdot 1^2$ satisfies the conditions of Theorem 14.2.3(a) as $h(\mathbb{Q}(\sqrt{-2})) = 1$, so that the only solutions in integers of the equation $y^2 = x^3 - 2$ are $(x, y) = (3, \pm 5)$. This result was first stated by Fermat.*

The values of k in the range $-200 < k < -2$ that satisfy the conditions of Theorem 14.2.3(a) are $k = -74 = 1 - 3 \cdot 5^2$ $(h(\mathbb{Q}(\sqrt{-74})) = 10)$ and $k = -146 = 1 - 3 \cdot 7^2$ $(h(\mathbb{Q}(\sqrt{-146})) = 16)$. Hence, by Theorem 14.2.3(a), the only solutions in integers of $y^2 = x^3 - 74$ are $(x, y) = (99, \pm 985)$ and the only solutions in integers to $y^2 = x^3 - 146$ are $(x, y) = (195, \pm 2723)$.

Example 14.2.2 *The smallest integer k in absolute value satisfying the conditions of Theorem 14.2.3(b) is $k = -13 = -1 - 3 \cdot 2^2$ as $h(\mathbb{Q}(\sqrt{-13})) = 2$. Hence, by Theorem 14.2.3(b), the only solutions in integers x and y of the equation $y^2 = x^3 - 13$ are $(x, y) = (17, \pm 70)$. In the range $-200 < k < -1$ there is only one other value of k that satisfies the conditions of Theorem 14.2.3(b), namely, $k = -193 = -1 - 3 \cdot 8^2$ $(h(\mathbb{Q}(\sqrt{-193})) = 4)$. The only solutions in integers of the equation $y^2 = x^3 - 193$ are $(x, y) = (257, \pm 4120)$.*

Example 14.2.3 *The integer $k = -5$ satisfies the conditions of Theorem 14.2.3(c) as $h(\mathbb{Q}(\sqrt{-5})) = 2$. Hence the equation $y^2 = x^3 - 5$ has no solutions in integers. We note that $k = -5$ was also covered by Theorem 14.1.1.*

Similarly, we find that $y^2 = x^3 + k$ is not solvable in integers x and y for

$$k = -6, -10, -14, -17, -21, -22.$$

In the next theorem we find a result similar to that of Theorem 14.2.3(c) in the case when k is positive.

Theorem 14.2.4 *Let k be an integer such that*

$$k > 0,$$
$$k \text{ is squarefree},$$
$$k \equiv 2, 3 \pmod{4},$$
$$h(\mathbb{Q}(\sqrt{k})) \not\equiv 0 \pmod 3.$$

Let $T + U\sqrt{k}$ be the fundamental unit of $K = \mathbb{Q}(\sqrt{k})$ of norm 1. If

$$k \equiv 4 \pmod 9, \quad U \equiv 0 \pmod 9$$

or

$$k \equiv 7 \pmod 9, \quad U \equiv \pm 3 \pmod 9$$

or

$$k \equiv 4 \,(\mathrm{mod}\ 7), \quad U \equiv 0 \,(\mathrm{mod}\ 7)$$

then the equation $y^2 = x^3 + k$ has no solutions in integers x and y.

Proof: Exactly as in the proof of Theorem 14.2.3 we obtain

$$y + \sqrt{k} = \epsilon (a + b\sqrt{k})^3,$$

where ϵ is a unit of O_K. Let η be the fundamental unit of O_K so that

$$\epsilon = \pm \eta^l$$

for some $l \in \mathbb{Z}$. As the cubes $-1 = (-1)^3$ and $\eta^{3m} = (\eta^m)^3$ can be absorbed into the cube $(a + b\sqrt{k})^3$, we have $y + \sqrt{k} = \epsilon (a + b\sqrt{k})^3$, where $\epsilon = 1, \eta$, or η^2. Further, as $\eta = \eta^3/\eta^2$ and $\eta^2 = \eta^3/\eta$, we have

$$y + \sqrt{k} = \epsilon (a + b\sqrt{k})^3, \quad \text{where } \epsilon \in \left\{1, \eta, \frac{1}{\eta}\right\} \text{ or } \left\{1, \frac{1}{\eta^2}, \eta^2\right\}.$$

We choose $\epsilon \in \{1, \eta, 1/\eta\}$ if η has norm 1 and $\epsilon \in \{1, 1/\eta^2, \eta^2\}$ if η has norm -1. Thus in both cases

$$\epsilon \in \{1, T + U\sqrt{k}, T - U\sqrt{k}\},$$

where $T + U\sqrt{k}$ is the fundamental unit (> 1) of O_K of norm 1. If $\epsilon = 1$, equating the coefficients of $\sqrt{k}$ we obtain $1 = 3a^2 b + kb^3$, so that $b \mid 1$ and thus $b = \pm 1$. Hence $\pm 1 = b = 3a^2 b^2 + kb^4 = 3a^2 + k \geq k > 1$, a contradiction. Thus $\epsilon = T \pm U\sqrt{k}$. Then

$$
\begin{aligned}
y + \sqrt{k} &= (T \pm U\sqrt{k})(a + b\sqrt{k})^3 \\
&= (T \pm U\sqrt{k})((a^3 + 3kab^2) + (3a^2 b + kb^3)\sqrt{k}) \\
&= (T(a^3 + 3kab^2) \pm Uk(3a^2 b + kb^3)) \\
&\quad + (T(3a^2 b + kb^3) \pm U(a^3 + 3kab^2))\sqrt{k}
\end{aligned}
$$

so that

$$1 = T(3a^2 b + kb^3) \pm U(a^3 + 3kab^2). \tag{14.2.7}$$

Case (i): $k \equiv 4 \,(\mathrm{mod}\ 9)$, $U \equiv 0 \,(\mathrm{mod}\ 9)$. As $U \equiv 0 \,(\mathrm{mod}\ 9)$, from $T^2 - kU^2 = 1$ we obtain $T \equiv \pm 1 \,(\mathrm{mod}\ 81)$, say

$$T \equiv \epsilon \,(\mathrm{mod}\ 81), \quad \epsilon = \pm 1.$$

Then from (14.2.7) modulo 9, we deduce that

$$1 \equiv \epsilon (3a^2 b + 4b^3) \,(\mathrm{mod}\ 9). \tag{14.2.8}$$

Clearly this congruence implies that $b \not\equiv 0 \,(\text{mod } 3)$. Hence $b \equiv \pm 1 \,(\text{mod } 3)$, say

$$b \equiv \lambda \,(\text{mod } 3), \quad \lambda = \pm 1.$$

Thus

$$b^3 \equiv \lambda \,(\text{mod } 9).$$

Then from (14.2.8) we deduce that

$$1 \equiv \epsilon\lambda(3a^2 + 4) \,(\text{mod } 9),$$

so that

$$3a^2 + 4 \equiv \epsilon\lambda \equiv \pm 1 \,(\text{mod } 9),$$

giving

$$3a^2 \equiv 4 \text{ or } 6 \,(\text{mod } 9),$$

both of which are impossible.

Case (ii): $k \equiv 7 \,(\text{mod } 9)$, $U \equiv \pm 3 \,(\text{mod } 9)$. In this case we have $U^2 \equiv 0$ $(\text{mod } 9)$. Then from $T^2 - kU^2 = 1$ we deduce that $T^2 \equiv 1 \,(\text{mod } 9)$, so that

$$T \equiv \epsilon \,(\text{mod } 9), \quad \epsilon = \pm 1.$$

Next, from (14.2.7) modulo 3, we obtain

$$1 \equiv \epsilon b^3 \,(\text{mod } 3),$$

so that

$$b \equiv \epsilon \,(\text{mod } 3), \quad b^3 \equiv \epsilon \,(\text{mod } 9).$$

Then from (14.2.7) modulo 9 we have

$$1 \equiv 3a^2 + 7 \pm 3a^3 \,(\text{mod } 9).$$

Clearly this implies $a \not\equiv 0 \,(\text{mod } 3)$, so $a \equiv \pm 1 \,(\text{mod } 3)$, $a^2 \equiv 1 \,(\text{mod } 3)$, and $a^3 \equiv a \,(\text{mod } 3)$. Hence

$$1 \equiv 1 \pm 3a \,(\text{mod } 9),$$

giving $a \equiv 0 \,(\text{mod } 3)$, a contradiction.

Case (iii): $k \equiv 4 \,(\text{mod } 7)$, $U \equiv 0 \,(\text{mod } 7)$. From

$$y + \sqrt{k} = (T \pm U\sqrt{k})(a + b\sqrt{k})^3$$

we deduce that

$$y - \sqrt{k} = (T \mp U\sqrt{k})(a - b\sqrt{k})^3,$$

so that

$$x^3 = y^2 - k = (y + \sqrt{k})(y - \sqrt{k})$$
$$= (T \pm U\sqrt{k})(a + b\sqrt{k})^3 (T \mp U\sqrt{k})(a - b\sqrt{k})^3$$
$$= (T^2 - kU^2)\left(a^2 - kb^2\right)^3$$
$$= \left(a^2 - kb^2\right)^3$$

and hence

$$x = a^2 - kb^2.$$

Now

$$x^3 \equiv 0, 1, 6 \pmod 7$$

and

$$y^2 \equiv 0, 1, 2, 4 \pmod 7,$$

so

$$y^2 - x^3 = k \equiv 4 \pmod 7$$

gives

$$y^2 \equiv 4 \pmod 7, \quad x^3 \equiv 0 \pmod 7.$$

Thus

$$x \equiv 0 \pmod 7$$

and so

$$a^2 - 4b^2 \equiv 0 \pmod 7;$$

that is,

$$a \equiv \pm 2b \pmod 7.$$

From $U \equiv 0 \pmod 7$ and $T^2 - kU^2 = 1$ we deduce that

$$T^2 \equiv 1 \pmod{49}$$

so that

$$T \equiv \pm 1 \pmod{49}.$$

Then from (14.2.7) we obtain

$$1 \equiv \pm 2b^3 \pmod 7,$$

which is impossible.

This completes the proof that $y^2 = x^3 + k$ is insolvable in integers x and y in all three cases. ∎

Example 14.2.4 *We choose $k = 58$ so that $k \equiv 2 \,(\text{mod } 4)$ and $k \equiv 4 \,(\text{mod } 9)$. In this case $h(\mathbb{Q}(\sqrt{58})) = 2$ and the fundamental unit of $\mathbb{Q}(\sqrt{58})$ is $99 + 13\sqrt{58}$ of norm -1. Thus the fundamental unit of norm 1 is*

$$(99 + 13\sqrt{58})^2 = 19603 + 2574\sqrt{58}$$

so that

$$U = 2574 \equiv 0 \,(\text{mod } 9).$$

Thus, by Theorem 14.2.4, the equation $y^2 = x^3 + 58$ is not solvable in integers x and y. We note that this equation is also covered by Theorem 14.1.3.

Example 14.2.5 *We choose $k = 7$. Here $k \equiv 3 \,(\text{mod } 4)$ and $k \equiv 7 \,(\text{mod } 9)$. Also, $h(\mathbb{Q}(\sqrt{7})) = 1$ and the fundamental unit of $\mathbb{Q}(\sqrt{7})$ of norm 1 is $8 + 3\sqrt{7}$ so that $U = 3 \equiv 3 \,(\text{mod } 9)$. Hence, by Theorem 14.2.4, the equation $y^2 = x^3 + 7$ has no solutions in integers x and y. This equation is also covered by Theorem 14.1.2.*

Example 14.2.6 *We choose $k = 158$. Here $k \equiv 2 \,(\text{mod } 4)$ and $k \equiv 4 \,(\text{mod } 7)$. Also, $h(\mathbb{Q}(\sqrt{158})) = 1$ and the fundamental unit of $\mathbb{Q}(\sqrt{158})$ of norm 1 is $7743 + 616\sqrt{158}$ so that $U = 616 \equiv 0 \,(\text{mod } 7)$. Thus, by Theorem 14.2.4, the equation $y^2 = x^3 + 158$ has no solutions in integers x and y.*

We conclude this section by giving an example where $h(\mathbb{Q}(\sqrt{k})) \equiv 0 \,(\text{mod } 3)$.

Theorem 14.2.5 *The equation*

$$y^2 = x^3 - 31 \tag{14.2.9}$$

has no solutions in integers x and y.

Proof: Suppose that $y^2 = x^3 - 31$ has a solution in integers x and y.

First we note that $31 \nmid y$, for if $31 \mid y$ then $31 \mid x$ and so $31^2 \mid x^3 - y^2 = 31$, a contradiction.

Next we show that x must be even. Suppose that x is odd. If $x \equiv 1 \,(\text{mod } 4)$ then $x^3 \equiv 1 \,(\text{mod } 4)$ so that $y^2 \equiv 2 \,(\text{mod } 4)$, which is impossible. If $x \equiv 3 \,(\text{mod } 4)$ then $x^2 + 3x + 9 \equiv 3 \,(\text{mod } 4)$. Also $x^2 + 3x + 9 > 1$.

Hence $x^2 + 3x + 9$ has a prime factor $p \equiv 3 \pmod 4$. Now

$$y^2 + 4 = x^3 - 27 = (x - 3)(x^2 + 3x + 9),$$

so that $y^2 + 4 \equiv 0 \pmod p$, which is impossible. This proves that x is even and y is odd.

An integral basis for $K = \mathbb{Q}(\sqrt{-31})$ is $\left\{1, \frac{1+\sqrt{-31}}{2}\right\}$. The prime ideal factorization of 2 in O_K is given by

$$\langle 2 \rangle = \langle 2, \frac{3 + \sqrt{-31}}{2} \rangle \langle 2, \frac{3 - \sqrt{-31}}{2} \rangle \qquad (14.2.10)$$

(see Theorem 10.2.1). We show that $\langle 2, \frac{3+\sqrt{-31}}{2} \rangle$ is not a principal ideal. Suppose on the contrary that $\langle 2, \frac{3+\sqrt{-31}}{2} \rangle$ is a principal ideal. Then there exist rational integers a and b such that

$$\langle 2, \frac{3 + \sqrt{-31}}{2} \rangle = \langle a + b \left(\frac{1 + \sqrt{-31}}{2} \right) \rangle.$$

Taking norms we obtain

$$2 = N \left(\langle 2, \frac{3 + \sqrt{-31}}{2} \rangle \right) = N \left(\langle a + b \left(\frac{1 + \sqrt{-31}}{2} \right) \rangle \right)$$

$$= | N(\frac{2a + b + b\sqrt{-31}}{2}) | = \frac{(2a + b)^2 + 31b^2}{4}$$

so that

$$(2a + b)^2 + 31b^2 = 8,$$

which is clearly impossible.

Next, appealing to (14.2.9) and (14.2.10), we deduce that

$$\langle \frac{y + \sqrt{-31}}{2} \rangle \langle \frac{y - \sqrt{-31}}{2} \rangle = \langle 2, \frac{3 + \sqrt{-31}}{2} \rangle \langle 2, \frac{3 - \sqrt{-31}}{2} \rangle \langle \frac{x}{2} \rangle^3. \quad (14.2.11)$$

We show that the two ideals $\langle \frac{y+\sqrt{-31}}{2} \rangle$ and $\langle \frac{y-\sqrt{-31}}{2} \rangle$ are coprime. If not, then there exists a prime ideal P such that

$$P \mid \langle \frac{y + \sqrt{-31}}{2} \rangle, \quad P \mid \langle \frac{y - \sqrt{-31}}{2} \rangle.$$

Then

$$\frac{y + \sqrt{-31}}{2} \in P, \quad \frac{y - \sqrt{-31}}{2} \in P,$$

so

$$\sqrt{-31} = \left(\frac{y + \sqrt{-31}}{2} \right) - \left(\frac{y - \sqrt{-31}}{2} \right) \in P.$$

Thus

$$P \mid \langle \sqrt{-31} \rangle.$$

But P and $\langle \sqrt{-31} \rangle$ are both prime ideals so that

$$P = \langle \sqrt{-31} \rangle.$$

Hence

$$\langle \sqrt{-31} \rangle \mid \left\langle \left(\frac{y + \sqrt{-31}}{2} \right) \right\rangle$$

so that $\frac{y + \sqrt{-31}}{2} \in \langle \sqrt{-31} \rangle$. This shows that there exist integers u and v such that

$$\frac{y + \sqrt{-31}}{2} = \sqrt{-31} \left(\frac{u + v\sqrt{-31}}{2} \right).$$

Hence $u = 1$ and $y = -31v$, contradicting $31 \nmid y$. This proves that the ideals $\left\langle \left(\frac{y + \sqrt{-31}}{2} \right) \right\rangle$ and $\left\langle \left(\frac{y - \sqrt{-31}}{2} \right) \right\rangle$ are coprime. Thus, replacing y by $-y$ if necessary, we see from (14.2.11) that there exists an ideal A of O_K such that

$$\begin{cases} \left\langle \left(\dfrac{y + \sqrt{-31}}{2} \right) \right\rangle = \langle 2, \dfrac{3 + \sqrt{-31}}{2} \rangle A^3, \\[4mm] \left\langle \left(\dfrac{y - \sqrt{-31}}{2} \right) \right\rangle = \langle 2, \dfrac{3 - \sqrt{-31}}{2} \rangle \bar{A}^3, \\[4mm] \langle \dfrac{x}{2} \rangle = A\bar{A}, \end{cases} \tag{14.2.12}$$

where $\bar{A}$ denotes the conjugate ideal of A. Since $h(\mathbb{Q}(\sqrt{-31})) = 3$ the ideal A^3 is principal. Then, from the first equality in (14.2.12), we deduce that the ideal $\langle 2, \frac{3+\sqrt{-31}}{2} \rangle$ is principal, a contradiction. This completes the proof that the equation $y^2 = x^3 - 31$ has no solutions in integers x and y. ∎

We conclude this section by giving two short tables (Tables 15 and 16) of solutions of $y^2 = x^3 + k$.

14.3 The Diophantine Equation
$y(y+1) = x(x+1)(x+2)$

In this section we use the arithmetic of the cubic field

$$K = \mathbb{Q}(\theta), \quad \theta^3 - 4\theta + 2 = 0, \tag{14.3.1}$$

Table 15.   *Solutions $(x, y) \in \mathbb{Z}^2$ of*
$$y^2 = x^3 + k, \quad -20 \leq k < 0$$

k	Solutions (x, y) of $y^2 = x^3 + k$
-1	$(1, 0)$
-2	$(3, \pm 5)$
-3	insolvable
-4	$(2, \pm 2)$, $(5, \pm 11)$
-5	insolvable
-6	insolvable
-7	$(2, \pm 1)$, $(32, \pm 181)$
-8	$(2, 0)$
-9	insolvable
-10	insolvable
-11	$(3, \pm 4)$, $(15, \pm 58)$
-12	insolvable
-13	$(17, \pm 70)$
-14	insolvable
-15	$(4, \pm 7)$
-16	insolvable
-17	insolvable
-18	$(3, \pm 3)$
-19	$(7, \pm 18)$
-20	$(6, \pm 14)$

to determine all the solutions in integers x and y of the equation

$$y(y + 1) = x(x + 1)(x + 2); \tag{14.3.2}$$

that is, we determine all those integers that are simultaneously a product of two consecutive integers and a product of three consecutive integers. This problem was proposed by Edgar Emerson to Burton W. Jones (1902–1983) and was first solved by Louis J. Mordell (1888–1972) in 1963. We follow the solution given by Mordell in his paper [6].

We need the following facts about the field K and its ring of integers O_K:

$$O_K = \mathbb{Z} + \mathbb{Z}\theta + \mathbb{Z}\theta^2, \tag{14.3.3}$$

O_K is a unique factorization domain, $\tag{14.3.4}$

a fundamental system of units of O_K is $\{\epsilon, \eta\}$,

where $\epsilon = \theta - 1$ and $\eta = 2\theta - 1$. $\tag{14.3.5}$

Result (14.3.3) is Exercise 13 of Chapter 7. Result (14.3.4) is Exercise 14 of Chapter 12. For result (14.3.5) see Example 13.6.1. By Dirichlet's unit theorem every unit of O_K is given by $\pm \epsilon^m \eta^n$ for integers m and n.

Table 16. *Solutions* $(x, y) \in \mathbb{Z}^2$ *of*
$$y^2 = x^3 + k, \ 0 < k \le 20$$

k	Solutions (x, y) of $y^2 = x^3 + k$
1	$(-1, 0)$, $(0, \pm 1)$, $(2, \pm 3)$
2	$(-1, \pm 1)$
3	$(1, \pm 2)$
4	$(0, \pm 2)$
5	$(-1, \pm 2)$
6	insolvable
7	insolvable
8	$(-2, 0)$, $(1, \pm 3)$, $(2, \pm 4)$, $(46, \pm 312)$
9	$(-2, \pm 1)$, $(0, \pm 3)$, $(3, \pm 6)$, $(6, \pm 15)$, $(40, \pm 253)$
10	$(-1, \pm 3)$
11	insolvable
12	$(-2, \pm 2)$, $(13, \pm 47)$
13	insolvable
14	insolvable
15	$(1, \pm 4)$, $(109, \pm 1138)$
16	$(0, \pm 4)$
17	$(-2, \pm 3)$, $(-1, \pm 4)$, $(2, \pm 5)$, $(4, \pm 9)$, $(8, \pm 23)$, $(43, \pm 282)$, $(52, \pm 375)$, $(5234, \pm 378661)$
18	$(7, \pm 19)$
19	$(5, \pm 12)$
20	insolvable

If we set

$$X = 2x + 2, \ Y = 2y + 1, \tag{14.3.6}$$

the equation (14.3.2) becomes

$$2Y^2 = X^3 - 4X + 2. \tag{14.3.7}$$

Clearly any solution of (14.3.7) must have X even and Y odd. We will show that the only solutions of (14.3.7) are

$$(X, Y) = (-2, \pm 1), \ (0, \pm 1), \ (2, \pm 1), \ (4, \pm 5), \ (12, \pm 29).$$

Thus all the solutions of (14.3.2) are

$$(x, y) = (0, 0), \ (0, -1), \ (-1, 0), \ (-1, -1), \ (-2, 0), \ (-2, -1), \ (1, 2),$$
$$(1, -3), \ (5, 14), \ (5, -15).$$

This proves that the only integers that are simultaneously a product of two consecutive integers as well as a product of three consecutive integers are 0, 6, and 210.

All such solutions are given by

$$0 = (-1)(0) = (0)(1) = (-2)(-1)(0) = (-1)(0)(1) = (0)(1)(2),$$
$$6 = (2)(3) = (-3)(-2) = (1)(2)(3),$$
$$210 = (14)(15) = (-15)(-14) = (5)(6)(7).$$

We let θ, θ', $\theta'' \in \mathbb{C}$ be the three roots of $x^3 - 4x + 2 = 0$ so that

$$\begin{cases} \theta + \theta' + \theta'' = 0, \\ \theta\theta' + \theta'\theta'' + \theta''\theta = -4, \\ \theta\theta'\theta'' = -2. \end{cases} \qquad (14.3.8)$$

We need a number of lemmas.

Lemma 14.3.1 θ *is a prime in* O_K.

Proof: From (14.3.8) we deduce that $|N(\theta)| = |\theta\theta'\theta''| = 2$, which is a rational prime, so that θ is a prime in O_K. ∎

Lemma 14.3.2 $4\theta - 3$ *is a prime in* O_K.

Proof: We have by (14.3.8)

$$\begin{aligned} N(4\theta - 3) &= (4\theta - 3)(4\theta' - 3)(4\theta'' - 3) \\ &= 64\theta\theta'\theta'' - 48(\theta\theta' + \theta'\theta'' + \theta''\theta) + 36(\theta + \theta' + \theta'') - 27 \\ &= 64(-2) - 48(-4) + 36(0) - 27 \\ &= -128 + 192 - 27 \\ &= 37, \end{aligned}$$

which is a rational prime, so that $4\theta - 3$ is a prime in O_K. ∎

Lemma 14.3.3 $2 = \rho\theta^3$, *where* $\rho \in U(O_K)$.

Proof: From $\theta^3 - 4\theta + 2 = 0$ we deduce that

$$\frac{\theta^3}{2} = 2\theta - 1 \in O_K.$$

Further,

$$N\left(\frac{\theta^3}{2}\right) = \frac{N(\theta)^3}{8} = \frac{(-2)^3}{8} = -1.$$

Hence

$$\frac{\theta^3}{2} \in U(O_K).$$

Thus

$$\frac{2}{\theta^3} \in U(O_K)$$

and so

$$\frac{2}{\theta^3} = \rho$$

for some $\rho \in U(O_K)$.   ∎

Lemma 14.3.4 *If $(X, Y) \in \mathbb{Z}^2$ is a solution of (14.3.7) then*

$$(X - \theta)(X^2 + \theta X + (\theta^2 - 4)) = \rho\theta^3 Y^2.$$

Proof: We have by (14.3.1), (14.3.7), and Lemma 14.3.3

$$\begin{aligned}
(X - \theta)(X^2 + \theta X + (\theta^2 - 4)) &= X^3 - 4X - \theta^3 + 4\theta \\
&= X^3 - 4X + 2 \\
&= 2Y^2 \\
&= \rho\theta^3 Y^2.
\end{aligned}$$ ∎

Lemma 14.3.5 *The only possible primes in O_K dividing both $X - \theta$ and $X^2 + \theta X + (\theta^2 - 4)$ are θ and $4\theta - 3$.*

Proof: Let π be a prime of O_K dividing both $X - \theta$ and $X^2 + \theta X + (\theta^2 - 4)$. Then π divides

$$\begin{aligned}
(X^2 + \theta X + (\theta^2 - 4)) &- (X + 2\theta)(X - \theta) \\
= 3\theta^2 - 4 &= \frac{3\theta^3 - 4\theta}{\theta} = \frac{3(4\theta - 2) - 4\theta}{\theta} = \frac{8\theta - 6}{\theta} \\
&= \frac{2}{\theta}(4\theta - 3) = \rho(4\theta - 3)\theta^2,
\end{aligned}$$

by (14.3.1) and Lemma 14.3.3. As ρ is a unit this shows that the only possibilities for π are $\pi = \theta$ and $\pi = 4\theta - 3$.   ∎

Lemma 14.3.6 *θ is a common factor of $X - \theta$ and $X^2 + X\theta + (\theta^2 - 4)$ such that $\theta^2 \nmid X - \theta$.*

Proof: By Lemma 14.3.3 we have $\theta^3 \mid 2$ in O_K. Hence, as X is even, we deduce that $\theta^3 \mid X$ in O_K. Hence $\theta \mid X$ and so $\theta \mid X - \theta$ and $\theta \mid X^2 + X\theta + (\theta^2 - 4)$. Finally, as $\theta^2 \mid X$ and $\theta^2 \nmid \theta$, we have $\theta^2 \nmid X - \theta$.   ∎

We are now ready to prove the main result of this section.

Theorem 14.3.1 *The solutions in integers X and Y of the equation*

$$2Y^2 = X^3 - 4X + 2$$

are

$$(X, Y) = (-2, \pm 1), \ (0, \pm 1), \ (2, \pm 1), \ (4, \pm 5), \ (12, \pm 29).$$

Proof: We define the nonnegative integer n by

$$(4\theta - 3)^n \mid X - \theta, \quad (4\theta - 3)^{n+1} \nmid X - \theta. \tag{14.3.9}$$

Then, as O_K is a unique factorization domain, from (14.3.3), (14.3.5), Lemma 14.3.4, Lemma 14.3.5, and (14.3.9), we deduce that

$$X - \theta = \pm \theta (4\theta - 3)^n \epsilon^l \eta^m (a + b\theta + c\theta^2)^2 \tag{14.3.10}$$

for some integers l, m, a, b, c. By absorbing squares into the square $\left(a + b\theta + c\theta^2\right)^2$, we may rewrite (14.3.10) as

$$X - \theta = \pm \theta (4\theta - 3)^N \epsilon^L \eta^M \left(A + B\theta + C\theta^2\right)^2, \tag{14.3.11}$$

where

$$L, M, N \in \{0, \ 1\}. \tag{14.3.12}$$

Taking norms of both sides of (14.3.11), we obtain

$$X^3 - 4X + 2 = \pm 2 \cdot 37^N \cdot Z^2 \tag{14.3.13}$$

for some $Z \in \mathbb{Z}$. As X is even we may set $X = 2X_1$, where $X_1 \in \mathbb{Z}$, in (14.3.13), and obtain

$$4X_1^3 - 4X_1 + 1 = \pm 37^N \cdot Z^2. \tag{14.3.14}$$

Reducing (14.3.14) modulo 8, we obtain

$$1 \equiv \pm 5^N \pmod{8},$$

showing that $N \neq 1$. Hence by (14.3.12) we have $N = 0$. Thus (14.3.11) becomes

$$X - \theta = \pm \theta \epsilon^L \eta^M \left(A + B\theta + C\theta^2\right)^2, \tag{14.3.15}$$

where $L, M \in \{0, \ 1\}$. Expanding the square in (14.3.15) and making use of $\theta^3 = 4\theta - 2$, we obtain

$$X - \theta = \pm \theta \epsilon^L \eta^M ((A^2 - 4BC) + (2AB + 8BC - 2C^2)\theta \\ + (2AC + B^2 + 4C^2)\theta^2). \tag{14.3.16}$$

We now consider each of the four possibilities

$$(L, M) = (0, 0), \ (0, 1), \ (1, 0), \ \text{and} \ (1, 1).$$

(i): $(L, M) = (0, 0)$. In this case (14.3.16) becomes

$$X - \theta = \pm\theta((A^2 - 4BC) + (2AB + 8BC - 2C^2)\theta$$
$$+ (2AC + B^2 + 4C^2)\theta^2). \qquad (14.3.17)$$

Using $\theta^3 = 4\theta - 2$, and equating terms in 1, θ, and θ^2 on both sides of (14.3.7), we obtain

$$4AC + 2B^2 + 8C^2 = \mp X, \qquad (14.3.18)$$
$$A^2 - 4BC + 4B^2 + 8AC + 16C^2 = \mp 1, \qquad (14.3.19)$$
$$AB + 4BC - C^2 = 0. \qquad (14.3.20)$$

Taking the Eq. (14.3.19) modulo 4, we see that the plus sign holds. Then (14.3.18)–(14.3.20) can be written as

$$X = 4AC + 2B^2 + 8C^2, \qquad (14.3.21)$$
$$(A + 4C)^2 + 4B^2 - 4BC = 1, \qquad (14.3.22)$$
$$B(A + 4C) = C^2. \qquad (14.3.23)$$

When $B = 0$, (14.3.23) gives $C = 0$. Then from (14.3.21) we obtain $X = 0$. If $B \neq 0$, from (14.3.22) and (14.2.23) we deduce that

$$\frac{C^4}{B^2} + 4B^2 - 4BC = 1. \qquad (14.3.24)$$

Now for all $x \in \mathbb{R}$ we have

$$(x - B)^2((x + B)^2 + 2B^2) \geq 0.$$

Hence

$$x^4 \geq 4B^3x - 3B^4$$

and so

$$\frac{x^4}{B^2} + 4B^2 - 4Bx \geq B^2.$$

Taking $x = C$ in this inequality, and appealing to (14.3.24), we deduce that

$$1 \geq B^2,$$

so that $B = \pm 1$. Then from (14.3.24) we have

$$C^4 + 4 \mp 4C = 1,$$

so that $C = \pm 1$. Hence from (14.3.23) we obtain $A = C^2/B - 4C = \mp 3$. Finally, from (14.3.21) we obtain

$$X = 2 - 12 + 8 = -2.$$

(ii): $(L, M) = (0, 1)$. In this case (14.3.15) becomes

$$X - \theta = \pm\theta(2\theta - 1)\left(A + B\theta + C\theta^2\right)^2.$$

Multiplying by θ, and absorbing θ^2 into the square, we obtain with a slight change of notation

$$\mp(X\theta - \theta^2) = (1 - 2\theta)((A^2 - 4BC) + (2AB + 8BC - 2C^2)\theta$$
$$+ (2AC + B^2 + 4C^2)\theta^2),$$

and so equating coefficients of $1, \theta, \theta^2$ we have

$$0 = A^2 - 4BC + 4(2AC + B^2 + 4C^2), \tag{14.3.25}$$
$$\mp X = 2AB + 8BC - 2C^2 - 2(A^2 - 4BC) - 8(2AC + B^2 + 4C^2), \tag{14.3.26}$$
$$\pm 1 = 2AC + B^2 + 4C^2 - 2(2AB + 8BC - 2C^2). \tag{14.3.27}$$

From (14.3.25) we see that A is even and from (14.3.27) that B is odd. Then, considering (14.3.27) modulo 4, we deduce that the $+$ sign holds in (14.3.27). Hence the $-$ sign holds in (14.3.26). Thus (14.3.25)–(14.3.27) become

$$0 = (A + 4C)^2 + 4B(B - C), \tag{14.3.28}$$
$$1 = A(2C - 4B) + B^2 - 16BC + 8C^2, \tag{14.3.29}$$
$$X = 2A^2 + 8B^2 + 34C^2 - 2AB + 16AC - 16BC. \tag{14.3.30}$$

Suppose first that $C = 2B$. Then from (14.3.29) we obtain $B^2 = 1$. Since solutions (A, B, C) and $(-A, -B, -C)$ give the same value for X, we need only take $B = 1$. Hence $C = 2$. Then from (13.3.28) we obtain $(A + 8)^2 = 4$ so that $A = -6, -10$. Then from (14.3.30) with $(A, B, C) = (-6, 1, 2)$ we obtain $X = 4$ and with $(A, B, C) = (-10, 1, 2)$ we obtain $X = 12$.

Suppose next that $C \neq 2B$. Then from (14.3.29) we have

$$A = \frac{B^2 - 16BC + 8C^2 - 1}{4B - 2C}. \tag{14.3.31}$$

Thus

$$A + 4C = \frac{B^2 - 1}{4B - 2C}. \tag{14.3.32}$$

Then, from (14.3.28), we deduce that

$$\left(\frac{B^2 - 1}{4B - 2C}\right)^2 + 4B(B - C) = 0,$$

so that

$$B^4 - 2B^2 + 1 + 16B(2B - C)^2(B - C) = 0. \tag{14.3.33}$$

This shows that $B \mid 1$, so that $B = \pm 1$. Then (14.3.33) gives (as $C \neq 2B$) $C = B = \pm 1$. Next from (14.3.32) we obtain $A = -4C = \mp 4$ and finally from (14.3.30)

$X = 2$. All solutions of $2Y^2 = X^3 - 4X + 2$ have now been found.

(iii): $(L, M) = (1, 0)$. In this case (14.3.16) becomes

$$\mp(X - \theta) = (\theta - \theta^2)((A^2 - 4BC) + (2AB + 8BC - 2C^2)\theta$$
$$+ (2AC + B^2 + 4C^2)\theta^2).$$

Equating coefficients of θ and θ^2, we obtain

$$\pm 1 = (A^2 - 4BC) + 6(2AC + B^2 + 4C^2) - 4(2AB + 8BC - 2C^2),$$
$$0 = (2AB + 8BC - 2C^2) - (A^2 - 4BC) - 4(2AC + B^2 + 4C^2).$$

The first equation shows that A is odd and the second that A is even. This case cannot occur.

(iv): $(L, M) = (1, 1)$. In this case (14.3.15) becomes

$$\pm(X - \theta) = \theta(1 - \theta)(1 - 2\theta)\left(A + B\theta + C\theta^2\right)^2.$$

On multiplying by θ, and absorbing θ^2 into the square, we obtain with a slight change of notation

$$\pm\theta(X - \theta) = (1 - 3\theta + 2\theta^2)((A^2 - 4BC) + (2AB + 8BC - 2C^2)\theta$$
$$+ (2AC + B^2 + 4C^2)\theta^2).$$

Equating coefficients of 1 and θ^2, we obtain

$$0 = (A^2 - 4BC) + 6(2AC + B^2 + 4C^2) - 4(2AB + 8BC - 2C^2),$$
$$\mp 1 = 9(2AC + B^2 + 4C^2) - 3(2AB + 8BC - 2C^2) + 2(A^2 - 4BC).$$

The first equation shows that A is even and then that B is even since $6B^2 \equiv 0 \pmod 4$. The second equation shows that B is odd. This case cannot occur.

This completes the proof of the theorem. ∎

As an immediate consequence of Theorem 14.3.7 we have the main result of this section.

Theorem 14.3.2 *The only solutions in integers x and y of the equation*

$$y(y + 1) = x(x + 1)(x + 2)$$

are

$$(x, y) = (0, 0), (0, -1), (-1, 0), (-1, -1), (-2, 0), (-2, -1), (1, 2), (1, -3),$$
$$(5, 14), (5, -15).$$

Proof: This follows immediately from Theorem 14.3.1 by using the transformation (14.3.6).   ∎

Exercises

1. Determine all integers k in the range $|k| < 200$ to which Theorem 14.1.1 applies.
2. Determine all integers k in the range $|k| < 200$ to which Theorem 14.1.2 applies.
3. Determine all integers k in the range $|k| < 200$ to which Theorem 14.1.3 applies.
4. Determine all integers k in the range $|k| < 200$ to which Theorem 14.1.4 applies.
5. Let M and N be integers such that

$$M \equiv 4 \,(\text{mod } 8), \quad N \equiv 1 \,(\text{mod } 2),$$
$$p \text{ (prime)} \mid N \implies p \equiv 1 \text{ or } 3 \,(\text{mod } 8).$$

   Set

$$k = M^3 - 2N^2.$$

   Prove that the equation $y^2 = x^3 + k$ has no solutions in integers x and y.
6. Determine all integers k in the range $|k| < 200$ to which the result of Exercise 5 applies.
7. Let M and N be integers such that

$$M \equiv 3 \,(\text{mod } 4), \quad N \equiv \pm 2 \,(\text{mod } 6),$$
$$p \text{ (prime)} \mid N \implies p \equiv \pm 1 \,(\text{mod } 12).$$

   Set

$$k = M^3 + 3N^2.$$

   Prove that the equation $y^2 = x^3 + k$ has no solutions in integers x and y.
8. Determine all integers k in the range $|k| < 200$ to which the result of Exercise 7 applies.
9. Formulate and prove a result analogous to that of Exercise 7 when k has the form $M^3 - 3N^2$.
10. Determine all integers k in the range $|k| < 200$ to which the result of Exercise 9 applies.
11. Prove that the equation

$$y^2 = x^3 + 45$$

   has no solutions in integers x and y.
12. Determine a class of integers k containing $k = 45$ for which the equation

$$y^2 = x^3 + k$$

   has no solutions in integers x and y.
13. Let M and N be integers such that

$$M \equiv 2 \,(\text{mod } 6), \quad N \equiv \pm 1 \,(\text{mod } 6),$$
$$p \text{ (prime)} \mid M \implies p \equiv 2 \,(\text{mod } 3).$$

   Set

$$k = 4M^3 - 3N^2.$$

   Prove that the equation $y^2 = x^3 + k$ has no solutions in integers x and y.

14. Determine all integers k in the range $|k| < 200$ to which the result of Exercise 13 applies.
15. Show that the condition $M \equiv 2 \,(\text{mod } 6)$ can be replaced by $M \equiv 0 \,(\text{mod } 6)$, $M \neq 0$ in Exercise 13 without affecting the result.
16. Determine all integers k in the range $|k| < 200$ to which the result of Exercise 15 applies.
17. Formulate and prove an analogous result to that of Exercise 13 for k of the form $4M^3 + 3N^2$.
18. Prove that $y^2 = x^3 + 13$ has no solutions in integers.
19. Prove that $y^2 = x^3 + 51$ has no solutions in integers.

Suggested Reading

1. W. W. Rouse Ball, *A Short Account of the History of Mathematics*, Dover, New York, 1960.

   A brief discussion of Bachet's work is given on pages 305 and 306.
2. E. Brown and B. T. Myers, *Elliptic curves from Mordell to Diophantus and back*, American Mathematical Monthly 109 (2002), 639–649.

   In this beautifully written article the authors discuss the number of integer points (x, y) on the elliptic curve $y^2 = x^3 - x + m^2$, where m is a nonnegative integer.
3. H. M. Edgar, *Classes of equations of the type $y^2 = x^3 + k$ having no rational solutions*, Nagoya Mathematical Journal 28 (1966), 49–58.

   Conditions are given for the equation $y^2 = x^3 + k$ to have no rational solutions.
4. T. Heath, *A History of Greek Mathematics*, Volume 1: *From Thales to Euclid*, Volume 2: *From Aristarchus to Diophantus*, Dover, New York, 1981.

   Chapter 20 in Volume 2 contains an interesting discussion of the work of Diophantus including his methods for finding solutions in integers of equations of degrees 1, 2, and 3.
5. H. London and R. Finkelstein, *On Mordell's Equation $y^2 - k = x^3$*, Bowling Green State University Press, Bowling Green, Ohio, 1973.

   The authors provide a comprehensive treatment of the equation $y^2 = x^3 + k$ with many references.
6. L. J. Mordell, *On the integer solutions of $y(y + 1) = x(x + 1)(x + 2)$*, Pacific Journal of Mathematics 13 (1963), 1347–1351.

   Section 13.3 is based on this beautifully written paper of Mordell.
7. L. J. Mordell, *Diophantine Equations*, Academic Press, London and New York, 1969.

   Mordell's book is a very readable standard reference text on Diophantine equations.

Biographies

1. V. Bjerknes, *Axel Thue*, Nordisk Matematisk Tidskrift 4 (1922), 33–46.

   A biography of Axel Thue is given.
2. J. W. S. Cassels, *L. J. Mordell*, Bulletin of the London Mathematical Society 6 (1974), 69–96.

   A biography of L. J. Mordell (1888–1972) is given.
3. H. Davenport, *L. J. Mordell*, Acta Arithmetica 9 (1964), 3–12.

   Another biography of L. J. Mordell is presented.

4. L. J. Mordell, *Reminiscences of an octogenarian mathematician*, American Mathematical Monthly 78 (1971), 952–961.

This article is based on a talk given to the Fellows of St. John's College, Cambridge, in 1968.

5. The website

http://www-groups.dcs.st-and.ac.uk/~history/

has biographies of Bachet, Diophantus, Mordell, and Thue.

List of Definitions

Location of Theorems

Location of Theorems

Location of Lemmas

Bibliography

[1] Z. I. Borevich and I. R. Shafarevich, *Number Theory*, Academic Press, New York and London, 1966.

[2] H. Cohen, *A Course in Computational Algebraic Number Theory*, Springer-Verlag, Berlin/Heidelberg/New York, 1996.

[3] H. Cohn, *Advanced Number Theory*, Dover, New York, 1980.

[4] H. Cohn, *A Classical Invitation to Algebraic Numbers and Class Fields*, Springer-Verlag, New York/Heidelberg/Berlin, 1978.

[5] D. A. Cox, *Primes of the Form $x^2 + ny^2$*, Wiley, New York, 1989.

[6] R. Dedekind, *Theory of Algebraic Integers*, Cambridge University Press, Cambridge, UK, 1996.

[7] J. Esmonde and M. Ram Murty, *Problems in Algebraic Number Theory*, Springer-Verlag, New York, 1999.

[8] G. J. Janusz, *Algebraic Number Fields*, Second Edition, Graduate Studies in Mathematics Volume 7, American Mathematical Society, Providence, Rhode Island, 1996.

[9] H. Koch, *Number Theory: Algebraic Numbers and Functions*, Graduate Studies in Mathematics Volume 24, American Mathematical Society, Providence, Rhode Island, 2000.

[10] S. Lang, *Algebraic Number Theory*, Springer-Verlag, New York, 1986.

[11] R. L. Long, *Algebraic Number Theory*, Dekker, New York, 1977.

[12] H. B. Mann, *Introduction to Algebraic Number Theory*, Ohio State University Press, Columbus, Ohio, 1955.

[13] D. A. Marcus, *Number Fields*, Springer-Verlag, New York/Heidelberg/Berlin, 1977.

[14] R. A. Mollin, *Algebraic Number Theory*, Chapman and Hall/CRC Press, London/Boca Raton, Florida, 1999.

[15] R. Narasimhan, S. Raghavan, S. S. Rangachari, and S. Lal, *Algebraic Number Theory*, Tata Institute of Fundamental Research, Bombay, India, 1966.

[16] W. Narkiewicz, *Elementary and Analytic Theory of Algebraic Numbers*, Springer-Verlag, Berlin/Heidelberg/New York, 1989.

[17] T. Ono, *An Introduction to Algebraic Number Theory*, Plenum, New York, 1990.

[18] M. E. Pohst, *Computational Algebraic Number Theory*, Birkhäuser Verlag, Basel/Boston/Berlin, 1993.

[19] M. Pohst and H. Zassenhaus, *Algorithmic Algebraic Number Theory*, Cambridge University Press, Cambridge, UK, 1989.

[20] H. Pollard and H. G. Diamond, *The Theory of Algebraic Numbers*, Mathematical Association of America, Washington, DC, 1975.

[21] P. Ribenboim, *Classical Theory of Algebraic Numbers,* Springer-Verlag, New York, 2001.
[22] P. Samuel, *Algebraic Theory of Numbers,* Kershaw, London, 1972.
[23] I. Stewart and D. Tall, *Algebraic Number Theory and Fermat's Last Theorem,* A. K. Peters, Natick, Massachusetts, 2002.
[24] E. Weiss, *Algebraic Number Theory,* McGraw-Hill, New York, 1963.
[25] H. Weyl, *Algebraic Theory of Numbers,* Princeton University Press, Princeton, New Jersey, 1940.

Index

Lightning Source UK Ltd.
Milton Keynes UK
UKOW021118240412

191362UK00003B/15/P

9 780521 540117